Advanced Drug Formulation Design to Optimize Therapeutic Outcomes

DRUGS AND THE PHARMACEUTICAL SCIENCES
A Series of Textbooks and Monographs

Advanced Drug Formulation Design to Optimize Therapeutic Outcomes

Edited by

Robert O. Williams III
University of Texas at Austin
Austin, Texas, USA

David R. Taft
Long Island University
Brooklyn, New York, USA

Jason T. McConville
University of Texas at Austin
Austin, Texas, USA

informa

healthcare

New York London

Informa Healthcare USA, Inc.
52 Vanderbilt Avenue
New York, NY 10017

International Standard Book Number-10: 1-4200-4387-0 (Hardcover)
International Standard Book Number-13: 978-1-4200-4387-7(Hardcover)

Library of Congress Cataloging-in-Publication Data

Advanced drug formulation design to optimize therapeutic outcomes / edited by Robert O. Williams, III, David R. Taft, Jason T. McConville.
    p. ; cm. – (Drugs and the pharmaceutical sciences; v. 172)
    Includes bibliographical references.
    ISBN-13: 978-1-4200-4387-7 (hardcover : alk. paper)
    ISBN-10: 1-4200-4387-0 (hardcover : alk. paper) 1. Drug–Design. 2. Drug delivery systems. 3. Pharmaceutical chemistry. I. Williams, Robert O., 1956- II. Taft, David R. III. McConville, Jason T. IV. Series.
    [DNLM: 1. Chemistry, Pharmaceutical. 2. Drug Delivery Systems. WI DR893B v. 172 2007/QV 744 A2444 2007]
    RS420.A42 2007
    615'.19–dc22   2007023345

Visit the Informa web site at
www.informa.com

and the Informa Healthcare Web site at
www.informahealthcare.com

Preface

Organizational and economic changes in the pharmaceutical industry have led to a marked reduction in the supply of new chemical entities. This combined with an increase in the expiration of patents and tighter regulatory control has lead many companies to rethink existing product design. The growing trend to improve the efficacy of existing drug products, to both extend patent life and allow competition with opportunistic generic companies, has allowed for a greater understanding of formulation design. This trend is set to continue for some time as economic and political pressure increases.

There have been many improvements in understanding the theoretical and practical aspects of drug delivery. Numerous therapeutic compounds have been investigated to treat a variety of different illnesses. There are many methods of drug delivery available, requiring a variety of formulation options. Advanced formulation design may be thought of as a way to optimize the mode of drug delivery to a patient that will ultimately lead to a more effective therapy and approach an optimal therapeutic outcome for a particular drug. This book is designed to take the reader through various cutting-edge areas of advanced formulation design by describing therapeutic categories that may be beneficial. The reader will be introduced to some novel technologies and updated on the state of the art in areas designed to overcome poor efficacy issues.

In each chapter, the reader is taken through an introduction providing background information relating to a variety of disease states and is then invited to review particular therapeutic categories and the treatments and outcomes for specific disease states. Where appropriate, each contributing author has indicated cutting-edge technologies related to formulation designs that have the potential to target specific diseases. Pre-clinical and/ or clinical evidence is often presented showing an enhanced therapy, achieved using advanced formulation design.

Chapters 1 through 4 focus on the area of respiratory drug delivery research. This particular branch of research has led to many innovations in recent years. The reader is introduced to a variety of advances in pulmonary delivery, covering a wide area of dry powder and liquid propellant-based delivery systems. Following the strategies for advancement in pulmonary delivery, the reader is informed about the rapidly emerging global healthcare issue of invasive pulmonary aspergillosis. This chapter focuses on a common nosocomial pulmonary infection that has historically been treated using ineffective therapeutic regimens. The reader is taken through a journey describing a scientific rationale that can lead to improvements in efficacy using advancements in nanoparticle formulation design.

An overview of pulmonary administration of anticancer agents is addressed in Chapter 3. The reasoning behind aerosol delivery to treat lung cancer is addressed, and a case study for the aerosol delivery of camptothecin analogues is described. Additionally, the future perspective of pulmonary administration of anticancer agents is discussed.

The recent approval of the first pulmonary delivery product of insulin for the treatment of diabetes mellitus is covered in Chapter 4, with emphasis on its clinical translation. Throughout the first four chapters, the impact of lung pharmacokinetics of inhalation aerosols is important, whether the advanced formulation design is related to local or systemic active pharmaceutical ingredient administration.

Chapter 5 is related to the therapeutic opportunities that can be exploited for ocular delivery. A variety of infectious pathogens are described in the text along with current therapeutic strategies. Throughout the chapter, it is indicated where advanced formulation design can be more effective than conventional dose administration.

In Chapter 6 the reader is presented with a comprehensive overview of cancer drug therapy as well as emerging cancer therapies. The authors describe what potential problems must be overcome for effective drug delivery to cancer cells. A strategic approach to advances in formulation design is indicated for optimizing therapeutic outcomes in cancer and ultimately improving patient mortality.

There are a variety of diseases that are clinically significant infections in the gastrointestinal (GI) tract, as indicated in Chapter 7. The reader is taken on a journey through the GI tract looking at infections such as *herpes simplex* (cold sores) and oral thrush caused by the *Candida* species. in the mouth, through to colonic infections. Current therapeutic options are described, as well as the use of advanced therapeutic options and their future perspective.

Delivery of drugs across the blood-brain barrier is described in Chapter 8. The blood-brain barrier poses a challenge for the treatment of a variety of disease states including brain cancer and neurodegenerative diseases. The authors describe potential of intracranial-controlled drug delivery systems to allow new therapeutic strategies and/or to improve existing ones for optimized therapeutic outcomes.

Chapter 9 introduces the reader to a novel approach involving nasal delivery systems that are applicable for a variety of disease states. The authors describe the importance of understanding appropriate nasal anatomy and physiology. The method for enhancing nasal absorption is discussed as well as the operation of an advanced formulation design of nasal insert is fully described with pre-clinical data.

Chapter 10 describes the problems associated with central nervous system drug delivery to treat diseases such as with Alzheimer's and Parkinson's disease. Effective methods to properly assess the impact of central nervous system drug delivery are indicated. Additionally, strategies and tools are described that relate to the optimization of dosage forms. The authors also look at the future of formulation design in this area.

Chapter 11 is related to the therapeutic opportunities that can be exploited for cardiovascular delivery. The spectrum of cardiovascular disease states are described in detail in the text along with current therapeutic strategies. The chapter indicates where some advancement in formulation design has been shown to be more effective than conventional dose administration.

Advances in immunosuppression therapy are described in Chapter 12. The authors detail the delivery of glucocorticoids, immunophilin binding compounds, and cytostatics. In addition, the reader is introduced to advancements in drug delivery that clearly demonstrate optimized therapeutic outcomes.

Chapter 13 deals with solid dispersion technologies that may be used for a variety of patient needs from pain management through to treatment of cardiovascular disorders, cancer therapy, and immunosuppression.

Each of the chapters has been authored by scientists who are recognized as leaders in their field. The authors are truly diverse and are selected from a global community of leading healthcare scientists who dedicate their research time to providing better health care across a broad spectrum of therapeutic areas. As a whole, this book is concerned with therapeutic categories or specific therapeutic needs, which have been the focus of recent research in the field of advanced formulation design to optimize therapeutic outcomes.

Jason T. McConville
David R. Taft
Robert O. Williams III

Contents

Contributors

Carlos A. Alvarez   Texas Tech School of Pharmacy, Dallas, Texas, U.S.A.

Hannah Batchelor   Medicines Research Unit, School of Life and Health Sciences, Aston University, Birmingham, U.K.

Barbara Conway   Medicines Research Unit, School of Life and Health Sciences, Aston University, Birmingham, U.K.

Martin Donovan   College of Pharmacy, Health Sciences Center, University of New Mexico, Albuquerque, New Mexico, U.S.A.

Stephen Edge   Inhalation Device and Technology, Novartis Pharma AG, Basel, Switzerland

Tariq Javed   Research, Enterprise and Regional Affairs, University of Greenwich, London, U.K.

Jason T. McConville   Divisions of Pharmaceutics and Pharmacotherapy, College of Pharmacy, University of Texas, Austin, Texas, U.S.A.

James W. McGinity   College of Pharmacy, University of Texas at Austin, Austin, Texas, U.S.A.

Fiona McInnes   Strathclyde Institute of Pharmacy and Biomedical Sciences, University of Strathclyde, Glasgow, U.K.

Dave A. Miller   College of Pharmacy, University of Texas at Austin, Austin, Texas, U.S.A.

Gyan Prakash Mishra   Division of Pharmaceutical Sciences, School of Pharmacy, University of Missouri-Kansas City, Kansas City, Missouri, U.S.A.

Ashim K. Mitra   Division of Pharmaceutical Sciences, School of Pharmacy, University of Missouri-Kansas City, Kansas City, Missouri, U.S.A.

Kirk A. Overhoff   Schering-Plough Research Institute, Kenilworth, New Jersey, U.S.A.

Troy Purvis   College of Pharmacy, University of Texas at Austin, Austin, Texas, U.S.A.

Imran Saleem   College of Pharmacy, Health Sciences Center, University of New Mexico, Albuquerque, New Mexico, U.S.A.

Ghassan F. Shattat   Health and Life Sciences Department, Coventry University, Coventry, U.K.

Florence Siepmann   College of Pharmacy, University of Lille, Lille, France

Juergen Siepmann   College of Pharmacy, University of Lille, Lille, France

Prapasri Sinswat   Department of Pharmacy, Chulalongkorn University, Bangkok, Thailand

Hugh D.C. Smyth   College of Pharmacy, Health Sciences Center, University of New Mexico, Albuquerque, New Mexico, U.S.A.

Howard N.E. Stevens   Strathclyde Institute of Pharmacy and Biomedical Sciences, University of Strathclyde, Glasgow, U.K.

David Taft   Division of Pharmaceutical Sciences, Long Island University, Brooklyn, New York, U.S.A.

Ravi S. Talluri   Division of Pharmaceutical Sciences, School of Pharmacy, University of Missouri-Kansas City, Kansas City, Missouri, U.S.A.

Panna Thapa   Department of Pharmacy, Kathmandu University, Dhulikhel, Kavre, Nepal

Daniela Traini   Faculty of Pharmacy, University of Sydney, Sydney, Australia

Curtis Triplitt   Division of Diabetes, Texas Diabetes Institute, University of Texas Health Science Center at San Antonio, San Antonio, Texas, U.S.A.

Claire F. Verschraegen   Cancer Research and Treatment Center, Health Sciences Center, University of New Mexico, Albuquerque, New Mexico, U.S.A.

Nathan P. Wiederhold   Divisions of Pharmaceutics and Pharmacotherapy, College of Pharmacy University of Texas, Austin, Texas, U.S.A.

Robert O. Williams III   College of Pharmacy, University of Texas at Austin, Austin, Texas, U.S.A.

Paul M. Young   Faculty of Pharmacy, University of Sydney, Sydney, Australia

Xudong Yuan   Division of Pharmaceutical Sciences, Long Island University, Brooklyn, New York, U.S.A.

1

Advances in Pulmonary Therapy

Paul M. Young and Daniela Traini

Faculty of Pharmacy, University of Sydney, Sydney, Australia

Stephen Edge

Inhalation Device and Technology, Novartis Pharma AG, Basel, Switzerland

INTRODUCTION

Pharmacologically active substances have been administered to humans for thousands of years. However, it is only in relatively recent times that the true nature of drug delivery to the human body has become an exact science. Today, drugs are delivered via many portals into the body using a variety of dosage forms which allow medicament administration, for example, orally or parenterally. One relatively recent pharmaceutical development is delivery via the lung. Although the breathing of "vapors" has been used since ancient times as a way of relieving respiratory problems, the effective acceptance and mass commercialization of inhaled respiratory medicines was not achieved until 1948 when Abbot Laboratories developed the Aerohaler for inhaled penicillin-G powder and then revolutionized the field in 1955 with the advent of the pressurized metered dose inhaler (pMDI) (1,2). Since this inception, the range of inhaler products and medicaments has grown and expanded to encompass alternative drug delivery systems, namely those based on the dry powder inhaler (DPI), and combination products.

The lung offers a unique and challenging route for drug delivery for the treatment of respiratory diseases, such as asthma, chronic obstructive pulmonary disease (COPD), and cystic fibrosis. The organ offers a high surface area, circa $100 \, m^2$ (3), offering the possibility of high absorption and non-hepatic drug delivery, with obvious pharmacological advantages. Whilst inhalation pharmaceutical products may appear to operate on simplistic assumptions, the challenge for pharmaceutical technologists, formulators, and device engineers has been how to produce a product, which meets the demands expected by the pharmaceutical regulatory authorities and, importantly, patient acceptability. As well as developing such a product there are challenges in pharmaceutical development in terms of material and formulation characterization and scale-up.

The methods used to characterize inhalation products, as with other solid dosage forms such as tablets, are relatively simple and have been developed in recognition of the particular challenges for characterizing products for administration to the body. All pharmaceutical products must exhibit efficacy, safety, stability, and conform to regulatory drug product requirements. In addition, solid dosage forms tend to have their own set of characterization methods, for example, tablets are often tested for their hardness and disintegration time. The characterization methods for inhalation products have developed to include tests for performance based on, for example, fine particle fraction (FPF) and emitted dose. The development of such tests has progressed, in part, due to the complexity of the inhalation drug delivery process: the tests have to mimic the inhalation process from emission from the device to deposition in various regions in the lung.

In general, the practicable delivery of an effective respiratory dose of any medicament is achieved if the therapeutic component has an aerodynamic diameter of less than 6 µm (4). Although the deposition, adsorption, diffusion, clearance, and residence of particles in the lung is still not fully understood, it is generally accepted that this particle size range will produce a therapeutic outcome, with minimized upper bronchial impaction and expulsion due to limited sedimentation time in the alveoli.

The evaluation of the deposition profiles of particulates in the lungs and airways has been extensively reported (4–7). However, it is generally accepted that the inspirational airflow which inhalation aerosols and dry powders are exposed to affects their lung deposition distribution. Understanding particulate deposition in the lung is a complex issue. A physical basis for the deposition of aerosol particles in the human lung became available when Findeisen published the first lung deposition model (8). Important parameters for lung deposition are: the particle size distribution; the inhalation and exhalation flow rates, tidal volume, and the shape of the upper airways. The structure of the bronchial and alveolar regions in normal lungs is considered to be of minor influence; however, this is probably not the case for diseased lungs. It is generally agreed that particles smaller than 0.1 µm in diameter will deposit due to a diffusion processes (9), if the residence time is sufficient. Furthermore, the distance a particle travels by diffusion transport increases with decreasing particle size and increasing respiratory rate. Diffusional deposition therefore decreases with increasing particle size up to about 1 µm and becomes negligible for larger particles.

Sedimentation processes become dominant for particles larger than 0.1 μm diameter, and deposition increases with particle size, particle density, and respiratory rate. Consequently, in the size range of 0.1–1 μm, particles are simultaneously deposited by gravitational and diffusional transport. Inertial transport is the effective mechanism for larger particles, and deposition due to impaction increases with particle size, particle density, and airflow rate. Thus, in the size range above 1 μm, particles are deposited due to impaction and sedimentation. In general it can be concluded that deposition in the extra-thoracic and upper bronchial airways, through which the inhaled air passes at high velocity, is a result of impaction mechanisms while sedimentation processes dominate deposition in the lower bronchial airways and the gas exchange region where the residence time of the inhaled air is large (Fig. 1).

As the size of physical objects is reduced, their bulk properties and behavior are greatly modified. For example, sugar is available as granulated sugar and icing sugar. However, icing sugar (∼15 μm) and granulated sugar (∼500 μm) have very different cohesive properties. This phenomenon is similar to particulates for inhalation. Part of the reason is that the ratio of surface area to mass ratio increases as particle size decreases, leading to an increased chemical and physical (increased surface energy) activity. This creates difficulties for the processing, handling and aerosolization of particulates in

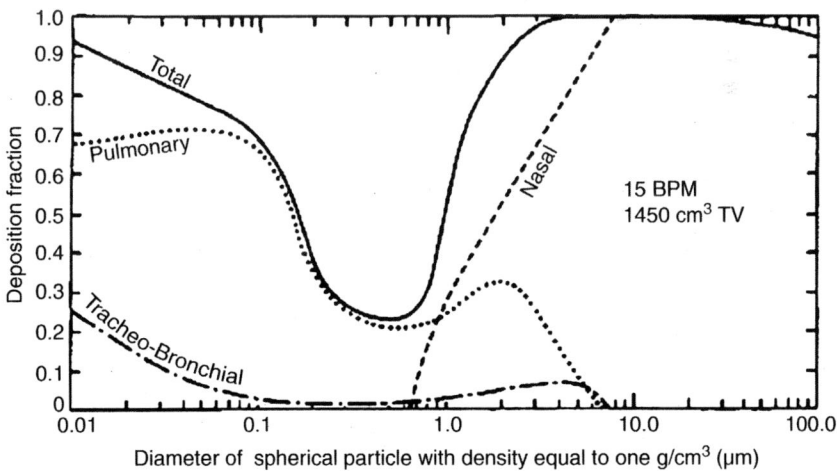

Figure 1 Deposition as a function of various sizes of inhaled particles in the human respiratory tract as calculated by the ICRP Task Group on Lung Dynamics for nose breathing at a rate of 15 BPM and a tidal volume of 1450 cm³. *Abbreviation*: BPM, breaths per minute. *Source*: From Ref. 10.

the dry state and non-aqueous liquid suspended states since the powders often form agglomerates, hindering the aerosolization of primary respirable particles. The intrinsic cohesiveness of such fine powders makes the accurate delivery of particulate medicaments to the lung a real challenge. A variety of approaches are currently utilized to overcome this challenge and optimize formulations for achieving a therapeutically effective delivery to the lungs.

This chapter will review the current state of the art of the three main inhalation delivery platforms (DPIs, pMDIs, and nebulizers), including novel exploratory delivery systems. Please note that this chapter will not extensively review the multitude of patents filed in this field.

DRY POWDER INHALATION

The Montreal protocol of 1989 banning the use of chlorofluorocarbons (CFCs) was a laudable achievement for the protection of the environment. However, while this ban was aimed at industries that used large volumes of these harmful materials, it also affected highly beneficial medicinal products, such as pMDIs, which had traditionally contained low levels of these CFCs. The consequence of this has been that existing pMDI products have had to be reformulated to contain hydrofluroalkane systems and new formulations must be developed to contain the replacement propellants. The challenges associated with such efforts have led to an increased popularity in alternative inhalation technologies, namely dry powder inhalers. As the name implies, DPIs contain and deliver the active medicament as a dry powder of suitable aerodynamic size for respiratory therapy. Dry powder particles of suitable size range, generally considered between 1 and 6 μm (4), can be readily produced. However, such particles have high surface area to mass ratio, making them highly cohesive/adhesive in nature. Consequently, the drug must be formulated in such a way that the energy input during inhalation is sufficient to overcome the contiguous adhesive and cohesive particle forces and aerosolize the powder for respiratory deposition. Although, in principle, this approach may appear straightforward, and there exist many DPI products which achieve this, the physico-chemical nature and interactive mechanisms of the components in a DPI system are still relatively poorly understood. Many commercial products have, by pharmaceutical standards, relatively poor efficiencies, with often less than 20% drug being delivered to the lung (11). This has generated a significant amount of research activity in the fields of pharmaceutics, powder technology, surface, aerosol and colloid science which has focused on understanding the interactions in DPI systems.

In simple terms, DPI technology can be categorized into four areas; Formulation; active pharmaceutical ingredient (API) powder, Device; and Manufacturing. It is clear that understanding and engineering the physicochemical properties of the materials and processes in DPI technology will allow us to influence, and control, the aerodynamic efficiency and thus therapeutic efficacy. The DPI device will play a pivotal role in efficient API aerosolization, since its geometry and configuration will influence the flow and sheer forces acting on the formulation. Many design approaches have been used to achieve and improve drug liberation, as can be seen from any cursory survey of the patent and product literature. Currently, marketed devices can be generally classified according to their method of energy input, formulation approach and dosing regime, and how these products achieve reproducible delivery of respirable API particulates. Common approaches employed to assure such reproducible delivery of the API (delivered dose) are by formulation of a micron sized drug with a larger inert carrier, agglomeration with an excipient of similar particle size or by simple agglomeration of the drug (where metering dose constraints allow). These formulation approaches may be termed as carrier and agglomeration based systems, respectively. Both systems result in improved powder flow, ease of metering and consistent drug mass entrainment. Examples of commercially available formulations which have been developed using these approaches are represented in Figure 2.

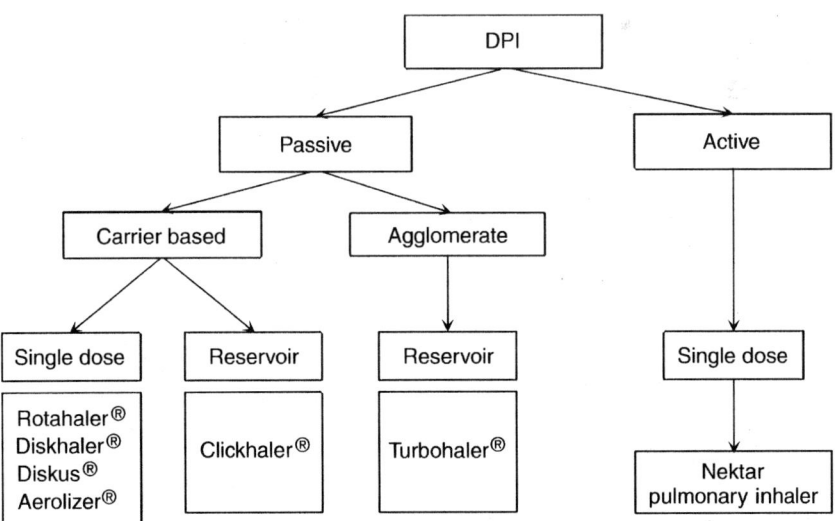

Figure 2 Conventional marketed device approaches. *Abbreviation*: DPI, dry powder inhaler.

The delivery of the drug particulates to the lung from passive inhaler products is achieved after liberation of the drug from the formulation during the inhalation actuation, as represented in Figure 3.

In addition, recent technological developments in spray drying and particle engineering have resulted in co-processed single dose DPI formulations combining, for example, a mixture of excipients (12) in the commercial insulin product, Exubra® (Pfizer/Nektar) (13), or by the addition of excipients which promote aerosolization efficiency through altering the particle physical morphology (14,15).

Carrier-Based Systems

The majority of DPI products use the carrier-based formulation system. This approach has proved successful for a range of dosing regimes from high dose, that is, up to several hundred micron, for example, 400 μg for albuterol sulphate to low dose, for example, ≤12 μg for formoterol. In the 1970s, Hersey theorized that when small particles were mixed with larger particles (such as conventional carrier material) a so-called ordered mix was formed (16). Although these observations were deemed relevant to low dose tablet formulations, this theory was later expanded in the 1980s and 1990s by Staniforth (17), and has since become an underlying theory for carrier based DPI powder formulations. In general, this approach has been

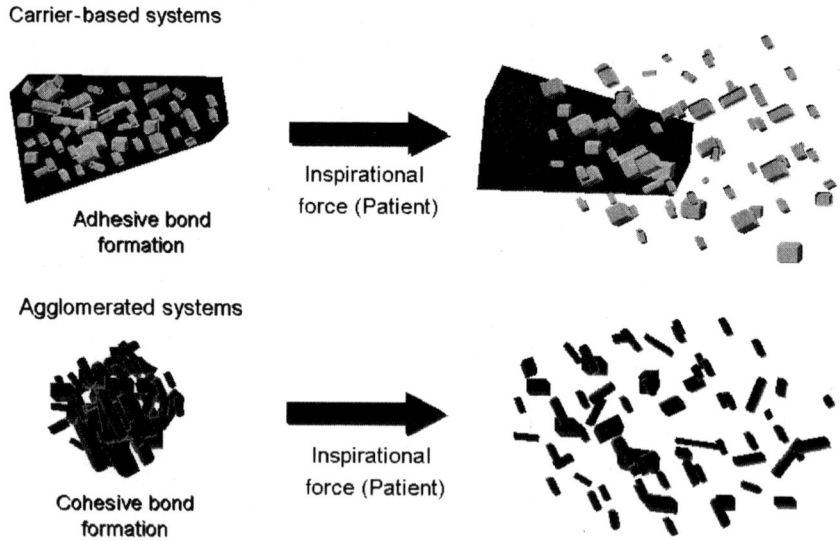

Figure 3   Conventional DPI formulation approaches

adopted by the pharmaceutical industry to produce many successful inhalation products, and, interestingly, as a way of preparing low dose blends for tableting applications. These simple DPI formulations are typically prepared by mixing micronized drug with larger lactose carrier particles. The drug adheres to the larger particles in such a way that results in good blend uniformity and flow. Additionally, and importantly, when the formulation is delivered to the patient via a device, the drug particles are liberated to provide a reliable efficacious dose to the patient. This technique has been proven to be versatile and has been used to develop products which contain low dose drugs products, for example, Novartis® 12 μg formoterol Foradil® product, equivalent to a c.a. 0.05% w/w drug/carrier blend. To put this achievement in powder blend and emitted dose uniformity into perspective, an image of a low dose Foradil blend is shown in Figure 4. As can be seen from Figure 4, the formulation consists of a mixture of larger carrier particles together with associated lactose fines and drug particulates. The relative surface distribution of drug particles to carrier material will be very low and yet the product provides good dose uniformity. The same is true for higher dose formulations which contain several hundred micrograms of drug which contain higher ratios of drug to carrier. However, the important point with drug loading in DPIs is that the drug will be associated with the surface of the carrier, making the surface ratios much higher than mass ratios. Obviously, for all DPI products, carriers and drugs with consistent physiochemical characteristics are required to maintain product performance. This dependency on material characteristics has implications for drug aerosolization (liberation) performance if variations in API adhesion exist throughout the carrier system.

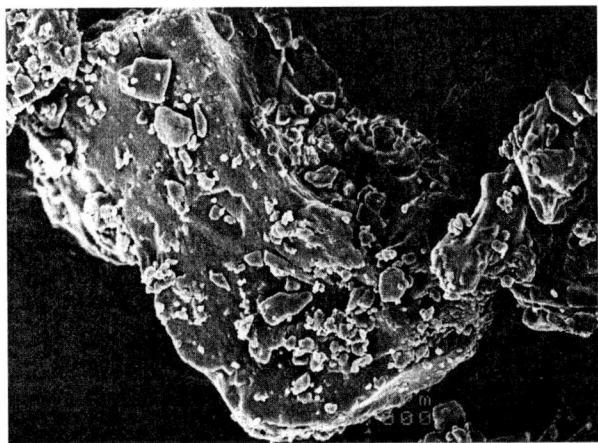

Figure 4 SEM of a commercial Foradil® formulation.

Agglomerated Systems

Agglomerated systems are not as commonly used as the carrier-based systems but the technique can be successfully employed to produce DPI products. Agglomerates containing pure drug, or a mixture of drug and fine particle excipients, can be prepared by spheronization under controlled conditions. The key requirement of the agglomerates is that they have sufficient strength to survive the manufacturing processes, filling, and metering, but still allow efficient de-agglomeration and aerosolization during delivery to the patient. The most obvious, and successful, example of this formulation approach is Astra Zeneca's Turbuhaler® where the agglomerated formulation contains only the drug (e.g. 100, 200, and 400 μg dose budesonide Pulmicort®) or with the addition of micronized lactose excipient (6 and 12 μg dose formoterol Oxis®).

As with carrier based systems, these formulation approaches result in improved reproducibility in emitted dose and respiratory deposition efficiency. However, both the efficiency of respiratory deposition will be dependent upon the energy imparted into the system upon inhalation (i.e. by the inspiration energy of the patient), where in general, agglomerated based systems require a higher pressure drop across the device to achieve this goal (18). For all DPIs, this is the real challenge for formulators and engineers: how to balance the mixing of particulate materials so they form a stable processable formulation but ensure that the interactions are weak enough to allow liberation of the drug particulates during inhalation.

Although these principles of formulation seem simplistic, it is possible to put this technological and scientific challenge into perspective when considering the forces involved in a typical DPI system. For example, it is estimated that the force acting on a 1-μm spherical particle (adhered in a carrier system) in a 10 m/s laminar air flow, comparable to the axial velocity in an Aerolizer® DPI (19), may be estimated as 2 nN while the adhesion force, using empirical calculation, is of the order of 100 nN (7). Clearly, with such disparity in expected and predicted particle exposure and detachment forces, efficient aerosolization efficiency is a scientific and technological challenge. Subsequently, efforts to improve the performance of DPIs have focused on the engineering of excipient and API physico-chemical properties to overcome these forces and, modifications in device design, to improve energy translation and efficiencies.

DPI Devices

To date, devices have been primarily passive, in that they rely on the patients inspiratory flow to generate the energy for drug entrainment and liberation. Such devices often contain engineered components which can facilitate drug liberation. Historically, these devices aerosolized a powder as a fixed single formulation dose (e.g. Rotahaler® GlaxoSmithkline®) or by aerosolization of a single formulation dose from a rotating capsule (e.g. Aerolizer Novartis), where drug particle liberation is presumably achieved by exposure of the formulation to the airstream, and in some cases, a baffle or grid (18,20). To improve the aerosolization efficiency of these early devices many attempts have been made to increase aerosolization efficiency, including more tortuous flow pathways and increased pressure drops (resulting in increased particle acceleration and turbulence), for example, the Turbuhaler (11,18,20). However, in terms of delivered dose, it is fair to say that there has been no great leap in efficiency improvements. In addition to these advances, the incorporation of multiple blister or reservoir systems has resulted in improvements in the ease of use by the patient. More recently, active devices, utilizing pressurized gas, have been marketed that potentially increase aerosol efficiency and remove the requirement for patient energy input (e.g. as used in Nektar's Pulmonary inhaler) (21). However, such systems have a unit price and ease of use consideration. Ultimately, whichever approach is adopted, advantages and disadvantages will exist and the efficiency of the system will be related to the formulation and design ingenuity.

Engineering a DPI Powder to Improve Therapeutic Outcomes

As previously discussed, the efficient respiratory deposition of the API powder will be dependent on the physico-chemical nature of the particulate system and, in addition, the physical and environmental conditions to which they are exposed. At this point it is important to state that there are limited products presently available which contain engineered excipients or drugs. However, there has been much research activity into the modification of the powder and material components of DPIs.

Carrier Modification

One of the most commonly studied areas in DPI formulation design is that of carrier modification. Difficulties (both technical and regulatory) in altering the physical properties of micron sized drug particulates has led to the more popular approach of modifying excipient carrier materials to achieve improvements in blend uniformity and delivered drug dose. It must be stated, however, that in general, this approach is still only led to limited empirical observations concerning a material descriptor and formulation performance, and it is unclear to what degree such observations have impacted pharmaceutical manufacturing processes. A series of approaches have indicated that the optimum performance characteristics of a DPI product require a formulation which exhibits adequate flow (for processing and dosing) and a certain level of excipient fines.

The effect of surface morphology: It is reasonable to expect that the macro-, micro-, and nanoscopic morphology of an excipient carrier will play a major role in efficient drug blending and aerosolization since micron sized particles will regularly encounter adhesive forces that are greater than those required for their liberation. Subsequently, the most common investigated approaches to improve drug removal from carrier particles during inhalation are to (1) reduce the contact area between drug particle and carrier, (2) alter the particle geometry of the carrier, and (3) reduce the surface energy of the contiguous surfaces. Of these three approaches, by far the greatest attention has been focused on modifying the carrier morphology in order to achieve changes in the contact geometry between drug and carrier. A simplistic representation of the ranges of surface geometries which may be encountered is shown in Figure 5. In simple terms, by varying the rugosity of a carrier it becomes possible to vary the degree of adhesion and, thus, alter the efficiency of drug liberation (Fig. 5).

Although roughness is believed to be one of the dominating factors for the performance of such systems, it is generally very difficult to investigate the relationships between one material factor, in this case roughness, and performance without altering another potentially influencing factor, such as particle shape or quantity of intrinsic carrier fines. This makes quantifiable interpretation of any observations difficult. Furthermore, the potential

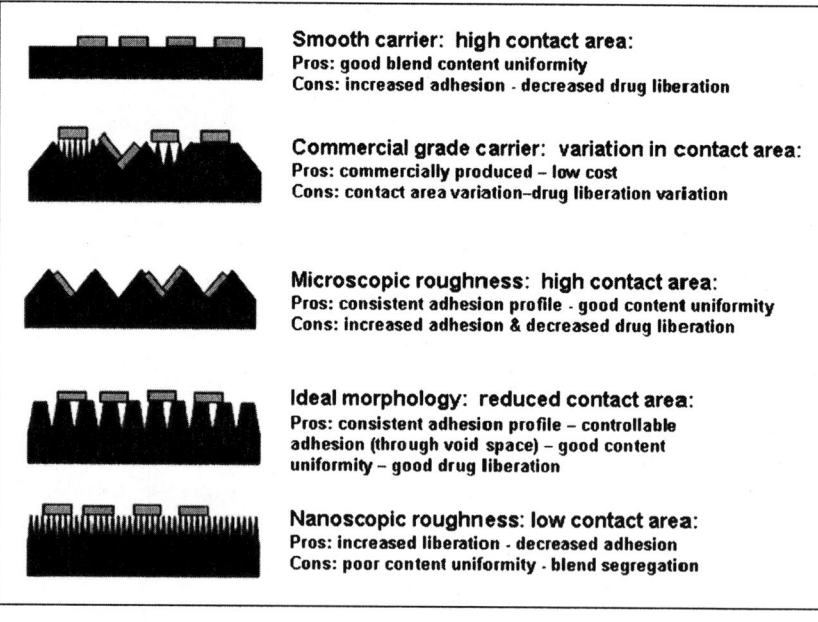

Figure 5   Influence of carrier morphology on drug particle adhesion.

therapeutic improvement of altering the carrier roughness has, in general, been investigated via comparison of off-the-shelf lactose excipients with materials produced by some variation in batch or process parameters. However, in general, the hypothesized influence of carrier morphology on aerosolization performance (as represented in Fig. 5) is generally in good agreement with recent literature. For example, recent work by Flament et al. (22) has suggested a linear reduction in the FPF of micronized terbutaline sulphate with increased roughness, measured in this case by microscopic luminescence, of a 63–90 µm sieve fractioned lactose samples. Earlier work, by Kawashima et al. (23), also related the surface area and roughness parameters of a series of lactose carriers (approximated median diameter 60–65 µm with similar geometric standard deviations) to the aerosolization performance of micronized pranlukast hydrate (Fig. 6).

In general, as can be seen from Figure 6, an increased carrier roughness and surface area resulted in a reduction in drug aerosolization efficiency. Interestingly, a deviation from linearity was observed for lactose type F, and was attributed to a variation in amorphous content and/or increased drug-carrier contact area (e.g. as in the smooth carrier example shown in Figure 5. Other recent studies have reported similar observations. For example, Zeng et al. (24) reported that when the surface structure of a re-crystallized lactose carrier was modified (by etching with a 95% w/w

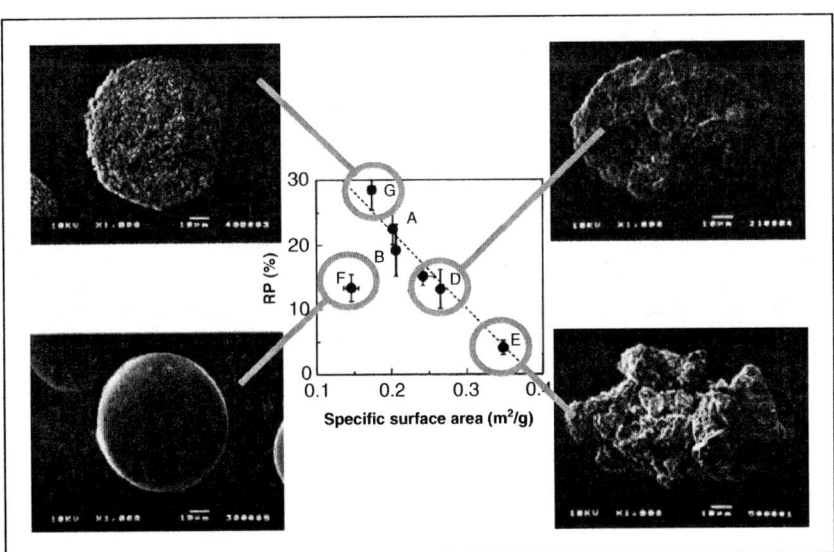

Figure 6 Influence of carrier surface area on the aerosolization performance of pranlukast hydrate. SEM images represent the carrier used for the measurements circled. *Abbreviation*: RP, respirable particle. *Source*: From Ref. 23.

ethanol solution), the induced surface cavities resulted in a decrease in FPF of blended albuterol sulphate. It is interesting to note, however, that the addition of fine lactose, approximated as 5% (between 5 and 10 µm), rectified this reduction. In other recent studies, the macroscopic etching of commercial grade lactose via mixing in an ethanolic solution (25–27) or via temperature controlled surface dissolution (28) has been shown to improve the aerosolization of various micronized APIs. As with the studies presented previously, it is envisaged that the improvement is, in part, due to a reduction in macroscopic roughness, limiting the contact geometry between drug and carrier surfaces. This potential for particle entrapment in large pits, crevices or craters is decreased as the macroscopic roughness is reduced, as can be seen by atomic force microscopy (AFM) topographical images of a commercial lactose carrier pre- and post-etching (Fig. 7).

In addition to direct surface modification, significant research has focused on the relationships between carrier geometry, for example, crystal habit, dimensions, and performance. In recent studies, Zeng et al. (29) have suggested that the crystal shape of lactose carriers can have significant impact on the aerosolization performance of attached drug particulates. In general, an increase in elongation (aspect) ratio was reported to increase the aerosolization performance (FPF) of albuterol sulphate, although a decrease in blend uniformity was observed (29). Again in these studies, it is suggested that a reduced carrier roughness resulted in increased fine particle drug. Furthermore, it is interesting to note that during these investigations, another important factor influencing aerosolization performance was noticed; the presence of fines.

Apart from direct morphological and surface roughness parameters, one of the key factors that have been suggested to influence aerosolization performance of adhered drug particulates is the presence of similar sized excipient fines, for example, particles with a volume median diameter < 5 µm.

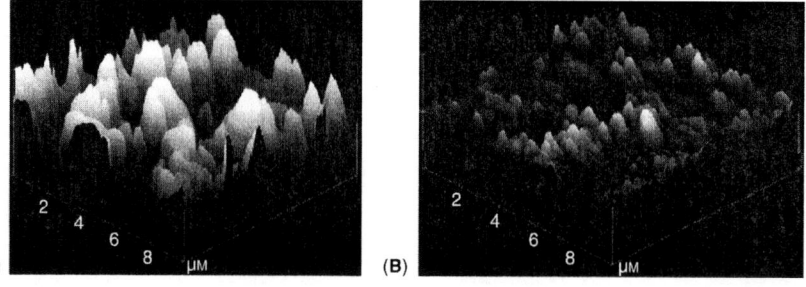

Figure 7 AFM topographical images of (**A**) a commercial grade lactose carrier surface and (**B**) surface-etched carrier. *Source*: From Ref. 26.

The "positive" effect of excipient fines for drug aerosolization would also be expected since the surface of a commercial grade lactose carrier, as represented in Figure 8, will contain a distribution of "energy" sites that will promote particle adhesion to different degrees (16), and thus affect drug particle liberation in DPI carrier based systems (30). These sites of high adhesion may be attributed to a combination of "particle entrapping" crevices in the surface morphology of the carrier, crystalline phases with higher surface energy and the presence of amorphous material. Recent research, investigating the presence of these "active sites" in carrier systems, has suggested that an energy distribution exists for such sites, which, for low dose formulations, significantly influences drug aerosol performance (31,32). This proposed variation in drug adhesion can be observed when directly measuring the adhesion forces experienced by a single micronized drug particle, over the surface of a commercial lactose carrier (Fig. 8).

Clearly, the distribution of such sites could lead to increased particle adhesion and thus variability in content uniformity and drug aerosolization efficiency. The addition of excipient fines to a simple carrier formulation, may improve the aerosol performance due to "filling" these "active sites"

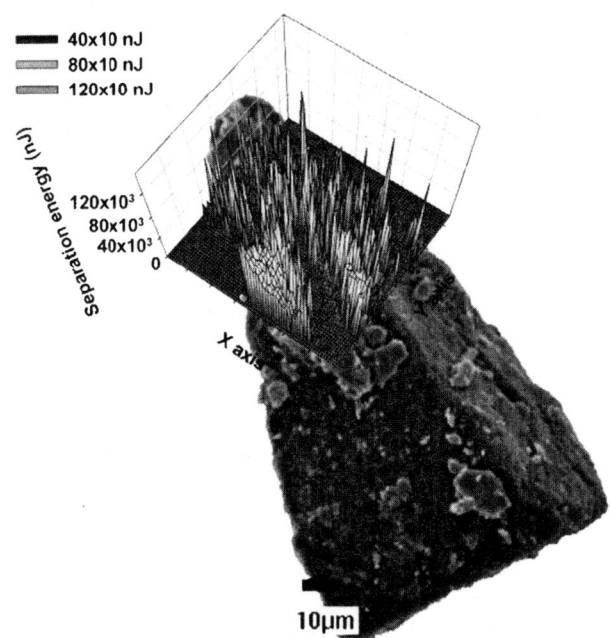

Figure 8 SEM of lactose carrier with an overlaid adhesion map of the interaction between a single micronized albuterol sulphate and carrier surface.

(by displacing existing drug particles), and/or by improving aerosol performance by the formation of fine-drug agglomerates. The presence of both these morphological features can be observed via high resolution Scanning electron micrograph (SEM) of a carrier particle surface (Fig. 9).

The mechanism for improved performance in these systems is dependent on both the previously described formulation factors, and the relative cohesive/adhesive balance between the components involved. Recently, direct measurement of inter-particulate forces in conventional DPI systems has shown that a clear balance may be observed between the adhesion and cohesion profiles of each component in the formulation (33,34), and these factors may be used to predict the performance of the formulation. Such observations are also in good correlation with previous studies that predict ordered mixing in tableting ingredients (35).

Regardless of the mechanism, it has been shown, as discussed previously, that the addition of fine excipient material to an inhalation formulation, results in improvement in performance outcomes (i.e. improved aerosolization efficiency). Recent work by several groups has suggested that the presence of carrier fines, significantly improve drug aerosolization performance by either filing these "active sites" or forming discrete agglomerates (24,36–42). For example, studies by Lucas et al. (42) indicated that, regardless of the mixing order, the addition of fine lactose particulates to a formulation increased the aerosol performance (Fig. 10).

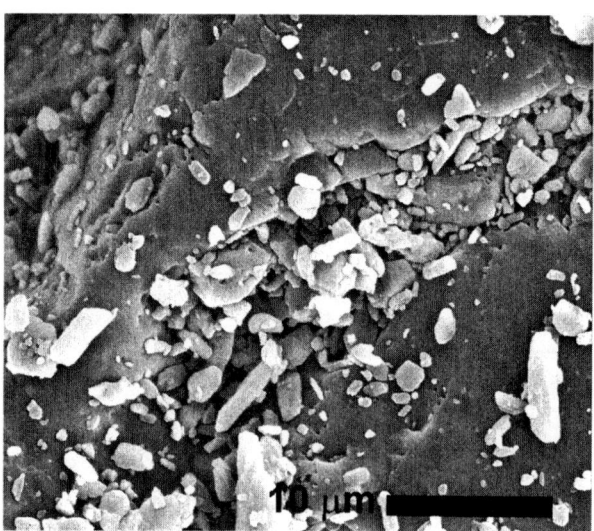

Figure 9 SEM of a fine particulate system in the crevice of a lactose carrier surface (albuterol sulphate–lactose carrier). *Source*: From Ref. 32.

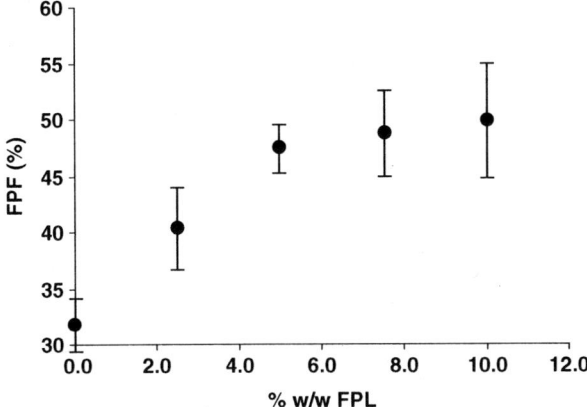

Figure 10 The influence of fine-particle lactose (FPL) additive on the aerosolization performance of a protein-carrier formulation. *Source*: From Ref. 42.

Similar observations were later described by Louey et al. Islam et al., and Adi et al., (37,38,43,44). Of note, Islam et al. (37,38) suggested that, while the carrier size and surface features had an influence on aerosol performance, the addition of fines into these formulations dominated performance. More recently, similar studies, investigating the influence of carrier milling, on aerosolization efficiency of blended APIs, again suggested that the introduction of fines dominated the performance compared to other factors, such as particle size and amorphous content (40,41).

Although not fully understood, the presence of a certain level of fines clearly results in improved drug aerosolization performance and the influence of these small particulates on active site filling and/or spontaneous agglomeration is still under investigation. For a more in-depth review of the influence of fines on DPI performance, the reader is referred to a recent review by Jones and Price (45).

Another approach to improve the aerosolization efficiency of blended drug particulates from carrier based systems is to alter the surface chemistry of the parent carrier. Work by Tee et al. (46) and Steckel and Bolzen (47) have investigated the influence of various sugars and sugar alcohols (from polyols, such as mannitol, to more complex disaccharides, such as lactose) on the aerosol performance of different model drugs. The theory behind this approach is that any particular crystalline system has a finite surface energy. Thus, by changing the nature of the crystalline habit, morphology or surface chemistry, the relative particle adhesion of an API will be altered. For example, Tee et al. (46) reported that, unsurprisingly, different sugar alcohols, such as mannitol, altered the aerosol performance of a micronized albuterol sulphate formulation. Interestingly, these studies also highlighted the positive effect of adding different amounts of sugar or sugar alcohol fines to these formulations. Other studies have investigated the influence of different crystalline carrier materials on drug aerosol performance and have reported that materials such as sugars and sugar alcohols can increase/ decrease aerosol performance (47–49). However, it is important to note, that such observations are difficult to compare, since different drugs, carrier size distributions and analytical techniques are used. Furthermore, other compounding factors may ultimately influence performance when trying to compare alternative carriers. For example, authors such as Steckel et al. (47) have reported significant variability in drug aerosolization performance when evaluating a single carrier material (e.g. lactose or mannitol) from multiple batches or different suppliers (47,50). That is not to say, it is impossible to make comparisons between materials, when certain carrier descriptors are altered, as long as most variables are retained constant. In practice, however, this is very difficult since carbohydrates have different moisture sorption characteristics, important for DPI performance, and it is challenging to produce sieve fractions with identical descriptors to that of the comparator material, namely lactose. One alternative to using different carbohydrate type materials for modifying performance is to alter the surface chemistry of a conventional lactose carrier. This may be achieved by varying the crystallization conditions, to alter the crystal habit or poly-morphic form (51), or by the addition of ternary agents that act to control the force of interaction (26,52–55). For example, materials such as magnesium stearate (25,26,52,53,56) and leucine (54) are reported to improve the aerosolization efficiency of carrier based system by modifica-tion of the adhesive forces between the drug and the carrier. This change in adhesive behavior can be observed in formulations of beclometasone dipropionate which suggested that the addition of magnesium stearate into the surface of a smoothed lactose surface increased the respirable dose of by a factor of 3.5 (29 and 102 μg doses for the smoothed and magnesium stearate treated lactose carriers, respectively) (26). Similarly, investigation of the cumulative adhesion energies (measured by colloid probe microscopy, indicated a decrease in the 90th percentile adhesion energy by the same factor (3.5) (from a 112×10^{-9} nJ to 32×10^{-9} nJ for magnesium stearate treated systems). More recent work by Iida et al. (52,57), has reported similar improvements, and, improved performance after storage of lactose-magnesium based formulations at elevated humidities. Recent technologies that incorporate this formulation strategy are Vectura's PowderHale[®] (58), and Chiesi's beclometasone Pulvinal[®] product (56).

In general, however, regardless of direct carrier adhesion or agglomeration formation, the ultimate aerosol performance of a DPI powder will be based on the inter-particulate forces acting between drug and carrier before, during and after processing and after aerosolization. As can be seen form the overview above, significant advances have been made in understanding these forces (e.g. recent work using colloid microscopy has been utilized to predict carrier adhesion and/or agglomerate formulation (26,33,34,59–61), and in general, the aerosol efficiencies of recent DPI systems are greatly improved compared to the early devices.

Modification of the API to Improve Respiratory Deposition

As previously discussed, in order to achieve efficient respiratory delivery, the drug powder requires an aerodynamic diameter of $< 6\,\mu m$ to avoid impaction and/or sedimentation in the upper airways (4). Historically, the method of producing particles in this size range is through high energy milling or micronization of a larger crystalline starting material. However, this method of production leads to irregular particle morphology and the potential for the formation of amorphous regions in the sample surface, as represented in Figure 11. Such regions can be highly unstable, and for small pharmaceutical molecules, this may result in spontaneous re-crystallization when exposed to environments that facilitate a lowering of the glass transition (such as elevated humidity) to ambient conditions (62).

Since these micronized particulates have a high surface area, the instability in any surface amorphous regions may lead to variations in surface energetics and, where re-crystallization occurs, the potential for particle fusion, which may affect the particle size distribution, and subsequent FPF. Consequently, there has been an impetus in the physical sciences to gain a greater understanding of surface induced amorphous material and to detect its presence at the low levels (63,64). In a drive to reduce surface instability, amorphous content, and high interfacial energies, in DPI drug, a significant amount of research into particle engineering has occurred.

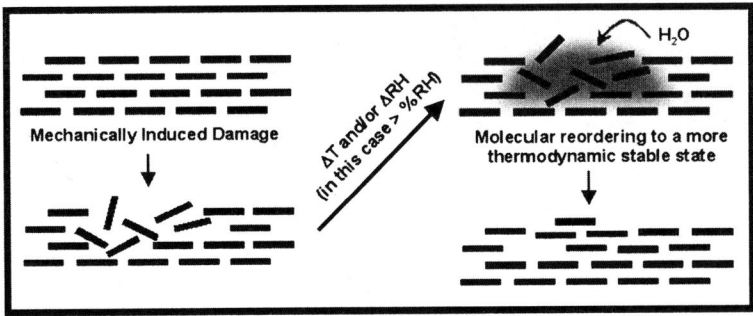

Figure 11 Schematic of mechanically induced molecular surface damage (amorphous regions) and recrystallization.

There are several approaches which have been successfully used to prepare respirable drug particles that have shown improved therapeutic outcomes, when compared to conventional micronized drugs. The most common approaches are to either precipitate material of the required size range (i.e. controlled crystallization), or to flash evaporate the drug from a solution with/without excipients, that facilitate a stable particulate system with chosen physical attributes.

Precipitation of Particles to Improve Respiratory Therapy

As previously discussed, the conventional bulk precipitation of small molecule type drugs will result in a product which exhibits a particle size distribution which is unsuitable for respiratory delivery (65). In order to precipitate respiratory sized particles from solution in a single-step, a high degree of micro-mixing and molecular diffusion is required. Many techniques have been developed recently to overcome the growth kinetics, and generally either involve high energy input, for rapid micro-mixing, and/ or rapid diffusion of solvent and anti-solvent. One area that has received considerable recent attention is the use of supercritical fluids to precipitate particles of respirable size (66). Supercritical fluids are at a temperature and pressure that result in them behaving like both a gas and liquid (i.e. they can diffuse through materials easily, like a gas, whilst maintaining a high solvation capacity, as a liquid). Furthermore, by modifying the temperature and/or pressure within this supercritical region, large variations in properties such as density can be controlled. Since physical properties, such as density can be directly related to the solubility (or anti-solvent power), a range of technologies have been patented that use supercritical fluids to prepare particulates of a respirable size range (66). One of the most successful approaches, in applying this technology to dry powder inhalation was the research group at Bradford University (UK) (67–69) whose spin-off company, Bradford Particle Design (later acquired by Inhale (now Nektar)) (70), utilized a process referred to as Solution Enhanced Dispersion by Supercritical Fluids (SEDS) (Fig. 12). The controlled precipitation of particles using supercritical fluids has been used to produce different drug polymorphs (68) and powders with more suitable morphologies for efficient aerosolization (69). In addition recent studies, comparing the aerosolization efficiency of SEDS and micronized drug powders from lactose carrier blends, has suggested, that while the aerosol performance of similarly sized SEDS and micronized powders were similar, the physical stability of the SEDS material was improved (71).

The difference in aerosolization performance between the micronized and SEDS sample was attributed to mechanically induced amorphous content, present in the micronized sample, re-crystallizing, and causing particle fusion (71). This would be expected to occur when the

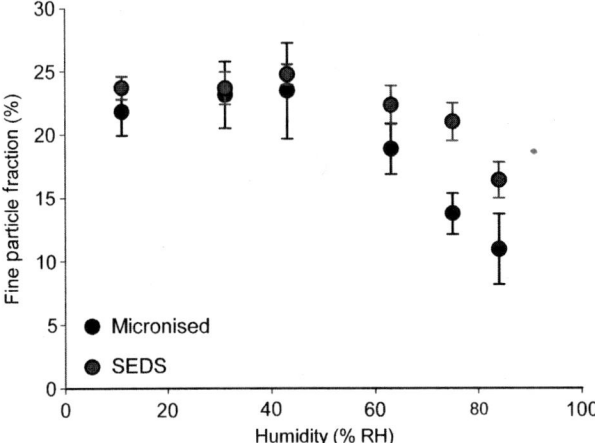

Figure 12 Aerosolization efficiency of micronized and SEDS-produced albuterol sulphate from a lactose carrier formulation as a function of relative humidity. *Abbreviation:* SEDS, solution-enhanced dispersion by supercritical fluids. *Source:* From Ref. 71.

environmental storage humidity was increased sufficiently to reduce the glass transition to ambient temperatures (albuterol sulphate glass transition humidity at 25°C is estimated to be approximately 50% RH (72)) and could thus be observed in investigations conducted after storage at higher humidities (>60% RH).

Other methods capable of producing discrete particle precipitation, rapid nucleation, and high rates of micro-mixing include sonication (73), high gravity precipitation (74–76), and impinging jet methodology (77). Recent work by Kaerger and Price (73) has suggested that aerosol droplets produced via electro-hydrodynamic or an air pressure atomizer could be captured in a anti-solvent and crystallized via sonic energy. This method of atomization and crystallization via sonication (SAXS) was shown to produce uniform particles with nanoscopic morphology within a suitable size range for respiratory delivery (1–5 μm).

High gravity precipitation has also been recently investigated (74,76) and relies on rapid mixing, in a high gravity rotating packed bed (HGRPB) rector, to produce micron and sub-micron particulates. As with all these high energy systems, such as SAX described above, the HGRPB relies on a high energy input (in this case intensified mass and heat transfer) to control the nucleation and crystallization of small particulates at a size suitable for respiratory therapy. Recently, this method has been successfully utilized to produce respiratory sized albuterol sulphate crystals, which depending on the operating conditions, can exhibit FPFs up to 55% (76).

Particle Formation via Rapid Drying

An alternative to precipitation and crystal formation of APIs with required size distributions for respiratory delivery is the rapid evaporation of the particles by spray drying (78). Spray drying has conventionally been used in the production of foodstuffs and pharmaceutical excipients, however, in recent years it has gained popularity as a means of preparing APIs in a suitable form for respiratory deposition. The most obvious commercial example of API spray drying is the recent Exubera® formulation (13).

By modifying the drug solution concentration and operating conditions it becomes possible to, with relative ease, alter the particle size or morphology of the final product. This has obvious implications, as reported by Chew et al. (79), where the altering of the spray drying conditions resulted in the control of the degree of particle corrugation of an serum albumin (BSA), and thus, the aerosolization efficiency. An example of the surface morphologies of these particles is shown in Figure 13, and the image suggests that the degree of particle interactions between contiguous surfaces will be altered.

Additionally, the aerosol performance can be improved by the addition of excipients so as to vary the particle density, aerodynamic size, contact area, and physico-chemical stability of the formulation. For example, a mixture of excipients is employed during the spray drying process for the preparation of the insulin formulation for the inhaled product Exubera (13). The excipients act as bulking agents and aid stabilization of the formulation (12).

The addition of other excipients, such as "blow-out" fluorocarbons, during the spray drying of formulations results in porous particles with reduced density and contact area (81). This approach was used in Nektars

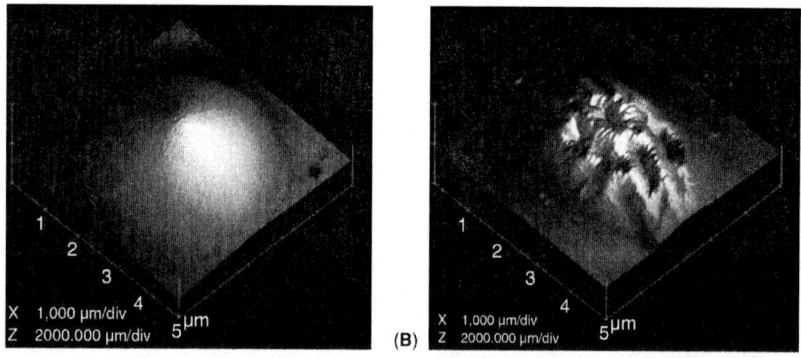

Figure 13   AFM topography image of spray-dried bovine serum albumin particles prepared using methodologies described by Chew et al. (**A**) Smooth particulates and (**B**) corrugated particulates. *Source*: From Refs. 79,80.

PulmoSphere® technology. In a recent study, dry powder budesonide PulmoSpheres were reported to exhibit improved in vivo respiratory deposition when compared to a conventional micronized formulation (14). These hollow, porous systems are designed to have reduced contact area and lower density. The culmination of these two factors will result in improved powder aerosolization efficiency and reduced aerodynamic diameter (82). However, it is important to note, such systems will still have a relatively high surface area to mass ratio and therefore will presumably be highly cohesive/ adhesive in nature. An approach to overcoming this is to produce larger particles, which have volume diameters of the order 20 μm but with densities that result in aerodynamic diameters of around 5 μm (15). An example of this approach is Alkermes Air® technology.

Improvement in Device Design and Packaging Optimizes Delivery Performance

The performance of any DPI product will be dependent upon the relationships between the device design characteristics, the formulation and the airflow generated by the patient (and in the case of active systems, the device). The basic role of the device is two-fold; to facilitate patient manipulation and to provide a path for the formulation from the reservoir or storage unit to the patient. Additionally, in the case of reservoir based products, it should provide a stable environment for the formulation and an appropriate metering system. Therefore, the device component of a DPI product offers opportunities to modify the performance of the formulation by design engineering, for example, incorporating gauzes and meshes and modifying the nature of the airflow and aerosol generation of the formulation pre and post emission from the dose container system.

The original DPI device, the Spinhaler®, consists of a simple capsule which contained the medicament, a single drug, disodiumcromglycate. The capsule is inserted into the device, pierced and a combination of the airflow generated by the patient and the Spinhaler device ensured the delivery of the drug to the patient. This "simplistic" concept of using a piercable formulation-containing a capsule in device has continued to be successfully used in DPI products, with particular application to the so-called "once-a-day" medications. For example, the Spinhaler concept was the basis for the development of GlaxoSmithKline's Rotohaler® and, more recently, the HandiHaler® (Boehringer-Ingelheim) and the Aerolizer (Novartis) and a multi-capsule product, the Aerohaler® (Boehringer-Ingelheim). As well as successfully producing single dose capsule based DPI products the industry has also invested in developing multiple dose and multi-drug products. Perhaps the best known multiple dose products are Astra Zeneca's Turbuhaler and GlaxoSmithKline's Diskhaler® and Diskus®. Whilst they are termed multiple dose products, their achievement of dose multiplicity is very different. The Turbuhaler is based on a device containing a reservoir of powder, which can be pure drug or a blend with lactose, which is metered immediately prior to delivery to the patient. The Diskhaler and Diskus multiple unit dose products employ discs and strips of individual blisters of medicament formulation respectively. The blister is pierced (Diskhaler) or removed (Diskus) immediately prior to delivery to the patient. In the case of the Diskhaler, discs of blistered medicament can be replaced in the device, whereas the Diskus cannot be replenished when the blister strip is empty. Since the introduction of these products, many pharmaceutical companies have continued to design more elaborate devices for delivery of generic drugs, drug combinations, and new chemical entities. In particular, a cursory review of the literature reveals that there have been developments in marketed multiple dose devices such as Chiesi's Pulvinal and Innovata's Clickhaler® (11). Often such devices are based on the principles of the first generation products but have evolved to include patient friendly features such as dose counting, for example, in the Turbuhaler, the Clickhaler and the Twisthaler®. In addition, the concept of replenishable single dose capsule formulations has been extended to multiple dose DPIs with the marketing of the Novolizer®, which has a replaceable multi-dose cartridge.

For a more in-depth review and discussion of conventional passive DPI systems, the reader is referred to recent reviews (11,83).

Perhaps one of the most interesting applications of DPI technologies is the delivery of high molecular mass materials, such as proteins, to the body, offering systemic drug delivery. One such recent development is Inhale's pulmonary Exubera system. In addition to being an active device, the product allows the delivery of insulin to the lung via a respirable mixture of excipients and insulin. The obvious advantage is that this avoids the parenteral route for insulin treatment. Active devices also remove the need for patient energy input, and provide a means of reproducibly aerosolizing the powder, which maybe inhaled via a spacer unit or coordinated with the patient's inspiration. For example, Vectura's Aspirair® (84) utilizes a compressed air reservoir (manually charged by the patient) to aerosolize the powder when the patient inhales. Activation in this system is achieved via a breath sensor, and the aerosol pathway is designed so the powder is de-agglomerated in a vortex type assembly (84).

Another approach, employed by Britannia Pharmaceuticals, utilized a pressurized air or a CO_2 canister to aerosol high dose (25–250 mg), formulations through a venturi (85), which could be inhaled via steady or tidal breathing. As an alternative to coordinated breathing or automated activation of powder aerosolization, the use of a holding chamber may be utilized. An example of this approach, is Nektar's Pulmonary inhaler, which generates a standing cloud of aerosolized powder, via pressurized gas (manually charged by the patient), which can be inhaled via controlled or slow tidal inhalation.

LIQUID AND PROPELLANT SYSTEMS

Liquid and propellant aerosol systems can be rudimentarily sub-classified into nebulizer and pMDI technologies. The main difference between these two systems is that nebulizers utilize an external energy source to produce aerosolized fine particulate droplets of the formulation whilst, pMDI systems incorporate a supercritical propellant into the formulation which provides the energy for aerosolization. Both of these approaches have advantages and disadvantages can contain either solution or suspension based formulation strategies and have undergone technological advances in recent years.

Recent Advances in pMDI Technology

Although the respiratory tract has been a therapeutic target for many centuries, the pMDI is generally considered the first generation of marketed inhalation products that adhere to the vigorous regulatory requirements. Marketed since their introduction by Riker Laboratories, Santa Barbara (now 3M Healthcare) (86) in 1956 and many authors have described their basic form and functionality (87,88). Fundamentally, pMDIs rely on aerosol propellant technology, in which a high vapor pressure gas (typically a liquid propellant) is contained in a pressurized canister, in a super-cooled state. By metering a given volume of propellant containing formulation (typically 25–100 µl) within an integral metering valve, a reproducible dose of drug can be emitted and delivered. The high, internal energy of the propellant results in rapid expansion of formulation through the metering stem and vaporization via an actuator orifice. Subsequently, micron sized droplets are produced containing the formulation that are suitable for respiratory delivery. In pMDI based systems, the drug can either be homogeneously suspended (colloidal system) or solubilized in a propellant (either with or without the addition of co-solvents).

Until 1995, all marketed pMDI products contained CFCs as the delivery propellant. However, concerns over the possible detrimental ozone depleting effects of CFCs (89), reported in 1974, resulted in 150 nations signing the Montreal protocol in 1987, which committed the signatories to cease CFC production by 1996. In order to continue to use and develop pMDIs, pharmaceutical companies have subsequently committed significant resources to the development of CFC-free pMDI systems (90). The candidates that emerged from this research as potentially suitable CFC replacements were short-chain hydrofluoroalkanes (HFAs) (91): specifically, 1,1,1,2,3,3,3 Heptafluoropropane (HFA-227) and 1,1,1,2 Tetrafluoroethane (HFA-134a). Although these materials were accepted as reasonable replacements for the previously used CFCs, the HFA propellants possessed different physical and chemical proprieties and thus made a simple formulation propellant exchange impossible, spawning a whole new series of challenges for the formulation of pMDIs. Even though these challenges still exist, significant time and effort has been invested in overcoming these formulation hurdles, and many HFA formulations are currently available as marketed products or are currently under development (Table 1).

Table 1   Marketed and In-Development pMDI Products Based on Suspension Formulations

Trade name	Active compound	Manufacturer	Therapeutic use
Ventolin®/ Proventil® -HFA	Albuterol	GlaxoSmithKline	Bronchodilator
Azmacort®	TAA	Kos Pharmaceuticals Inc..	Prophylactic therapy of asthma
Flovent® HFA	Fluricasone propionate	GlaxoSmithKline	Corticosteroid bronchodilator
Combivent®	Ipratropium bromide and albuterol sulphate	Boehringer Ingelheim Pharmaceuticals, Inc.	Anticholinergic bronchodilator in patients with COPD
Aerobid® inhaler system	Flunisolide	Forest Pharmaceuticals, Inc.	Corticosteroid anti-inflammatory, antiallergic
Tilade®	Nedocromil sodium	Rhone-Poulenc Rorer Pharmaceuticals Inc.	Anti-inflammatory
Azmacort	TAA	Aventis Pharma (in development)	Cortoicosteroid
Symbicort®	Budesonide and formoterol	AstraZeneca (in development)	Corticosteroid and selective beta2-agonist
Pulmicort®	Budesonide	AstraZeneca (in development)	Corticosteroid
Serevent® Evohaler	Salmeterol xinafoate	GlaxoSmithKline	Bronchodilator
Xopenex®HFA	Levalbuterol tartrate	Sepracor Inc..	Selective beta2-agonist
	Mometasone furoate	Schering–Plough Research Institute	Corticosteroid
Brycanil®	Terbutaline sulphate	AstraZeneca	Short-acting beta2-agonists

Abbreviations: COPD, chronic obstructive pulmonary disease; HFA, hydrofluoroalkanes; TAA, triamcinolone acetonide.

Formulation of pMDI Systems

Advances in Suspension-Based pMDI Systems

Due to the poor solvent properties of HFAs, suspension based systems are a popular formulation route for current pMDIs. In general, research in this area has focused on two main themes over the past 5 years: particle engineering and the application physical stabilizers.

As with DPI systems, the drug particles used in suspension based systems should ideally have a diameter of less than 6 μm to achieve a therapeutic affect upon administration. As previously discussed, these particulate systems have a high surface area to mass ratio and therefore tend to be highly cohesive/adhesive in nature. Although, effects such as humidity plays a minor role in pMDI systems, the presence of solvation, van-der Waals, and electrical double layer forces, as well as density effects, may result in suspension instability and uncontrolled agglomeration and/or caking. Until recently, suspension pMDIs were mostly comprised of micronized drug particles. As previously discussed, micronization is a high energy process that leads to a broad particle size distribution and allows poor control over particle morphology and density. Although this approach is still used in current pMDI suspensions, research has focused on improving internal stability by particle engineering, especially in light of increasingly stringent regulatory standards proposed for pMDIs. Since many examples of particle engineering technologies exist only a limited few that are specifically related to pMDI suspension technology are discussed here.

Media milling is a concept that has found some application in biopharmaceutical processing and may be applicable to pMDI suspensions. Media milling incorporates the suspending medium (i.e. HFA) into the processing step, such that the final production step is conducted in situ and the final product can be extracted at the mill output. The adaptation of this technique to in situ pressurized milling has had some success. For example, work conducted by Lizio et al. (92) suggested that this methodology could be successfully used to produce a pMDI suspended peptide formulation containing particles with a volumetric mean diameter of 3.1 μm, without generating degradation products or contaminants from the milling process. Similarly, DuPont workers, reported increased formulation stability of in situ media milled budesonide formulations when compared to conventionally micronized particulate suspensions (93).

Another approach for improving stability via modified particulate milling is to process the drug with pharmaceutical stabilizers by co-grinding. For example, Williams et al. (94) investigated the effect of co-grinding a model drug, triamcinolone acetonide (TAA), with a polymeric surfactant (Pluronic F77). The co-processed material exhibited reduced propellant solubility, increased formulation stability and improved aerosol dispersion (resulting in increased FPF) when compared to a TAA control.

Apart from co-processing methodologies, specific particle engineering techniques may be employed to improve suspension stability. In general, any of the particle engineering methodologies discussed in the DPI section of this chapter may be utilized. However, of note is the use of porous particles (15,81). As with DPI formulations, the control of the density, particle size distribution, and porosity of the powder can affect the relative surface area to mass ratio thus modifying the adhesion characteristics. However, in addition, for pMDI systems the rate of sedimentation (or creaming), will be directly related to the densities of the media and the particulates. Thus, by modifying the density of the API powder it may become possible to "density match" the formulation and thus create a stable system. This approach was employed in Nektar's PulmoSphere technology (81). Furthermore, by incorporating a porous surface topography and spherical morphology, contact area effects are minimized.

The monitoring of particle surface energy is another possibility that has been explored in order to improve the suspension stability in pMDIs. For example, the atomic force microscope colloidal probe technique (as described previously) may be used to measure the forces between individual micron sized particulates in model propellants, similar to those used in pMDIs, namely, pressurized HFAs (95,96). Furthermore, these techniques can be expanded to investigate the relationship between drug–drug interactions, surface free energy, and in vitro performance (97,98).

An alternative option to particle engineering is the use of HFA soluble surfactants and/or stabilizing agents. The molecules in these materials reduce particulate interactions by accumulating at the interface of the colloidal particles and act to either reduce interfacial tension, produce steric hindrance between particulates, or form colloid bridges to increase flocculation. An example of the influence of polymeric additives, such as polyethelene glycol (PEG) on drug cohesive forces is represented in Figure 14. As can be seen from Figure 14, particle cohesion in this model system is dependent on both the concentration and the molecular weight of the polymer used (99).

It is not surprising that there has been a significant focus on HFA stabilizing agents for pMDI applications since they may provide a more immediate, and less costly, alternative to particle engineering. Indeed, historically, CFC formulations were stabilized using surfactants such as oleic acid. Unfortunately, these molecules were incompatible with the replacement HFAs and alternatives have been sought. Such excipients include long chain polymers, such as PEG (100) and polyvinylpirrolidone (101); fluorinated carboxylic acids or ester surfactants (102) and hydrophilic surfactants (103). Furthermore, the solubility of conventional CFC pMDI surfactants (e.g. oleic acid, lecithin) may be utilized by using specific co-solvents (104).

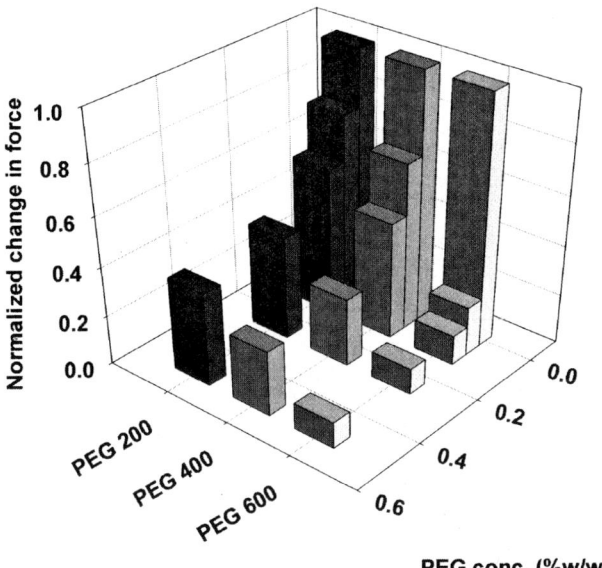

PEG conc. (%w/w)

Figure 14 Influence of PEG molecular weight and concentration on drug–drug interactions in a model pMDI system. *Abbreviations*: PEG, polyethelene glycol; pMDI, pressurized metered dose inhaler.

Solution-Based pMDI Systems

The most obvious approach for overcoming suspension stability issues in pMDIs is to solubilize the drug in the propellant, with or without co-solvents. This approach has obvious advantages. Since the drug is effectively present as a molecular dispersion any issues with flocculation, sedimentation, Oswald ripening, particle-device/component adhesion, and metering consistency are overcome. However, it is important to note, new challenges are faced with solution pMDI systems. For example, in most cases, the drug will not be directly soluble in the non-aqueous propellant and will require a considerable amount co-solvent for solublization. Furthermore, the formulation may be chemically less stable (and require stabilizers to reduce degradation) or may be more prone to temperature cycle instabilities and precipitation.

Apart from the improvement in physical stability and metering consistency, a key advantage of using solution pMDI technology is the potential to accurately control the particle size distribution of the resulting aerosol. For suspension based systems, the aerosol particle size will ultimately be governed by the drug particulate size and adhesive/cohesive interactions. In comparison, the spray pattern and final aerosol particle size distribution of a solution based system will be governed by the vapor pressure of the propellant, the proportion and volatility of co-solvent and the dimensions of the actuator orifice. By controlling these parameters in a system containing volatile co-solvents, the final particle size distribution will depend on the drug concentration in the aerosol droplets.

The most successful example of pMDI solution based systems is 3Ms beclomethosone dipropionate formulation; Qvar[®] (105) which utilizes ethanol as a co-solvent. As discussed above, in comparison to suspension based pMDIs, the FPF of the aerosol cloud is not be dependent on micronized API size (and interactions) but on the size of the evaporating propellant/co-solvent droplets. Subsequently, the aerosol mass median aerodynamic diameter of Qvar beclometasone is reported as 1.1 μm (less than most conventional suspension pMDI formulations), resulting in improved respiratory deposition (between 50% and 60% as measured in human studies) (106).

As with the Qvar system, the use of volatile co-solvents in pMDI solution formulations may yield particles with small aerodynamic diameters (of the order 0.8–1.2 μm). Although this size range is clearly suitable for respiratory delivery, it will generally result in a product targeted largely to the alveoli, which for generic API formulations, may not be comparable to existing deposition patterns and clinical efficacy.

One method of controlling the aerosol particle size of solution based pMDI formulations is by addition of a non-volatile component into the formulation (107,108). The presence of a non-volatile additive in a soluiblized HFA-volatile-co-solvent system will result in an increase in the final droplet size. Since evaporation is dependent on the initial solution composition, the integration of a non-volatile component, such as glycerol, into the primary pMDI solution will result in the final aerosol containing both drug and non-volatile additive. Subsequently, after evaporation of the volatile components the resultant particulate size will be dependent on the concentration of the non-volatile component which will dominate the particle size. This approach has been successfully employed in the Modulite[®] technology developed by Chiesi, which, depending on the non-volatile concentration of glycerol or polyethylene glycol, has resulted in the ability to modify the aerodynamic diameter of the resultant aerosol (Fig. 15) (107).

In addition to simple co-solvents, and the addition of non-volatile agents which modify aerosol particle size, significant advances have been made using alternative solubilizing agents. For example, hydrophilic and hydrophobic excipients that are HFA-soluble have been recently developed. Examples of these novel excipients include oligolactic acids (OLAs). These acids, first introduced as suspension aids, are short-chain polylactides that, in the absence of co-solvents, degrade to endogenous lactic acid. The OLAs, with an average of 6–15 repeat units, interact with the API to form API/excipient complexes that are highly soluble in HFA propellants (> 5% w/w). These OLAs have no detectable toxicity, as measured by radio-labelled

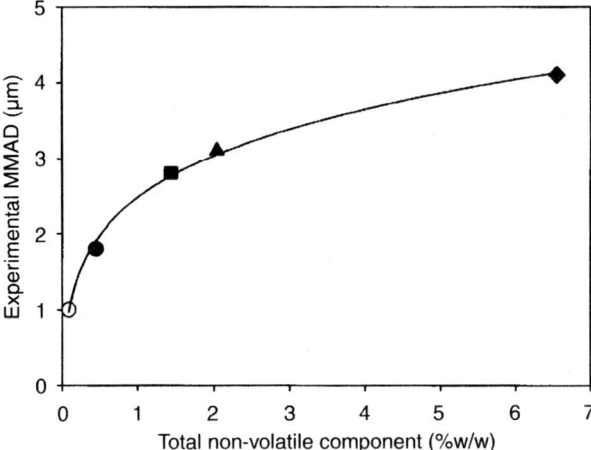

Figure 15 Effect of total content of non-volatile component on MMAD of BDP pMDI solution formulations in HFA 134a. *Abbreviations*: MMAO, mass median aerodynamic diameter; BDP, beneficiary database prototype; pMDI, pressurized metered dose inhaler. *Source*: From Ref. 108.

distribution, and inhalation safety studies (109). A further advantage of this new excipient class is the potential for OLA-drug matrix formation to provide the potential for sustained drug release in a very simple, economical manner, potentially increasing the efficacy of many compounds (Fig. 16).

One drawback of OLA excipients is that they are water insoluble, making the pre-formulation and solubilization of polar drugs difficult. Consequently, another group of hydrophilic excipients based on functionalized PEGs are available for use with biopharmaceuticals or small molecules in HFA pMDIs (111). These compounds, generally used at low concentrations relative to the drug, are also readily soluble in both HFA 134a and 227 (alone, or with minimal amounts of ethanol). Furthermore, these functionalized PEGs also have the potential to form ion-pairs with drug salts. For example, 3M workers examined the ability to prepare HFA-soluble ion-pairs and reported significant improvements in solubility in HFA/ethanol systems (111). Other solubility improving compounds include mono-amides and mono-esters, two ubiquitous substances in the body that degrade to endogenous or known biocompatible compounds. These new mono-functionalized excipient-drug complexes, when combined with ethanol solvents, have recently been reported to increase solubility of a drug to 1.5 wt%, which corresponds to an ex-valve dose of over a milligram (111).

In addition to solution based systems, other "molecular dispersions" may be formed using reverse micelles and microemulsion technology. These systems have the same advantages in terms of aerosol performance, as pure

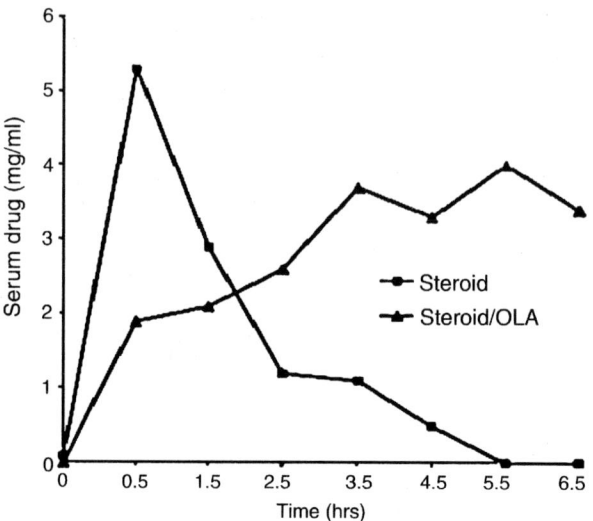

Figure 16 Sustained release of Butixicot (steroid) using OLA as a sustained-release matrix following pMDI aerosolization. *Abbreviation*: OLA, oligolactic acids. *Source*: From Ref. 110.

solution based systems, but may have the potential for incorporation of molecules that will not be soluble using conventional co-solvents. Although these formulation strategies are not currently utilized in pMDI products, a limited number of studies have investigated these systems in HFA propellants. For example, Butz et al. (112) reported that water soluble compounds could be emulsified in a water-in-fluorocarbon emulsions (using a perfluorooctyl bromide dispersed phase and perfluoroalkylate-dimorpholinophosphate stabilizing surfactant). The subsequent emulsion was readily dispersible in all proportions in both HFA 134a and 227 (112). Other molecular dispersants include cyclodextrins. Cyclodextrins have been used extensively to form inclusion complexes with many substances since the complex exhibits higher aqueous solubility and improved chemical stability. This approach can be utilized in pMDI systems, as reported by Williams and Liu (113), who investigated a novel technique of incorporating aspirin in a hydroxypropyl-β-cyclodextrin inclusion complex in HFA 134a.

Advances in pMDI Device Design

Apart from the pMDI formulation approaches discussed above, product efficiency, patient compliance, and usability will also be governed by the functionality of the device itself. In simple terms, pMDI devices contain 5 key components, namely, the pressurized canister, the actuator device/

canister housing, the valve components, crimp seal components, and actuator (Fig. 17).

Dose counting: In comparison to multi-dose DPI systems, which integrated dose counting mechanisms into the earliest devices, pMDIs have, until recently, had no convenient way for patients to track the number of doses remaining in the canister at any given time. Indeed, in early pMDI systems, patients could keep track of pMDI usage by recording each dose taken manually on a record sheet, and subtracting the total from the labeled number of doses. Alternatively, patients could test the fill volume by a rudimentary float-test.

Such unreliable methods often lead to patients throwing away a product which still contained an acceptable number of doses or using a product beyond the recommended number of doses, the latter being potentially dangerous as patients could be inhaling sub-therapeutic doses. A recent survey of 342 adult asthmatics conducted in 2004 by Dr. Bradley E. Chipps, of Capital Allergy and Respiratory Disease Centre in Sacramento (California), reported that 62% of patients had no idea that they were supposed to keep track of the dose status of their pMDI (114). Amongst those who knew, only 24% were aware of what was left in their pMDI, 25% of subjects found their MDI empty when they needed it, and 8% of those people ended up calling 911 emergency for help (114).

In 2003, the FDA issued guidance for the industry concerning the incorporation of dose counting technologies in pMDI devices. This announcement prompted the industry to accelerate development of several dose-counter technologies, resulting in a surge in activity and the filing of

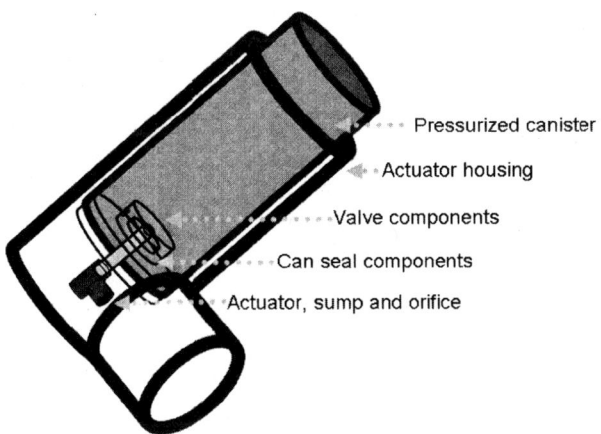

Figure 17 Schematic representation of a conventional pMDI.

more than 30 patents relating to counting systems by several drug delivery device companies. These included Bespak, KOS, Trudell Medical, and Valois Pharmaceutical Divisions. These companies offered proprietary technologies such as the requiring 50% less force to release a dose. This is especially important for the elderly and young children. Commercially, in 2004, GlaxoSmithKline led the way in applying these new FDA dose counting guidelines by launching the Seretide® Evohaler® in the United Kingdom.

Overcoming patient coordination/improving the patient experience: Many patients find it difficult to coordinate the actuation of their pMDI with inhalation maneuvers. Overcoming this potential hurdle was one of the key attractive features that have been promoted as an advantage for passive DPI systems. Furthermore, inspiratory flow rate can influence the dose emitted from an inhaler, the amount inhaled, the oropharyngeal deposition, and the regional lung deposition of inhaled medications (115). Consequently, the effective use of conventional pMDIs, as shown in Figure 17, is technique dependent (116,117).

In the late 1960s, development began on the first breath-actuated inhaler. Having seen the difficulty that some patients experienced when coordinating pressing and breathing, an opportunity was identified for a device that would actuate as the patient breathed in. This would eliminate the need for coordinated actuation or the use of add-on devices such as aerosol-holding chambers.

The first Autohaler® device was launched as the Duohaler® in 1970. The Autohaler was pocket-sized and easy for patients to use, however some complained that the loud "click" that sounded when the mechanism fired was disconcerting. The device did not gain instant popularity and was shelved until work began on an improved device in the 1980s. Subsequently, pirbuterol (Maxair, 3M, St Paul, Minnesota) and albuterol HPA (IVAX Laboratories, Miami, Florida) are now available in the Autohaler. Similar breath actuated pMDI device designs include the Baker Norton Easyhaler®, the more recent Xcelovent® (Meridica/SkyePharma) and CCLs Integrated Breath Actuated Inhaler with dose counter. In general, these devices have a mechanical flow trigger that, after priming, actuates the release of the drug when the patient's inhalation airflow reaches a required flow rate (e.g. 30 l/min). As expected, this approach has been shown to improve drug delivery in adults and children with poor coordination (118–120).

Just as coordination influences patient compliance, so does the aerosol plume velocity and temperature. Since pMDI formulations are based on the volatile evaporation of a unit dose, the resultant aerosol may travel at high velocity and be "cold". This relationship can be visualized when comparing the plume velocities of commercially available pMDIs, where plume forces have been reported as varying between 29 and 117 mN and temperatures between $-32°C$ and $+8°C$ (when measured 5 cm from the end of a actuator mouthpiece) (121). These patient "feel" issues may result in a reluctance and/or difficulty for effective patient inhalation, since the Freon effect, in combination with particle impaction, is felt at the back of the throat; possibly causing a gagging effect.

Developments in the design of actuators, notably the use of actuators with smaller orifice diameters which produce a much slower, "warmer" spray, may improve patient co-ordination and compliance. By using actuators with small orifice diameters, it is possible to produce a relatively slow and "warm" spray from HFA pMDIs, compared with traditional CFC products (122). This approach makes it easier for a patient to coordinate the act of firing the pMDI with an inhalation maneuver. Furthermore, for solution based systems, if the unit dose remains constant, and the orifice diameter is reduced, the reduction in actuator orifice diameter will result in increased FPF (123) (Fig. 18). For solution based systems, the relationships between, for example, co-solvent, non-volatile component and actuator orifice were developed by Lewis et al. (107,108,124) into a set of empirical equations that could describe the performance of a pMDI based on formulation variables.

Improvements to internal pMDI components to increase therapeutic outcomes: Other advances in pMDI device technology have been a result of the modification of canister components, including valve, crimp assembly and internal canister materials.

For example, conventional metering valves are designed in such a way, that a metered volume of liquid is retained, by surface tension, in the chamber after an actuation. The nature of such valve designs results in priming issues, and where suspensions are used, homogeneity issues. To overcome this, valves are being developed with either more "open valves" or

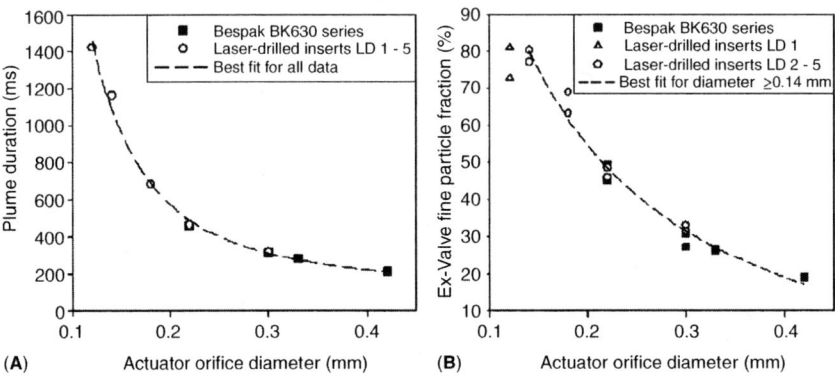

Figure 18 Relationship between actuator orifice diameter and (**A**) plume and (**B**) FPF for a solution-based pMDI. *Source*: From Ref. 125.

virtual metering tanks, which only form a closed system on actuation. These improvements increase dose reproducibility, remove priming effects, and potentially increase therapeutic outcomes (126). Furthermore, these developments in valve technology, particularly the "Easi-fill" valve (BK361 Bespack, Fig. 19), can be incorporated into new pMDI designs with relative ease.

Since pMDI formulations are pressurized systems, the crimp assembly and internal seals are important components since they should, for obvious reasons, avoid propellant leakage, and moisture ingress. As with the formulation issues that arose with the conversion of CFCs to HFAs, the simple switch using conventional canister components was not straight forward, since the solvent properties of HFA propellants (and co-solvents) affected the degree of elastomer swelling (or shrinkage) and increased propellant leakage when compared to the CFC based formulations (127). In order to overcome such issues, alternative elastomer materials with improved HFA compatibility were required. The Spraymiser™ Valve (3M) (128) is a new generation of valve that incorporates a number of EPDM (Ethylene Propylene Diene) elastomers that can offer additional benefits to the performances of HFA pMDI products, that is, improved sealing, elimination of valve sticking, etc. Furthermore, the new materials contained reduced levels of extractable materials (129). This was to become increasingly important since the FDA endorses for more stringent guidance for levels of acceptable extractables (130). For a pMDI, the PQRI (Product Quality Research Institute, Virginia, USA) leachable and extractable working group recommends: "that AETs (Analytical Evaluation Threshold) for MDI leachables profiles be based on the Safety Concern Threshold of 0.15 µg/day for an individual organic leachable. This recommendation includes potential organic leachables derived from critical

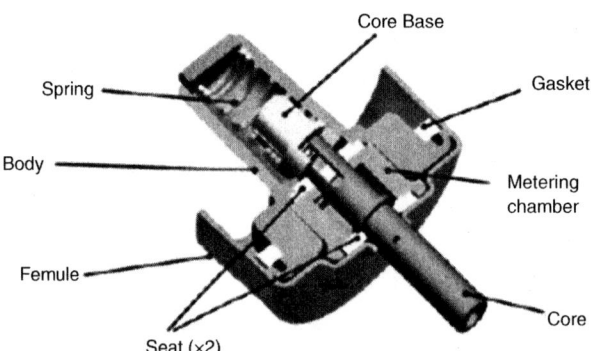

Figure 19 Bespak BK361 "Easi-fill" valve, schematic cross-sectional drawing. The larger flow path in comparison with a standard metering valve enables an easier fill and drain. *Source*: www.bespak.com/ddel_resp_easifill.asp.

components of the dose metering valve, canister inner surface, and inner surface coating if present." Subsequently, there are numerous patents that include the uses of novel materials or describe new methods for the removal of sources of polynuclear aromatic compounds, but this aspect of the pMDI valve development is beyond the scope of this chapter.

As previously discussed, the surface area to mass ratio of suspended particles used in pMDI inhalation formulations results in a high adhesion potential. Consequently, exposed material surfaces inside the pressurized canister may facilitate adhesion and thus reduce emitted doses throughout the lifespan of a device. Although this may not be critical for conventional dose pMDIs, where small variations in drug loss may not influence the required pharmacopoeial tolerances, it may be critical for lower dose or more adhesive APIs. An example of the relative differences in the median adhesion forces between single micronized albuterol sulphate particles and three materials used in pMDI canisters is shown in Figure 20, which shows that there is a difference of two orders of magnitude in the drug adhesion to the components. In view of these potential issues, a series of coatings have been developed, namely fluorinated polymers (131), which are now utilized in the Ventolin® HFA product (132). These coatings are generally made of relatively inert organic materials, such as perfluoroalkoxyalkane, epoxy-phenol resin, and fluorinated-ethylene-propylene polyether sulfone (133),

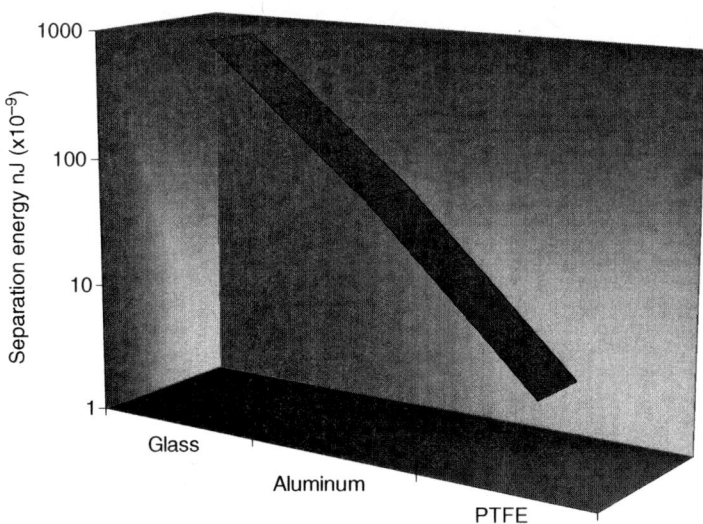

Figure 20 Median adhesion between an albuterol particle and three surfaces used in pMDI canister construction collected in a model HFA propellant using colloid probe microscopy. *Source*: From Ref. 95.

which have low surface energies and are thus are less likely to cause particle adhesion (134) (Fig. 21).

The major developments in pMDI formulation and technology have been reviewed and an overview provided as to how the latest advances in aerosol technology have been used in order to improve upon existing inhaler performance. There are a multitude of formulation and device factors to consider when developing a pMDI. For these reasons a multiplicity of approaches have been applied for the determination of the effect of such variables on performances. As a result, although much is known about pMDIs at the empirical level, a systematic approach between in vitro measurements and in vivo clinical outcomes has been clearly missing.

Recent Advances in Nebulizer Technology

Nebulizers have been used for asthma therapy for many years and are used for the delivery of fine droplets of drugs to the lungs. For this reason they have been optimized for aerosol delivery in the upper range of inhalation

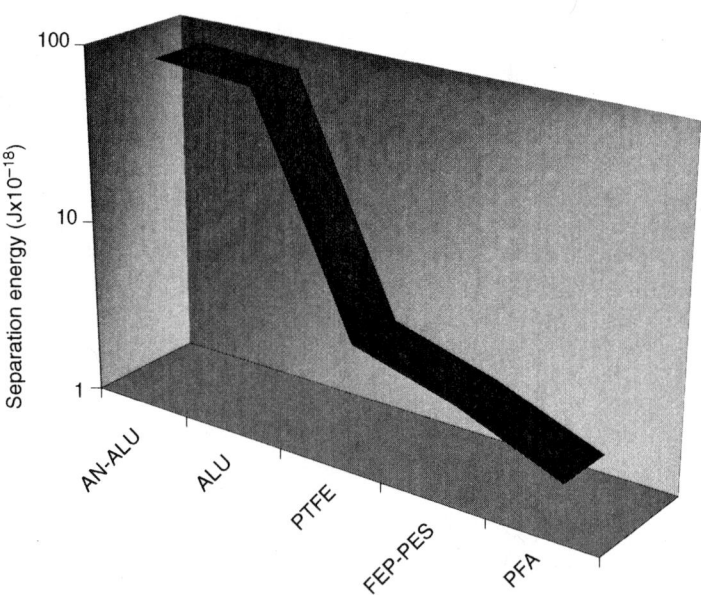

Figure 21 Median adhesion between an albuterol particle and the internal walls of five pMDI canisters in model propellant using colloid probe microscopy. Canister coatings were as follows: Aluminum (ALU), anodized aluminum (AN-ALU), polytetrafluoroethylene (PTFE), fluorinated ethylene propylene–polyether sulphone (FEP-PES) and perfluoroalkoxy (PFA). *Source*: From Ref. 134.

delivery, that is, 1–6 μm. Nebulizers can produce a constant stream of aerosol particles, from both solutions and/or suspensions which can be inhaled via tidal breathing. Furthermore, the aerosol droplets produced may be of a smaller range than pMDIs, making them capable of easily penetrating the small airways (135). Nebulizers enable treatment of patients that require higher doses of drugs (i.e. antibiotics), that can, until now, only be formulated to be delivered via a nebulizer (i.e. DNAse for the treatment of cystic fibrosis), or have difficulties with conventional inhalers, such as the very young or elderly patients. However, Nebulizer systems have many drawbacks. For example, historically, nebulizers have been cumbersome, often being used at home or in a clinical setting. Furthermore, they deliver medication over a long timescale and have a complexity of device assembly with low efficiency and high variability in drug delivery (136–138).

The most common type of nebulizer is the air-jet nebulizer (139). Jet nebulizer devices utilize a pressurized gas to create the aerosol particles (140). The gas stream is passed over the solution, creating a liquid film. This film breaks apart due to breakage of surface tension forces, forming aerosol particles, with the aerosol droplet size being dependent on the compressed air pressure applied (7). Subsequently, respiratory sized droplets can easily be produced once the large droplets (>10 um) have been filtered out by impaction on the device surfaces or baffles (141).

The major disadvantage of jet-nebulizers is their constant aerosol output during the inhalation, exhalation, and breath-hold, which is not matched to the patient's respiratory effort (142,143). Other disadvantages include: a residual volume (0.5–1.0 ml), a long therapy time (10–15 min for completion), a variable mass output throughout nebulization and an increase in osmolarity of the aerosol/liquid with nebulization time (139,144). Of all the jet nebulizers on the market, the most popular systems are probably those produced by Pari GmbH, for example, the Pari LC Plus® and the Pari LC Star® products. These systems belong to the so called jet enhanced nebulizers. Conventional jet nebulizers aerosolize continuously, resulting in higher losses of dose and shorter nebulization times. Since the aim is to be able to deliver the drug only during inhalation, novel breath-operated nebulizers have been developed. These novel nebulizers have been shown to increase the in vivo mass of the inhaled drug compared with conventional jet nebulizers (145).

Breath-synchronized nebulizers have been reported to effectively reduce drug wastage during the expiratory phase, both in vivo and in vitro (146,147). Examples of breath operated jet nebulizers include the AeroEclipset (Trudell Medical International, London, Ontario, Canada) and the Smartstream® (Medic-Aid, West Sussex, UK). These devices are designed to deliver drug during the whole inspiration, whereas new adaptive electronic dosimetric jet nebulizers, such as the HaloLite® (Medic-Aid) are operated using a different principle, for example, delivering drug during the first ~50% of each inspiration, or the delivery of aerosol only during inhalation, as in the conventional jet nebulizer, the AKITA® (InAMed, Germany). Furthermore, improvements in constant output nebulizers have also been made with the advent of the breath-enhanced nebulizers, for example, the PARI LC Star (PARI, Germany) and the Ventastream® nebulizer (Medic-Aid, UK). In these new devices the aerosol is produced at a higher rate during inhalation than during exhalation, using valves to control air flow and mixing (144). In general, the level of waste of the nebulized dose, in these systems, is reduced by at least 50% in comparison with the original assisted open vent nebulizers. Since droplet size distribution and output rate are also influenced by the physical properties of the drug solution (suspension) and air flow rate from the compressor, it can, be concluded that there are high variations in the performance of different types of such nebulizers (143,148,149).

Adaptive aerosol delivery technology (150) is an approach developed to address the highly variable drug delivery characteristic of conventional delivery systems. These novel types of nebulizers adapt to the individual breathing pattern of the patient (determining the shapes of the inspiratory and expiratory flow pattern) and deliver the aerosol only during inhalation. Such systems pulses aerosol delivery only during inhalation and each pulse is matched to the previously determined inspiration time. The breathing pattern is continuously monitored and the system adapts to the changes in pattern. Furthermore, such systems are programmed to deliver a preset metered dose. An example of this new technology can be found in the HaloLite AAD system (Profile Therapeutics, UK). In a recent study (151), utilizing a modified venturi Venstream® nebulizer (Profile Respiratory Systems, UK) which incorporated an AAD system, nebulized budesonide was administered to 125 Spanish children with moderate to severe asthma for 24 weeks in a double-blind, randomized, parallel group study. A total of 75% of the children received 100% of the programmed dose. Although there was no statistically significant difference between the treatment regimens, there was a clear improvement in the overall health score. Similarly, in another study involving 47 children aged less than 3 years old, a higher number of patients achieved successful treatment with the HaloLite AAD system (81%) compared with the Ventstream nebulizer (66%) (152). Overall the AAD technology appears to be a promising approach for the challenge of delivering reproducible doses of aerosolized drug to the lungs (Fig. 22).

An alternative popular type of nebulizer is the ultrasonic nebulizer. Ultrasonic nebulizers generate high frequency ultrasonic waves, produced by a rapidly vibrating piezoelectric crystal. A given solution presented to the piezoelectric crystal surface is broken into fine droplets where the aerosol droplet size is inversely proportional to the power of the acoustic frequency. In a similar way to jet nebulizers, baffles within the nebulizer remove large droplets. Ultrasonic nebulizers have many advantageous over compressed

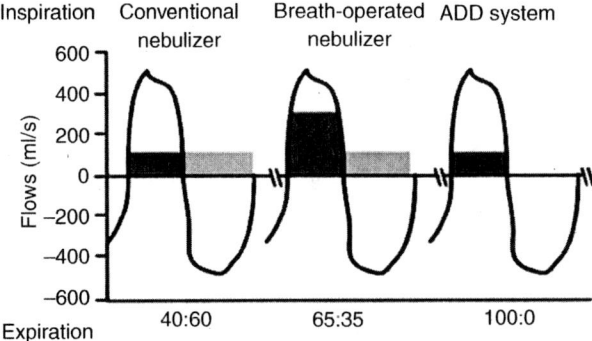

Figure 22 Comparison of drug output during inspiration and expiration for a conventional jet nebulizer, active venture nebulizer, and an ADD system. The shaded areas represent drug output. *Source*: From Ref. 153.

air systems, including the small size of device, the minimal patient coordination required, and rapid drug delivery (138). However, it is important to note, that these systems also have disadvantages, including the high device costs, susceptibility to mechanical breakdowns and potential for contamination. Depending on the method (direct or indirect) used to produce the aerosol droplets the ultrasonic class of nebulizers can be sub-classified into passive and active vibrating devices. Passive vibrating mesh ultrasonic devices [e.g. the Omron MicroAir NE-U22 (Omron Healthcare, Ltd., Milton Keynes, UK)] are small battery-operated nebulizers which utilize a piezoelectric crystal in an ultrasonic horn to force drug solutions through a mesh of hundreds (or thousands) of micron-sized holes to creating an aerosol (154). The example given here (the MicroAir), in comparison to many conventional devices, results in primary droplets that are sufficiently small so that no baffles are required to filter the spray, thus allowing the use of a smaller volume fill.

Active vibrating mesh ultrasonic nebulizers (155) ensure the mesh in contact with the reservoir fluid is vibrated directly by a piezoelectric crystal to generate the aerosol cloud. These devices use a perforate membrane and a micro-pumping action to draw jets of fluid though the holes in the membrane, dispersing the jets into a drug cloud. The size of the aerosol droplets is controlled by the shape/size of the holes and the surface chemistry and composition of the drug solution. Examples of such types of nebulizers are: the AeroNeb®Pro and AeroNebGo (Nektar Therapeutics) (156), the e-Flow™ (Pari GmbH) (157,158) and the TouchSpray™ (ODEM Ltd.) (159).

In comparison, a novel class of nebulizer device which has been investigated by Battelle Pharma, Ohio, is the Electrodynamic Aerosol Devices (EAD) (160). This novel device uses a powerful electric field to create near monodisperse, low velocity charged aerosols from a liquid formulation. As the solvent in the droplets evaporate in entrained air, the surface charges on each droplet repel each other and cause the droplets to separate, producing an aerosol cloud with a small particle size. The limitation of this technique is that electric charge is known to affect aerosol deposition, both in vitro and in vivo (161,162). To the authors' knowledge no devices with the EHD technology are on the market as yet, however, early performance results have shown promise (156).

In general, there have been few recent advances in nebulizer technology. In 2003 Boehringer Ingelheim introduced a new nebulizer, the Respimat® Soft Mist™ Inhaler which is a propellant-free, multidose system (163). This nebulizer system uses an innovative approach for inhalation therapy combining a mixed form of nebulizer and pMDI. The medication is stored in a sealed plastic container inside of the cartridge. When the dose-release button is pressed, the energy released from the spring forces the solution through a "uniblock", releasing a slow-moving aerosol mist. Observations concerning the use of this kind of system with suspensions has not been reported. Furthermore, recently, Aradigm launched AERx® (164) Pulmonary Delivery System in 2000. This nebulizer generates an aerosol by mechanically pressurizing a disposable blister pack, forcing the liquid through an array of micromachined holes (165). This device has been used with a vast array of drug molecules with various successes including: morphine, rhDNase, insulin, and fentanyl. Finally, Chrysalis Technologies (a division of Philip Morris, USA) has developed a novel proprietary capillary aerosol generation system in which the aerosol is formed by pumping the drug formulation through a small, electrically heated capillary (166,167). This device enables thermally stable drugs and vehicles to be evaporated and subsequently condensed in entrained air in a controlled fashion, producing a soft aerosol. Although this device has the advantage of being able to exhibit very high delivery efficiencies (fraction of milliliter), one envisaged disadvantage is that it is unlikely that all drugs can be aerosolized in this way without some degree of drug degradation occurring.

At the present, there is a great deal of debate between health professionals in how to decide which nebulizer to use for a particular application or situation. Pharmaceutical companies still supply drugs to be used for nebulization independently of which device is going to be used for its aerosol delivery. This approach may seriously affect the intended therapeutic benefit. At the moment it is considered to be sufficient to merely characterize nebulizer performance using a simple saline solution at or near physiological concentration (168,169). It seems reasonable that in view of this testing methodology and the examples given here, that there is a need for a more stringent evaluation standardization test that will improve the future health care of patients.

CONCLUSIONS

In recent times the number of DPI products on the market has increased significantly, with many more technologies still in development. Additionally, there have been numerous patents filed and a significant body of research literature published. In general, devices and/or technologies are available, or in development, which deliver generic drugs, new chemical entities and combination products, such as the Advair Diskus®, which contains two therapeutic agents. It seems likely that the industry will continue to develop new products based on these approaches. In addition, efforts will continue for the delivery of larger molecules such as proteins and peptides to the lung. However, as the potential applications of DPI expands, for example, by the delivery of complex molecules to the lung, further studies into all aspects of DPIs must be undertaken in order to gain a deeper understanding of this drug delivery system.

The evaluation of the published literature and patents confirms that there have been many advances in nebulizer technology over the past 10 years. Recognition of the importance of breathing pattern in aerosol delivery and deposition has led to the design of devices that allow targeting of deposition to specific areas of the airways and novel systems incorporating patient feedback. Such systems have provided control of factors that affect drug deposition and dose and have resulted in the development of new products that improve therapeutic outcomes.

In the early 1960s, inhaled therapy was regarded as a specialized therapeutic route, limited to the treatment of severe lung disease. This opinion has been "rewritten" in the past 50 years, with the perceptions regarding inhalation therapy changing to, not only the treatment of respiratory lung diseases, but also systemic delivery. Consequently, pulmonary delivery is becoming an increasingly popular delivery route for both small molecules and the delivery of peptide and proteins. With increased knowledge concerning the safety and consequent acceptance of delivery via the pulmonary route, there has recently been a significant level of research activity in this area undertaken by pharmaceutical companies and academic institutions. This approach has already revolutionized inhaled drug therapy, with many different therapeutic approaches via inhalation already possible and others that may be developed for clinical use in the near future. Advances in formulation/device technology may aid the development of novel inhalational delivery systems, but a great deal of research is still required to make this delivery route truly profitable and effective in future.

REFERENCES

1. Versteeg HK, Hargrave G. Fundamentals and resilience of the original MDI actuator design. Proc Respir Drug Deliv 2006; 1:91–100.

2. Anderson PJ2005. History of aerosol therapy: liquid nebulization to MDIs to DPIs. Respir Care 2005; 50:1139–50.

3. Patton JS. Mechanisms of macromolecule absorption by the lungs. Adv Drug Deliv Rev 1996; 19:3–36.

4. Pritchard JN2001. The influence of lung deposition on clinical response 1. J Aerosol Med Deposit Clearance Eff Lung 2001; 14:S19–26.

5. Yeh HC, Schum GM. Models of human-lung airways and their application to inhaled particle deposition. Bull Math Biol 1980; 42:461–80.

6. Schum M, Yeh HC. Theoretical evaluation of aerosol eeposition in anatomical models of mammalian lung airways. Bull Math Biol 1980; 42:1–15.

7. Hinds WC. Aerosol Technology, 2nd ed. New York: John Willey & Sons; 1999.

8. Findeisen W. The precipitation of small air-suspended particles in human lungs during respiration. Pflugers Arch Die Gesamte Physiol Menschen Tiere 1935; 236:367–79.

9. Ferron GA. Aerosol properties and lung deposition. Eur Respir J 1994; 7:1392–4.

10. Task Group on Lung Dynamics. Deposition and retention models for internal dosimetry of the human respiratory tract. Health Phys 1966; 12: 173–207.

11. Smith IJ, Parry-Billings M. The inhalers of the future? A review of dry powder devices on the market today. Pulm Pharmacol Ther 2003; 16:79–95.

12. Franks F1999. Thermomechanical properties of amorphous saccharides: Their role in enhancing pharmaceutical product stability. Biotechnol Genet Eng Rev 1999; 16:281–92.

13. White S, Bennett DB, Cheu S, et al. 2005. EXUBERA®: Pharmaceutical development of a novel product for pulmonary delivery of insulin. Diabetes Technol Ther 2005; 7:896–906.

14. Duddu SP, Sisk SA, Walter YH, et al. Improved lung delivery from a passive dry powder inhaler using an engineered PulmoSphere (R) powder. Pharm Res 2002; 19:689–95.

15. Edwards DA, Hanes J, Caponetti G, et al. Large porous particles for pulmonary drug delivery. Science 1997; 276:1868–71.

16. Hersey JA. Ordered Mixing–New Concept in Powder Mixing Practice 17. Powder Technol 1975; 11:41–4.

17. Staniforth JN. Order out of chaos. J Pharm Pharm 1987; 39:329–34.

18. Timsina MP, Martin GP, Marriott C, Ganderton D, Yianneskis M1994. Drug-delivery to the respiratory-tract using dry powder inhalers. Int J Pharm 1994; 101:1–13.

19. Coates MS, Fletcher DF, Chan HK, Raper JA. Effect of design on the performance of a dry powder inhaler using computational fluid dynamics. Part 1: Grid structure and mouthpiece length. J Pharm Sci 2004; 93:2863–76.

20. Srichana T, Martin GP, Marriott C. Dry powder inhalers: The influence of device resistance and powder formulation on drug and lactose deposition in vitro. Eur J Pharm Sci 1998; 7:73–80.

21. Owens DR, Zinman B, Bolli G. Alternative routes of insulin delivery. Diabet Med 2003; 20:886–98.

22. Flament MP, Leterme P, Gayot A. The influence of carrier roughness on adhesion, content uniformity and the in vitro deposition of terbutaline sulphate from dry powder inhalers 19. Int J Pharm 2004; 275:201–9.

23. Kawashima Y, Serigano T, Hino T, Yamamoto H, Takeuchi H. Effect of surface morphology of carrier lactose on dry powder inhalation property of pranlukast hydrate. Int J Pharm 1998; 172:179–88.

24. Zeng XM, Martin GP, Marriott C, Pritchard J. Lactose as a carrier in dry powder formulations: The influence of surface characteristics on drug delivery. J Pharm Sci 2001; 90:1424–34.

25. Ferrari F, Cocconi D, Bettini R et al. 2004. The surface roughness of lactose particles can be modulated by wet-smoothing using a high-shear mixer. AAPS PharmSciTech 2004; 5:1–6.

26. Young PM, Cocconi D, Colombo P, et al. 2002. Characterization of a surface modified dry powder inhalation carrier prepared by "particle smoothing". J Pharm Pharm 2002; 54:1339–44.

27. Iida K, Hayakawa Y, Okamoto H, Danjo K, Leuenberger H. Preparation of dry powder inhalation by surface treatment of lactose carrier particles. Chem Pharm Bull 2003; 51:1–5.

28. El-Sabawi D, Price R, Edge S, Young PM. Novel temperature controlled surface dissolution of excipient particles for carrier based dry powder inhaler formulations 1. Drug Dev Ind Pharm 2006; 32:243–51.

29. Zeng XM, Martin GP, Marriott C, Pritchard J. The influence of carrier morphology on drug delivery by dry powder inhalers. Int J Pharm 2000; 200: 93–106.

30. Staniforth JN. Pre-formulation Aspects of Dry Powder Aerosols. Proc Respir Drug Deliv 1996; 1:65–73.

31. El-Sabawi D, Edge S, Price R, Young P. Continued investigation into the influence of loaded dose on the performance of dry powder inhalers: surface smoothing effects. Drug Dev Ind Pharm 2006; 32:1135–8.

32. Young PM, Edge S, Traini D, et al. The influence of dose on the performance of dry powder inhalation systems 1. Int J Pharm 2005; 296:26–33.

33. Begat P, Morton DAV, Staniforth JN, Price R. The cohesive-adhesive balances in dry powder inhaler formulations II: Influence on fine particle delivery characteristics. Pharm Res 2004; 21:1826–33.

34. Begat P, Morton DAV, Staniforth JN, Price R. The cohesive-adhesive balances in dry powder inhaler formulations I: Direct quantification by atomic force microscopy. Pharm Res 2004; 21:1591–7.

35. Ahfat NM, Buckton G, Burrows R, Ticehurst MD. Predicting mixing performance using surface energy measurements. Int J Pharm 1997; 156:89–95.

36. Louey MD, Razia S, Stewart PJ. Influence of physico-chemical carrier properties on the in vitro aerosol deposition from interactive mixtures 3. Int J Pharm 2003; 252:87–98.

37. Islam N, Stewart P, Larson I, Hartley P. Effect of carrier size on the dispersion of salmeterol xinafoate from interactive mixtures. J Pharm Sci 2004; 93:1030–8.

38. Islam N, Stewart P, Larson I, Hartley P. Lactose surface modification by decantation: Are drug-fine lactose ratios the key to better dispersion of salmeterol xinafoate from lactose-interactive mixtures? 2. Pharmaceutical Research 2004; 21:492–9.

39. Adi H, Larson I, Chiou H, Young P, Traini D, Stewart P. Agglomerate strength and dispersion of salmeterol xinafoate from powder mixtures for inhalation. Pharm Res 2006; 23:2556–65.

40. Steckel H, Markefka P, teWierik H, Kammelar R. Effect of milling and sieving on functionality of dry powder inhalation products. Int J Pharm 2006; 309:51–9.

41. Young PM, Traini D, Chan HK, Chiou H, Edge S, Tee T. The influence of mechanical processing dry powder inhaler carriers on drug aerosolisation performance. J Pharm Sci 2007; in press.

42. Lucas P, Anderson K, Staniforth JN. Protein deposition from dry powder inhalers: Fine particle multiplets as performance modifiers. Pharm Res 1998; 15:562–9.

43. Louey MD, Stewart PJ. Particle interactions involved in aerosol dispersion of ternary interactive mixtures 5. Pharm Res 2002; 19:1524–31.

44. Handoko A, Ian L, Herbert C, Paul Y, Daniela T, Peter S. Agglomerate strength and dispersion of salmeterol xinafoate from powder mixtures for inhalation. Pharm Res 2006; V23:2556–65.

45. Jones MD, Price R. The influence of fine excipient particles on the performance of carrier-based dry powder inhalation formulations. Pharm Res 2006; 23:1665–74.

46. Tee SK, Marriott C, Zeng XM, Martin GP. The use of different sugars as fine and coarse carriers for aerosolised salbutamol sulphate. Int J Pharm 2000; 208: 111–23.

47. Steckel H, Bolzen N. Alternative sugars as potential carriers for dry powder inhalations. Int J Pharm 2004; 270:297–306.

48. Hooton JC, Jones MD, Price R. Predicting the behavior of novel sugar carriers for dry powder inhaler formulations via the use of a cohesive-adhesive force balance approach. J Pharm Sci 2006; 95:1288–97.

49. Traini D, Young PM, Jones MD, Edge S, Price R. Comparative study of erythritol and lactose monohydrate as carriers for inhalation: Atomic force microscopy and in vitro correlation. Eur J Pharm Sci 2006; 27:243–51.

50. Steckel H, Markefka P, teWierik H, Kammelar R. Functionality testing of inhalation grade lactose 12. Eur J Pharm Biopharm 2004; 57:495–505.

51. Zeng XM, Martin GP, Marriott C, Pritchard J. The use of lactose recrystallised from carbopol gels as a carrier for aerosolised salbutamol sulphate. Eur J Pharm Biopharm 2001; 51:55–62.

52. Iida K, Hayakawa Y, Okamoto H, Danjo K, Luenberger H. Effect of surface layering time of lactose carrier particles on dry powder inhalation properties of salbutamol sulfate. Chem Pharm Bull 2004; 52:350–3.

53. Begat P, Price R, Harris H, Morton DAV, Staniforth JN. The influence of force control agents on the cohesive-adhesive balance in dry powder inhaler formulations. KONA 2005; 23:109–21.

54. Lucas P, Anderson K, Potter UJ, Staniforth JN. Enhancement of small particle size dry powder aerosol formulations using an ultra low density additive. Pharm Res 1999; 16:1643–7.

55. Kumon M, Suzuki M, Kusai A, Yonemochi E, Terada K. Novel approach to DPI carrier lactose with mechanofusion process with additives and evaluation by IGC. Chem Pharm Bull 2006; 54:1508–14.

56. Anon. Beclometasone Pulvinal product data sheet., http://www.trinity-chiesi. co.uked.; 2007.

57. Iida K, Hayakawa Y, Okamoto H, Danjo K, Luenberger H. Influence of storage humidity on the in vitro inhalation properties of salbutamol sulfate dry powder with surface covered lactose carrier. Chem Pharm Bull 2004; 52:444–6.

58. Begat P, Green M, Morton D, Whittock A, Staniforth JN. PowderHale: A novel high performance dry powder inhaler formulation technology for targeted and systemic drug delivery. Proc Drug Deliv Drug Deliv Lung 2001; XII119.

59. Louey MD, Mulvaney P, Stewart PJ. Characterisation of adhesional properties of lactose carriers using atomic force microscopy 1. J Pharm Biomed Anal 2001; 25:559–67.

60. Berard V, Lesniewska E, Andres C, Pertuy D, Laroche C, Pourcelot Y. Affinity scale between a carrier and a drug in DPI studied by atomic force microscopy 1. Int J Pharm 2002; 247:127–37.

61. Young PM. Characterisation of particle-particle interactions using the atomic force microscope. U 2002; 1–187.

62. Ward GH, Schultz RK. Process-induced crystallinity changes in albuterol sulfate and its effect on powder physical stability. Pharm Res 1995; 12:773–9.

63. Shah B, Kakumanu VK, Bansal AK. Analytical techniques for quantification of amorphous/crystalline phases in pharmaceutical solids. J Pharm Sci 2006; 95:1641–65.

64. Buckton G, Darcy P. Assessment of disorder in crystalline powders—a review of analytical techniques and their application. Int J Pharm 1999; 179:141–58.

65. Handbook of Industrial Crystallization, 2nd ed., Boston, MA: Butterworth-Heinemann; 2002.

66. Jung J, Perrut M. Particle design using supercritical fluids: Literature and patent survey. J Supercrit Fluids 2001; 20:179–219.

67. Tong HHY, Shekunov BY, York P, Chow AHL. Influence of polymorphism on the surface energetics of salmeterol xinafoate crystallized from supercritical fluids. Pharm Res 2002; 19:640–8.

68. Tong HHY, Shekunov BY, York P, Chow AHL. Characterization of two polymorphs of salmeterol xinafoate crystallized from supercritical fluids. Pharm Res 2001; 18:852–8.

69. Shekunov BY, Feeley JC, Chow AHL, Tong HHY, York P. Aerosolisation behaviour of micronised and supercritically-processed powders. J Aerosol Sci 2003; 34:553–68.

70. Anon. Inhale announces that offer to acquire Bradford particle design plc has met conditions to close. Archived press releasesed; 2007.

71. Young PM, Price R. The influence of humidity on the aerosolisation of micronised and SEDS produced salbutamol sulphate. Eur J Pharm Sci 2004; 22:235–40.

72. Burnett DJ, Thielmann F, Booth J. Determining the critical relative humidity for moisture-induced phase transitions. Int J Pharm 2004; 287:123–33.

73. Kaerger JS, Price R. Processing of spherical crystalline particles via a novel solution atomization and crystallization by sonication (SAXS) technique. Pharm Res 2004; 21:372–81.

74. Chen JF, Zhou MY, Shao L, et al. Feasibility of preparing nanodrugs by high-gravity reactive precipitation. Int J Pharm 2004; 269:267–74.

75. Chan HK. Dry powder aerosol drug delivery—Opportunities for colloid and surface scientists. Colloids Surfaces A Physicochem Eng Aspects 2006; 284: 50–5.

76. Chiou H, Li T, Hu T, Chan HK, Chen J-F, Yon J. Production of salbutamol sulfate for inhalation by high-gravity controlled antisolvent precipitation. Int J Pharm 2007; 331:93–8.

77. Johnson BK, Prud'homme RK. Chemical processing and micromixing in confined impinging jets. AICHE J 2003; 49:2264–82.

78. Broadhead J, Rouan SKE, Rhodes CT. The spray drying of pharmaceuticals. Drug Dev Indl Pharm 1992; 18:1169–206.

79. Chew NYK, Tang P, Chan HK, Raper JA. How much particle surface corrugation is sufficient to improve aerosol performance of powders? Pharm Res 2005; 22:148–52.

80. Chew NYK, Chan HK. Use of solid corrugated particles to enhance powder aerosol performance. Pharm Res 2001; 18:1570–7.

81. Dellamary LA, Tarara TE, Smith DJ, et al. Hollow porous particles in metered dose inhalers. Pharm Res 2000; 17:168–74.

82. Weers JG. Dispersible powders for inhalation applications. Innov Pharm Tech 2000; 1:111–16.

83. Newman SP, BUSSE WW. Evolution of dry powder inhaler design, formulation, and performance. Respir Med 2002; 96:293–304.

84. Tobyn M, Staniforth JN, Morton D, Harmer Q, Newton ME. Active and intelligent inhaler device development. Int J Pharm 2004; 277:31–7.

85. Price R, Staniforth JN, Woodcock D, Young PM. Delivery device for a dry powder aerosol. PCT/GB2004/002490 ed. 2004; 1–49.

86. Thiel C. From Susie's question to CFC free: An inventor's perspective on forty years of MDI development and regulation. In: Dalby RN, ed. Respiratory Drug Delivery V, Phenix, Arizona: Interpharm Press, 1996; 115–24.

87. Moren F, Dolovich M, Newhouse M, Newman S. Aerosols in Medicine: Principles, Diagnosis and Ttherapy. Amsterdam: Elsevier Science Publishers; 1993.

88. Hickey A. Inhalation Aerosols, Physical and Biological Basis for Therapy, New York: Marcel Dekker Inc.; 1996.

89. Molina MJ, Rowland FS. Stratospheric sink for chlorofluoromethanes–chlorine atomic-catalysed destruction of ozone. Nature 1974; 249:810–12.

90. Vervaet C, Byron PR. Drug-surfactant-propellant interactions in HFA-formulations. Int J Pharm 1999; 186:13–30.

91. Dalby RN, Byron PR, Shepherd HR, Papadoupoulos E. CFC propellant substitution: P-134a as a potential replacement for P-12 in MDIs. Pharm Technol 1990; 14:26–33.

92. Lizio R, Damm M, Sarlikiotis AW, Bauer HH, Lehr CM. Low-Temperature Micronization of a Peptide Drug in Fluid Propellant : Case Study Cetrorelix. AAPS Pharm Sci Tech 2007; 2:1–7.

93. Green J, Gommeren E. Pharmaceutical aerosols–enhancing the metered dose inhaler. 2007.

94. Williams RO, Repka MA, Barron MK. Application of co-grinding to formulate a model pMDI suspension. Eur J Pharm Biopharm 1999; 48:131–40.
95. Young PM, Price R, Lewis D, Edge S, Traini D. Under pressure: predicting pressurized metered dose inhaler interactions using the atomic force microscope. J Colloid Interface Sci 2003; 262:298–302.
96. Ashayer R, Luckham PF, Manimaaran S, Rogueda P. Investigation of the molecular interactions in a pMDI formulation by atomic force microscopy 2. Eur J Pharm Sci 2004; 21:533–43.
97. Traini D, Rogueda P, Young PM, Price R. Surface energy and interparticle forces correlations in model pMDI formulations. Pharm Res 2005; 22:816–25.
98. Traini D, Young PM, Rogueda P, Price R. In vitro investigation of drug particulates interactions and aerosol performance of pressurised metered dose inhalers. Pharm Res 2007; 24:125–35.
99. Traini D, Young PM, Rogueda P, Price R. Investigation into the influence of polymeric stabilizing excipients on inter-particulate forces in pressurised metered dose inhalers. Int J Pharm 2006.
100. Blondino FE, Byron PR. Surfactant dissolution and water solubilization in chlorine-free liquified gas propellants. Drug Dev Ind Pharm 1998; 24:935–45.
101. Griffiths P, Paul A, Rogueda P. Advance in non-aqueous colloids. London, UK: Royal Society of Chemistry; 2007.
102. Schultz RK, Quessy SN. Use of soluble fluorosurfactants for the preparation of metered-dose aerosol formulations. 668597 ed. 1991.
103. Byron P, Blondino F. Metered dose inhaler fomulations which include the ozone-friendly propellant HFC 134a and a pharmaceutically acceptable suspending, solubilizing, wetting, emulsifying or lubricating agent. 217012 ed. 1994.
104. Purewal TS, Greenleaf DJ. Medicinal aerosol formulations. 471618 ed. 1995.
105. Stefely J, Schultz D, Schallinger L, Perman C, Leach C, Duan D. Biocompatible compounds for pharmaceutical drug delivery systems. In Properties Company MI, ed. 797803 ed. 1997.
106. Leach CL. Targeting inhaled steroids. Int J Clin Pract1998; 23–27.
107. Ganderton D, Lewis D, Davies R, Meakin B, Brambilla G, Church T. Modulite (R): a means of designing the aerosols generated by pressurized metered dose inhalers. Respir Med 2002; 96:S3–8.
108. Brambilla G, Ganderton D, Garzia R, Lewis D, Meakin B, Ventura P. Modulation of aerosol clouds produced by pressurised inhalation aerosols. Int J Pharm 1999; 186:53–61.
109. Leach C, Hameister M, Tomai M, Hammerbeck D, Stefely J. Oligolactic acid (OLA) biomatrices for sustained release of asthma therapeutics. In Dalby RN, Byron PR, Farr SJ, Peart JP, eds. Respiratory Drug Delivery VII, 1st edn. Florida: Serentec Press; 2000; 75–82.
110. Stefely JS. Excipients for inhalation drug delivery: Expanding the capability of the MDI. drug delivery technology 2002; 2:62–9.
111. Stefely J, Brown B, Hammerbeck D, Stein S. Equipping the MDI for the 21st century by expanding its formulation options. In Dalby RN, Byron P, Peart JP, Farr SJ, eds. Respiratory Drug Delivery VIII, 1st ed., NC, USA: Davis Horwood International. 2002; 207–14.

112. Butz N, Porte C, Courrier H, Krafft MP, Vandamme TF. Reverse water-in-fluorocarbon emulsions for use in pressurized metered-dose inhalers containing hydrofluoroalkane propellants. Int J Pharm 2002; 238:257–69.

113. Williams RO, Liu J. Influence of formulation technique for hydroxypropyl-beta-cyclodextrin on the stability of aspirin in HFA 134a. Eur J Pharm Biopharm 1999; 47:145–52.

114. Chipps BE. Determinants of asthma and its clinical course. Ann Allergy Asthma Immunol 2004; 93:309–16.

115. Dolovich M. Influence of inspiratory flow rate, particle size, and airway caliber on aerosolized drug delivery to the lung. Respir Care J 2000; 45:597–608.

116. Deblaquiere P, Christensen DB, Carter WB, Martin TR. Use and misuse of metered-dose inhalers by patients with chronic lung-disease–a controlled, randomized trial of 2 instruction methods. Am Rev Respir Dis 1989; 140: 910–6.

117. Cochrane MG, Bala MV, Downs KE, Mauskopf J, Ben-Joseph RH. Inhaled corticosteroids for asthma therapy–Patient compliance, devices, and inhalation technique. Chest 2000; 117:542–50.

118. Ruggins NR, Milner AD, Swarbrick A. An assessment of a new breath actuated inhaler device in acutely wheezy children. Arch Dis Childhood 1993; 68:477–80.

119. Hampson NB, Mueller MP. Reduction in patient timing errors using a breath-activated metered-dose inhaler. Chest 1994; 106:462–5.

120. Newman SP, Weisz AWB, Talaee N, Clarke SW. Improvement of drug delivery with a breath actuated pressurized aerosol for patients with poor inhaler technique. Thorax 1991; 46:712–16.

121. Gabrio BJ, Stein SW, Velasquez DJ. A new method to evaluate plume characteristics of hydrofluoroalkane and chlorofluorocarbon metered dose inhalers. Int J Pharm 1999; 186:3–12.

122. Ross DL, Gabrio BJ. Advances in metered dose inhaler technology with the development of a chlorofluorocarbon-free drug delivery system. J Aerosol Med Deposition Clearance Eff Lung 1999; 12:151–60.

123. Newman SP. Principles of metered dose inhalers. Respir Care 2005; 50: 1177–87.

124. Lewis D, Ganderton D, Meakin B, Brambilla G. Theory and practice with solution systems. In Dalby RN, Byron PR, Peart J, Suman JD, Farr S, eds., 1st edn. Illinois: Davis Healthcare International 2004; 109–15.

125. Lewis D, Meakin B, Brambilla G. New actuators versus old: reasons and results for actuator modifications for HFA Solution MDIs. InDalby R, Byron PR, Peart J, Suman JD, Farr S, eds., 1 edn. Illinois: Davis Healthcare International Publishing. 2006; 101–10.

126. Wilby M. Increasing dose consistency of pMDIs. Drug Deliv Technol 2005; 5: 59–65.

127. Tiwari D, Goldman D, Dixit S, Malick WA, Madan PL. Compatibility evaluation of metered-dose inhaler valve elastomers with tetrafluoroethane (P134a), a non-CFC propellant. Drug Dev Ind Pharm 1998; 24:345–52.

128. Bradley L, Hunt K, Fenn T. The Performance of the EPDM Spraymiser Valve. Drug Delivery to the Lungs XV, Edimborough, UK: The Aerosol Society. 2005; 101–4.

129. Howlett D, Colwell J. Improvements in extractables from pMDI elastomer systems. Drug Delivery to the Lungs VIII The Aerosol Society. 1997; 36–8.

130. Cummings RH. Pressurized metered dose inhalers: chlorofluorocarbon to hydrofluoroalkane transition-valve performance. J Allergy Clin Immunol 1999; 104:S230–6.

131. Jinks P, Marsden S. The Development and Performance of a Fluoropolymer Lined Can for Suspension Metered Dose Inhaler Products., Portisghead, UK: The Aerosol Society. 2007; 177–80.

132. Herman CS, Riebe MT, Li-Bovet Li, Ashurst IC. Metered dose inhaler for albuterol. 831268 ed. 2000.

133. Lewis D, Ganderton D, Meakin B, Ventura P, Brambilla G, Garzia R. Pressurised metered dose inhalers (MDI). 612072 ed. 2004.

134. Traini D, Young PM, Rogueda P, Price R. The use of AFM and surface energy measurements to investigate drug-canister material interactions in a model pressurized metered dose inhaler formulation. Aerosol Sci Tech 2006; 40:227–36.

135. Gupta PK, Hickey AJ. Contemporary Approaches in Aerosolized Drug Delivery to the Lung. J Control Release 1991; 17:127–47.

136. Selroos O, Pietinalho A, Riska H. Delivery devices for inhaled asthma medication-Clinical implications of differences in effectiveness. Clin Immunotherapeut 1996; 6:273–99.

137. Newman S, Pavia D2085. Aerosol deposition in man. InNewhouse M, Dolovich M, editors. Aerosols in Medicine. Amsterdam: Elsevier. 2005; 193–218.

138. O'Callaghan C, Barry PW. The science of nebulised drug delivery. Thorax 1997; 52(Suppl 2):S31–44.

139. Dalby R, Hickey A, Tiano S. Medical Devices for the Delivery of Therapeutic Aerosols to the Lungs. New York: Marcel Dekker Inc. 1996.

140. Mercer TT, Tillery MI, Chow HY. Operating characteristics of some compressed-air nebulizers. Am Ind Hygiene Assoc J 1968; 29:66.

141. Nerbrink O, Dahlback M, Hansson HC. Why do medical nebulizers differ in their output and particle-size characteristics. J Aerosol Med 1994; 7: 259–76.

142. Hess D, Fisher D, Williams P, Pooler S, Kacmarek RM. Medication nebulizer performance–Effects of diluent volume, nebulizer flow, and nebulizer brand. Chest 1996; 110:498.

143. Loffert DT, Ikle D, Nelson HS. A comparison of commercial jet nebulizers. Chest 1994; 106:1788–92.

144. Knoch M, Sommer E. Jet nebulizer design and function. Eur Resp Rev 2000; 10:183–6.

145. Coates AL, Ho SL. Drug administration by jet nebulization. Pediat Pulm 1998; 26:412–23.

146. Nikander K, Bisgaard H. Impact of constant and breath-synchronized nebulization on inhaled mass of nebulized budesonide in infants and children. Pediat Pulm 1999; 28:187–93.

147. Pelkonen AS, Nikander K, Turpeinen M. Jet nebulization of budesonide suspension into a neonatal ventilator circuit: Synchronized versus continuous nebulizer flow. Pediat Pulm 1997; 24:282–6.

148. LeBrun PP, deBoer AH, Gjaltema D, Hagedoorn P, Heijerman HG, Frijlink HW. Inhalation of tobramycin in cystic fibrosis. Part 1: the choice of a nebulizer. Int J Pharm 1999; 189:205–14.

149. Smith EC, Denyer J, Kendrick AH. Comparison of twenty three nebulizer/compressor combinations for domiciliary use. Eur Respir J 1995; 8:1214–21.

150. Denyer J1997. Adaptive aerosol delivery in practice. Eur Respir Rev 1997; 7: 388–9.

151. Nikander K, Denyer J, Cobos N. Compliance with nebulized budesonide in Spanish children with asthma using adaptive aerosol delivery (AAD). Am J Respir Crit Care Med 1999; 159:A119.

152. Denyer J, Ritson S, Everard ML. Aerosol delivery systems acceptable to young children improve drug delivery. Thorax 1998; 53:A55.

153. Nikander K. Drug-delivery systems. J Aerosol Med 1994; 7:S19-S24.

154. Takano H, Nakazawa M., Asai L, Itoh M. Performance of a new clinical nebuliser for drug administration based on the mesh-type ultrasonic atomisation by elastic surface waves. J Aerosol Med 1999; 12:98.

155. Dhand R. Nebulizers that use a vibrating mesh or plate with multiple apertures to generate aerosol. Respir Care 2002; 47:1406–16.

156. Simon M, Gopalakrishnan V. AeroDose inhaler feasibility studies with the aerosol generator. In Dalby R, Byron PR, Farr S, Peart J, ed., 1 edn., North Carolina: Davis Horwood. 2000; 311–14.

157. Keller M, Lintz F, Walther E. Novel liquid formulation technologies as a tool to design the aerosol performance of nebulisers., London, UK: The Aerosol Society. 2001; 99–102.

158. Stangl R. The PARI eFlow: tool in the development process of a new inhaler for topic and systemic pulmonary drug delivery. Drug Delivery to the Lungs 12, London, UK: The Aerosol Society.; 2001; 107–10.

159. Smart J, Stangl R, Halsall I, Chrystyn H. TouchSpray technology: In vitro evaluation of pharmaceutical compounds. Drug Delivery to the Lungs 12, London, UK: The Aerosol Society. 2001; 103–6.

160. Grace JM, Marijnissen JCM. A Review of Liquid Atomization by Electrical Means. J Aerosol Sci 1994; 25:1005–19.

161. Bailey AG. The inhalation and deposition of charged particles within the human lung. J Electrostat 1997; 42:25–32.

162. Bailey AG, Hashish AH, Williams TJ. Drug delivery by inhalation of charged particles. J Electrostat 1998; 44:3–10.

163. Dalby R, Spallek M, Voshaar T. A review of the development of Respimat ((R)) Soft Mist (TM) Inhaler. Int J Pharm 2004; 283:1–9.

164. Schuster J, Rubsamen R, Lloyd P, Lloyd J. The AER(X)(TM) aerosol delivery system. Pharm Res 1997; 14:354–7.

165. Gonda I, Cipolla D, Farr S, et al. AERxTM: Synthesis of technologies to provide therapeutic benefits via pulmonary drug delivery. Abstracts Papers Am Chem Soc 1999; 217:U572.

166. Hong JN, Hindle M, Byron PR. Control of particle size by coagulation of novel condensation aerosols in reservoir chambers. J Aerosol Med 2002; 15: 359–68.

167. Gupta R, Hindle M, Byron P, Cox KA, McRae DD. Investigation of a novel condensation aerosol generator: solute and solvent effects. Aerosol Sci Tech 2003; 37:672–81.

168. [Anon] Current best practice for nebuliser treatment—Introduction. Thorax 1997; 52:S1.

169. [Anon] Current best practice for nebuliser treatment—Summary of guidelines. Thorax 1997; 52:S2–3.

2

Invasive Pulmonary Aspergillosis: Therapeutic and Prophylactic Strategies

Jason T. McConville and Nathan P. Wiederhold

Divisions of Pharmaceutics and Pharmacotherapy, College of Pharmacy, University of Texas, Austin, Texas, U.S.A.

INTRODUCTION

Invasive aspergillosis is a common cause of infectious disease related to morbidity and mortality in immunocompromised patients, with the lungs serving as a primary point of entry and site of infection for invasive infections caused by *Aspergillus* species. Over the last few decades the incidence of invasive aspergillosis has increased. While our antifungal armamentarium has also increased with newer agents having broader spectrums of in vitro activity and improved safety profiles compared to older antifungal drugs, response rates in the treatment of invasive aspergillosis remain suboptimal. Because of poor outcomes associated with the treatment of invasive mycoses in immunocompromised patients, antifungal prophylaxis in populations at high risk for these infections has gained increased interest. However, this strategy is often limited by toxicities and adverse effects associated with systemic exposure. To overcome these limitations many investigators have begun to study the use of aerosolized antifungal agents as prophylaxis with the goal of achieving high localized concentrations at the primary site of infection.

EPIDEMIOLOGY

The incidence of invasive infections caused by *Aspergillus* species has increased significantly in recent decades. At some large cancer centers, invasive aspergillosis cases have increased three- to four-fold over the past 20 years (1–3). Unfortunately, these infections are associated with high morbidity and mortality in patients undergoing hematopoietic stem cells transplantation and those with hematologic malignancies (4,5). Invasive pulmonary aspergillosis is also observed in patients undergoing solid organ transplantation. However, the risk of these infections differs depending on the type of organ transplantation, with lung transplant recipients having the highest risk. Similarly, the risk of such infections is higher among patients undergoing allogeneic hematopoietic stem cell transplantation versus autologous stem cell transplants (4,6). Classic risk factors associated with invasive pulmonary aspergillosis include neutropenia, defects in cell-mediated immunity, the use of corticosteroids, and graft-versus-host disease (7,8). For allogeneic stem cell recipients the incidence of infection has been shown to be bimodal. Both autologous and allogeneic stem cell recipients are at increased risk during the neutropenic period prior to engraftment. During this period risk is associated with the degree and extent of neutropenia. However, allogeneic stem cell recipients are also at increased risk post-engraftment in the absence of neutropenia (2). Risk factors for later disease include graft-versus-host disease, the use of corticosteroids, and concurrent infection with cytomegalovirus.

MICROBIOLOGY/PATHOPHYSIOLOGY

Aspergillus species are ubiquitous fungi in nature. In the laboratory these fungi can be easily grown on routine culture media. The saprophytic growth stage consists of hyphal stalks with sporulating conidial heads. These conidial heads release thousands of conidida (spores) that are capable or remaining buoyant on air currents for prolonged periods of time (6). While over 180 *Aspergillus* species have been identified, only four species are commonly associated with infections in humans, including *A. fumigatus, A. terreus, A. flavus,* and *A. niger.* Of these, *A. fumigatus* is the species most commonly associated with invasive infections. The exact reason for this remains unclear. However, it has been speculated that the small size of the conidia (2–3 μm) and the ability for rapid mycelial growth at 37°C may favor this species (9).

Due to their small size and ability to remain in the air for prolonged periods of time, conidia are the means by which *Aspergillus* is spread in the environment and inhaled into the lungs. The majority of conidia are excluded from the lungs by the ciliary action after entrapment on the mucous epithelium (6). However, some are able to penetrate to the distal alveolar spaces due to their small size. In addition, *A. fumigatus* may produce toxins that inhibit ciliary activity (10). Pulmonary macrophages represent the first line of cellular defense against invasive disease and are capable of ingesting and killing conidia as well as the coordination of cellular defenses through the secretion of cytokines and chemokines (6,11). However, this process may be inefficient and delayed as germination of conidia into hyphae within macrophages may continue (9). Growth of angioinvasive hyphae are responsible for much of the pathologic features observed in patients with invasive pulmonary aspergillosis including local tissue damage, hemorrhage, infarction, coagulative necrosis, and potential hematogenous dissemination (7). To combat conidia that escape phagocytosis and hyphae, polymorphonuclear neutrophils and monocytes are recruited to the lungs by chemotactic factors. The neutrophils contact and adhere to the surface of hyphae, resulting in a respiratory burst, the secretion of reactive oxygen intermediates, and release of granules containing myeloperoxidase (11,12). These lines of cellular defense (pulmonary macrophages and neutrophils), along with the physical barriers and ciliary action of the mucous epithelium, are capable of preventing disease in healthy individuals. However, impairment of these defenses with the use of corticosteroids and cytotoxic chemotherapy place the host at risk for invasive disease (2,13,14).

Increases in the numbers of infections caused by non-*fumigatus* species has been reported to be increasing at large cancer centers (15,16). The causative organism can influence the choice of therapy as infections caused by *A. terreus* and *A. flavus* have been shown to be less susceptible to amphotericin B with clinical failures reported in the literature (17–19). However, the species responsible for infection is often unknown due to the difficulties in the diagnosis and identification of the causative organism. Isolation of *Aspergillus* species from bronchial alveolar lavage fluid, bronchial washings, or expectorated sputum has low sensitivity for the diagnosis of invasive pulmonary aspergillosis (7,8). Furthermore, the clinical manifestations of invasive disease, which may include fever, hemoptysis, cough, or pleuritic chest pain, are neither sensitive nor specific (7).

TREATMENT

Response rates in the treatment of invasive pulmonary aspergillosis are highly individualized. Factors that influence outcomes include the immune status of the host, the extent of disease, and patient's ability to tolerate antifungal therapy (5,20). For many years amphotericin B deoxycholate was the standard of care. By binding to ergosterol within the cell membrane and allowing leakage of a variety of small molecules, amphotericin B is able to inhibit a wide spectrum of fungi in vitro. However, use of this agent is often limited by drug toxicities, including nephrotoxicity and infusion related reactions. Nephrotoxicity is the principal adverse affect associated with the prolonged courses and cumulative doses need for the treatment of invasive aspergillosis. The availability of newer therapies, including lipid formulations of amphotericin B, second-generation triazoles (i.e. voriconazole and posaconazole), and the echinocandins, have improved the overall effectiveness of antifungal therapy for these infections, principally through increased tolerability. Response rates from various antifungal studies in the treatment of invasive aspergillosis are shown in Figure 1.

AMPHOTERICIN B

Lipid formulations of amphotericin B, including liposomal amphotericin B, amphotericin B lipid complex, and amphotericin B colloidal dispersion, have improved safety profiles compared to amphotericin B deoxycholate. Lower rates of infusion related reactions and nephrotoxicity are observed with these agents. However, while these toxicities are reduced they are not eliminated (21–23). Furthermore, the extent to which these formulations have improved outcomes is unclear. Most available efficacy data for the lipid formulations for the treatment of invasive aspergillosis come from small, open-label studies that have enrolled heterogeneous patient populations (Fig. 1). Only amphotericin B colloidal dispersion has been compared directly to amphotericin B deoxycholate with no differences in response rates (51% and 52%, respectively) reported in a randomized double-blind study (24). The other lipid formulations have been compared to amphotericin B deoxycholate against invasive fungal infections not limited to those caused by *Aspergillus* species. In one study involving 55 patients with document or suspected invasive aspergillosis, a trend in favor of liposomal amphotercin B compared to amphotericin B deoxycholate was reported (25). However, the number of patients with documented infection was small and the difference in response rates did not reach significance.

Despite the availability and extensive use of lipid formulations for the treatment of invasive fungal infections the optimal dose of these agents still remains in question. Standard doses in clinical trials have ranged between 3 and 5 mg/kg/day. Higher doses (up to 10 mg/kg/day) have been advocated for the treatment of fungal infections involving the central nervous system

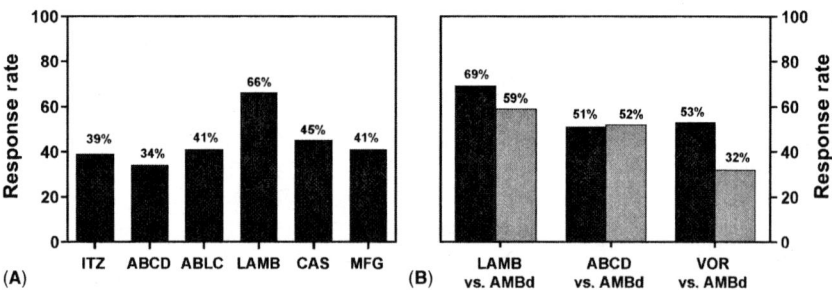

Figure 1   Response rates to antifungal therapy from (**A**) open-label studies and (**B**) randomized controlled trials. *Abbreviations*: ABCD, amphotericin B colloidal dispersion; ABLC, amphotercin B lipid complex; CAS, caspofungin; ITZ, itraconazole; LAMB, liposomal amphotericin B; MFG, micafungin; VOR, voriconazole.

and those caused by Zygomycetes (26,27). However, this strategy has not consistently demonstrated improved outcomes in patients with invasive pulmonary aspergillosis. Similar response rates were reported in patients with invasive aspergillosis who received liposomal amphotericin B doses of either 1 or 4 mg/kg/day in one randomized multicenter trial (28). Other uncontrolled studies have suggested that dose of up to 7.5 mg/kg/day of lipid formulations of amphotericin B are relatively well tolerated (29,30). In contrast, a recent study demonstrated that doses of 10 mg/kg/day of liposomal amphotericin B did not improve outcomes compared to doses of 3 mg/kg/day but were associated with increased toxicity (31).

Itraconazole

The azoles are an alternative class of antifungal agents to amphotericin B that inhibit the biosynthesis of ergosterol. Currently, three members of this class with activity against *Aspergillus* species are available in the United States: itrazonazole, voriconazole, and posaconazole. One of the advantages of these agents is the availability of oral formulations, thus eliminating the necessity for intravenous access required for amphotericin B. Of these azoles, itraconazole has been available longest. However, its role in the treatment of invasive aspergillosis remains poorly defined. Open-label studies have suggested response rates ranging between 30 and 60% (32,33). However, many of the early studies were conducted when itraconazole was only available in capsule form, thus potentially biasing the results in favor of this agent since only patients well enough to tolerate oral medications were included.

In addition to the limited clinical data for the treatment of invasive aspergillosis, itraconazole oral capsules are also hampered by low and erratic bioavailability. An oral solution and intravenous formulation have since become available, both of which utilize hydroxy-propyl-β-cyclodextrin to overcome the insolubility of itraconazole in aqueous solution. The oral solution has improved the bioavailability and consistency of plasma levels of itraconazole (34). However, high rates of gastrointestinal toxicity, including nausea, vomiting, and diarrhea, have been reported with the oral solution which can often limit therapy. One potential strategy to overcome the variable bioavailability of the itraconazole capsules involves loading patients with the intravenous formulation followed by administration of the oral capsules with close monitoring of itraconazole plasma concentrations. However, published data regarding the efficacy of this strategy in the treatment of invasive pulmonary aspergillosis are limited. A response rate of 38.7% after 14 weeks of therapy (2 weeks intravenous itraconazole followed by 12 weeks of oral capsules) was reported in one study (35). However, the lack of a comparator group makes it difficult to evaluate the results of this trial and compare to other studies for the treatment of for invasive pulmonary aspergillosis.

Voriconazole

Voriconazole is another triazole with activity against *Aspergillus* species and is considered by many experts to be the drug of choice for the primary treatment of invasive aspergillosis. Voriconazole is available as both an intravenous formulation and as tablets for oral delivery, which have significantly improved bioavailbility (~90%) and tolerability compared to oral formulations of itraconazole. In the largest randomized clinical trial evaluating the treatment of invasive aspergillosis to date, voriconazole was shown to be superior to amphotericin B deoxycholate and other licensed antifungal therapy (Fig. 1B) (36). In this study, patients were randomized to receive either voriconazole (6 mg/kg IV every 12 hours for 2 doses followed by 4 mg/kg IV every 12 hours for at least 7 days, then 200 mg po (every 12 hour) or amphotericin B deoxycholate (1 mg/kg IV once daily). Patients could be switched to other licensed antifungal therapy after the initial randomized therapy. After 12 weeks, 52.8% of patients treated with voriconazole experienced a complete or partial response compared to 31.6% of patients randomized to receive amphotericin B deoxycholate. Among patients with pulmonary disease, 54.5% responded to voriconazole therapy versus 34.2% for amphotericin B deoxycholate. Voriconazole was superior to amphotericin B regardless of the site of infection, underlying condition, and neutropenia status. Survival was also significantly higher among patients randomized to voriconazole (70.8%) compared to amphotericin B deoxycholate (57.9%). However, response rates remained poor in heavily immunocompromised patients. Only 32% of allogeneic stem transplant recipients randomized to voriconazole had a complete or partial response, further underscoring the importance of the immune response in patients with invasive pulmonary aspergillosis.

While voriconazole has been a significant advancement in the treatment of invasive aspergillosis, the use of this agent is associated with a number of clinically significant drug interactions as well as the potential for hepatotoxicity. In addition, recent attention has focused on the variable bioavailability of orally administered voriconazole. A potential cause for this variability may be genetic polymorphisms in the cytochrome P450 enzyme 2C19, an influential covariant for exposure and clearance of this drug (37,38). This variability may influence response to therapy as a recent study has suggested that voriconazole trough plasma concentrations below 1 mcg/mL may be associated with poor outcomes in patients with invasive fungal infections (39).

Posaconazole

Posaconazole is the most recent triazole antifungal to become available. Structurally similar to itraconazole, this agent has potent activity against a number of filamentous fungi include *Aspergillus* species. In a small salvage study, response rates for the treatment of invasive aspergillosis ranged between 31 and 37% for allogeneic stem cell transplant patients and those with hematologic malignancies, respectively, to as high as 86% for autologous stem cell transplant recipients (40). However, no data are currently published evaluating the use of posaconazole as primary therapy. In addition, posaconazole is currently only available as an oral suspension. Moderate interpatient variability in trough concentrations reported in clinical trials (41–43), and this agent must be administered two to four times per day to achieve reliable plasma concentrations (44). Even with multiple doses per day moderate interpatient variability has been observed with posaconazole trough concentrations in patients with hematologic malignancies and autologous/ allogeneic stem cell recipients.

Echinocandins

One of the most important advances in the treatment of invasive fungal infections has been the development and availability of the echinocandin class of antifungal agents. Currently, three agents are available for use in the United States: anidulafungin, caspofugnin, and micafungin. By inhibiting the fungal specific target beta-1,3-glucan synthase, resulting in the reduction of beta-1,3-glucan in the cell walls of *Candida* and *Aspergillus* species, these agents avoid toxicities to mammalian cells associated with amphotericin B as well as the drug interactions commonly caused by triazoles due to inhibiton of cytochrome P450 enzymes (45–47). While the excellent safety profile and therapeutic efficacy for the treatment of invasive candidiasis have been well documented in clinical trials (48–50), the efficacy of the echinocandins for the treatment of infections caused by *Aspergillus* species is less established. The majority of clinical data for the treatment of invasive aspergillosis come from small noncomparative studies that included patients who were unresponsive to or intolerant of other antifungal therapy. In these trials response rates for caspofungin and micafungin have ranged between 40% and 50% (Fig. 1A) (51,52). Currently, there is considerable interest in combined use of echinocandins with other antifungal therapies, such as lipid formulations of amphotericin B and voriconazole, to potential improve response rates for invasive aspergillosis. This strategy has been supported by in vitro and in vivo studies (53). However, the available clinical data for combination therapy are from retrospective salvage studies (54,55).

PROPHYLAXIS

Strategies to prevent invasive mycoses have gained wide interest due to the poor response rates observed in the treatment of these infections in heavily immunosuppressed patients. The prophylactic use of antifungal agents has been shown in randomized controlled trials to reduce the incidence of invasive infections and improve survival (Fig. 2) (56,57). Fluconazole prophylaxis has been endorsed as an appropriate infection-control measure during the pre-engraftment period in patients undergoing hematopoietic stem cell transplantation in consensus guidelines (58). Two double-blind, placebo controlled studies have demonstrated decreases in invasive fungal infections during the pre-engraftment period with the use of this strategy (Fig. 2) (56,57). In both studies patients randomized to fluconazole had lower rates of invasive fungal infections as well as reductions in fungal-related mortality. Reductions in invasive fungal infections was primarily due decreases in the incidence of invasive canidiasis. Unfortunately, the use of fluconazole prophylaxis is not without consequences. An autopsy study including 355 allogeneic transplant recipients, half who received fluconazole and half no prophylaxis, reported an increase in the rates of *Aspergillus* and *Zygomycete* infections in patients who received prophylaxis (59). Whether this was due to selective pressure resulting from the lack of activity that fluconazole has against these organisms or increased risk of invasive mould infections due to prolonged survival is unknown.

Itraconazole Prophylaxis

Although prophylactic use of fluconazole is safe and effective in preventing invasive candidiasis, this strategy is ineffective against invasive aspergillosis. Many transplant centers have instituted prophylactic strategies utilizing agents with activity against *Aspergillus* species such as itraconazole secondary to the high morbidity and mortality rates associated with invasive aspergillosis in high-risk immunocompromised patients. In a

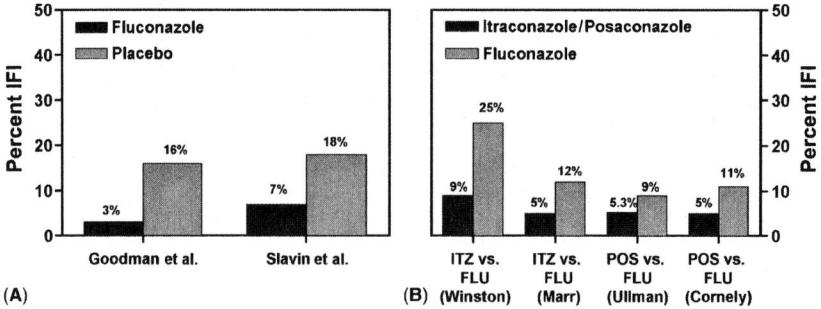

Figure 2 Percent of patients who developed invasive fungal infections in studies evaluating antifungal prophylaxis. (**A**) Fluconazole versus placebo and (**B**) itraconazole or posaconazole versus fluconazole. *Abbreviations*: FLU, fluconazole; ITZ, itraconazole; POS, posaconazole.

prospective, open-label multi-center trial of antifungal prophylaxis in allogeneic hematopoietic stem cell transplants recipients itraconazole did reduce the number of invasive fungal infections compared to fluconazole (9% vs. 25%, respectively; Fig. 2B) (60). A second open-label study conducted at a single-center reported significantly more proven and probable invasive fungal infections in patients randomized to fluconazole versus itraconazole (61). This difference was due to increases in the number of invasive mould infections (primarily aspergillosis). Interestingly, this study was amended after 197 patients were enrolled due to concerns with the safety of itraconazole when administered with cyclophosphamide during the conditioning regimen. Increases in serum bilirubin and a doubling of serum creatinine were observed in patients due a previously unrecognized drug interaction between these two agents when administered concurrently (62). Following this interim analysis, the protocol was changed to start prophylaxis the day of stem cell transplant after completion of the conditioning regimen. In addition, in both studies a significant portion (24–36%) of patients experienced gastrointestinal adverse effects, which resulted in approximately one quarter of the patients discontinuing prophylaxis with itraconazole in one study (61).

Posaconazole Prophylaxis

Posaconazole is also an attractive agent for use as prophylaxis in patients at high risk for invasive fungal infections due to its potent activity against *Aspergillus* species and availability as an oral formulation. Prospective clinical trials have demonstrated that posaconazole is effective in preventing invasive fungal infections and reducing fungal related mortality in high-risk patients. In a randomized multicenter study posaconazole prophylaxsis was compared to fluconazole or itraconazole in patients with neutropenia secondary to chemotherapy for acute myelogenous leukemia or myelodysplastic syndrome (41). Significantly fewer invasive fungal infections were observed in patients randomized to posaconazole (5%) compared to those who received fluconazole or itraconazole (11%; $P = 0.003$) within 100 days following randomization. This difference was primarily due to a lower incidence of invasive aspergillosis in the posaconazole group (1% vs. 7%, respectively; $P < 0.001$; Fig. 2B). A second prophylaxis study compared posaconazole to fluconazole in allogeneic stem cell transplant patients with graft-versus-host disease who were receiving immunosuppressive therapy (42). Similar rates of invasive fungal infections were observed in the posaconazole recipients compared to those randomized to fluconazole (5.3% vs. 9%; $P = 0.07$). However, posaconazole was superior to fluconazole in preventing proven or probable invasive aspergillosis (1% vs. 5.9%; $P = 0.001$). Adverse events related to study drug were similar between prophylaxis groups in both studies.

However, variable trough concentrations were reported in patients who received posaconazole (coefficients of variation ranging between 57% and 68%), raising concerns about the bioavailability of this agent.

Echinocandin Prophylaxis

An alternative strategy for the prevention of invasive fungal infections is the use of the echinocandins due to their excellent safety profile and activity against *Candida* and *Aspergillus* species. The results of a clinical study comparing micafungin to fluconazole as prophylaxis have been reported (63). This prospective, double-blind, multicenter trial enrolled both autologous and allogeneic stem cell transplant recipients randomized to either micafungin or fluconazole and continued until approximately 5 days after engraftment (up to 42 days following transplant). Treatment success, defined as the absence of proven, probable, or suspected systemic fungal infections through the end of prophylaxis and the absence of proven or probable systemic fungal infections the the end of the 4-week post-treatment period, was higher for those who received micafungin (80%) compared to fluconazole (73.5%; $P = 0.03$). This difference was primarily due to breakthrough aspergillosis, which occurred in one patient randomized to micafungin and 7 patients who received fluconazole ($P = 0.071$), and suspected fungal infections during the prophylaxis period (15.1% vs. 21.4%, respectively; $P = 0.024$). Patients randomized to micafungin also received less empirical antifungal therapy compared to fluconazole (15.1% vs. 21.4%, respectively; $P = 0.024$). Despite the success of micafungin in this trial, the median duration of prophylaxis was only 18 days. This is of particular concern in the setting of allogeneic stem cell transplants where observational studies have demonstrated a bimodal distribution of invasive infections caused by *Aspergillus* species (3). This highlights one of the limitations of the echinocandins in that these agents are only available in intravenous formulations making it difficult to continue daily prophylaxis for a prolonged period of time.

AEROSOLIZED ANTIFUNGAL PROPHYLAXIS

A major concern with prophylaxis is exposure of a large group of patients, many of whom will not develop an invasive fungal infection even without prophylaxis, to the associated systemic toxicities and drug interactions that occur with currently available antifungal agents. Recently, attention has been focused on the pulmonary delivery of antifungal agents for the prevention of invasive aspergillosis (64). This strategy is gaining favor for the prevention of invasive mycoses that primarily affect the upper and lower respiratory tract. Aerosolized delivery results in high concentrations of antifungal agents localized at the site of infection, and systemic toxicities and drug interactions may be minimized due to low systemic exposures. However, to date wide spread use of this strategy has not been well studied in clinical trials.

Aerosolized Amphotericin B

Amphotericin B has been the most widely studied aerosolized antifungal agent thus far. One survey reported use of aerosolized amphotericin B as prophylaxis in 61% of transplant centers (65). However, variability exists between centers as to the formulation and dose administered and the aerosolization procedures (i.e. nebulizer) used. Several centers have reported the aerosolization of amphotericin B deoxycholate. Nebulized amphotericin B lipid complex has also been used and shown to be safe in lung transplant recipients with few adverse effects (66,67). Adverse effects that have been reported include cough, dyspnea, bronchospasm, nausea, and vomiting (66–68). In a comparative safety study, adverse effects were more common with the deoxycholate formulation compared to the amphotericin B lipid complex (66). However, decreases in FEV_1 of greater than 20% in addition to cough, shortness of breath, and wheezing were reported following the administration of these formulations (32,69). The significance of these adverse effects in patients with compromised pulmonary function or structural lung disease may not be fully recognized.

Aerosolized administration of amphotericin B as prophylaxis has not been consistently shown to reduce the incidence of invasive fungal infections in clinical trials (70–72). Unfortunately, most clinical studies of the prophylactic use of aerosolized amphotericin B have been underpowered due to the small number of patients enrolled. In addition, these studies often have included heterogenous patients with different risk factors for invasive aspergillosis, and have been short in duration thus preventing evaluation of the effectiveness of this strategy in allogeneic stem cell transplant recipients at risk in the late post-engraftment period secondary to graft-versus-host disease. Recently, the use of inhaled liposomal amphotericin B has been evaluated as prophylaxis against invasive aspergillosis (73). In this study 271 patients who were expected to have profound and prolonged neutropenia were randomized to receive either inhaled liposomal amphotericin B, administered as 12 mg doses 2 days per week, or placebo. Significantly fewer patients randomized to inhaled liposomal amphotericin B (4%) were reported to have developed proven or probable invasive aspergillosis compared to placebo (14%).

The formulations of amphotericin B often used include intravenous preparations not specifically designed for aerosolized administration. Recently, a dry-powder amphotericin B formulation specifically designed for pulmonary delivery was reported to be effective in preventing invasive aspergillosis due to *A. fumigatus* in a persistently neutropenic rabbit model

(74). These results agree with previous animal studies demonstrating improvements in survival with aerosolized administration of amphotericin B deoxycholate as well as lipid amphotericin B formulations (75–77). However, as previously noted, concerns regarding the activity and efficacy of amphotericin B against non-*fumigatus Aspergillus* species exist. In particular, decreased in vitro activity and clinical failures with the use amphotericin B have been reported for both *A. flavus* and *A. terreus* (17–19,78,79).

Aerosolized Nanostructured Itraconazole

Recent attention has been focused on new technologies capable of improving the bioavailability of poorly water-soluble drugs and enhancing aerosolized delivery to the lungs. Evaporative precipitation of aqueous solution (EPAS) and spray-freeze into liquid (SFL) are two such technologies utilized to improve the dissolution and bioavailability of poorly water-soluble agents, and can produce nanostructured particles capable of drug delivery to the alveolar space (80,81). The utility of nanostructured itraconazole particles produced by these engineering processes has recently been evaluated in a series of in vitro and in vivo studies.

Manufacture of Nanostructured Itraconazole Particles

Improving the solubility of poorly water soluble drugs is an increasingly important aspect for increased efficacy of treatment regimens. Approximately 40% of investigational active pharmaceutical ingredients (APIs) are estimated to have poor aqueous solubility. The Biopharmaceutics Classification System indicates that class II APIs display poor water solubility with high mucosal permeability (an example of this class of compound is itraconazole). For APIs in this class, solubility and subsequent dissolution rate are rate limiting steps for absorption. Particle size reduction through milling to form micronized particles of API with a high surface area have been shown to improve dissolution and bioavailability (82).

Spray freezing into liquid: Spray freezing into liquid (SFL) is a technique used to formulate APIs to have surface areas from 12 to 83 m^2/g, compared to approximately 4 or 5 m^2/g for unprocessed APIs. Figure 3 indicates the difference in bulk density of an unprocessed itraconazole formulation compared to that of an SFL processed formulation of the same composition. Following processing using the SFL technique there is a marked reduction in bulk density (Fig. 3), which is directly proportional to in increase in surface area, compared to the low bulk density of the raw unprocessed formulation ingredients (Fig. 3).

As there is such an increase in surface area, there is a marked improvement in dissolution rates. During this particle engineering technique (as shown in Fig. 4), an API and stabilizing and/or dissolution enhancing

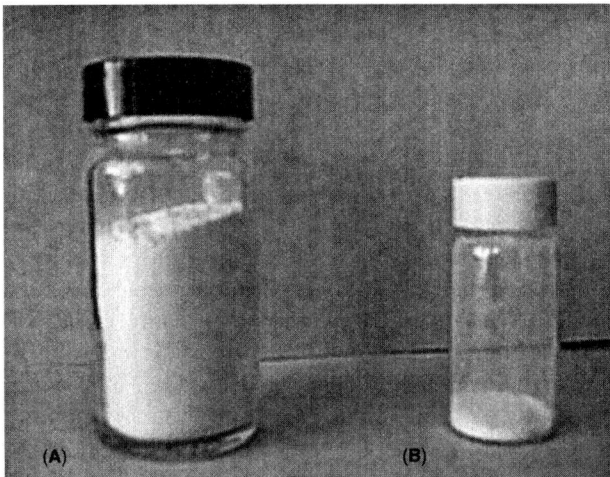

Figure 3 Powder formulations composed of: 0.75 g Itraconazole, 0.75 g Poloxamer 407, and 0.25 g Polysorbate 80. **(A)** Processed using the SFL technique; **(B)** raw unprocessed powder formulation components.

component (pharmaceutical excipients selected from a variety of surfactants and polymers) are dissolved in an organic solvent (e.g. acetonitrile). The organic solution is atomized directly into a cryogenic liquid (e.g. liquid nitrogen (83), or CO_2 (84)), through an insulated nozzle (a 63–126 μm inner diameter polyether ether ketone (PEEK) nozzle). During this process the atomized organic solution is instantly frozen (as frozen amorphous microparticles) (83,85–89), and a molecular dispersion is achieved during this rapid freezing, thus preventing phase separation.

The frozen molecularly dispersed microparticles are then placed into a lyophilizer to remove the frozen organic solvent (by sublimation). After the lyophilization process is complete, the resulting dried powder is composed of discrete microparticles where the API is molecularly dispersed with a polymer in a porous matrix (90) (Fig. 5).

Previous studies have indicated that rapid dissolution, of SFL formulations, is due to the amorphous morphology (molecular dispersion), high surface area, and enhanced wetting (related to the high surface area and inclusion of surfactants and/or hydrophilic stabilizing polymers) (91–94).

Evaporative precipitation into aqueous solution: Evaporative precipitation into aqueous solution (EPAS) is another technique whereby an API can be altered to have an increased surface area, and like the SFL technique results in an improvement dissolution rate (90). During this

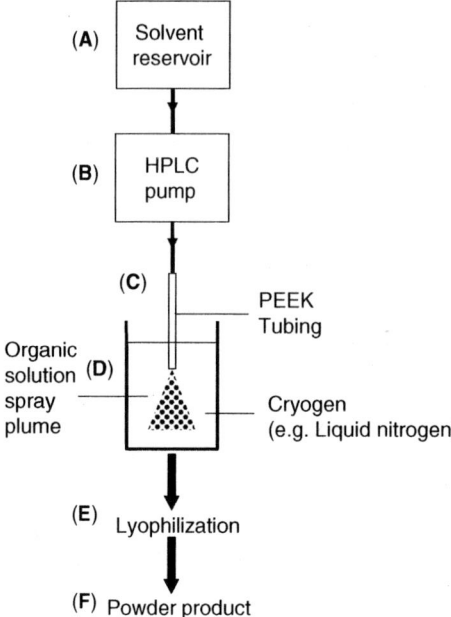

Figure 4 A typical setup for the spray-freezing into liquid technique. **(A)** Solvent reservoir containing drug with stabilizing excipient(s) co-dissolved in an organic solvent; **(B)** pump; **(C)** insulating tubing consisting of PEEK; **(D)** cryogen (e.g. liquid nitrogen); **(E)** lyophilization process to obtain **(F)** the final dry powder product.

particle engineering technique (as shown in Fig. 6), an API and stabilizing dissolution enhancing component (pharmaceutical excipients selected from a variety of surfactants and polymers) are dissolved in a heated organic solvent (e.g. dichloromethane). The organic solution is atomized directly into a heated aqueous solution bath, which may contain a surfactant, through a crimped tube nozzle (90,95,96). At this stage the volatile solvent rapidly evaporates and the API rapidly precipitates. Depending on the selection of stabilizing excipients, the particle size as determined by crystal growth is inhibited. During this manufacture a colloidal dispersion of crystalline drug is produced. This could be used directly for nebulization or lyophilized (after flash-freezing in a cryogen) to obtain a dry powder product.

Nebulization of Colloidal Itraconazole

Inhalation and deposition of submicron particles (particles below 1 µm in size) in the lung is a challenge. Submicron particles have a high charge to mass ratio and as a result may easily agglomerate (97) to form large particles that have the potential to impinging in the upper airways of the lung. Conversely, submicron particles that do not agglomerate will essentially posses a very small inertia during inhalation and so will not impinge on the lung wall, they will then be subsequently exhaled. This problem has been

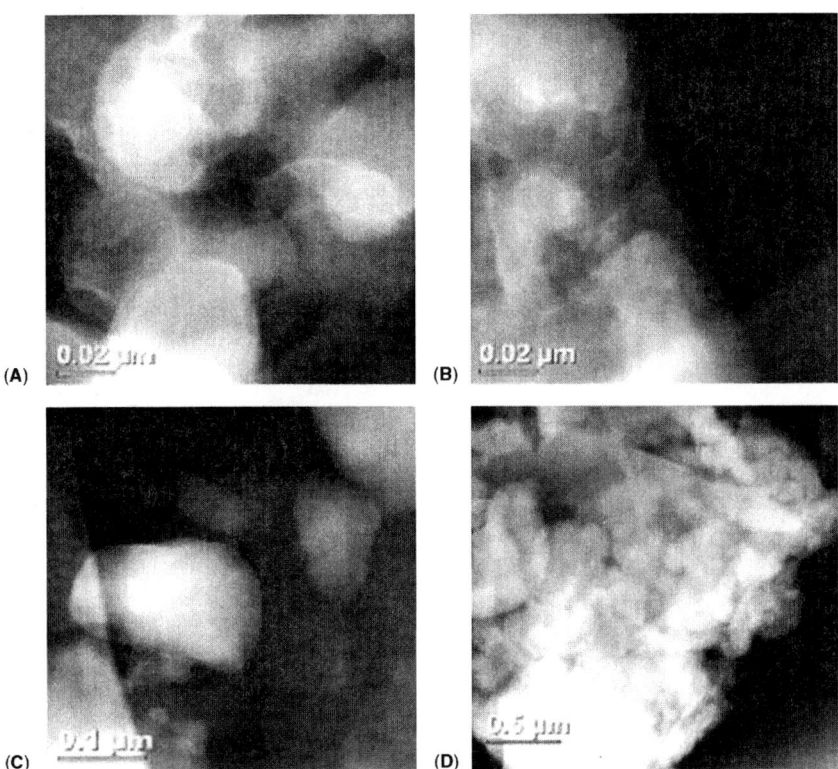

Figure 5 Scanning tunneling electron microscopy micrographs of the SFL (**A** and **B**) processed drug (example shown is the poorly water-soluble drug danazol) and the EPAS (**C** and **D**) processed danazol. *Source*: From Ref. 90.

overcome by embedding the submicron particles within carrier materials that are suitably sized to deposit in the deep airways of the lung (98). Another approach is to disperse the submicron material within an aqueous medium and nebulize (95). This nebulization approach effectively removes a processing step, as the lung deposition is controlled by the functionality of the nebulizer (Fig. 7).

In the example of aerosolized itraconazole nanoparticles prepared by either EPAS or SFL technologies (95), comparable emitted doses and respirable fractions were observed when using a micropump nebulizer (Table 1).

Animal Dosing with Nebulized Colloidal Dispersions of Itraconazole

Due to the fact that itraconazole colloidal dispersions could be nebulized within a recognized droplet size range that was consistent with peripheral

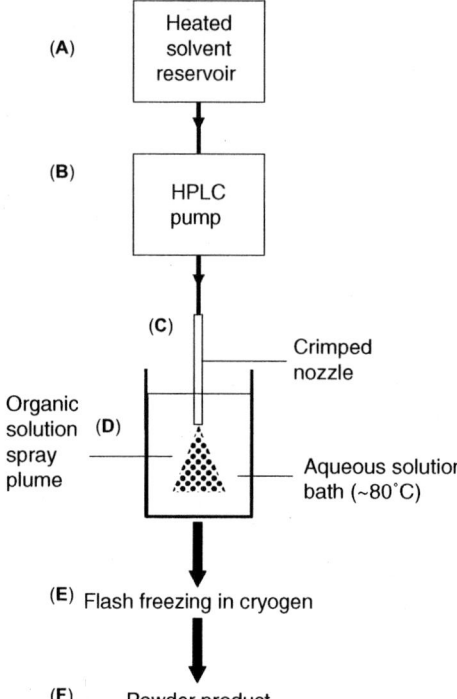

(A) Heated solvent reservoir

(B) HPLC pump

(C) Crimped nozzle

Organic solution spray plume **(D)**

Aqueous solution bath (~80°C)

(E) Flash freezing in cryogen

(F) Powder product

Figure 6 A typical setup for the EPAS technique. (**A**) Solvent reservoir containing drug with stabilizing excipient(s) co-dissolved in an organic solvent (which may be heated); (**B**) pump; (**C**) crimped narrow bore nozzle manufactured from HPLC tubing; (**D**) aqueous solution bath (this may contain further stabilizing surfactants to prevent aggregation and crystal growth); (**E**) lyophilization process to obtain (**F**) the final dry powder product.

lung deposition (99), dosing of a mammalian model was considered possible (100). Previously a small rodent whole body exposure unit had been evaluated for use with an Aeroneb Pro® micro-pump nebulizer (Nektar Therapeutics, formerly Aerogen Inc., Mountain View, CA), using the model drug caffeine (101). This dosing apparatus had been designed to allow up to 14 mice to be housed simultaneously. The apparatus was re-evaluated using a colloidal dispersion of itraconazole which had been prepared using the SFL technique (95). The study indicated that the simultaneous dosing of 10 mice resulted in only a 13% relative standard deviation in itraconazole mouse lung concentration.

Lung Pharmacokinetics of Nebulized Colloidal Dispersions of Itraconazole

Dosing of the nebulized colloidal dispersions of itraconazole was performed in three groups of mice (14 per group) for a 20 min nebulization period in the specially designed dosing chamber (indicated above) (95). Interestingly the data revealed that irrespective of the processing technique used to prepare itraconazole (EPAS or SFL, which result in either crystalline or amorphous

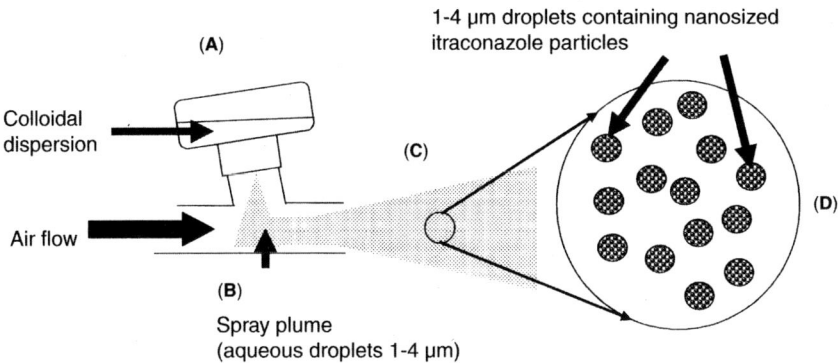

Figure 7 The scheme of nebulization of a colloidal dispersion of a poorly water-soluble drug: (**A**) A colloidal dispersion of drug is contained within a nebulizer reservoir (e.g., an Aeroneb® Pro micropump nebulizer); (**B**) the drug is nebulized; and (**C**) a nebulized spray plume is produced with aqueous droplets between 1 and 4 μm (suitably sized for deep lung deposition); (**D**) nanoparticles of drug are dispersed within nebulized aqueous droplets.

material respectively) (96), approximately the same level of lung deposition was shown to occur, when the emitted dose from the nebulizer was approximately equivalent. Pharmacokinetic analysis of the EPAS and SFL formulations indicated that the retention time of the drug was similar too, with similar elimination rates. The retention time of the itraconazole in the lung was of key importance in these studies, as it was an indication of how long the drug would remain above the minimum lethal concentration (MLC) for preventing the germination of *Aspergillus* spp. Data reported by Johnson et al. indicated that the 90 percentile MLC (MLC_{90}) for *A. flavus* (the concentration at which 90% of *A. flavus* is terminated), was 0.5 μg/mL of blood serum (102). From this information it was inferred that the MLC_{90} for lung tissue would be 0.5 μg/g of wet lung tissue weight. The pharmacokinetic profiles obtained for the lung retention time for the

Table 1 Cascade Impactor Data

Formulation	TED (μg/min)	FPF* (%)	MMAD (μm)	GSD
EPAS-ITZ	1743	76	2.70	1.9
SFL-ITZ	1134	85	2.82	1.7

Note: The data was obtained for the nebulization of colloidal dispersions of itraconazole prepared from powders manufactured using either the EPAS or SFL techniques.
Abbreviations: FPF, fine particle fraction; GSD, geometric standard deviation; MMAD, mass median aerodynamic diameter; TED, total emitted dose.
Source: From Ref. 95.

nebulized EPAS and SFL colloidal dispersion of itraconazole indicated that the drug remained MLC_{90} for at least 24 h following a single dose administration (Fig. 8).

In order to obtain lung levels of itraconazole greater than the MLC_{90} of *A. flavus* in the lung tissue using conventional dosing regimens (i.e. by oral administration of the drug), it would mean that high and sustained doses of itraconazole would be required. This high and sustained itraconazole dosing regimen needed may be considered higher than that indicated for a typical human dosing regimen (103). The strain and degree of a given fungal infection, treatment may vary. Typically values of 0.5 µg/mL blood serum are considered to be therapeutically important (104). Human doses of orally administered itraconazole vary from 200 to 800 mg itraconazole b.i.d. or t.i.d. (104–110).

Steady State Dosing of Nebulized Itraconazole

In a follow up study for dosing itraconazole to the lung, a repeat single dose study was conducted using the itraconazole prepared using SFL technique, and the steady state dosing kinetics was determined (111). Using the pharmacokinetic data from the single dose study, it was predicted that the steady state lung concentration for nebulized itraconazole (prepared using the SFL technique) would be obtained within four half lives when administered at 12 h intervals (b.i.d.). With this information Vaughn et al. designed a study to dose twelve mice b.i.d. using the nebulized SFL itraconazole, and 12 mice b.i.d. using a conventional orally administered Sporanox® formulation. The study was run for 12 days, and 3 mice were euthanized from each group on days 3, 8, and 12. From the 3 mice euthanized

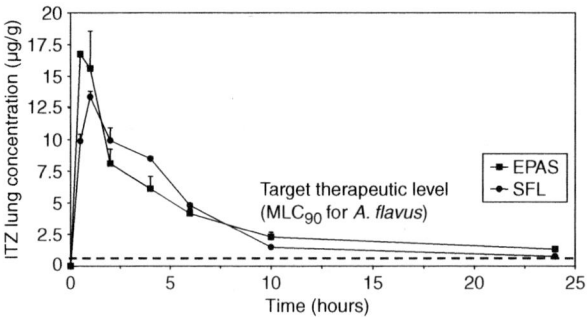

Figure 8   Lung deposition in mice for nebulized ITZ formulations (prepared either using the EPAS or SFL techniques) by single dose administration (an equivalent dose exposure of 30 mg/kg by aerosolization over a 20-min period). *Source*: From Ref. 95.

at each of the time intervals, the lung and serum concentrations could be determined for both pulmonary and orally administered itraconazole. The data showed conclusively that the amount of itraconazole present in the lung at the dose trough level following b.i.d. administration was much greater for the pulmonary administered SFL formulation. Furthermore, the levels obtained were shown to be consistently above the MLC_{90} for *A. flavus*. For the orally administered drug this was shown not to be the case. Additionally, information was obtained on the histopathology of the lung. This data indicated that there was no apparent tissue damage as a result of the inhaled SFL itraconazole formulation (112). Further, a qualitative assessment of alveolar macrophage uptake of itraconazole, performed using a bronchoalveolar lavage (BAL) method, in conjunction with mass spectroscopy, was described (112). The BAL study indicated the presence of itraconazole in isolated alveolar macrophages. Interestingly the studies were able confirm that previously reported incidences of significant side-effects (113,114) were apparent, due to the occurrence of 2 dose related mortalities in the orally administered Sporanox group. The studies concluded that pulmonary dosing with itraconazole could an effective method for delivery of antifungal therapy for the treatment and prophylaxis of invasive fungal infections. Using the nebulized itraconazole formulations high and sustained lung tissue concentrations were achieved via inhalation of an amorphous nanoparticulate ITZ-pulmonary composition while maintaining serum levels which are within the MLC range measured for of *A. flavus* (102). The lung:serum ratio was significantly greater in pulmonary dosed ITZ and could prove to enhance treatment while reducing side effects, by decreasing the systemic Cmax and maintaining serum levels above the MLC range demonstrated for itraconazole. By enhancing the local delivery and reducing the potential for side effects of ITZ delivery, this delivery method can greatly improve morbidity and mortality for patients who are at risk of or infected with life-threatening fungal infections.

Nanostructured Itraconazole for the Prevention of Invasive Pulmonary Aspergillosis

A collaborative study between the University of Texas at Austin and the University of Texas Health Science Center in San Antonio was undertaken using a prophylaxis strategy for the pulmonary administration of itraconazole. Animal models of invasive pulmonary aspergillosis demonstrated that the strategy is effective as prophylaxis compared to the commercially available oral solution (Fig. 9) (115,116). In mice immunocompromised with high dose cortisone acetate and challenged with *A. flavus*, both aerosolized EPAS and SFL formulations of itraconazole significantly prolonged and increased survival (up to 60% survival at the end of the study) compared to placebo and mice administered a commercially available

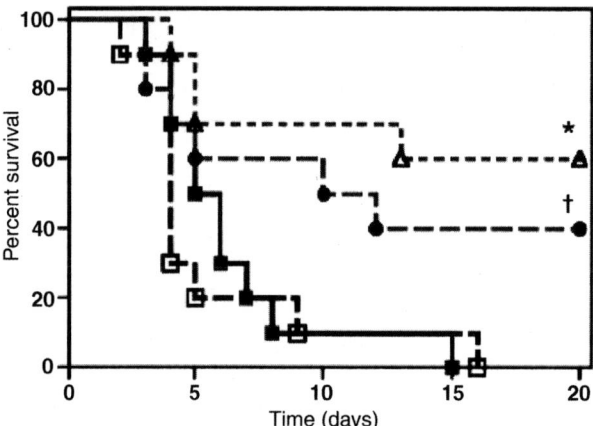

Figure 9 Survival curves for mice that received prophylaxis with aerosolized itraconazole prepared by EPAS (•) and SFL (Δ), or orally administered Sporanox® Oral Liquid (□) and control group (■). *Source*: From Ref. 115.

solution of itraconazole by oral gavage (both with a 0% survival level at the termination of the study) (115).

Similar effects of aerosolized SFL itraconazole on survival were observed in a subsequent study in mice immunosuppressed with cortisone acetate and cyclophosphamide and challenged with the more prevalent and virulent species *A. fumigatus* (116). In addition, aerosolized SFL itraconazole markedly reduced lung damage as observed by histopathology. In contrast to the extensive damage observed in the lower airways of mice administered placebo or oral itraconazole solution, the epithelial disruption and necrosis observed in the SFL itraconazole group was mainly superficial and isolated more proximally at branch points of the airways with less vascular congestion. Despite these promising data in vitro and in vivo, the clinical utility of nanostructured itraconazole is unknown as these formulations have not been tested in healthy volunteers or patients.

CONCLUSIONS AND FUTURE PERSPECTIVES

Pulmonary administration of a variety of anti-infective agents is likely to play an important role in the future. In the example of pulmonary fungal infection a local therapeutic effect is observed. This locally acting administration is one approach that could be exploited for an extensive list of infectious airborne pathogens whether they be fungal, bacterial, or viral infectious diseases (e.g. aspergillosis, tuberculosis, hantavirus, etc.).

Additionally, it is hypothesized that with the emergence of new processing technologies (such as cryogenic and supercritical fluid techniques, as well as controlled precipitation techniques) and next-generation delivery devices a variety of compounds may be administered systemically to treat disease states that may have disseminated from the lung throughout the body. Furthermore, new classes of compounds are now able to be administered to the lung using combinations of the aforementioned breakthroughs in processing and delivery.

REFERENCES

1. Baddley JW, Stroud TP, Salzman D, Pappas PG. Invasive mold infections in allogeneic bone marrow transplant recipients. Clin Infect Dis 2001; 32:1319–24.
2. Marr KA, Carter RA, Crippa F, Wald A, Corey L. Epidemiology and outcome of mould infections in hematopoietic stem cell transplant recipients. Clin Infect Dis 2002; 34:909–17.
3. Wald A, Leisenring W, vanBurik JA, Bowden RA. Epidemiology of Aspergillus infections in a large cohort of patients undergoing bone marrow transplantation. J Infect Dis 1997; 175:1459–66.
4. Lin SJ, Schranz J, Teutsch SM. Aspergillosis case-fatality rate: systematic review of the literature. Clin Infect Dis 2001; 32:358–66.
5. Denning DW. Therapeutic outcome in invasive aspergillosis. Clin Infect Dis 1996; 23:608–15.
6. Latge JP. *Aspergillus fumigatus* and aspergillosis. Clin Microbiol Rev 1999; 12: 310–50.
7. Marr KA, Patterson T, Denning D. Aspergillosis. Pathogenesis, clinical manifestations, and therapy. Infect Dis Clin North Am 2002; 16:875–94, vi.
8. Wiederhold NP, Lewis RE, Kontoyiannis DP. Invasive aspergillosis in patients with hematologic malignancies. Pharmacotherapy 2003; 23:1592–610.
9. Latge JP. The pathobiology of *Aspergillus fumigatus*. Trends Microbiol 2001; 9:382–89.
10. Amitani R, Taylor G, Elezis EN, et al. Purification and characterization of factors produced by *Aspergillus fumigatus* which affect human ciliated respiratory epithelium. Infect Immun 1995; 63:3266–71.
11. Schaffner A, Douglas H, Braude A. Selective protection against conidia by mononuclear and against mycelia by polymorphonuclear phagocytes in resistance to Aspergillus. Observations on these two lines of defense in vivo and in vitro with human and mouse phagocytes. J Clin Invest 1982; 69:617–31.
12. Allen JE, Liu LX. Immunity to parasitic and fungal infections. In Pier GB, Lyczak JB, Wetzler LM, eds. Immunology, Infection, and Immunity. ASM Press, 2004; 469–96.
13. Schaffner A. Therapeutic concentrations of glucocorticoids suppress the antimicrobial activity of human macrophages without impairing their responsiveness to gamma interferon. J Clin Invest 1985; 76:1755–64.
14. Schaffner A, Schaffner T. Glucocorticoid-induced impairment of macrophage antimicrobial activity: mechanisms and dependence on the state of activation. Rev Infect Dis 1987; 9(Suppl 5):S620–9.

15. Chamilos G, Luna M, Lewis RE, et al. Invasive fungal infections in patients with hematologic malignancies in a tertiary care cancer center: an autopsy study over a 15-year period (1989–2003). Haematologica 2006; 91:986–9.

16. Lionakis MS, Lewis RE, Torres HA, Albert ND, Raad, II, Kontoyiannis DP. Increased frequency of non-fumigatus Aspergillus species in amphotericin B- or triazole-pre-exposed cancer patients with positive cultures for aspergilli. Diagn Microbiol Infect Dis 2005; 52:15–20.

17. Hachem RY, Kontoyiannis DP, Boktour MR, et al. Aspergillus terreus: an emerging amphotericin B-resistant opportunistic mold in patients with hematologic malignancies. Cancer 2004; 101:1594–600.

18. Paterson PJ, Seaton S, Prentice HG, Kibbler CC. Treatment failure in invasive aspergillosis: susceptibility of deep tissue isolates following treatment with amphotericin B. J Antimicrob Chemother 2003; 52:873–6.

19. Steinbach WJ, Benjamin DK, Jr., Kontoyiannis DP, et al. Infections due to Aspergillus terreus: a multicenter retrospective analysis of 83 cases. Clin Infect Dis 2004; 39:192–8.

20. Denning DW. Invasive aspergillosis. Clin Infect Dis 1998; 26:781–803; quiz 804-785.

21. Wingard JR. Lipid formulations of amphotericins: are you a lumper or a splitter? Clin Infect Dis 2002; 35:891–5.

22. Wingard JR, Kubilis P, Lee L, et al. Clinical significance of nephrotoxicity in patients treated with amphotericin B for suspected or proven aspergillosis. Clin Infect Dis 1999; 29:1402–7.

23. Wingard JR, Leather H. A new era of antifungal therapy. Biol Blood Marrow Transplant 2004; 10:73–90.

24. Bowden R, Chandrasekar P, White MH, et al. A double-blind, randomized, controlled trial of amphotericin B colloidal dispersion versus amphotericin B for treatment of invasive aspergillosis in immunocompromised patients. Clin Infect Dis 2002; 35:359–66.

25. Leenders AC, Daenen S, Jansen RL, et al. Liposomal amphotericin B compared with amphotericin B deoxycholate in the treatment of documented and suspected neutropenia-associated invasive fungal infections. Br J Haematol 1998; 103: 205–12.

26. Kontoyiannis DP, Lewis RE. Invasive zygomycosis: update on pathogenesis, clinical manifestations, and management. Infect Dis Clin North Am 2006; 20: 581–607, vi.

27. Ostrosky-Zeichner L, Marr KA, Rex JH, Cohen SH. Amphotericin B: time for a new "gold standard". Clin Infect Dis 2003; 37:415–25.

28. Ellis M, Spence D, dePauw B, et al. An EORTC international multicenter randomized trial (EORTC number 19923) comparing two dosages of liposomal amphotericin B for treatment of invasive aspergillosis. Clin Infect Dis 1998; 27: 1406–12.

29. Kontoyiannis DP, Andersson BS, Lewis RE, Raad, II. Progressive disseminated aspergillosis in a bone marrow transplant recipient: response with a high-dose lipid formulation of amphotericin B. Clin Infect Dis 2001; 32:E94–6.

30. Walsh TJ, Goodman JL, Pappas P, et al. Safety, tolerance, and pharmaco-kinetics of high-dose liposomal amphotericin B (AmBisome) in patients

infected with Aspergillus species and other filamentous fungi: maximum tolerated dose study. Antimicrob Agents Chemother 2001; 45:3487–96.

31. Cornely OA, Maertens J, Bresnik M, Herbrecht R. Liposomal amphotericin B as initial therapy for invasive filamentous fungal infections: A randomized, prospective trial of a high loading regimen vs. standard dosing (AmBiLoad Trial). In Amercian Society of Hematology 47th Annual Meeting, Atlanta, GA: 2005.

32. Denning DW, Lee JY, Hostetler JS, et al. NIAID Mycoses Study Group Multicenter Trial of Oral Itraconazole Therapy for Invasive Aspergillosis. Am J Med 1994, 97:135–44.

33. Stevens DA, Lee JY. Analysis of compassionate use itraconazole therapy for invasive aspergillosis by the NIAID Mycoses Study Group criteria. Arch Intern Med 1997; 157:1857–62.

34. Prentice HG, Caillot D, Dupont B, Menichetti F, Schuler U. Oral and intravenous itraconazole for systemic fungal infections in neutropenic haematological patients: meeting report. London, United Kingdom, 20 June 1998. Acta Haematol 1999; 101:56–62.

35. Caillot D, Bassaris H, McGeer A, et al. Intravenous itraconazole followed by oral itraconazole in the treatment of invasive pulmonary aspergillosis in patients with hematologic malignancies, chronic granulomatous disease, or AIDS. Clin Infect Dis 2001; 33:e83–90.

36. Herbrecht R, Denning DW, Patterson TF, et al. Voriconazole versus amphotericin B for primary therapy of invasive aspergillosis. N Engl J Med 2002; 347:408–15.

37. Mikus G, Schowel V, Drzewinska M, et al. Potent cytochrome P450 2C19 genotype-related interaction between voriconazole and the cytochrome P450 3A4 inhibitor ritonavir. Clin Pharmacol Ther 2006; 80:126–35.

38. Trifilio S, Ortiz R, Pennick G, et al. Voriconazole therapeutic drug monitoring in allogeneic hematopoietic stem cell transplant recipients. Bone Marrow Transplant 2005; 35:509–13.

39. Pascual AA, Bolay S, Marchetti O. Documentation of low voriconazole blood levels followed by dose adjustment in patients with invasive fungal infections not responding to therapy. In 46th Interscience Conference on Antimicrobial Agents and Chemotherapy; San Francisco, CA; 2006.

40. Walsh TJ, Raad I, Patterson TF, et al. Treatment of invasive aspergillosis with posaconazole in patients who are refractory to or intolerant of conventional therapy: an externally controlled trial. Clin Infect Dis 2007; 44:2–12.

41. Cornely OA, Maertens J, Winston DJ, et al. Posaconazole vs. fluconazole or itraconazole prophylaxis in patients with neutropenia. N Engl J Med 2007; 356:348–59.

42. Ullmann AJ, Lipton JH, Vesole DH, et al. Posaconazole or fluconazole for prophylaxis in severe graft-versus-host disease. N Engl J Med 2007; 356: 335–47.

43. Gubbins PO, Krishna G, Sansone-Parsons A, et al. Pharmacokinetics and safety of oral posaconazole in neutropenic stem cell transplant recipients. Antimicrob Agents Chemother 2006; 50:1993–9.

44. Ullmann AJ, Cornely OA, Burchardt A, et al. Pharmacokinetics, safety, and efficacy of posaconazole in patients with persistent febrile neutropenia or

refractory invasive fungal infection. Antimicrob Agents Chemother 2006; 50: 658–66.

45. Douglas CM, D'Ippolito JA, Shei GJ, et al. Identification of the FKS1 gene of Candida albicans as the essential target of 1,3-beta-D-glucan synthase inhibitors. Antimicrob Agents Chemother 1997; 41:2471–9.

46. Kurtz MB, Heath IB, Marrinan J, Dreikorn S, Onishi J, Douglas C. Morphological effects of lipopeptides against *Aspergillus fumigatus* correlate with activities against (1,3)-beta-D-glucan synthase. Antimicrob Agents Chemother 1994; 38:1480–9.

47. Wiederhold NP, Lewis RE. The echinocandin antifungals: an overview of the pharmacology, spectrum and clinical efficacy. Expert Opin Investig Drugs 2003; 12:1313–33.

48. Mora-Duarte J, Betts R, Rotstein C, et al. Comparison of caspofungin and amphotericin B for invasive candidiasis. N Engl J Med 2002; 347:2020–9.

49. Betts RF, Rotstein C, Talwar D, et al. Comparison of micafungin and caspofungin for candidemia or invasive candidiasis. In 46th Interscience Conference on Antimicrobial Agents and Chemotherapy; San Francisco, CA; 2006.

50. Reboli A, Rotstein C, Pappas P, Schranz J, Krause D, Walsh T. Anidulafungin versus fluconazole for treatment of candidemia and invasive candidiasis. In 45th Interscience Conference on Antimicrobial Agents and Chemotherapy December 16–19;Washington, D.C.; 2005.

51. Maertens J, Raad I, Petrikkos G, et al. Efficacy and safety of caspofungin for treatment of invasive aspergillosis in patients refractory to or intolerant of conventional antifungal therapy. Clin Infect Dis 2004; 39:1563–71.

52. Denning DW, Marr KA, Lau WM, et al. Micafungin (FK463), alone or in combination with other systemic antifungal agents, for the treatment of acute invasive aspergillosis. J Infect 2006; 53:337–49.

53. Petraitis V, Petraitiene R, Sarafandi AA, et al. Combination therapy in treatment of experimental pulmonary aspergillosis: synergistic interaction between an antifungal triazole and an echinocandin. J Infect Dis 2003; 187:1834–43.

54. >Kontoyiannis D, Hachem R, Lewis R, Rivero G, Kantarjian H, Raad I. Efficacy and Toxicity of the Caspofungin/Liposomal Amphotericin B Combination in Documented or Possible Invasive Aspergillosis in Patients with Hematologic Malignancies (abstract). In 42nd Interscience Conference on Antimicrobial Agents and Chemotherapy September 27–30;San Diego, CA; 2002.

55. Marr KA, Boeckh M, Carter RA, Kim HW, Corey L. Combination antifungal therapy for invasive aspergillosis. Clin Infect Dis 2004; 39:797–802.

56. Goodman JL, Winston DJ, Greenfield RA, et al. A controlled trial of fluconazole to prevent fungal infections in patients undergoing bone marrow transplantation. N Engl J Med 1992; 326:845–51.

57. Slavin MA, Osborne B, Adams R, et al. Efficacy and safety of fluconazole prophylaxis for fungal infections after marrow transplantation–a prospective, randomized, double-blind study. J Infect Dis 1995; 171:1545–52.

58. Dykewicz CA: Guidelines for preventing opportunistic infections among hematopoietic stem cell transplant recipients: focus on community respiratory virus infections. Biol Blood Marrow Transplant 2001; 7(Suppl):19S–22S.

59. van Burik JH, Leisenring W, Myerson D, et al. The effect of prophylactic fluconazole on the clinical spectrum of fungal diseases in bone marrow transplant recipients with special attention to hepatic candidiasis. An autopsy study of 355 patients. Medicine (Baltimore) 1998; 77:246–54.

60. Winston DJ, Maziarz RT, Chandrasekar PH, et al. Intravenous and oral itraconazole versus intravenous and oral fluconazole for long-term antifungal prophylaxis in allogeneic hematopoietic stem-cell transplant recipients. A multicenter, randomized trial. Ann Intern Med 2003; 138: 705–13.

61. Marr KA, Crippa F, Leisenring W, et al. Itraconazole versus fluconazole for prevention of fungal infections in patients receiving allogeneic stem cell transplants. Blood 2004; 103:1527–33.

62. Marr KA, Leisenring W, Crippa F, et al. Cyclophosphamide metabolism is affected by azole antifungals. Blood 2004; 103:1557–9.

63. van Burik J, Ratanatharathorn V, Lipton J, Miller C, Bunin N, Walsh TJ. Randomized, Double-Blind Trial of Micafungin versus Fluconazole for Prophylaxis of Invasive Fungal Infections in Patients undergoing Hematopoietic Stem Cell Transplant, NIAID/BAMSG Protocol 46. In 42nd Interscience Conference on Antimicrobial Agents and Chemotherapy September 27–30; San Diego, CA; 2002.

64. Perfect JR, Dodds Ashley E, Drew R. Design of aerosolized amphotericin b formulations for prophylaxis trials among lung transplant recipients. Clin Infect Dis 2004; 39(Suppl 4):S207–10.

65. Dummer JS, Lazariashvilli N, Barnes J, Ninan M, Milstone AP. A survey of anti-fungal management in lung transplantation. J Heart Lung Transplant 2004; 23:1376–81.

66. Drew RH, Dodds Ashley E, Benjamin DK, Jr., DuaneDavis R, Palmer SM, Perfect JR: Comparative safety of amphotericin B lipid complex and amphotericin B deoxycholate as aerosolized antifungal prophylaxis in lung-transplant recipients. Transplantation 2004; 77:232–7.

67. Palmer SM, Drew RH, Whitehouse JD, et al. Safety of aerosolized amphotericin B lipid complex in lung transplant recipients. Transplantation 2001; 72:545–8.

68. Dubois J, Bartter T, Gryn J, Pratter MR. The physiologic effects of inhaled amphotericin B. Chest 1995; 108:750–3.

69. Alexander BD, Dodds Ashley ES, Addison RM, Alspaugh JA, Chao NJ, Perfect JR. Non-comparative evaluation of the safety of aerosolized amphotericin B lipid complex in patients undergoing allogeneic hematopoietic stem cell transplantation. Transpl Infect Dis 2006; 8:13–20.

70. Schwartz S, Behere G, Heniemann V, et al. Aerosolized amphotericin B inhalation is not effective prophylaxis of invasive aspergillus infections during prolonged neutropenia in patients after chemotherapy or autologous bone marrow transplantation. Evidence-based Oncol 2000; 1:87–8.

71. Erjavec Z, Woolthuis GM, deVries-Hospers HG, et al. Tolerance and efficacy of Amphotericin B inhalations for prevention of invasive pulmonary aspergillosis in haematological patients. Eur J Clin Microbiol Infect Dis 1997; 16:364–8.

72. Conneally E, Cafferkey MT, Daly PA, Keane CT, McCann SR. Nebulized amphotericin B as prophylaxis against invasive aspergillosis in granulocytopenic patients. Bone Marrow Transplant 1990; 5:403–6.

73. Rijnders BJ, Slobbe L: Aerosolized liposomal amphotericin B to prevent invasive aspergillosis during prolonged neutropenia. In 46th Interscience Conference on Antimicrobial Agents and Chemotherapy; San Francisco, CA; 2006.

74. Kugler AR, Sweeney TD, Eldon MA. Prophylaxis of invasive aspergillosis in neutropenic rabbits using an inhaled, novel, dry-powder amphotericin B formulation. In 45th Interscience Conference on Antimicrobial Agents and Chemotherapy December 16–19; Washington, D.C.: 2005.

75. Allen SD, Sorensen KN, Nejdl MJ, Durrant C, Proffit RT. Prophylactic efficacy of aerosolized liposomal (AmBisome) and non-liposomal (Fungizone) amphotericin B in murine pulmonary aspergillosis. J Antimicrob Chemother 1994; 34:1001–13.

76. Ruijgrok EJ, Vulto AG, Van Etten EW. Efficacy of aerosolized amphotericin B desoxycholate and liposomal amphotericin B in the treatment of invasive pulmonary aspergillosis in severely immunocompromised rats. J Antimicrob Chemother 2001; 48:89–95.

77. Gavalda J, Martin MT, Lopez P, et al. Efficacy of nebulized liposomal amphotericin B in treatment of experimental pulmonary aspergillosis. Antimicrob Agents Chemother 2005; 49:3028–3030.

78. Abraham OC, Manavathu EK, Cutright JL, Chandrasekar PH. In vitro susceptibilities of Aspergillus species to voriconazole, itraconazole, and amphotericin B. Diagn Microbiol Infect Dis 1999; 33:7–11.

79. Lewis RE, Wiederhold NP, Klepser ME. In vitro pharmacodynamics of amphotericin B, itraconazole, and voriconazole against Aspergillus, Fusarium, and Scedosporium spp. Antimicrob Agents Chemother 2005; 49:945–51.

80. Muir DC, Davies CN. The deposition of 0.5 microns diameter aerosols in the lungs of man. Ann Occup Hyg 1967; 10:161–74.

81. Edwards DA, Ben-Jebria A, Langer R. Recent advances in pulmonary drug delivery using large, porous inhaled particles. J Appl Physiol 1998; 85:379–85.

82. Liversidge GG, Cundy KC. Particle-Size Reduction for Improvement of Oral Bioavailability of Hydrophobic Drugs .1. Absolute Oral Bioavailability of Nanocrystalline Danazol in Beagle Dogs. Int J Pharm 1995; 125:91–7.

83. Rogers TL, Nelsen AC, Hu JH, et al. A novel particle engineering technology to enhance dissolution of poorly water soluble drugs: spray-freezing into liquid. Eur J Pharm Biopharm 2002; 54:271–80.

84. Young TJ, Mawson S, Johnston KP, Henriksen IB, Pace GW, Mishra AK. Rapid expansion from supercritical to aqueous solution to produce submicron suspensions of water-insoluble drugs. Biotechnol Prog 2000; 16:402–7.

85. Rogers TL, Johnston KP, Williams RO. Solution-based particle formation of pharmaceutical powders by supercritical or compressed fluid CO_2 and cryogenic spray-freezing technologies. Drug Dev Ind Pharm 2001; 27:1003–15.

86. Rogers TL, Nelsen AC, Sarkari M, Young TJ, Johnston KP, Williams RO. Enhanced aqueous dissolution of a poorly water soluble drug by novel particle engineering technology: Spray-freezing into liquid with atmospheric freeze-drying. Pharm Res 2003; 20:485–93.

87. Rogers TL, Overhoff KA, Shah P, et al. Micronized powders of a poorly water soluble drug produced by a spray-freezing into liquid-emulsion process. Eur J Pharm Biopharm 2003; 55:161–72.

88. Hu JH, Rogers TL, Brown J, Young T, Johnston KP, Williams RO. Improvement of dissolution rates of poorly water soluble APIs using novel spray freezing into liquid technology. Pharm Res 2002; 19:1278–84.

89. Yu ZS, Rogers TL, Hu JH, Johnston KP, Williams III RO. Preparation and characterization of microparticles containing peptide produced by a novel process: spray freezing into liquid. Eur J Pharm Biopharm 2002; 54: 221–8.

90. Vaughn JM, Gao X, Yacaman M-J, Johnston KP, Williams III RO. Comparison of powder produced by evaporative precipitation into aqueous solution (EPAS) and spray freezing into liquid (SFL) technologies using novel Z-contrast STEM and complimentary techniques. Eur J Pharm Biopharm 2005; 60:81–9.

91. Hu JH, Johnston KP, Williams RO. Stable amorphous danazol nano-structured powders with rapid dissolution rates produced by spray freezing into liquid. Drug Dev Ind Pharm 2004; 30:695–704.

92. Hu JH, Johnston KP, Williams RO. Nanoparticle engineering processes for enhancing the dissolution rates of poorly water soluble drugs. Drug Dev Ind Pharm 2004; 30:233–45.

93. Hu JH, Johnston KP, Williams RO. Rapid dissolving high potency danazol powders produced by spray freezing into liquid process. Int J Pharm 2004; 271:145–54.

94. Williams RO, Johnston KP, Hu JH. Improvement of dissolution rates of poorly water soluble drugs using a new particle engineering technology – Spray Freezing into Liquid. Abstracts Papers Am Chem Soc 2003; 226:U527.

95. McConville JT, Overhoff KA, Sinswat P, et al. Targeted high lung concentrations of itraconazole using nebulized dispersions in a murine model. Pharm Res 2006; 23:901–11.

96. Vaughn JM, McConville JT, Crisp MT, Johnston KP, Williams RO. Supersaturation produces high bioavailability of amorphous danazol particles formed by evaporative precipitation into aqueous solution and spray freezing into liquid technologies. Drug Dev Ind Pharm 2006; 32:559–67.

97. Baxter J, Abou-Chakra H, Tuzun U, Lamptey BM. A DEM simulation and experimental strategy for solving fine powder flow problems. Chem Eng Res Design 2000; 78:1019–25.

98. Sham JOH, Zhang Y, Finlay WH, Roa WH, Lobenberg R. Formulation and characterization of spray-dried powders containing nanoparticles for aerosol delivery to the lung. Int J Pharm 2004; 269:457–67.

99. Hickey AJ. Inhalation Aerosols-Physical and Biological Basis for Therapy. New York: Marcel Dekker; 1996.

100. Miller FJ, Mercer RR, Crapo JD. Lower respiratory-tract structure of laboratory-animals and humans – dosimetry implications. Aerosol Sci Technol 1993; 18:257–71.

101. McConville JT, Carvalho TC, Iberg AN, et al. Design and Evaluation of a Restraint-Free Small Animal Inhalation Dosing Chamber. Drug Dev Ind Pharm 2005; 31:35–42.

102. Johnson EM, Szekely A, Warnock DW. In-vitro activity of voriconazole, itraconazole and amphotericin B against filamentous fungi. J Antimicrob Chemother 1998; 42:741–5.
103. Allendoerfer R, Loebenberg D, Rinaldi MG, Graybill JR. Evaluation of Sch51048 in an experimental-model of pulmonary aspergillosis. Antimicrob Agents Chemother 1995; 39:1345–8.
104. Sobel JD. Practice guidelines for the treatment of fungal infections. Clin Infect Dis 2000; 30:652.
105. Galgiani JN, Ampel NM, Catanzaro A, Johnson RH, Stevens DA, Williams PL. Practice guidelines for the treatment of coccidioidomycosis. Clin Infect Dis 2000; 30:658–61.
106. Kauffman CA, Hajjeh R, Chapman SW. Practice guidelines for the management of patients with sporotrichosis. Clin Infect Dis 2000; 30:684–7.
107. Kauffman CA, Hedderwick SA. Treatment of systemic fungal infections in older patients – Achieving optimal outcomes. Drugs Aging 2001; 18:313–23.
108. Pappas PG, Rex JH, Sobel JD, et al. Guidelines for treatment of candidiasis. Clin Infect Dis 2004; 38:161–89.
109. Rex JH, Walsh TJ, Sobel JD, et al. Practice guidelines for the treatment of candidiasis. Clin Infect Dis 2000; 30:662–78.
110. Wheat J, Sarosi G, McKinsey D, et al. Practice guidelines for the management of patients with histoplasmosis. Clin Infect Dis 2000; 30:688–95.
111. Vaughn JM, McConville JT, Burgess D, et al. Single dose and multiple dose studies of itraconazole nanoparticles. Eur J Pharm Biopharm 2006; 63:95–102.
112. Vaughn JM, Wiederhold NP, McConville JT, et al. Murine airway histology and intracellular uptake of inhaled amorphous itraconazole. Int J Pharm 2007; 338:219–224.
113. Vandewoude K, Vogelaers D, Decruyenaere J, et al. Concentrations in plasma and safety of 7 days of intravenous itraconazole followed by 2 weeks of oral itraconazole solution in patients in intensive care units. Antimicrob Agents Chemother 1997; 41:2714–18.
114. Winston DJ, Maziarz RT, Chandrasekar PH, et al. Intravenous and oral itraconazole versus intravenous and oral fluconazole for long-term antifungal prophylaxis in allogeneic hematopoietic stem-cell transplant recipients – A multicenter, randomized trial. Ann Int Med 2003; 138:705–13.
115. Hoeben BJ, Burgess DS, McConville JT, et al. In vivo efficacy of aerosolized nanostructured itraconazole formulations for prevention of invasive pulmonary aspergillosis. Antimicrob Agents Chemother 2006; 50:1552–54.
116. Alvarez CA, Wiederhold NP, McConville JT, et al. Aerosolized nanostructured itraconaozle as prophylaxis against invasive pulmonary aspergillosis. J Infect 2007; 55:68–74.

3

Pulmonary Delivery of Anti-Cancer Agents

Hugh D.C. Smyth, Imran Saleem, and Martin Donovan

College of Pharmacy, Health Sciences Center, University of New Mexico, Albuquerque, New Mexico, U.S.A.

Claire F. Verschraegen

Cancer Research and Treatment Center, Health Sciences Center, University of New Mexico, Albuquerque, New Mexico, U.S.A.

INTRODUCTION

There are several compelling reasons why pulmonary delivery of aerosolized chemotherapy for the direct local treatment of lung tumors is likely to provide advancements in therapy. For other diseases affecting the lungs, significant improvements in therapy have been achieved using inhalation aerosols. Asthma was the first example where the pharmacokinetic and pharmacodynamic advantage of aerosol drug delivery resulted in critical improvements in therapy. These improvements include rapid onset therapy (i.e. bronchodilators) and significant decreases in systemic side effects (i.e. corticosteroids). Subsequently, inhalation aerosols have been applied to other local diseases of the lungs including cystic fibrosis, chronic obstructive pulmonary disease, tuberculosis, pneumocystis carinii pneumonia, and pulmonary hypertension. It is surprising therefore, that scant attention has been paid to reapplying these same basic principles of pharmacokinetic advantage to the leading cause of cancer death, lung cancer. In this chapter, we outline the rationale, potential limitations, and previous investigations of pulmonary delivery of chemotherapy. Those studies that have investigated regional administration of chemotherapy agents using aerosols are reviewed with specific emphasis placed on the results of some of the first clinical trials

81

in this area that we have performed. Based on this data, we also explore the potential advantages and limitations for the aerosol delivery of anti-cancer agents. Finally, issues are examined for future clinical translation of therapies in development.

LUNG CANCER

Despite the decades of awareness, and the introduction of newer therapies and treatment regimens, overall lung cancer survival rates remain low, and the urgency of replacing conventional approaches with novel or combination therapies exhibiting enhanced anti-cancer efficacy is now being recognized. The most telling statistic regarding lung cancer is that the annual number of deaths attributed to this disease continues to rise with each passing year (Fig. 1). This dual lack of success in both lung cancer treatment and prevention has garnered increasing attention with not only the popular media, but also among policy makers, recently prompting the U.S. Congress to declare lung cancer a national public health priority and eliciting a call for an inter-agency attack on the primary cause of cancer death.

Apart from the inexorable increase in overall lung cancer deaths, another consideration is the shift in the frequency among the different types of lung cancer. Adenocarcinoma has emerged as the most recurrent histologic type (50%) while squamous, previously the most common, presently accounts for approximately one-third of lung cancers, and 15% of cases are attributed to small-cell lung cancer (SCLC) (1). Furthermore, non–small cell lung cancer (NSCLC) consists of varying histologies; however, the approaches to diagnosis, staging, prognosis, and treatment

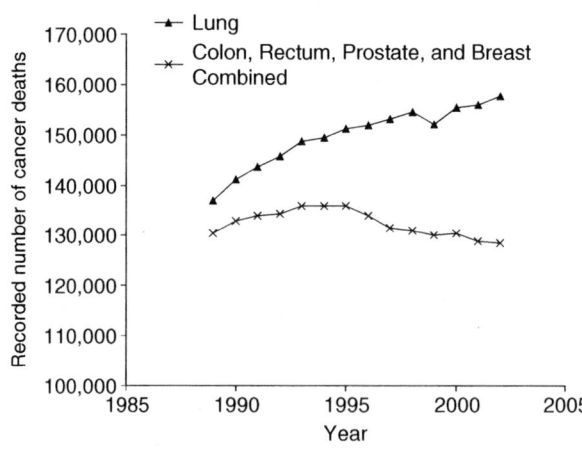

Figure 1 No declines in lung cancer deaths have occurred in the past 15 years.

remain similar between the different types. Current guidelines from the National Cancer Institute of the National Institutes of Health indicate the following approaches and associated prognosis may be expected: patients with resectable disease may be cured by either surgery or surgery with adjuvant chemotherapy; local control can be achieved with radiation therapy in a large number of patients with unresectable disease, though cure is seen only in a small number of patients. Patients with locally advanced, unresectable disease may have long-term survival with radiation therapy coupled with chemotherapy. Patients with advanced metastatic disease may achieve improved survival and palliation of symptoms with chemotherapy. Unfortunately, diagnosis is often made late, by which time surgical resection is no longer a viable option. Chemotherapy has produced short-term improvement in disease-related symptoms (2). However, chemotherapy for the treatment of NSCLC has been the source of much heated debate, with issues ranging from the toxicity to patients and the economic burden imposed upon the community (2). Second-line chemotherapy for NSCLC is also under discussion as most cases of NSCLC tend to relapse after first-line chemotherapy. The major issue is, again, not only whether second-line chemotherapy is effective, but perhaps equally important for individual patient care, health providers, and funding agencies, whether it is cost-effective. It has been suggested that no current evidence exists in support of second-line treatment of patients with poor performance status (3). In SCLC, the positive impact of chemotherapy is well documented, and compared to supportive care for extensive SCLC, chemotherapy prolongs mean survival (between 66 and 79 days) in patients with advanced SCLC, although a detailed study on the effect on the patients' quality of life remains largely unexplored (4).

Unfortunately, chemotherapy in advanced lung cancer, the most commonly presented staging, has encountered a plateau, and there exist few differences among the various combinations of drugs currently employed. Administered via intravenous injection, the current treatment of advanced SCLC and NSCLC with combination chemotherapy is characterized with the highly undesirable triumvirate of being nonspecific, nonselective, and toxic (5,6). The primary issue is the limitation of the dose that can be successfully delivered to the patient per treatment. Due to non-specific interactions with all tissues following systemic administration, severe toxicity limits the dose that a patient may be given, resulting in low concentrations of the chemotherapy agent at the target tumor site, and yielding ineffective killing of tumor cells. Second, when administered as an infusion, chemotherapy agents are only present at locally high concentrations for a very brief time period. Considerable evidence suggests that this pharmacokinetic profile leads to poor tumor penetration, resulting in rapid repopulation of tumor cells during periods where the cytotoxic agent is absent (7–10). These events are reflected in the unsuccessful and ineffective treatment of lung tumors, accompanied by poor patient prognosis. Thus, unlike many other diseases of the respiratory tract, virtually no drug delivery technology or targeting capacity has been developed for these therapeutic agents to optimize therapy outcomes, and as a result, current therapeutic regimens exhibit poor pharmacokinetic profiles against an often advanced and aggressive solid tumor. Clearly, the development of systems capable of enhancing the targeting of chemotherapy administered to lung tumors is urgently needed to improve drug pharmacokinetics and pharmacodynamics. Pulmonary delivery of anti-cancer agents is not a well studied area, but based upon the few preliminary preclinical and clinical trials, the potential and feasibility of this approach in patients has become increasingly evident. These studies will be discussed in light of a critical analysis of both the opportunities and potential limitations of pulmonary delivery of anti-cancer agents.

INHALATION AEROSOLS

Modern inhalation aerosol therapies have been in existence since the 1950s, but only in recent times have considerable efforts been devoted to producing inhaled technologies with increased capabilities over their predecessors. For example, we have recently witnessed the FDA approval of the first non-injectable insulin product: inhaled insulin, Exubera™ (11). Due to these efforts, understanding the dynamics of aerosols and their interactions with the lungs has recently borne witness to major advancements. As the main barriers to the targeted deposition of aerosols to the lung are aerodynamic in nature, the successful delivery of therapeutic aerosols to the desired region of the airways is directly dependent upon a combination of aerodynamic characteristics of inhaled particles (e.g. density, diameter, etc.), patient inhalation dynamics, and lung physiology/disease (12). Particles less than 1 μm in aerodynamic diameter tend to possess reduced lung deposition efficiency since they are exhaled before diffusive transport to the epithelia can occur. Conversely, aerosol particles greater than approximately 5 μm tend to be impacted within the mouth and throat via inertial mechanisms, where they are subsequently swallowed. Lung deposition efficiencies can be modulated somewhat by controlling both particle size distributions of the aerosol and the inspiratory flow rates that the patient can control. As an example, consider that lung deposition may be as high as 80–90% if 1–2 μm particles are slowly inhaled followed by a breath-hold. In these cases the majority of the aerosol deposits within the alveoli, the principle gas exchange region of the lung. While these lung deposition patterns have been related to particle aerodynamic characteristics for many years, the mechanisms for efficiently and reproducibly generating these aerosols for drug delivery remains the focus of intense research. Several recent reviews are available detailing the various modes of aerosol generation and administration (12–15).

RATIONALE FOR PULMONARY DELIVERY OF ANTI-CANCER AGENTS

As with other respiratory diseases treated locally through the use of inhalation aerosols, there are numerous pharmacokinetic and pharmacodynamic arguments favoring this delivery route in cancer therapy. In general, chemotherapy is characterized by a dose dependent response (cell apoptosis) coupled to a high degree of non-specificity, so that despite the introduction of several newer generations of chemotherapy agents, toxicity remains the principle limitation for effective anti-tumor response. Accordingly, the rationale for pulmonary delivery of these agents primarily focuses on the ability to increase regional targeting and the associated benefits arising from this pharmacokinetic advantage. These additional potential benefits are briefly described here, and are specific to the nature of the tumor microenvironment to which inhalation aerosols are targeted.

Overcoming Severe Systemic Side-Effects: Pharmacokinetic Advantage

The severe systemic side-effects experienced by patients during lung cancer chemotherapy have the dual disadvantage of limiting the effectiveness of therapeutic regimens while significantly decreasing the patient's quality of life (16,17). During therapy, the delicate balance between patient tolerability and effectiveness of intravenously infused chemotherapy must be rigorously maintained, and any deviations can result in sub-optimal treatment. As a means of overcoming this obstacle, it would be highly desirable to deliver significant increases in local concentrations while minimizing, if not circumventing entirely, exposure to the systemic circulation.

Direct aerosol administration possesses the capabilities necessary to achieve this goal and is corroborated by clinical studies that have been recently performed. The rationale for such an approach to lung cancer treatment is succinctly described by a variable termed pharmacokinetic advantage, or R_d. This variable has previously been described to elucidate the principle of regional therapy (18,19). For example, in inhaled drug delivery R_d can be defined (for drugs with linear pharmacokinetics) by the following equation:

$$R_d = \frac{\left(AUC_{lung}/AUC_{plasma}\right)_{inhalation}}{\left(AUC_{lung}/AUC_{plasma}\right)_{IV}} \tag{1}$$

where the numerator and denominators in this equation represent the ratios of the areas under the curves (AUCs) in lung and plasma with inhalational and IV administrations, respectively. Using this model, Sharma et al. estimated R_d for a drug with a relatively low $\log P$ (0.46), revealing an advantage of approximately 24 for the inhalational delivery of this compound (20). Essentially this indicates a 24-fold improvement in regional targeting, unambiguously demonstrating the potential for controlled release of chemotherapy in the lung, and indicating the likelihood of achieving R_d values greater than 24.

The pharmacokinetic advantage of inhaled chemotherapy is assisted not only by the direct deposition of drug in the tumor region, but also through the avoidance of hepatic metabolism, a critical drug delivery advantage that is especially important for drugs with high hepatic metabolism and/or toxic effects on the liver. Moreover, because of the specific targeting to the lung, the potential for increased exposure times and dosing frequency will, as discussed below, become mechanistically and therapeutically important.

Increasing Tumor Penetration

To maximize their efficacy, it is imperative that anticancer drugs penetrate tissue quickly and efficiently, arriving at the cancer cells in a concentration sufficient to exert a therapeutic effect (7). The efficient organization of the vasculature system, ensuring that cells are almost always within a few cell diameters of a blood vessel, allows the delivery of oxygen, nutrients, and drugs to the cells comprising the tissues of the body. However, it is well known that the homeostatic regulation of tissue and the growth of blood vessels break down in solid tumors, and thus cancer cells may be at significant distances from local blood supply (7,21), have irregular blood supply (22,23), have an absence of functional lymphatics (24), higher interstitial fluid pressure (25), and a modified extracellular matrix that can slow down the movement of molecules within the tumor (26). Collectively, these physical and physiological characteristics present formidable barriers impeding drug delivery to cells within solid tumors. Moreover, when drugs are administered into the blood stream by IV infusion, the time they spend in proximity to the tumor is quite brief, minimizing the opportunity for drug molecules to penetrate sufficiently into solid tumors and exert their intended therapeutic effect.

There now exists significant evidence that cells distant from blood vessels are resistant to conventional chemotherapy and though this observation may be attributed to several mechanisms, the most straight-forward explanation, and indeed one well supported by recent studies, is that cells distant from blood vessels are, due to limited drug access, exposed to insubstantial levels of therapeutic agents (7). Clearly, current methods of drug administration (e.g. i.v.) are poorly suited to overcome these barriers to tumor penetration. In marked contrast to these aforementioned treatments, direct aerosol delivery of chemotherapy to the tumor vicinity possesses the potential to significantly increase the concentration gradient for enhanced drug penetration, resulting in enhanced effectiveness and improved outcomes.

Preventing Accelerated Tumor Cell Repopulation

Administration of both radiotherapy and chemotherapy is performed in multiple doses, which are interrupted to allow for the recovery of normal tissues, such as the bone marrow and the GI tract, between treatments. Unfortunately, this interval between treatments presents surviving cancer cells an opportunity to proliferate, a process of repopulation that severely undermines the benefits of the therapy (9). There is significant evidence that not only does the repopulation of tumor cells limit the effectiveness of therapy, but that tumor–cell repopulation actually accelerates during the course of the treatment (9,27–30). While pulsatile delivery is necessary using standard chemotherapy regimens due to the non-specificity of cytotoxic agents, recent efforts have been directed toward delivering therapies with reduced intervals between doses. An example is the accelerated fractionation of radiotherapy, which seeks to minimize the overall treatment time, thereby providing decreased opportunity for tumor cell repopulation (31). With this approach, radiation treatments are administered as frequently as three times per day, and one of the most promising observations of this procedure is the enhanced recovery of normal tissues during the intervals between treatments. Furthermore, this approach has been beneficial in controlling Burkitt's lymphoma and NSCLC.

Sustained delivery of high local concentrations of cytotoxics would eliminate the propensity for the unbalanced cycle of treatment and recovery that favors tumor cell growth. Aerosol delivery, with its associated lack of systemic toxicity may potentially achieve these requirements. For instance, the low systemic toxicity of pulmonary delivery would eliminate the requirement for recovery of healthy tissues and systems. Furthermore, multiple doses may be able to be administered with greater frequency, as in fractionated radiotherapy, which would be ideal for preventing tumor cell resistance arising from increasing rates of repopulation.

AEROSOL DELIVERY TO TREAT LUNG CANCER

To date, reports of aerosolized chemotherapy for the treatment of lung cancer and pulmonary metastases have been quite limited. Table 1 summarizes all known studies that have investigated local lung delivery of anticancer agents including cytotoxics, gene therapy, chemopreventative agents,

(Text continues on page 94.)

Table 1 Local Lung Delivery of Anticancer Agents Including Cytotoxics, Gene Therapy, Chemopreventative Agents, Immunomodulating Agents, and Combinations Thereof

Drug	Delivery system	Model	Effect	Reference
Benzotaph, endoxan, thiophosphamid, cyclophosphan	Liquid aerosol	Human	8/58 showed tumor disappearance, 6/58 significant reduction	33
Mitomycin-C	Liquid aerosol	Human	1 of the 4 patients showed improvement	34
FU	Liquid aerosol	Human	FU alone 4/6 showed good response, FU aerosol + CMAF 5/6 patients showed a "remarkable" anti-tumor response. Side effects included glottitis	35
FU	Liquid aerosol	Human	Showed accumulation of 5-FU in lung after administration, including metabolites. Trace serum levels. "Satisfactory" anti-tumor response of 60%	36
BCG	Liquid aerosol	Human	There was no improvement in actuarial survival time	37
Cytosine arabinoside	Intratracheal instillation	Rat	Free cytosine arabinoside inhibited DNA synthesis in multiple organs (absorption from lung) while liposome-encapsulated drug inhibited DNA synthesis in the lung only	38
Ethanol	Vapor	Mice	Marked reduction of the pulmonary tumor growth	39
Interferon-alpha	Liquid aerosol	Human	Limited antitumor activity in locally advanced bronchioloalveolar carcinoma. 6/8 showed radiological stabilization of disease for 7–43 weeks (median 15)	40
Gamma-IFN and lipopolysaccharide	Liquid aerosol	Mice	Activated pulmonary alveolar macrophages showed enhanced cytotoxicity to tumor cells, also selectively killing tumor cells over 3T3 fibroblast cell line	41

Agent	Delivery	Species	Description	Ref.
Gamma-IFN	Liquid aerosol	Mice	Number of lung metastatic nodules was significantly reduced (by 50%; $P < 0.01$) after IFN-gamma aerosols, compared with controls	42
Doxorubicin	Isolated lung perfusion	Human	$N = 8$, There were two major complications and no objective responses	43
Doxorubicin	Isolated lung perfusion	Human	Isolated lung perfusion with high-dose doxorubicin is well tolerated and is associated with minimal cardiac and host toxicity	44
IL-2	Liquid aerosol	Human	Reversible airway irritation causing a nonproductive cough represented the dose-limiting toxicity. Out of 16 patients, one durable complete response, one partial response, and one mixed response were observed	45, 46
Gamma-IFN and muramyl tripeptide	Liquid aerosol	Mice	Organ-specific activation of alveolar macrophages was achieved	47
Budesonide	Liquid aerosol (in ethanol)	Mice	>80% inhibition of pulmonary tumor formation compared to the aerosol control and 90% or greater compared to mice not exposed to aerosol. The first published effort at the use of aerosol administration to prevent neoplasia of the respiratory tract	48–50
IL-2 Liposome	Liquid aerosol	Dogs	2/4 dogs with metastatic osteosarcoma had complete regression of metastases (for > 12, 20 months respectively). 1/3 dogs with lung carcinoma had stabilization > 8 months; the other had disease progression. Toxicity was minimal	51
Doxorubicin	Liquid aerosol	Dogs	Significant pharmacokinetic advantage of aerosol versus infusion therapies	52
9-NC, liposome	Liquid aerosol	Mice	Tumor growth was greatly reduced or tumors were undetectable after several weeks of treatment	53–56

(Continued)

Table 1 Local Lung Delivery of Anticancer Agents Including Cytotoxics, Gene Therapy, Chemopreventative Agents, Immunomodulating Agents, and Combinations Thereof (*Continued*)

Drug	Delivery system	Model	Effect	Reference
Paclitaxel or Doxorubicin	Liquid aerosols	Dogs	In 24 dogs, 6 responses were noted including 5 partial responses and 1 complete response. 4/18 responses to DOX and 2/15 responses to PTX. No systemic toxicities. Formulated in ethanol, or polyethylene glycol 200 and ethanol	57
GM-CSF	Liquid aerosol	Human	1/7 had progression of lung metastases. 5/7 received 2–6 months of intermittent aerosol therapy without side effects. 1/7 (Ewing's sarcoma) had complete response, 1/7 with melanoma had a partial response; 3/7 had stabilization of pulmonary metastases for 2–6 months	58
CPT, Liposome	Liquid aerosol	Mice	Rapid absorption from lung into systemic circulation	59
9-NC, liposome	Liquid aerosol	Human	Drug absorbed systemically (maximum at 2 h after end of aerosolization). Blood levels detectable at 24 h. Stabilization of disease was observed in 2/6 patients	60, 61
All-trans-retinoic acid	Propellant driven inhaler	Rats	Significantly longer pulmonary half-life, lower peak serum concentrations, and lower liver levels compared to IV	62
IL-2 liposomes	Liquid aerosol	Human	No significant toxicity was observed	63
PEI-p53 complexes	Liquid aerosol	Mice	PEI-p53 complexes inhibited the growth of lung metastasis. 50% increase in the mean length of survival	64–67
13-cis-Retinoic acid	Liquid aerosols	Rats	Epithelial delivery of retinoids in ethanol and PEG 300 to lung tissue was more efficacious in up-regulation of TGase II and retinoid receptors	68

Agent	Formulation	Model	Description	Reference
PEI-DNA	Liquid aerosol	Mice	Enhanced gene expression	69
Cationic liposome-DNA complexes	Liquid aerosol	Mice	Positive correlation between the in vitro p53 function and the in vivo antitumoral activities of liposome-p53 formulations	70
Paclitaxel liposome	Liquid aerosol	Mice	Significant reduction in lung weights and reduced number of visible tumor foci on the lung surfaces of mice treated with PTX aerosol were observed. Prolonged survival.	71, 72,66
MKK4	Liquid aerosol	Mice	MKK4 was selectively inhibited	73
Retinyl palmitate	Liquid aerosol	Human	Overall response rate (remission or partial remission) was 56%	74
Methotrexate	Propellant inhaler	HL-60 cells	Induction of apoptosis	75
Nimesulide and doxorubicin	Propellant inhaler	A549 cells	Enhanced cytotoxicity of combination treatment	76
PEI:IL-12 gene	Liquid aerosol	Mice	Selective gene expression and protein production in the tumor area. Number of lung metastases decreased significantly	77
Farnesol	Liquid aerosol	H460 and A549	Estimated airway surface liquid concentrations of the deposited farnesol reveal that the IC50 of the nebulized farnesol can be achieved over the entire tracheobronchial region (with polysorbate 80)	78
Vitamin E analogue	Liquid aerosol	Mice	Reduction in tumor volumes observed	79–81
TF-liposome-endostatin complex	Liquid aerosol	Mice	TF-liposome-mediated endostatin gene therapy strongly inhibited angiogenesis and the growth of mouse xenograft liver tumors	
Epigallocatechin gallate	Liquid aerosol	Mice	EGCG aerosol significantly reduced lung tumor multiplicity by 20–30% However, exposure to water solvent alone produced greater reduction (40%)	83

(Continued)

Table 1 Local Lung Delivery of Anticancer Agents Including Cytotoxics, Gene Therapy, Chemopreventative Agents, Immunomodulating Agents, and Combinations Thereof (*Continued*)

Drug	Delivery system	Model	Effect	Reference
DFMO	Liquid aerosol	Mice	Estimated pharmacokinetic advantage based on serum and lung AUCs was 13 in favor of inhalation.	84
Doxorubicin	Particles	S180 cells	Action of doxorubicin on growth of S180 murine sarcoma cells was enhanced in vitro by delivery in microparticles	85
PGA-TXL	Intratracheal injection	Mice	Decreased tumor burden and increased survival	86
Glucosylated conjugated PEI and plasmid DNA	Liquid aerosol	Mice	Apoptosis was detected in gene therapy delivered mouse lung	87
Vitamin E and 9-NC	Liquid aerosol	Mice	Combination treatments enhanced antiproliferative and proapoptotic activities in cell culture. The combination treatment showed a significant reduction in tumor volume in comparison to either treatment alone in murine model of cancer	88
CsA and PTX	Liquid aerosol	Mice	Co-administration of CsA with PTX demonstrated significant dose dependent anticancer effects against renal cell pulmonary metastases. Toxicity was weight loss at highest dose of CsA	89
DFMO and 5-FU	Liquid aerosol	Hamster	Both compounds resulted in a significant increase in the percent of cancer-free animals	90

Agent	Delivery	Model	Description	Ref.
GCB	Liquid aerosol	Mice	Aerosol GCB inhibited the growth of lung metastases. Also, aerosol GCB suppressed the growth of subcutaneous LM8 tumor	91
Celecoxib and docetaxel	Propellant inhaler	A459, H460 cell	Aerosolized celecoxib significantly enhances the in vitro cytotoxicity and apoptotic response of docetaxel against A549 and H460 cells	92
Gemcitabine	Liquid aerosol	Baboons, Mice	Inhibition of orthotopic tumour growth by aerosol delivery of gemcitabine. Demonstrated safety of nine weekly aerosol deliveries of gemcitibine in primates	93–95
Doxorubicin nanoparticles	Liquid aerosol	H460, A549	DOX-nanoparticles showed higher cytotoxicity at highest concentration compared to blank nanoparticles and free DOX.	96
5-FU LNPs	Liquid aerosol	Hamster	Within 24 h, greater than 99% of the LNPs were cleared from the respiratory tract and total 5-FU concentration mirrored the LNP concentration	97–99

Abbreviations: BCG, Bacillus Calmette-Guérin; CMAF, cis-platinum, mitomycin C, adriamycin and 5-FU combination therapy; CsA, Cyclosporin; DFMO, Difluoromethylornithine; FU, Fluorouracil; GCB, Gemcitabine; GM-CSF, Granulocyte macrophage-colony stimulating factor, IL-2, Interleukin 2; LNPs, lipid-coated nanoparticles; PTX, Paclitaxel; PEI-p53, Polyethyleneimine-p53; PGA-TXL, Poly (L-glutamic acid)-paclitaxel; TF, Transferrin.

immunomodulating agents, or any combination of these treatments. In almost all instances of in vitro, in vivo, or human clinical trials, inhaled chemotherapy resulted in positive outcomes relative to controls. Several broad generalizations may be garnered from a review of these studies performed over the course of nearly 40 years. First, while several different therapeutic agents have been investigated, very few studies have focused on chemotherapy agents that would currently be considered effective in lung cancer. Reasons for this discrepancy may include the relatively poor solubility and stability of many frontline chemotherapy agents readily available for aqueous systems. Also, these studies have occurred over a long time span, during which therapeutic guidelines have changed and newer agents become accepted. The second major observation regarding studies performed thus far is the overwhelming dominance of liquid aerosols as the delivery method. As previously discussed, nebulizers are readily accessible and require minimal expertise for operation. However, there are inherent disadvantages of nebulizers that are a specific concern for pulmonary delivery of chemotherapy agents: chemotherapy agents are generally lipophilic and poorly water soluble. Consequently, many of the studies described in Table 1 have employed novel formulation strategies such as liposomal encapsulation, where solubilization of the drug within these amphiphilic carrier systems permits the employment of liquid aerosolization technologies. The potential disadvantage of using liposomal systems is the inclusion of excipients that have not received extensive testing or evaluation in the respiratory tract. Their effects on drug absorption, toxicity, biodistribution, and metabolism, among other factors, have been largely unexplored, and therefore the potential for developing products for regulatory approval (i.e. by the US Food and Drug Administration) may be hindered or delayed using this approach. The third significant observation from analysis of Table 1 reveals the mouse as the leading preclinical model for pulmonary delivery of chemotherapy. It is well recognized that aerosol deposition studies in small animal models are highly problematic, and the efficiency of delivery is dramatically reduced in these species (32). Given that the majority of these studies have been performed in obligate nose breathers with high efficiency of aerosol filtration in the upper respiratory tract, these findings must be taken with caution, and extensive studies must be performed prior to fully understanding how they may be extrapolated to larger animals and humans.

CASE STUDY: AEROSOL DELIVERY OF CAMPTOTHECIN ANALOGUES

During the replication of the genome, the double-helix must be unwound prior to the arrival of the DNA polymerase. As DNA helicase proceeds down the strand, unwinding the DNA ahead of the replication machinery, the double-stranded DNA in front of the replication fork becomes increasingly super-coiled. This increased coiling creates a stress that must be relieved or else the entire replication process would abruptly and prematurely terminate. The enzymes responsible for alleviating this super-helical tension are the topoisomerase family, which in humans consist of topo-I and topo-II. Topo-I relaxes DNA by cleaving a single strand of a duplex DNA, allowing passage of the other strand through the nick before religation, and it is this enzyme that is targeted by camptothecin (CPT) and its analogs. CPT destructively interferes with the replication process by stabilizing the normally transient covalent linkage between the topo-I and the DNA strand, exploiting the fact that malignant cells contain greater amounts of topo-I than normal cells, making them more sensitive to this cytotoxic effect (100,101).

CPT and its analogs exist in a pH-dependent equilibrium between two distinct chemical conformations: a closed lactone E ring, and an open carboxylate form. While the lactone ring is the active compound required for binding with the topo-I-DNA complex, at physiologic pH in human serum it is the carboxylate form that dominates (102,103). In human serum, the AUC of the lactone form is between 0 and 16% of the AUC of both forms combined. This observation may be the reason for the low therapeutic index observed with these compounds in humans (104). Preclinical and clinical data indicate that the cytotoxic activity of CPT resides in the E-ring lactone. Futhermore, correlations between the degree of neutropenia and plasma concentrations of lactone and open-ring forms have suggested that the open-ring carboxylate may contribute to myelosuppression (105,106).

When tested in vivo against human tumors (in nude mice models), CPT and its analogs have demonstrated the remarkable capacity to cure the affected animals, exhibiting potency at least one hundred times greater over their cytotoxic counterparts. Moreover, once tumors have been eradicated they usually do not recur (107). In these mice models, 50% of the drug present in serum is in the lactone form compared to only about 10% in the human serum, presumably attributing to the dramatically reduced efficacy in humans, where anticancer responses are observed in fewer than 20% of patients.

While these results are somewhat discouraging, it is surprising, considering the low amount of lactone that reaches the tumor in humans, that any significant responses are even observed (100). The clinical activity of topotecan and other analogues may be attributed to the higher concentration of lactone observed at physiologic pH (around 30% for topotecan), though their therapeutic actions are hindered by their rapid elimination from the plasma (108).

As attested by the preceding paragraphs, CPT and its analogs appear to possess significant potential for effective anticancer activity, though their efficacy is limited by pharmacokinetic and delivery issues. Importantly, within the human blood stream the active α-hydroxy-δ-lactone ring moiety is significantly hydrolyzed to its inactive charged carboxylate form. Moreover, when administered systemically, CPTs exhibit significant toxicity, especially involving the bone marrow and gastrointestinal tract, prompting the investigation of alternative administration routes for this set of compounds (100). For example, 9-nitrocamptothecin (9-NC) has been orally administered at doses of 1 mg/m^2/day safely for extended periods. However, similar to IV systemic administration of other analogues, dose-limiting toxicities, most notably myelosuppression, have been observed (109). Pulmonary delivery of 9-NC was then investigated as a possible method of mitigating the delivery and pharmacokinetic limitations.

9-NC, like many other chemotherapy agents, is water insoluble. Preclinical investigations of aerosol formulations therefore required the development of a dispersed liquid system to facilitate aerosol generation. Liposomal formulations of 9-NC were developed and studied in animal models (53). Dilauroylphosphatidylcholine (DLPC) liposome systems had been well characterized in terms of synthesis and safety in animal models (110,111). Rats exposed to 1 h of continuous aerosol for 28 consecutive days exhibited no effect of the phospholipid (112). Phase I/II studies in humans with DLPC aerosol have also demonstrated its safety and tolerability (113). Liposomes were prepared using DMSO to dissolve the lipophilic drug before addition to the phospholipids dissolved in butanol. The mixtures are snap frozen and then lyophilized. Reconstitution of the liposomes was a simple procedure, involving addition of water for injection and after which, aerosol formation by nebulization was readily achieved.

The pharmacokinetic studies of aerosolized CPT liposomes in mice showed that high concentrations of CPT in the lungs could be attained. However, the drug was rapidly distributed to the liver, brain, and other organs (59). As discussed below, the absorption and distribution from the lung may have significant effects on local and systemic exposures and may be interpreted from two points of view. First, treatment of tumors within the lung may require sufficient residence time in local tissues for clinical efficacy. Additionally, rapid absorption from the lung following inhalation results in significant systemic exposure that may give rise to unfavorable side effects. Alternatively, pulmonary delivery of chemotherapy agents that are rapidly absorbed may possess distinct advantages for treatment of tumors that do not reside within the respiratory tract.

In terms of pharmacokinetic advantage, 9-NC-DLPC liposome aerosols have demonstrated a favorable profile (significantly reduced tumor growth rate and shrinkage without serious side effects) in treating three different human cancer xenografts (breast, colon, and lung) in a nude mouse model (53). It is believed that the liposomal formulation protects the lactone-ring, increasing the concentration of this active conformation, thus allowing for enhanced penetration in tumor cells. In animal models, equivalent antitumor activity of liposomal 9-NC as compared to the non-liposomated drug is achieved at less than 20% of the dose. Toxicity profiles in mice indicate a maximum tolerated dose of approximately 4 mg/kg when administered orally. Accordingly, in these studies the maximum dose delivered via aerosol of 307 µg/kg/day was also non-toxic. Antitumor activity was noted in the absence of weight loss or other evident toxicity (54).

This same liposomal 9-NC-DLPC aerosol system was also studied in a phase I trial in patients with advanced pulmonary malignancies (60,61). Patients were eligible if they had primary or metastatic cancer in the lungs, had failed standard chemotherapy regimens for their disease, exhibited normal bone marrow function along with normal hepatic and renal functions, had no known respiratory disease other than cancer, and possessed acceptable pulmonary function. Treatment in the feasibility cohort consisted of 6.7 µg/kg/day 9-NC in aerosol reservoir (nebulizer) for 60 min per day for 5 consecutive days/week for 1,2,4, or 6 weeks, followed by observation for 2 weeks. For the phase I portion of the study, doses were increased stepwise from 6.7 up to 26.6 µg/kg/day for 5 consecutive days for 8 weeks followed by 2 weeks rest. Twenty five patients received treatment. Does-limiting toxicity was chemical induced pharyngitis seen in two of the patients at the highest dosage (26.6 µg/kg/day). At 20.0 µg/kg/day, a grade 2 and 3 fatigue required dose reduction in 2 of the 4 patients. A reversible decrease in forced expiratory volume was also noted in patients treated with the aerosolized drug. Significantly, there was no notable hematologic toxicity, while 9-NC plasma levels were similar to those observed after oral ingestion, though the mechanisms behind this different toxicity response are unknown. Based on these studies, the recommended dose for phase II studies was 13.3 µg/kg/day delivered as two consecutive 30 min nebulizations/day from a nebulizer reservoir with 4 mg 9-NC in 10 ml water, for 5 consecutive days for 8 weeks every 10 weeks. Partial remissions were observed in two patients with uterine cancer and stabilization occurred in three patients with primary lung cancer (Fig. 2). A partial remission of a liver metastasis was also observed in a patient with endometrial cancer, demonstrating the systemic potential of aerosol delivery (Fig. 2).

Pharmacokinetic studies in patients demonstrated that total 9-NC plasma concentration continued to increase for 2–3 h from the start of treatment reaching a mean (±SD) peak concentration of 37.7 ± 20.2 ng/ml at 2 h (Fig. 3). Mean (±SD) clearance was biphasic with a $T_{1/2\alpha}$ of 1.9 ± 1.4 h and a $T_{1/2\beta}$ of 16.4 ± 10.5 h. The AUC of the lactone form measured in two patients comprised 3.2% and 3.5% of the total 9-NC. 9-NC concentrations in bronchoalveolar lavage fluid were 4.2 to 10.6 times higher than those measured concurrently in plasma.

Several recent animal studies have focused on aerosol delivery of 9-NC in combination with other chemotherapeutic agents. A recent investigation in mice examined the anticancer properties of a vitamin E analogue and 9-NC delivered as an aerosol against mouse mammary tumor cells (88). This combination treatment significantly enhanced antiproliferative and

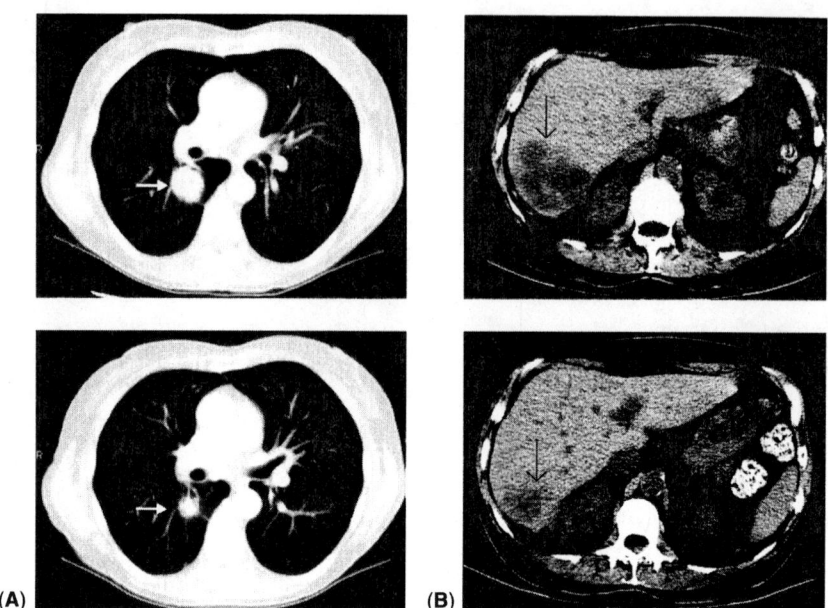

(A)   **(B)**

Figure 2 Response in lung and liver tumors after aerosol treatment (arrows). (**A**) partial response in a lung metastasis of endometrial cancer (arrow) and (**B**) partial response in a liver metastasis of endometrial cancer in another patient (arrow). *Source*: From Ref. 61.

proapoptotic activities both in cell culture, and when formulated in liposomes and delivered via aerosolization to treat metastatic murine mammary tumor. Administration of these agents in tandem exhibited a significant reduction in tumor volume when compared to either treatment alone.

PHARMACOKINETICS OF INHALATION AEROSOLS: IMPLICATIONS FOR CANCER TREATMENT

The absorption and distribution of drug following deposition of the aerosol particles in the airways is a complex process. However, it is of paramount importance to understand the dynamics of drug fate in order to achieve optimal efficacy of chemotherapeutic agents. Much information on the fate of inhaled drugs has been obtained from asthma medications that have been investigated extensively (114), though recently a great amount of research on this topic has arisen due to the interest in using the lungs as a portal to the systemic circulation (115).

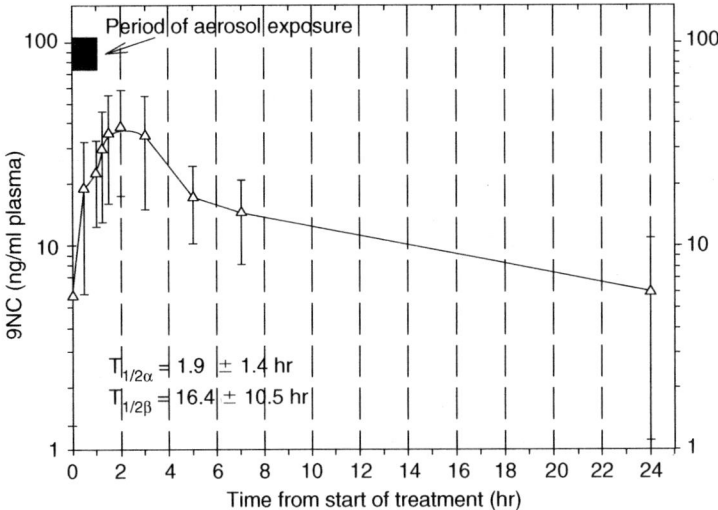

Figure 3 Pharmacokinetics of liposomal 9-NC. Mean (±SD) plasma levels in five cancer patients following treatment with 9-NC liposome aerosol by mouth-only breathing. *Source*: From Ref. 60.

The airways sequentially branch into multiple generations (usually modeled with 23 generations of bifurcations) and the morphological and cellular changes along these dividing structures can lead to significantly altered absorption and clearance rates depending on deposition patterns. The airway surface epithelial monolayer is a columnar epithelium populated by many mucus and ciliated cells that collectively form the mucociliary escalator. This barrier gradually thins as the airways branch. Particles that deposit on this portion of the airways may be efficiently cleared by the mucociliary escalator that is present along the conducting airways. Insoluble particles that deposit in the airways are efficiently swept up and removed from the lungs on this moving carpet of mucus over a time period of several hours (116).

The epithelial monolayer that comprises the alveoli is very different from the epithelia of the conducting airways. Alveolar Type 1 cells are thin and flattened, covering approximately 95% of the alveolar surface. Alveolar macrophages are a key cell in the maintenance of lung homeostasis and reside in the alveolar region, where they detect, engulf, and digest any foreign particles that have eluded the aerodynamic filters of the upper respiratory passages. Together these clearance mechanisms are well adapted to eliminating particles from the lungs in a matter of hours. Surmounting these mechanisms is particularly challenging and only a few reports of sustained release pulmonary delivery are found in the literature (117,118).

If a drug that is deposited in the lungs is soluble in the fluid lining the airways, the efficiency of the aforementioned clearance mechanisms will be significantly reduced, if not abrogated entirely. However, in general, residence time of small molecules deposited in the lungs is brief due to their rapid absorption into the systemic circulation. This absorption of small molecules from the respiratory tract into the systemic circulation is considered the fastest of any route of delivery other than intravenous (119), and stems from the exceptionally large surface area (between 50 and $100\,m^2$) of relatively high permeability epithelia and the highly dispersed nature of the drug dose as an aerosol. The resistance to systemic absorption of inhaled medicines from the airways appears to occur at the plasma membrane of the lung epithelium (120,121).

Absorption is dependent on both the location of drug deposition in the respiratory tract and the physicochemical nature of the drug. For example, small hydrophobic molecules are thought to be rapidly absorbed from the lungs regardless of deposition site via passive diffusion through the plasma membrane. Conversely, small hydrophilic molecules may be absorbed more slowly by specific transporters or across the tight junctions. The tightness of cell junctions, as measured by electrical resistance and conductivity, appears to decrease from a maximum value in the trachea to a minimum in the distal airways prior to once again increasing to a high value in the alveoli (122).

The distribution of pulmonary administered chemotherapy agents is important for clinical effectiveness. Thus, for practical purposes, chemotherapy absorption rates, lung residence time, and the potential for significant systemic exposure depend on the location of drug deposition and the physicochemical nature of the drug. Typical physicochemical properties of chemotherapeutics commonly employed for lung cancer chemotherapy generally possess low water solubility and high lipophilicity. For example, the octanol/water partition coefficient of paclitaxel is high (>100) and will therefore exhibit very rapid absorption from the lung. In fact, octanol/water partition coefficients greater than 100 generally lead to absorption half-lives of approximately 10 min (123,124). Thus, it is possible that aerosol administration of highly permeating chemotherapeutic agents will not provide appropriate pharmacokinetics for sustained local delivery and treatment of lung tumors using immediate release aerosols. Optimal tumor cell exposure may be enhanced by sustaining the release of the cancer drug so that prolonged concentrations capture all cells in different stages of the cell cycle, reducing the rate of tumor cell repopulation, and increasing tumor penetration of the cytotoxic.

POTENTIAL EFFECTS OF AIRWAY OBSTRUCTION AND CONCURRENT LUNG DISEASE

The lung function of patients with cancer is likely to be reduced due to the co-existence of other pulmonary diseases (i.e. obstructive pulmonary diseases, emphysema). Moreover, tumor masses within the airways are likely to cause changes in airflow patterns. Under these conditions, aerosol deposition in the airways and lungs deviates from that observed in healthy individuals, a scenario that is not unique to aerosol delivery for lung cancer treatment. Most marketed inhaler systems are designed for use in pulmonary diseases with varying degrees of obstruction and abnormal lung function (i.e. asthma, cystic fibrosis, etc.).

The presence of airway obstruction and tumors in the airway lumen will lead to increased turbulence and higher resistance to flow. This is generally known to result in the removal of more particles (specifically smaller particles) from the inhaled aerosol. Accordingly, for pulmonary delivery of chemotherapy aerosols, the particle size distributions and breathing parameters will need to be specifically engineered and controlled for these conditions. In fact, breathing parameters may be identified in which tumor targeting is maximized. Recently Kleinstreuer and Zhang modeled the effects of tumor in-growth into the airway lumen and the effects of different airflow conditions expected in lung cancer patients (129). They found that particle deposition near tumors could be modulated using particle size and breathing parameters such that deposition fractions of 30–90% could be achieved (Fig. 4). Most particles land on the front surface of the tumor and some deposit on the tubular wall. Clearly, particles landing on the tumor surface are most desirable for targeted drug aerosol delivery. For this relatively large-tumor-size case, very few particles enter and deposit at the downstream bifurcations due to the tumor blockage effect.

FUTURE AND CLINICAL TRANSLATION

Presently, lung cancer accounts for approximately one-third of cancer deaths, of which the vast majority of patients present or develop advanced lung cancer, and ultimately experience a rapid demise. Early trials with chemotherapy in NSCLC were quite disappointing, exhibiting a response rate of around 10–15%, and an approximate 5 week prolonged survival relative to supportive care. Moreover, in many cases chemotherapy regimens were found to be associated with worse results than supportive care alone (cyclophosphamide, methotrexate, and vinblastine). In more recent years, newer agents have been investigated that have demonstrated better median survival (1), although this median survival increased only to 7–10 months and fewer than 40% of patients remained alive at 1 year (130). As a result of these generally poor outcomes, it has been widely opined that chemotherapy in advanced lung cancer has reached a plateau, and newer approaches other than chemotherapy must be investigated (5,6). This opinion may be premature given the realization that chemotherapy for lung cancer has not been optimized using any advanced formulation or drug

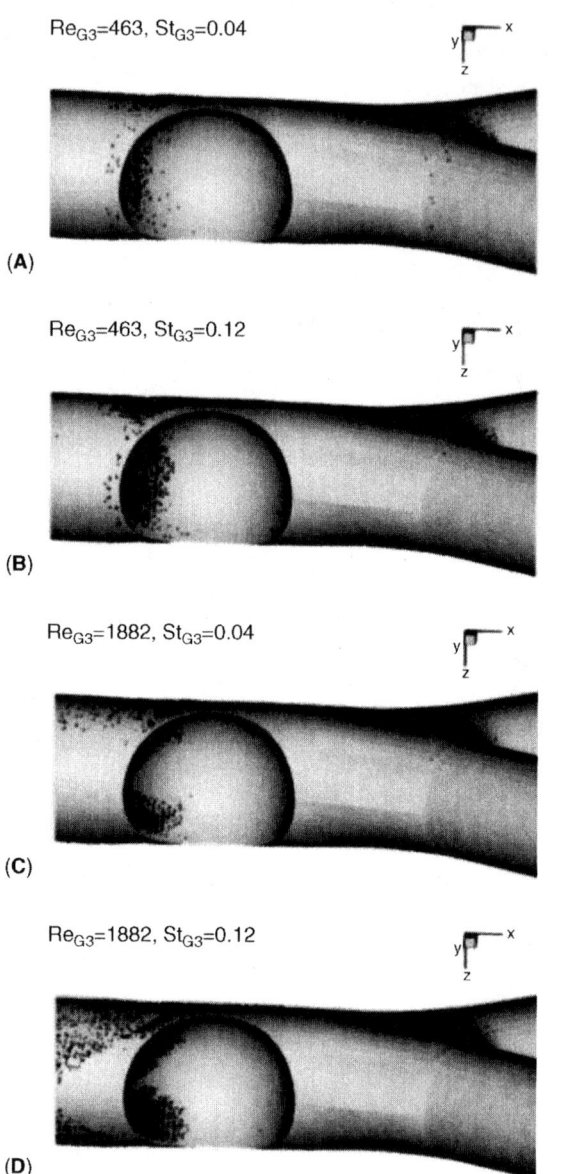

(A) $Re_{G3}=463$, $St_{G3}=0.04$

(B) $Re_{G3}=463$, $St_{G3}=0.12$

(C) $Re_{G3}=1882$, $St_{G3}=0.04$

(D) $Re_{G3}=1882$, $St_{G3}=0.12$

Figure 4 Three-dimensional views of local particle deposition patterns around an airway tumor modeled by Kleinstreuer and Zhang. Particle deposition on the tumor surface may be influenced by the local occlusion (i.e. tumor size and location), as well as inlet velocity profile and particle distribution as a function of Reynolds number (Re) and Stokes number (St). (A–D) indicate these different flow and particle regimes. In general, particle deposition, occurring mainly along the front surface of the tumor due to impaction, increases with both increasing Stokes and Reynolds numbers. *Source*: From Ref. 129.

delivery approaches. Using historical experiences with asthma and cystic fibrosis as examples, major advances in disease control, life expectancy, and quality of life may be gained in lung cancer with inhalation aerosols.

While it is evident that promising results are a shared feature of the few investigations of pulmonary delivery of chemotherapy, there are numerous obstacles that remain to be satisfactorily addressed, including formulation issues due to the poor water solubility of chemotherapy agents, improving regional targeting within the airways and minimizing esophageal deposition, and controlling the absorption of chemotherapy agents subsequent to their deposition in the airways. Rapid translocation into the systemic circulation may prove pharmacokinetically beneficial for tumors located outside the respiratory tract, but may provide a significant barrier to the treatment of lung tumors, an issue that may be addressed using novel sustained pulmonary release technologies that are currently under development. Existing and novel chemotherapy agents are equally likely candidates for Phases 1 and 2 clinical trials, as evidenced by data presented in Table 1.

The successful clinical translation of pulmonary delivery systems for anti-cancer agents will necessarily require a broad spectrum of expertise, including physicians experienced in aerosol delivery studies, and aerosol drug delivery scientists to address the unique formulation and drug delivery issues of chemotherapy, ensuring that optimal targeting, stability, and pharmacokinetic profiles are attained in both preclinical models and clinical trials. With these capabilities, researchers will possess the background and expertise ideally suited to transform these promising observations into a tangible and effective treatment that will restore the hope and improve the quality of life of those unfortunate enough to be diagnosed with lung cancer. The future of inhaled chemotherapy presents a promising direction in the treatment of lung cancer; a direction that if diligently pursued may one day aid in the discovery of the cure that presently resides beyond our grasp.

REFERENCES

1. Pirozynski M. 100 years of lung cancer. Respir Med 2006; 100(12):2073–84.
2. Yang QE, Fong SE, Li K, Gonda MA, Tobin GJ. Ex vivo expanded murine bone marrow cells with a multiple cytokine cocktail retain long-term hematopoietic reconstitution potentials. Med Sci Monit 2005; 11(6): BR154–61.
3. Bonfill X, Serra C, Sacristan M, Nogue M, Losa F, Montesinos J. Second-line chemotherapy for non-small cell lung cancer. Cochrane Database Syst Rev 2002(2):CD002804.
4. Agra Y, Pelayo M, Sacristan M, Sacristan A, Serra C, Bonfill X. Chemotherapy versus best supportive care for extensive small cell lung cancer. Cochrane Database Syst Rev 2003(4):CD001990.
5. Carney DN, Hansen HH. Non-small-cell lung cancer—stalemate or progress? N Engl J Med 2000; 343(17):1261–2.
6. Carney DN. Lung cancer—time to move on from chemotherapy. N Engl J Med 2002; 346(2):126–8.

7. Minchinton AI, Tannock IF. Drug penetration in solid tumours. Nat Rev Cancer 2006; 6(8):583–92.
8. Primeau AJ, Rendon A, Hedley D, Lilge L, Tannock IF. The distribution of the anticancer drug Doxorubicin in relation to blood vessels in solid tumors. Clin Cancer Res 2005; 11(24 Pt 1):8782–8.
9. Kim JJ, Tannock IF. Repopulation of cancer cells during therapy: an important cause of treatment failure. Nat Rev Cancer 2005; 5(7):516–25.
10. Davis AJ, Tannock IF. Tumor physiology and resistance to chemotherapy: repopulation and drug penetration. Cancer Treat Res 2002; 112:1–26.
11. Smyth HD, Leach CL. Alternative propellant aerosol delivery systems. Crit Rev Ther Drug Carrier Syst 2005; 22(6):493–534.
12. Smyth HDC, Garmise R, Cooney D, Garcia-Contreras L, Jones LD, Hickey AJ. Medical and Pharmaceutical Aerosols. In: Ruzer LH, N.H., ed. Aerosols Handbook: Measurement, Dosimetry, and Health Effects. New York: CRC Press; 2004.
13. Smyth HD. Propellant-driven metered-dose inhalers for pulmonary drug delivery. Expert Opin Drug Deliv 2005; 2(1):53–74.
14. Smyth HDC, Garcia-Contreras L. Dry-powder or liquid spray systems for inhaled delivery of peptides and proteins. Am J Drug Deliv 2005; 3(1): 29–45.
15. Smyth HDC, Hickey AJ. Carriers in drug powder delivery: implications for inhalation system design. Am J Drug Deliv 2005; 3(2):117–32.
16. Blackhall FH, Shepherd FA, Albain KS. Improving survival and reducing toxicity with chemotherapy in advanced non-small cell lung cancer: a realistic goal? Treat Respir Med 2005; 4(2):71–84.
17. Bahl A, Sharma DN, Julka PK, Rath GK. Chemotherapy related toxicity in locally advanced non-small cell lung cancer. J Cancer Res Ther 2006; 2(1): 14–16.
18. Collins JM. Pharmacologic rationale for regional drug delivery. J Clin Oncol 1984; 2:498–504.
19. Dedrick RL. Arterial drug infusion: Pharmacokinetic problems and pitfalls. J Natl Cancer Inst 1988; 80:84–89.
20. Sharma S, White DI, A. R., Placke ME, Vail DM, Kris MG. Development of inhalational agents for oncologic use. J Clin Oncol 2001; 19(6):1839–47.
21. Tomlinson RH, Gray LH. The histological structure of some human lung cancers and the possible implications for radiotherapy. Br J Cancer 1955; 9: 539–49.
22. Less JL, Skalak EM, Sevick EM, Jain RK. Micrvascular architecture in a mammary carcinoma: branching patterns and vessel dimensions. Cancer Res 1991; 51:265–73.
23. Brown JM, Giaccia AJ. The unique physiology of solid tumors: opportunities (and problems) for cancer therapy. Cancer Res 1998; 58:1408–16.
24. Leu AJ, Berk DA, Lymboussaki A, Alitalo K, Jain RK. Absence of functional lymphatics within a murine sarcoma: a molecular and functional evaluation. Cancer Res 2000; 60:4324–7.
25. Heldin CH, Rubin K, Pietras K, Ostman A. High interstitial fluid pressure— an obstacle in cancer therapy. Nature Rev Cancer 2004; 4:806–13.

26. Netti PA, Berk DA, Swartz MA, Grodzinsky AJ, Jain RK. Role of extracellular matrix assembly in interstitial transport in solid tumors. Cancer Res. 2000; 60:2497–503.

27. De Ruysscher D, Pijls-Johannesma M, Vansteenkiste J, Kester A, Rutten I, Lambin P. Systematic review and meta-analysis of randomised, controlled trials of the timing of chest radiotherapy in patients with limited-stage, small-cell lung cancer. Ann Oncol 2006; 17(4):543–52.

28. De Ruysscher D, Pijls-Johannesma M, Bentzen SM, et al. Time between the first day of chemotherapy and the last day of chest radiation is the most important predictor of survival in limited-disease small-cell lung cancer. J Clin Oncol 2006; 24(7):1057–63.

29. Brade AM, Tannock IF. Scheduling of radiation and chemotherapy for limited-stage small-cell lung cancer: repopulation as a cause of treatment failure? J Clin Oncol 2006; 24(7):1020–2.

30. Wu L, Tannock IF. Repopulation in murine breast tumors during and after sequential treatments with cyclophosphamide and 5-fluorouracil. Cancer Res 2003; 63(9):2134–8.

31. Maciejewski B, Majewski S. Dose fractionation and tumour repopulation in radiotherapy for bladder cancer. Radiother Oncol 1991; 21(3):163–70.

32. Nadithe V, Rahamatalla M, Finlay WH, Mercer JR, Samuel J. Evaluation of nose-only aerosol inhalation chamber and comparison of experimental results with mathematical simulation of aerosol deposition in mouse lungs. J Pharm Sci 2003; 92(5):1066–76.

33. Shevchenko IT, Resnik GE. Inhalation of chemical substances and oxygen in radiotherapy of bronchial cancer. Neoplasma 1968; 15(4):419–26.

34. Sugawa IT. A study on transairway carcinostatic dose for lung metastases. Ochanomizu Med J 1970; 18(3):103–14.

35. Tatsumura T, Yamamoto K, Murakami A, Tsuda M, Sugiyama S. [New chemotherapeutic method for the treatment of tracheal and bronchial cancers–nebulization chemotherapy]. Gan No Rinsho 1983; 29(7): 765–70.

36. Tatsumura T, Koyama S, Tsujimoto M, Kitagawa M, Kagamimori S. Further study of nebulisation chemotherapy, a new chemotherapeutic method in the treatment of lung carcinomas: fundamental and clinical. Br J Cancer 1993; 68 (6):1146–9.

37. Cusumano CL, Jernigan JA, Waldman RH. Aerosolized BCG (Tice strain) treatment of bronchogenic carcinoma: phase I study. J Natl Cancer Inst 1975; 55(2):275–9.

38. Juliano RL, McCullough HN. Controlled delivery of an antitumor drug: localized action of liposome encapsulated cytosine arabinoside administered via the respiratory system. J Pharmacol Exp Ther 1980; 214(2):381–7.

39. Batkin S, Tabrah FL. Ethanol vapour modulation of Lewis lung carcinoma, a murine pulmonary tumour. J Cancer Res Clin Oncol 1990; 116(2):187–9.

40. van Zandwijk N, Jassem E, Dubbelmann R, Braat MC, Rumke P. Aerosol application of interferon-alpha in the treatment of bronchioloalveolar carcinoma. Eur J Cancer 1990; 26(6):738–40.

41. Eisenberg BL, Taylor DD, Weese JL. Aerosolized gamma-interferon and lipopolysaccharide enhances cytotoxicity of murine pulmonary alveolar macrophages. J Immunother 1991; 10(1):51–6.

42. Kessler R, Dumont S, Bartholeyns J, Weitzenblum E, Poindron P. Antitumoral potential of aerosolized interferon-gamma in mice bearing lung metastases. Am J Respir Cell Mol Biol 1994; 10(2):202–6.

43. Johnston MR, Minchen RF, Dawson CA. Lung perfusion with chemotherapy in patients with unresectable metastatic sarcoma to the lung or diffuse bronchioloalveolar carcinoma. J Thorac Cardiovasc Surg 1995; 110: 368–73.

44. Ng B, Hochwald SN, Burt ME. Isolated lung perfusion with doxorubicin reduces cardiac and host toxicities associated with systemic administration. Ann Thorac Surg 1996; 61:969–72.

45. Lorenz J, Wilhelm K, Kessler M, et al. Phase I trial of inhaled natural interleukin 2 for treatment of pulmonary malignancy: toxicity, pharmacokinetics, and biological effects. Clin Cancer Res 1996; 2(7):1115–22.

46. Khanna C, Hasz DE, Klausner JS, Anderson PM. Aerosol delivery of interleukin 2 liposomes is nontoxic and biologically effective: canine studies. Clin Cancer Res 1996; 2(4):721–34.

47. Goldbach P, Dumont S, Kessler R, Poindron P, Stamm A. In situ activation of mouse alveolar macrophages by aerosolized liposomal IFN-gamma and muramyl tripeptide. Am J Physiol 1996; 270(3 Pt 1):L429–34.

48. Wattenberg LW, Wiedmann TS, Estensen RD, Zimmerman CL, Steele VE, Kelloff GJ. Chemoprevention of pulmonary carcinogenesis by aerosolized budesonide in female A/J mice. Cancer Res 1997; 57(24):5489–92.

49. Wattenberg LW, Wiedmann TS, Estensen RD, et al. Chemoprevention of pulmonary carcinogenesis by brief exposures to aerosolized budesonide or beclomethasone dipropionate and by the combination of aerosolized budesonide and dietary myo-inositol. Carcinogenesis 2000; 21(2): 179–82.

50. Estensen RD, Jordan MM, Wiedmann TS, Galbraith AR, Steele VE, Wattenberg LW. Effect of chemopreventive agents on separate stages of progression of benzo[alpha]pyrene induced lung tumors in A/J mice. Carcinogenesis 2004; 25(2):197–201.

51. Khanna C, Anderson PM, Hasz DE, Katsanis E, Neville M, Klausner JS. Interleukin-2 liposome inhalation therapy is safe and effective for dogs with spontaneous pulmonary metastases. Cancer 1997; 79(7):1409–21.

52. Trigg RL, Zutshi A. Absorption, distribution and elimination of 14C- labeled doxorubicin co-administered with 99mTc by inhalation to dogs. Battelle Data File; 1999.

53. Knight V, Koshkina NV, Waldrep JC, Giovanella BC, Gilbert BE. Anticancer effect of 9-nitrocamptothecin liposome aerosol on human cancer xenografts in nude mice. Cancer Chemother Pharmacol 1999; 44(3):177–86.

54. Knight V, Kleinerman ES, Waldrep JC, Giovanella BC, Gilbert BE, Koshkina NV. 9-Nitrocamptothecin liposome aerosol treatment of human cancer subcutaneous xenografts and pulmonary cancer metastases in mice. Ann NY Acad Sci 2000; 922:151–63.

55. Knight V, Koshkina N, Waldrep C, Giovanella BC, Kleinerman E, Gilbert B. Anti-cancer activity of 9-nitrocamptothecin liposome aerosol in mice. Trans Am Clin Climatol Assoc 2000; 111:135–45.

56. Koshkina NV, Kleinerman ES, Waidrep C, et al. 9-Nitrocamptothecin liposome aerosol treatment of melanoma and osteosarcoma lung metastases in mice. Clin Cancer Res 2000; 6(7):2876–80.

57. Hershey AE, Kurzman ID, Forrest LJ, et al. Inhalation chemotherapy for macroscopic primary or metastatic lung tumors: proof of principle using dogs with spontaneously occurring tumors as a model. Clin Cancer Res 1999; 5(9): 2653–9.

58. Anderson PM, Markovic SN, Sloan JA, et al. Aerosol granulocyte macrophage-colony stimulating factor: a low toxicity, lung-specific biological therapy in patients with lung metastases. Clin Cancer Res 1999; 5 (9):2316–23.

59. Koshkina NV, Gilbert BE, Waldrep JC, Seryshev A, Knight V. Distribution of camptothecin after delivery as a liposome aerosol or following intramuscular injection in mice. Cancer Chemother Pharmacol 1999; 44(3): 187–92.

60. Verschraegen CF, Gilbert BE, Huaringa AJ, et al. Feasibility, phase I, and pharmacological study of aerosolized liposomal 9-nitro-20(S)-camptothecin in patients with advanced malignancies in the lungs. Ann NY Acad Sci 2000; 922: 352–4.

61. Verschraegen CF, Gilbert BE, Loyer E, et al. Clinical evaluation of the delivery and safety of aerosolized liposomal 9-nitro-20(s)-camptothecin in patients with advanced pulmonary malignancies. Clin Cancer Res 2004; 10(7): 2319–26.

62. Brooks AD, Tong W, Benedetti F, Kaneda Y, Miller V, Warrell RP, Jr. Inhaled aerosolization of all-trans-retinoic acid for targeted pulmonary delivery. Cancer Chemother Pharmacol 2000; 46(4):313–8.

63. Skubitz KM, Anderson PM. Inhalational interleukin-2 liposomes for pulmonary metastases: a phase I clinical trial. Anticancer Drugs 2000; 11(7): 555–63.

64. Gautam A, Densmore CL, Waldrep JC. Inhibition of experimental lung metastasis by aerosol delivery of PEI-p53 complexes. Mol Ther 2000; 2(4): 318–23.

65. Densmore CL, Kleinerman ES, Gautam A, et al. Growth suppression of established human osteosarcoma lung metastases in mice by aerosol gene therapy with PEI-p53 complexes. Cancer Gene Ther 2001; 8(9):619–27.

66. Gautam A, Waldrep JC, Densmore CL, et al. Growth inhibition of established B16-F10 lung metastases by sequential aerosol delivery of p53 gene and 9-nitrocamptothecin. Gene Ther 2002; 9(5):353–7.

67. Gautam A, Densmore CL, Melton S, Golunski E, Waldrep JC. Aerosol delivery of PEI-p53 complexes inhibits B16-F10 lung metastases through regulation of angiogenesis. Cancer Gene Ther 2002; 9(1):28–36.

68. Wang DL, Marko M, Dahl AR, et al. Topical delivery of 13-cis-retinoic acid by inhalation up-regulates expression of rodent lung but not liver retinoic acid receptors. Clin Cancer Res 2000; 6(9):3636–45.

69. Gautam A, Densmore CL, Xu B, Waldrep JC. Enhanced gene expression in mouse lung after PEI-DNA aerosol delivery. Mol Ther 2000; 2(1):63–70.

70. Zou Y, Zong G, Ling YH, Perez-Soler R. Development of cationic liposome formulations for intratracheal gene therapy of early lung cancer. Cancer Gene Ther 2000; 7(5):683–96.

71. Koshkina NV, Waldrep JC, Roberts LE, Golunski E, Melton S, Knight V. Paclitaxel liposome aerosol treatment induces inhibition of pulmonary metastases in murine renal carcinoma model. Clin Cancer Res 2001; 7(10): 3258–62.

72. Koshkina NV, Knight V, Gilbert BE, Golunski E, Roberts L, Waldrep JC. Improved respiratory delivery of the anticancer drugs, camptothecin and paclitaxel, with 5% CO_2-enriched air: pharmacokinetic studies. Cancer Chemother Pharmacol 2001; 47(5):451–6.

73. Lee HY, Suh YA, Lee JI, et al. Inhibition of oncogenic K-ras signaling by aerosolized gene delivery in a mouse model of human lung cancer. Clin Cancer Res 2002; 8(9):2970–5.

74. Kohlhaufl M, Haussinger K, Stanzel F, et al. Inhalation of aerosolized vitamin a: reversibility of metaplasia and dysplasia of human respiratory epithelia—a prospective pilot study. Eur J Med Res 2002; 7(2):72–8.

75. Shaik MS, Haynes A, McSween J, Ikediobi O, Kanikkannan N, Singh M. Inhalation delivery of anticancer agents via HFA-based metered dose inhaler using methotrexate as a model drug. J Aerosol Med 2002; 15(3):261–70.

76. Haynes A, Shaik MS, Chatterjee A, Singh M. Evaluation of an aerosolized selective COX-2 inhibitor as a potentiator of doxorubicin in a non-small-cell lung cancer cell line. Pharm Res 2003; 20(9):1485–95.

77. Jia SF, Worth LL, Densmore CL, Xu B, Duan X, Kleinerman ES. Aerosol gene therapy with PEI: IL-12 eradicates osteosarcoma lung metastases. Clin Cancer Res 2003; 9(9):3462–8.

78. Wang Z, Chen HT, Roa W, Finlay W. Farnesol for aerosol inhalation: nebulization and activity against human lung cancer cells. J Pharm Pharm Sci 2003; 6(1):95–100.

79. Lawson KA, Anderson K, Menchaca M, et al. Novel vitamin E analogue decreases syngeneic mouse mammary tumor burden and reduces lung metastasis. Mol Cancer Ther 2003; 2(5):437–44.

80. Lawson KA, Anderson K, Simmons-Menchaca M, et al. Comparison of vitamin E derivatives alpha-TEA and VES in reduction of mouse mammary tumor burden and metastasis. Exp Biol Med (Maywood) 2004; 229(9):954–63.

81. Zhang S, Lawson KA, Simmons-Menchaca M, Sun L, Sanders BG, Kline K. Vitamin E analog alpha-TEA and celecoxib alone and together reduce human MDA-MB-435-FL-GFP breast cancer burden and metastasis in nude mice. Breast Cancer Res Treat 2004; 87(2):111–21.

82. Li X, Fu GF, Fan YR, et al. Potent inhibition of angiogenesis and liver tumor growth by administration of an aerosol containing a transferrin-liposome-endostatin complex. World J Gastroenterol 2003; 9(2):262–6.

83. Witschi H, Espiritu I, Ly M, Uyeminami D, Morin D, Raabe OG. Chemoprevention of tobacco smoke-induced lung tumors by inhalation of

an epigallocatechin gallate (EGCG) aerosol: a pilot study. Inhal Toxicol 2004; 16(11–12):763–70.

84. Liao X, Liang W, Wiedmann T, Wattenberg L, Dahl A. Lung distribution of the chemopreventive agent difluoromethylornithine (DFMO) following oral and inhalation delivery. Exp Lung Res 2004; 30(8):755–69.

85. Tian Y, Klegerman ME, Hickey AJ. Evaluation of microparticles containing doxorubicin suitable for aerosol delivery to the lungs. PDA J Pharm Sci Technol 2004; 58(5):266–75.

86. Zou Y, Fu H, Ghosh S, Farquhar D, Klostergaard J. Antitumor activity of hydrophilic Paclitaxel copolymer prodrug using locoregional delivery in human orthotopic non-small cell lung cancer xenograft models. Clin Cancer Res 2004; 10(21):7382–91.

87. Kim HW, Park IK, Cho CS, et al. Aerosol delivery of glucosylated polyethylenimine/phosphatase and tensin homologue deleted on chromosome 10 complex suppresses Akt downstream pathways in the lung of K-ras null mice. Cancer Res 2004; 64(21):7971–6.

88. Lawson KA, Anderson K, Snyder RM, et al. Novel vitamin E analogue and 9-nitro-camptothecin administered as liposome aerosols decrease syngeneic mouse mammary tumor burden and inhibit metastasis. Cancer Chemother Pharmacol 2004; 54(5):421–31.

89. Koshkina NV, Golunski E, Roberts LE, Gilbert BE, Knight V. Cyclosporin A aerosol improves the anticancer effect of paclitaxel aerosol in mice. J Aerosol Med 2004; 17(1):7–14.

90. Wattenberg LW, Wiedmann TS, Estensen RD. Chemoprevention of cancer of the upper respiratory tract of the Syrian golden hamster by aerosol administration of difluoromethylornithine and 5-fluorouracil. Cancer Res 2004; 64(7):2347–9.

91. Koshkina NV, Kleinerman ES. Aerosol gemcitabine inhibits the growth of primary osteosarcoma and osteosarcoma lung metastases. Int J Cancer 2005; 116(3):458–63.

92. Haynes A, Shaik MS, Chatterjee A, Singh M. Formulation and evaluation of aerosolized celecoxib for the treatment of lung cancer. Pharm Res 2005; 22(3): 427–39.

93. Gagnadoux F, Pape AL, Lemarie E, et al. Aerosol delivery of chemotherapy in an orthotopic model of lung cancer. Eur Respir J 2005; 26(4): 657–61.

94. Gagnadoux F, Leblond V, Vecellio L, et al. Gemcitabine aerosol: in vitro antitumor activity and deposition imaging for preclinical safety assessment in baboons. Cancer Chemother Pharmacol 2006; 58(2):237–44.

95. Gagnadoux F, LePape A, Urban T, et al. Safety of pulmonary administration of gemcitabine in rats. J Aerosol Med 2005; 18(2):198–206.

96. Azarmi S, Tao X, Chen H, et al. Formulation and cytotoxicity of doxorubicin nanoparticles carried by dry powder aerosol particles. Int J Pharm 2006; 319 (1–2):155–61.

97. Hitzman CJ, Wattenberg LW, Wiedmann TS. Pharmacokinetics of 5-fluorouracil in the hamster following inhalation delivery of lipid-coated nanoparticles. J Pharm Sci 2006; 95(6):1196–211.

98. Hitzman CJ, Elmquist WF, Wiedmann TS. Development of a respirable, sustained release microcarrier for 5-fluorouracil II: In vitro and in vivo optimization of lipid coated nanoparticles. J Pharm Sci 2006; 95 (5):1127–43.

99. Hitzman CJ, Elmquist WF, Wattenberg LW, Wiedmann TS. Development of a respirable, sustained release microcarrier for 5-fluorouracil I: In vitro assessment of liposomes, microspheres, and lipid coated nanoparticles. J Pharm Sci 2006; 95(5):1114–26.

100. Glaberman U, Rabinowitz I, Verschraegen CF. Alternative administration of camptothecin analogues. Expert Opin Drug Deliv 2005; 2(2):323–33.

101. Liebes L, Potmesil M, Kim T, et al. Pharmacodynamics of topoisomerase I inhibition: Western blot determination of topoisomerase I and cleavable complex in patients with upper gastrointestinal malignancies treated with topotecan. Clin Cancer Res 1998; 4(3):545–57.

102. Nabiev I, Fleury F, Kudelina I, et al. Spectroscopic and biochemical characterisation of self-aggregates formed by antitumor drugs of the camptothecin family: their possible role in the unique mode of drug action. Biochem Pharmacol 1998; 55(8):1163–74.

103. Gabr A, Kuin A, Aalders M, El-Gawly H, Smets LA. Cellular pharmacokinetics and cytotoxicity of camptothecin and topotecan at normal and acidic pH. Cancer Res 1997; 57(21):4811–6.

104. Verschraegen CF, Natelson EA, Giovanella BC, et al. A phase I clinical and pharmacological study of oral 9-nitrocamptothecin, a novel water-insoluble topoisomerase I inhibitor. Anticancer Drugs 1998; 9(1):36–44.

105. Hertzberg RP, Caranfa MJ, Holden KG, et al. Modification of the hydroxy lactone ring of camptothecin: inhibition of mammalian topoisomerase I and biological activity. J Med Chem 1989; 32(3):715–20.

106. Moertel CG, Schutt AJ, Reitemeier RJ, Hahn RG. Phase II study of camptothecin (NSC-100880) in the treatment of advanced gastrointestinal cancer. Cancer Chemother Rep 1972; 56(1):95–101.

107. Giovanella BC, Stehlin JS, Wall ME, et al. DNA topoisomerase I–targeted chemotherapy of human colon cancer in xenografts. Science 1989; 246(4933): 1046–8.

108. Herben VM, ten BokkelHuinink WW, Beijnen JH. Clinical pharmacokinetics of topotecan. Clin Pharmacokinet 1996; 31(2):85–102.

109. Natelson EA, Giovanella BC, Verschraegen CF, et al. Phase I clinical and pharmacological studies of 20-(S)-camptothecin and 20-(S)-9-nitrocamptothecin as anticancer agents. Ann N Y Acad Sci 1996; 803:224–30.

110. Gilbert BE, Wyde PR, Lopez-Berestein G, Wilson SZ. Aerosolized amphotericin B-liposomes for treatment of systemic Candida infections in mice. Antimicrob Agents Chemother 1994; 38(2):356–9.

111. Gilbert BE, Wyde PR, Wilson SZ. Aerosolized liposomal amphotericin B for treatment of pulmonary and systemic Cryptococcus neoformans infections in mice. Antimicrob Agents Chemother 1992; 36(7):1466–71.

112. Gilbert B, Black MB, Waldrep JC, Bennick J, Montgomery C, Knight CM. Cyclosporin A liposome aerosol:lack of acute toxicity in rats with a high incidence of underlying pneumonitis. Inhal Toxicol 1997; 9:717–30.

113. Waldrep JC, Gilbert BE, Knight CM, et al. Pulmonary delivery of beclomethasone liposome aerosol in volunteers. Tolerance and safety. Chest 1997; 111(2):316–23.

114. Lipworth BJ. Pharmacokinetics of inhaled drugs. Br J Clin Pharmacol 1996; 42(6):697–705.

115. Patton JS, Fishburn CS, Weers JG. The lungs as a portal of entry for systemic drug delivery. Proc Am Thorac Soc 2004; 1(4):338–44.

116. Sakagami M, Byron PR, Venitz J, Rypacek F. Solute disposition in the rat lung in vivo and in vitro: determining regional absorption kinetics in the presence of mucociliary escalator. J Pharm Sci 2002; 91(2):594–604.

117. Smyth HDC, Saleem I, Donovan MJ. Emerging carriers for pulmonary drug delivery, cover story. AAPS News Magazine 2006; 9(8):12–15.

118. Smyth HDC, Saleem I, Donovan M. Emerging carriers for pulmonary drug delivery. Am Assoc Pharm Scie News Mag 2006; 9 (8):Cover Story.

119. Patton JS, Fishburn CS, Weers JG. The lungs as a portal of entry for systemic drug delivery. Proc Am Thorac Soc 2004; 1(4):338–44.

120. Gorin AB, Stewart PA. Differential permeability of endothelial and epithelial barriers to albumin flux. J Appl Physiol 1979; 47(6):1315–24.

121. Wangensteen OD, Schneider LA, Fahrenkrug SC, Brottman GM, Maynard RC. Tracheal epithelial permeability to nonelectrolytes: species differences. J Appl Physiol 1993; 75(2):1009–18.

122. Boucher RC, Stutts MJ, Gatzy JT. Regional differences in bioelectric properties and ion flow in excised canine airways. J Appl Physiol 1981; 51 (3):706–14.

123. Byron PR. Determinants of drug and polypeptide bioavailability from aerosols delivered to the lung. Adv Drug Deliev Rev 1990; 5:107–32.

124. Enna SJ, Schanker LS. Absorption of drugs from the rat lung. Am J Physiol 1972; 223:1227–31.

125. Miller BJ, Rosenbaum AS. The vascular supply to metastatic tumors of the lung. Surg Gynecol Obstet 1967; 125:1009–12.

126. Milne EN, Noonan CD, Margulis AR, Stoughton JA. Vascular supply of pulmonary metastases: Experimental study in rats. Invest Radiol 1969; 4: 215–29.

127. Fontanini G, Calcinai A, Boldrini L, et al. Modulation of neoangiogenesis in bronchial preneoplastic lesions. Oncol Rep 1999; 6:813–17.

128. Keith RL, Miller YE, Gemmill RM, et al. Angiogenic squamous dysplasia in bronchi of individuals at high risk for lung cancer. Clin Cancer Res 2000; 6: 1616–25.

129. Kleinstreuer C, Zhang Z. Targeted drug aerosol deposition analysis for a four-generation lung airway model with hemispherical tumors. J Biomech Eng 2003; 125(2):197–206.

130. Schiller JH, Harrington D, Belani CP, et al. Comparison of four chemo-therapy regimens for advanced non-small-cell lung cancer. N Engl J Med 2002; 346(2):92–8.

4

Inhaled Insulin Therapy for Patients with Diabetes Mellitus

Carlos A. Alvarez

Texas Tech School of Pharmacy, Dallas, Texas, U.S.A.

Curtis Triplitt

Division of Diabetes, Texas Diabetes Institute, University of Texas Health Science Center at San Antonio, San Antonio, Texas, U.S.A.

INTRODUCTION

Diabetes mellitus is a devastating disease that is a leading cause of blindness, end stage renal disease, neuropathies, and peripheral vascular disease that lead to amputations. It is also a coronary heart disease equivalent (1,2) and is the sixth leading cause of death in the United States (3). Diabetes is a national epidemic that has been diagnosed in approximately 16 million Americans, and it is estimated that an additional 6 million Americans have undiagnosed diabetes (3). The incidence of diabetes is on the rise, and is expected to continue, especially type 2 diabetes mellitus (T2DM) (4). Diabetes can be simply classified into type 1 (insulin dependent) and type 2 (non-insulin dependent) diabetes mellitus (5). Type 1 diabetes (T1DM) is characterized by autoimmune destruction of insulin-producing β-cells in the pancreas causing hyperglycemia (6). These patients require exogenous insulin replacement to maintain glucose homeostasis. Type 2 diabetes is characterized by insulin resistance and progressive β-cell failure (7). Patients with T2DM can be managed with oral therapy initially (insulin sensitizers ± insulin secretagogues); however, a large number of patients will require exogenous insulin therapy along with insulin sensitizers due to progressive β-cell failure. Insulin is an excellent choice, as its glucose-lowering potential is limited only by the potential side effect of hypoglycemia. Insulin therapy may have additional benefits, as the early use of insulin may slow the loss of beta cell insulin secretory function in type 2 diabetic patients versus sulfonylureas.

Glycemic control is imperative to reduce the risk of complications associated with T1DM and T2DM (8–11). The American Diabetes Association recommends a hemoglobin A1c (A1C) level at a minimum below 7%, and as close to 6% as possible if it can be achieved without significant side effects (12). Glycemic control primarily reduces microvascular disease (retinopathy, neuropathy, and nephropathy) shows a trend to reduce macrovascular events. However, glycemic control is very difficult to achieve as evidenced in clinical trials (8–10) and in clinical practice (13,14). Insulin therapy is required to meet these goals in many patients, but multiple daily injections of insulin are perceived adversely by many patients and clinicians alike (15,16).

INHALED INSULIN

Insulin is a polypeptide that is denatured in the gastrointestinal tract when taken orally; therefore, parenteral administration is necessary. Patients are required to receive multiple doses of insulin either subcutaneously or intravenously. Since the introduction of insulin therapy, alternative delivery routes and systems have been investigated. Pulmonary delivery is attractive because insulin is permeable at the level of the alveoli.

In January 2006, an inhaled dry powder formulation of fast-acting insulin was approved in the United States and Europe for the treatment of type 1 and 2 diabetes in adults (17). There are several other insulin inhalational systems under various stages of development (Table 1). Differences between these systems include formulation, particle size, and delivery device. This chapter will review the ideal delivery, differing delivery systems, pharmacodynamic, pharmacokinetics, efficacy, and safety of inhaled insulin.

Ideal Pulmonary Delivery of Insulin

The concept of inhaled insulin was first introduced in 1925 where it was studied in five diabetic patients (18). Variability in bioavailability and system portability made pulmonary insulin delivery at the time unfeasible. Interest in pulmonary delivery resurfaced in the 1970s and 1980s in the clinical setting (19). With the advancements in drug delivery technology, the pulmonary delivery of insulin became more viable. Pulmonary delivery of insulin is ideal due to the large surface area of the lung available for absorption (up to $100\,m^2$); having high permeability at the alveolar surface, and a vast circulation. However, the lung prohibits deep penetration

Table 1 Available or Developing Inhaled Insulin Products

Trade name	Formulation	Particle characteristic	Delivery system
Exubera®	Dry-powder	< 5 µm aerodynamic diameter	Spacer device that propels dry powder via compressed air
AERx iDMS®	Liquid	< 5 µm particle diameter	Breath guidance system
AIR®	Dry-powder	10–20 µm geometric diameter; <5 m aerodynamic diameter	Breath-powered unit dose inhaler
Technosphere®	Cystalline dry powder	Ordered lattice array of 2 µm spherical particles	Commercially available asthma inhaler
Spiros®	Cystalline dry powder	2–3 µm; median size of lactose particles 150 µm	Handheld breath actuated device

through several protective mechanisms. Most airborne particles are deposited in the upper airways where the mucocilliary elevator expels them from the lungs into the digestive tract. In order for a substance to reach the alveoli it must possess certain characteristics such as appropriate size, particle density, and morphology (20,21). Particles sizes of 1–3 µm are ideal for penetration to the alveoli and minimally deposit in the oropharynx. Particles smaller than 1 µm are mostly exhaled, and sizes larger than 10 µm have little chance of making it past the oropharynx (20). Furthermore, substances that do not have the appropriate density or morphology will be removed by macrophages in the alveoli. Small nonporous particles are engulfed by macrophages quickly, whereas porous particles are removed at a slower rate (20,21). Mode of delivery to the agent's target can also hinder therapy via the pulmonary route.[20] Many marketed delivery systems propel particles with too much velocity depositing the substance in the oropharynx instead of in the lower airways (i.e. pressurized metered dose inhalers). New inhaler technologies have advanced considerably and are aimed to provide high levels of dosing while minimizing losses within the device and environment. Moreover, advances in particle technology have improved reproducibility and enhanced delivery to the deep lung.

PHARMACOKINETICS AND GLUCODYNAMICS

Several insulin delivery systems are in development or on the market. Each pulmonary insulin delivery system consists of three components: an insulin formulation, a unique insulin delivery "packaging," and a unique delivery device. Though they all may be characterized as inhaled insulin systems, these unique characteristics require us to individually scrutinize each product. The inhalers range from handheld breath actuated devices to meter dose inhalers similar to those used for asthma agents. Insulin delivery packaging ranges from dry-powder blister packs or capsules, to liquid blisters or liquid in asthma type canisters. Diverse particle technologies have also been investigated for the delivery of insulin. The Exubera and AERx® systems were first technologies investigated in the 1990s.

Exubera System

Exubera is the first Food and Drug Administration approved inhaled insulin product. It is approved for the treatment of adult patients with T1DM and T2DM (22). Exubera is currently available in a short acting formulation to control post prandial glucose excursions. In patients with T1DM, Exubera should be used in combination with long acting insulin, which provides basal insulin replacement. In patients with T2DM, Exubera can be used either as monotherapy or in combination with long acting insulin or insulin sensitizers. The delivery system administers a dry-powder through a device that provides a spacer into which particles are launched via a compressed air supply then slowly inhaled by the patient (Fig. 1). The spacer helps decouple the high speed of the jet needed to disaggregate particles from particle impaction in the back of the oropharnx. The aerosolized powder is contained in blister packets which are administered through the inhaler. Each blister pack contains either 1 or 3 mg of insulin. The 1-mg blister pack is equivalent to approximately 3 units of SC insulin and the 3-mg blister pack is equivalent to approximately 8 units. The Exubera system is designed to emit a specific dose of insulin from the mouthpiece (Table 2). Up to 45% of the 1-mg blister contents and 25% of the 3-mg blister contents may be retained within the blister well. Patients must take great care in dosing insulin with the Exubera system since three 1-mg blister packs do not equal one 3-mg blister pack. Three 1-mg blister packs have a 30–40% higher C_{max} and area under the curve (AUC) compared to one 3-mg blister pack (22).

In patients with diabetes, the serum insulin concentrations peak faster with Exubera (49 min, range 30–90 min) than SC regular insulin (105 min, range 60–240 min; Fig. 2) (22). In healthy volunteers, the onset of glucose lowering activity occurs in 31 min with the maximum glucose lowering effect occurring at approximately 110 min. The duration of action is 6 h, which is longer than that of SC insulin lispro and comparable to SC regular insulin (23). In another study in 17 healthy male volunteers, Exubera's® time action profile was compared to SC insulin lispro and SC regular insulin using the

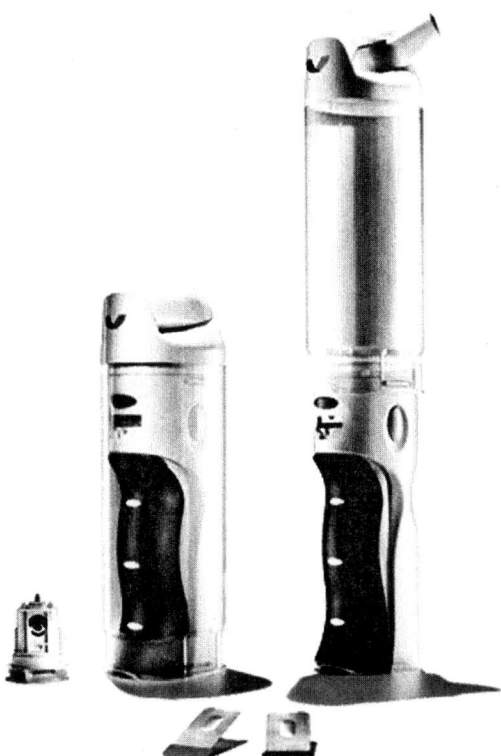

Figure 1   Exubera inhaler.

euglycemic insulin clamp method (24). When compared to SC regular insulin, Exubera had a faster onset of action (32 vs. 48 min, respectively; $P=0.001$). The same could be said when Exubera was compared to SC insulin lispro (32 vs. 41min, respectively; $P < 0.05$). The time to maximal

Table 2   Exubera Emitted Dosing Information

Fill mass[*]	Nominal dose[**]	Emitted dose[**a,c]	Fine particle dose[**b,c]
1.7	1.0	0.53	0.4
5.1	3.0	2.03	1.0

[*]mg of powder; [**]mg of insulin.
[a]Flow rate of 30 L/min for 2.5 s.
[b]Flow rate of 28.3 L/min for 3 s.
[c]Emitted dose and fine particle dose information are not intended to predict actual pharmacodynamic response.

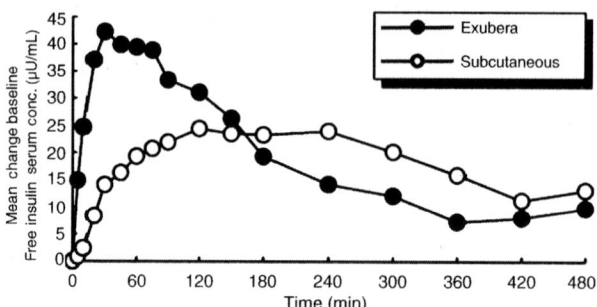

Figure 2 Change from baseline in free serum insulin concentration in healthy, non-diabetic subjects who received a single dose of Exubera® (6 mg) and a single dose of SC regular insulin (18 units) on separate days.

metabolic effect was comparable between Exubera and SC insulin lispro (143 vs. 137 min; $P = $ NS) but was shorter compared to SC regular insulin (193 min; $P < 0.01$). The duration of metabolic activity for Exubera was similar to SC insulin regular, but significantly longer than SC insulin lispro.

Exubera's insulin pharmacokinetic and glucokinetic reproducibility have been investigated in two studies (25,26). The reproducibility of Exubera was similar to SC regular insulin and the intrasubject differences between the two same-dose route of administration were small.

Special Populations and Influence of Absorption

Underlying lung disease: Exubera has not been studied in certain patient populations such as diabetic patients with asthma or chronic obstructive pulmonary disease (COPD). Since it has not been studied, the use of Exubera in these patients is not recommended. When studied in non-diabetic patients with COPD, the plasma insulin levels were twice those in subjects without COPD (22).

Smoking: Smoking increases insulin levels by more than two times compared to non-smokers. Therefore, it is recommended that patients who smoke or have stopped smoking in the previous 6 months not receive Exubera. Within 1 week of discontinuation of smoking the bioavailability of Exubera decreased by approximately 50% when compared to smokers, but was still higher than non-smokers. These rapid changes in absorption make dose prediction, and the potential risk of hypoglycemia, higher in smokers (22,27).

Asthma: Asthma can greatly affect the ability to move air and thus deliver particles to the deep lung. Studies have consistently reported reductions (approximately 20%) in absorption of inhaled insulin in subjects with mild to moderate asthma when compared to subjects without asthma. Additionally, bronchodilator therapy, if administered prior to inhaled insulin therapy, can increase absorption by 25–50%. These large absorption differences are the basis for the current recommendation against inhaled insulin use in subjects with asthma (22).

Intercurrent respiratory tract illness: Clinically, concurrent, uncomplicated upper respiratory illnesses (bronchitis, upper respiratory infection, and rhinitis) did not affect rates of glycemic control or hypoglycemia (22). About 1 in the 20 subjects temporarily discontinued Exubera therapy during the illness and alternative therapy would have to be implemented in these subjects (2).

AERx System

The liquid insulin based AERx system utilizes a microprocessor controlled inhaler device that monitors the breathing pattern of the patient and actuates when a combination of breathing parameters falls in a predetermined range (28,29). A fine mist is created by driving the insulin solution through a disposable nozzle, which prevents clogging. The inhaler is about the size of a paperback book and utilizes a slot for the insertion of the insulin strip. The insulin strip uses a one-time use nozzle which has hundreds of 1-μm holes that release the aerosolized insulin uniformly (28,29).

Dosing with the AERx system is in units, with each AERx unit providing the same glucose lowering effect as one unit of subcutaneously injected insulin. The patient can select to give between 2 and 10 units of insulin with each strip and can be adjusted by one unit increments (29,30).

The onset of action with the AERx system is approximately 10 min with peak insulin concentrations occurring approximately 1 h after inhalation (30). Dose effects the duration of action with a dose of 0.3 units/kg lasting 5 h and a dose of 1.8 units/kg lasting 8 h (30). Pharmacokinetic studies found that inhaled delivery using the AERx system resulted in rapid absorption of insulin and a corresponding faster decline in plasma glucose compared with subcutaneous (SC) administration (Table 3) (29). Intrapatient variability is similar in T1DM patients given equivalent doses of either SC regular insulin or inhaled insulin (31). The speed of absorption and resultant hypoglycemic activity is dependent on the volume and depth of breath, hence the inhaler design (29).

Studies in special populations with the AERx system have mirrored those with Exubera. This includes smoking, underlying lung disease, asthma, and intercurrent respiratory tract infection. Pharmacokinetic parameters are altered in smokers who use the AERx system. Insulin AUC was 60% higher, C_{max} was three times higher, and absorption was

Table 3 Pharmacokinetic and Pharmacodynamic Parameters (mean ± SD) after Inhaled and Subcutaneous Administration of Insulin Using the AERx System in Healthy Fasting Adults

Study number	Subcutaneous administration				AERx delivery			
	C_{max}	T_{max}	G_{min}	T_{Gmin}	C_{max}	T_{max}	G_{min}	T_{Gmin}
IN001	21 ± 6	66 ± 10	30 ± 10	112 ± 14	22 ± 5	15 ± 5	35 ± 10	66 ± 9
IN001	21 ± 6	54 ± 20	27 ± 10	115 ± 44	30 ± 11	7 ± 6	25 ± 11	66 ± 28

Abbreviations: C_{max}, maximum serum insulin concentration; G_{min}, minimum plasma glucose level expressed as percent decrease from fasting levels; $T_{G_{min}}$, time to minimum glucose concentration; T_{max}, time to maximum insulin concentration.

more rapid in smokers compared to non-smokers (Fig. 3) (27). Studies in asthmatic patients found that inhaled insulin absorption was decreased compared to those healthy non-asthmatic subjects (32). With the reduced absorption there were smaller reductions in blood glucose, possibly implying that patients with asthma may require higher doses of inhaled insulin to achieve glycemic control, or bronchodilator therapy prior to each dose of inhaled insulin. In patients with respiratory tract infections it was found that insulin pharmacokinetics were not significantly different during or immediately after acute infection (33).

AIR™ System

The AIR system uses a semi-disposable inhaler to deliver a dry-powder insulin particle that is characterized by large size but low density keeping the aerodynamic diameter small. Currently under development are the 2 and 6 unit capsules which contain 0.9 and 2.6 mg/capsule of insulin, respectively. The inhaler is approximately the size of a cell phone and utilizes breath actuated technology to deliver the dose (Fig. 4) (34,35). The insulin capsule is placed in the inhaler where it is punctured by the device (34). The inhalation of air through the device provides enough energy to make the capsule spin, forming an aerosol for inhalation.

When compared to SC insulin lispro in healthy subjects, the AIR system seems to be comparable in terms of pharmacokinetic and pharmacodynamic parameters (35). Subjects who were given doses of 2.6, 5.2, and 7.8 mg, the onset of action were 13, 13, and 15 min, respectively (Fig. 5). This was found to be a significantly faster onset compared to SC insulin lispro. Peak serum concentrations were reached in 45, 30, and 45 min for the 2.6, 5.2, and 7.8 mg doses, respectively. Mean AUC values were similar between the inhaled AIR system and SC insulin lispro. C_{max} values were similar for the 7.8 mg dose compared to 18 units of SC insulin lispro, however, the 2.6 and 5.2 mg had lower C_{max} values compared to the 6 and 12 units of SC insulin lispro,

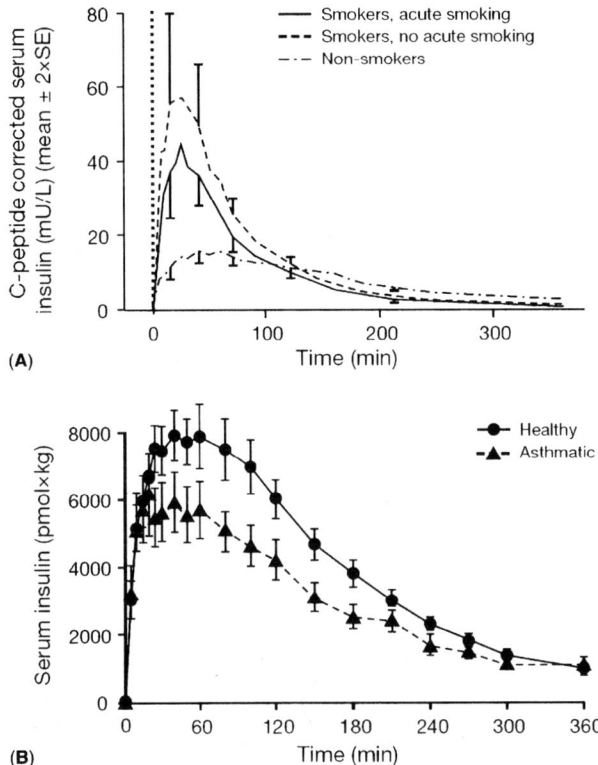

(A)

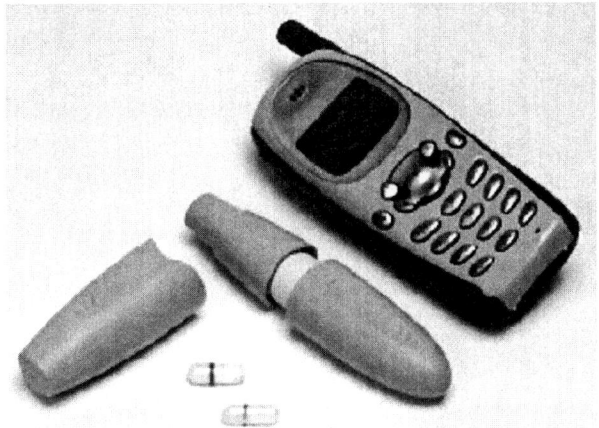

(B)

Figure 3 Insulin pharmacokinetics in smokers and asthmatics using the AERx® system.

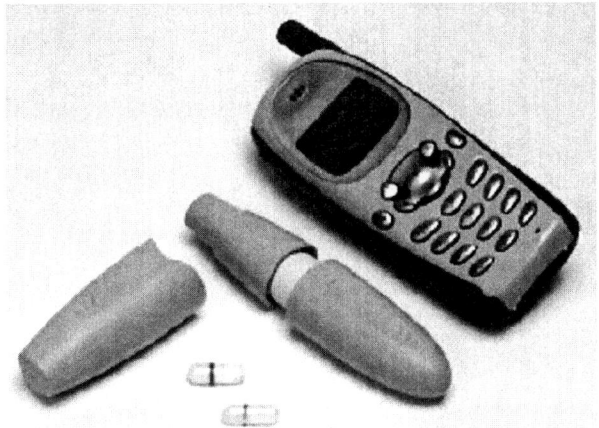

Figure 4 AIR™ system inhaler.

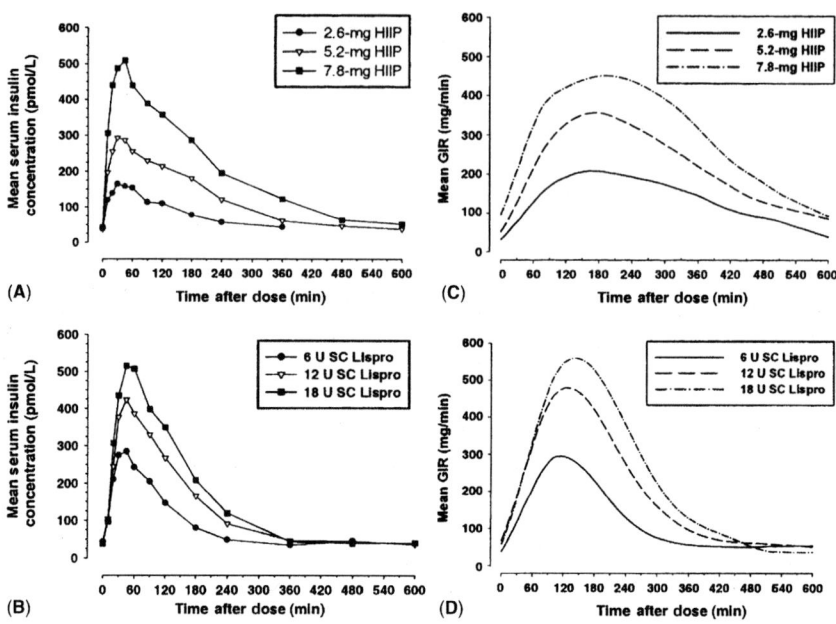

Figure 5   Pharmacokinetic and glucodynamic results after inhalation of insulin using the AIR product and SC lispro injection. (**A**) Mean serum insulin concentration of three doses of inhaled insulin. (**B**) Mean serum insulin concentration of three different SC lispro doses. (**C**) Plots of glucose infusion rates of three different doses of inhaled insulin. (**D**) Plots of glucose infusion rates of three different SC lispro doses.

respectively. Higher doses extends the duration of action. In subjects who received the 7.8 mg dose, the duration of action was 8 h compared to subjects who received the 2.6 and 5.2 mg dose whose duration of action was 6–7 h. Intrasubject variability of the inhaled product is similar to SC insulin lispro (35). A major difference between the Exubera product and the AIR product is the capsule dose equivalency. Unlike Exubera, 3 inhalations of 2 unit AIR capsules are equivalent to one inhalation of a 6-unit capsule.

One study evaluated the pharmacokinetics and glucodynamics of the AIR system in patient with COPD (36). The mean AUC and metabolic effect were reduced in patients with COPD compared to healthy subjects. Moreover, the intrasubject variability was greater in patients with chronic bronchitis. More extensive studies are needed in this population to further investigate the safety profile.

Technosphere™/Insulin

The Technosphere/Insulin technology is unique in that it captures and stabilizes peptides in small particles (37). A small organic molecule, 3,6-bis

[N-fumaryl-N-(n-butyl)amino-2,5-diketopiperazine] self-assembles in a mild acidic environment into microspheres of approximately 2 µm (Fig. 6) (38). Peptides in the reaction solution, such as insulin, are trapped in the microspheres during self assembly. The particles are then dried and suitable for pulmonary delivery. In the neutral pH of the deep lungs, the microspheres dissolve and facilitate the absorption of the insulin into the systemic circulation. The carrier molecules are excreted as ammonium salts in the urine within hours (37). The Technosphere/Insulin is provided in gelatin capsules that contain 50 IU of dry powder regular human insulin (38). In studies evaluating Technosphere/Insulin technology a specifically developed breath-powered unit dose dry powder inhaler was utilized (37). The inhaler does not require any external energy and subjects were instructed to take three deep breaths per capsule. Delivery of the drug is dependant on the breath effort of the patients (38).

In a pilot study of five healthy subjects, a euglycemic clamp technique was used to evaluate the biologic efficacy and pharmacokinetic properties of the Technosphere/Insulin technology (38). Subjects were randomized to one of three groups inhaled 100 IU of Technosphere/Insulin, 10 units of SC regular insulin and 5 units of intravenous (IV) regular insulin (Fig. 7). Technosphere/Insulin showed an onset of action similar to that of IV regular insulin (13 vs. 9 min) and much quicker than that of SC regular insulin (121 min). Insulin concentrations with the Technosphere/Insulin technology returned to baseline after 3 h. In a following dose ranging study, subjects were randomized to receive 25, 50, and 100 IU of Technosphere/ Insulin. This study found that there is a linear dose dependant increase in metabolic activity and systemic insulin uptake. Technosphere/Insulin

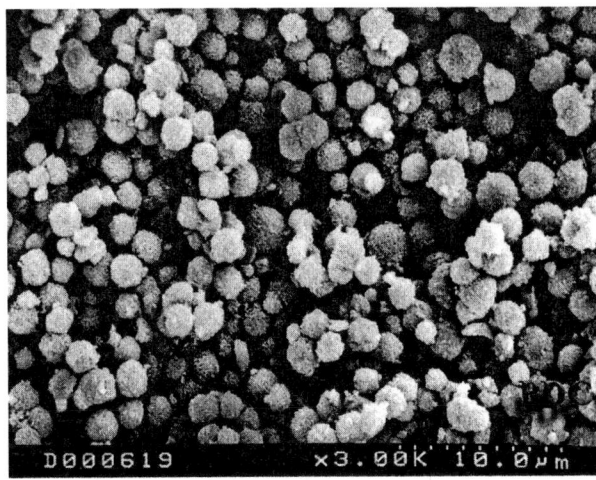

Figure 6 Scanning electron micrograph of precipitated Technosphere particles. Mean particle size = 2–3 µm.

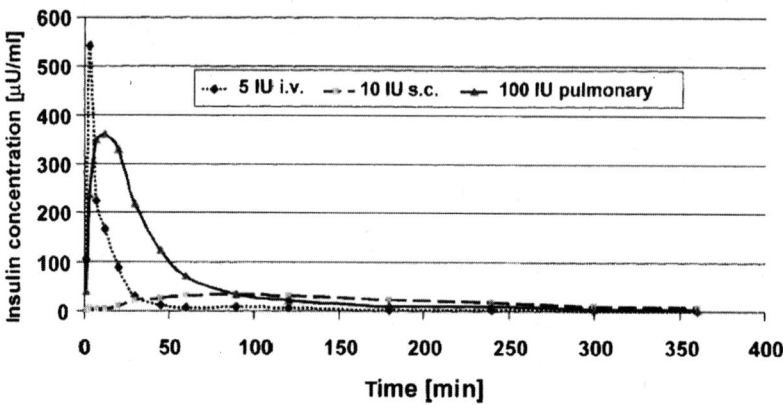

Figure 7 Serum insulin concentrations after administration of 5 IU of insulin IV, 10 IU of insulin SC, and 100 IU of Technosphere™/Insulin in five healthy volunteers in a euglycemic clamp study.

technology differs from the other inhaled technologies in that the onset is faster, although they have not been directly compared.

Spiros®

The Spiros insulin consists of a dry powder inhaler and blisterdisks. The blisterdisks contain a preparation of 11% micronized insulin crystals blended with lactose monohydrate NF. The blend is filled into blisterdisks at a target fill weight of 9.5 mg per well. This results in 1.05 mg of insulin in each blister well that is equivalent to 30 units. The median aerodynamic particle size of the insulin crystals is 2–3 μm and the median size of the lactose particles is 150 μm. The inhaler is a hand held breath activated device that delivers the dry powder crystals (Fig. 8). Inspiratory flow activates a battery-powered twin blade impeller that assisted in deagglomeration of the drug and carrier during inspiration (39).

In an open label euglycemic glucose clamp in healthy volunteers four doses of the Spiros system's (60, 90, 120, and 150 U) pharmacokinetics and glucodynamics were evaluated against those subjects receiving 2 of the 3 possible doses of SC regular insulin (8, 14, and 20 U). Serum insulin levels following inhalation of insulin peaked approximately 60 min earlier when compared to SC regular insulin ($P < 0.001$). The time to maximum glucose infusion rate was reached approximately 70 min earlier with inhaled insulin than with SC regular insulin. The Spiros system intra- and intersubject variability were found to be comparable to that of SC regular insulin (39). Smokers have increased serum insulin levels; however they have reduced glucodynamic effect due to increased insulin resistance (40).

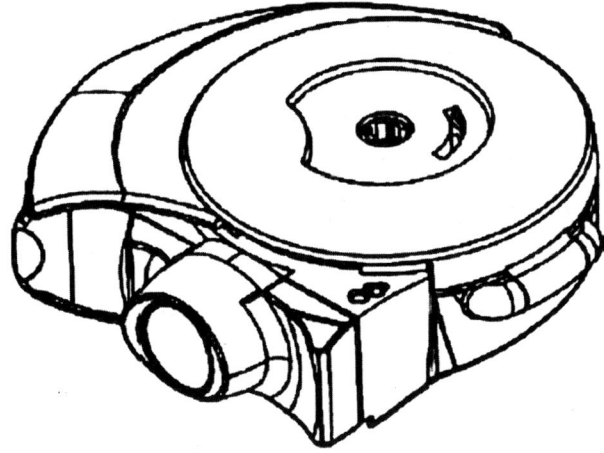

Figure 8 Spiros dry powder inhaler.

CLINICAL STUDIES AND EFFICACY

Exubera System

The efficacy and safety of Exubera has been evaluated in over 4000 individuals with T1DM (4 studies) and T2DM (7 studies).

Type 1 Diabetes Mellitus

Skyler reported on two trials in T1DM. In a proof-of-concept trial conducted in 72 T1DM patients (41), a similar reduction in A1C was achieved when randomized to twice daily isophane insulin plus Exubera before meals versus once daily ultralente insulin plus SC regular insulin before meals (Table 4). Second, 328 patients with T1DM (baseline A1C approximately 8.2%) received premeal Exubera or regular SC insulin in combination with Neutral Protamine Hagedorn (NPH) insulin before breakfast and at bedtime. After 24 weeks, the A1C was $7.7 \pm 0.1\%$ in the Exubera group and $7.8 \pm 1.2\%$ in the SC insulin group. The two treatment regimens were comparable for A1C reduction and number of subjects who reached an A1C $< 7.0\%$, but Exubera significantly lowered the fasting plasma glucose (FPG) greater than SC insulin (-35 vs. $4\,\mathrm{mg/dl}$ [CI -57.50 to -21.56]), respectively (42).

Quattrin et al. (43) randomized 335 T1DM subjects with a baseline A1C of approximately 8.3% to either premeal Exubera plus bedtime ultralente or NPH insulin twice daily (breakfast and evening or bedtime) in combination with two regular insulin doses prior to breakfast and the evening meal for 24 weeks. Exubera and SC insulin reduced the A1C to $7.9 \pm 1.1\%$ and $7.7 \pm 0.9\%$ (adjusted treatment group difference; 0.16% [95% CI, -0.01 to 0.32]), respectively. In addition, a comparable numbers

Table 4 Exubera Published Clinical Trials

Reference	n	Study duration (weeks)	Mean age (range or SD)	Women (%)	White (%)	Duration of diabetes (years)	Baseline A1C (%)	Design	Inhaled insulin regimen/additional therapy	Comparator regimen
Type 1 diabetes (inhaled vs. SQ insulin)										
41	72	12	37+/-9	47	81	14.5	8.5	Randomized, open-label	TID with meals, UL at bedtime	Reg BID/NPH BID
43	334	24	33 (11–63)	46	90	16	8.3	Randomized, open-label	TID with meals, UL at bedtime	Reg BID/NPH BID
44	45	24	36 (18–50)	64	NR	17	7.0	Randomized, open-label	TID with meals, NPH BID	Reg BID/NPH BID
42	328	24	30 (11–65)	77	90	13	8.2	Randomized, open-label	TID with meals, NPH BID	Reg TID/NPH BID
Type 2 diabetes (inhaled vs. oral agents)										
45	69	12	51 (33–65)	35	64	8	9.9	Randomized, open-label	TID with meals, MET and/or SU	Continue MET and/or SU
46	143	12	54 (28–80)	45	74	3.5	9.5	Randomized, open-label	Diet/Exercise TID with meals	Diet/Exercise, Rosiglitazone

49	298	12	57 (35–80)	34	80	9	9.5	Randomized, open-label	MET or TZD + SU. (1) Monotherapy D/C orals, TID with meals. (2) Continue orals, Add TID with meals	MET or TZD + SU continued
47	423	24	60 (35–80)	47	NR	9	9.7	Randomized, open-label	SU + TID with meals	SU + MET
48	470	24	55 (35–77)	43	NR	8	9.5	Randomized, open-label	MET + TID with meals	MET + SU
Type 2 diabetes (inhaled vs. SQ insulin)										
50	26	12	51 (35–65)	38	NR	11	8.7	Randomized, open-label	TID with meals, UL at bedtime	2–3 injections/ day
51	298	24	57 (23–80)	34	NR	13.5	8.5	Randomized, open-label	TID with meals, UL at bedtime	Reg BID/NPH BID

Abbreviations: MET, metformin; NR, not reported; PFT, pulmonary function tests; Reg, regular insulin; SD, standard deviation; SU, sulfonylurea; UL, ultralente insulin.

of subjects achieved A1C < 7% (15.9% and 15.5%), respectively. The FPG in patients receiving Exubera (baseline FPG = 194 mg/dl) was reduced greater than in patients receiving SC insulin (baseline FPG = 203 mg/dl), [−25.2 mg/dl (95% CI −143.4 to −7.0 mg/dl)].

Additionally, Heise et al. (44) examined 47 T1DM subjects given NPH twice daily in combination with pre-meal SC regular insulin or Exubera. A1C and FPG were not significantly different between regimens.

Overall, in T1DM, Exubera given pre-meal appears to be equally efficacious, as measured by the A1C and attainment of A1C < 7.0%, as SC regular insulin. Exubera has shown in several trials to significantly lower FPG greater than SC insulin, and this may result in the need to adjust the basal insulin, or at a minimum be cognizant of the potential for nocturnal hypoglycemia in a particular patient.

Type 2 Diabetes Mellitus

Exubera versus or in combination with oral agents: Weiss et al. (45) evaluated the effect of the addition of Exubera or continuation of baseline oral antihyperglycemics (baseline A1C approximately 9.8%) on glycemic control in 68 T2DM subjects on near maximal or maximal therapeutic doses of a sulfonylurea, metformin, or the combination. After 12 weeks, subjects randomized to Exubera three times daily prior to meals versus continued baseline oral therapy had an A1C reduction of 2.3% versus 0.1% (P < 0.001). Eleven subjects (34%) taking Exubera versus none continuing oral agents achieved an A1C < 7%.

DeFronzo et al. (46) randomized 145 T2DM patients with poor glucose control, A1C at baseline approximately 9.5%, despite diet and exercise, to receive premeal Exubera (average dose = 15.3 mg at end of study) or rosiglitazone 4 mg twice daily. After 12 weeks, the A1C was reduced greater in the Exubera group versus rosiglitazone (−2.3% vs. −1.4%, P < 0.0001), and the odds of reaching an A1C of either <7% or ≤6.5% was significantly higher with Exubera. Changes in FPG were comparable between treatments (−64 and −56 mg/dl) for Exubera and rosiglitazone.

Barnett et al. reported on two open-label, parallel, 24-week trials in T2DM subjects randomized to addition of pre-meal Exubera versus addition of metformin in subjects poorly controlled on sulfonylurea monotherapy (47), or versus addition of a sulfonylurea in poorly controlled subjects on metformin monotherapy (48). In 423 subjects taking near maximal effective doses failing sulfonylureas (47), addition of Exubera (average daily dose = 12.1 mg) lowered the A1C greater than addition of metformin (average dose = 1 g twice daily), (−2.06% vs. −1.83%, p = 0.014), respectively. The high A1C arm (baseline A1C≥9.5%) saw reductions of 2.17% versus 1.79%, p = 0.002, with Exubera and metformin, respectively. A1C ≤ 7.0% was achieved by 54 (25%) Exubera subjects and 45

(23%) of metformin subjects ($P =$ NS). 476 subjects failing metformin (> 1.5 grams/day for ≥ 2 months) were given Exubera (average daily dose $= 13.2$ mg) or glibenclamide (glyburide in United States, at average daily dose $= 7.6$ mg) (48). A1C was reduced (2.03% and 1.88%, $P =$ NS), with Exubera and glibenclamide, respectively, but in the high A1C arm, reductions were greater with Exubera (-0.37% [95% CI -0.62 to -0.12, $P = 0.004$]). Odds of achievement of an A1C $\leq 7.0\%$ were significantly better with Exubera if the baseline A1C was $> 9.5\%$. Overall, Exubera proved to be non-inferior to metformin or glibenclamide at an A1C $< 9.5\%$, and superior to addition of metformin or sulfonylurea if the baseline A1C is $> 9.5\%$.

In an open-label, multicenter study (baseline A1C approximately 9.3%) in 309 T2DM, subjects were randomized to one of three groups: (1) continued current oral antihyperglycemics; (2) premeal Exubera as monotherapy (discontinuation of all oral agents); or (3) current oral agent therapy plus Exubera. After 12 weeks, A1C reductions in the three groups were 0.2%, 1.4%, and 1.7%, respectively, with both Exubera monotherapy and Exubera addition to oral agents significantly lowering A1C versus continued oral agents. Odds of attainment of an A1C $< 7.0\%$ was significantly higher with both Exubera monotherapy (17%) and addition of Exubera to oral agents (32%) versus continued oral agents (1%). The mean dose of Exubera was higher with monotherapy (26.4 mg) than when oral agents are continued (13.1 mg) (49).

Exubera versus SC insulin regimens: In a small, three month trial (50), 26 insulin-treated T2DM not taking oral agents were randomized to continue their conventional insulin regimen (2–3 injections/day) or to start premeal Exubera with a single bedtime injection of ultralente insulin. In the Exubera group, A1C declined by 0.71% (baseline A1C 8.67%), and remained unchanged in the conventional insulin regimen group. Subjects were on 14.6 mg Exubera and 35.7 units of ultralente, compared to 19 units of regular insulin and 51 units of long-acting insulin at baseline. Thus, it appears Exubera may have the potential to lower the basal insulin requirements, despite being prandial insulin.

Hollander et al. (51) randomized 298 T2DM subjects, poorly controlled (baseline A1C = ~8.5%) on at least two daily injections of insulin, to receive premeal Exubera plus bedtime ultralente or at least two SC injections of mixed regular/NPH insulin for 6 months. The reduction in A1C in the Exubera (-0.7%) and mixed regular/NPH insulin (-0.8%) groups were similar. An A1C $\leq 7\%$ was achieved in 67 (47%) subjects receiving Exubera versus 46 subjects (32%) on the mixed regular/NPH regimen (adjusted odds ratio 2.27 [95% CI 1.24–2.14]). The FPG was significantly lower with Exubera, but did not show a significant difference in 2-h post-prandial glucose readings versus the mixed regular/NPH group.

Summary of efficacy: In all studies in T1DM and T2DM, Exubera as proven to be as effective as regular insulin in improving glycemic control. In addition, Exubera appears to lower the FPG when compared to regular insulin, which could result in the need to reduce the basal insulin dose and also supports the potential use of Exubera as monotherapy. Though Exubera can be used in monotherapy, combination with oral agents appears to be advantageous in that you can attain glycemic control with a lower dose of inhaled insulin (49). It should be noted that the majority of trials were started prior to the introduction of the basal insulin glargine or detemir, and intermediate-acting or ultralente acted as the basal insulin in these trials. The choice of basal insulin can be important, and most clinicians currently would recommend the use of insulin glargine or detemir. Insulin glargine and detemir have less risk of hypoglycemia and/or nocturnal hypoglycemia when compared to NPH, but trials do not support superior efficacy. Thus, if cost is paramount, NPH could be substituted, as lente and ultralente insulins have been discontinued by the manufacturer.

Dosing/Prescribing/Maintenance of Device

Guidelines for initial premeal Exubera dosing are based upon the patient's body weight or the appropriate cross-over dose when switching from regular SC insulin. (Table 5). Based upon clinical trials in which patients consumed three meals per day, an initial meal-time dose can also be calculated by: (body weight in kg) $\times$ (0.05 mg/kg), where tenths of a milligram are rounded down to the nearest whole milligram (22). These are only initial premeal doses, and further individualization based on glycemic control, timing in regards to meal, content, and size of meal are appropriate. For example, a patient with an A1C of 10% may start at this dose, but is likely to be controlled on a higher dose, whereas if the patient's starting A1C was 7.0%, post-prandial or fasting hypoglycemia may be of more concern if incorrectly dosed, base on the meal. It should be intuitive that T2DM patients are more insulin resistant, thus higher doses of Exubera are likely to be required in patients with T2DM versus those with T1DM. Also, Exubera is dosed in milligrams, and each milligram increase is equal to 2–3 units injected subcutaneously. Clinically, this is usually a large dosage increase for T1DM, and must be balanced with the current blood glucose readings.

Exubera is available in a Combination Pack 12 (90, 1-mg blister packs and 90, 3-mg blister packs, plus 2 insulin release units) and a Combination Pack 15 (180, 1-mg blister packs and 90, 3-mg blister packs, plus 2 insulin release units) for subjects who may use 12 or 15 mg/day, respectively. Blister packs in the foil can be used, if stored at room temperature, until the expiration date, but once the foil overlay has been opened they are good for 3 months (22).

The insulin release units within the Exubera device should be replaced every 2 weeks so as to lower the risk of insulin powder blockage where the

Table 5 Exubera®: Approximate Weight-Based Dosing and Equivalent SC Regular Dosing

Pt. weight (kg)	Pt. weight (lb)[a]	Initial dose/meal (mg)	Number of 1-mg blisters/meal	Number of 3-mg blisters/meal
30–39.9	66–87	1	1	–
40–59.9	88–132	2	2	–
60–79.9	133–176	3	–	1
80–99.9	177–220	4	1	1
100–119.9	220–264	5	2	1
120–139.9	265–308	6	–	2

Approximate Equivalent SC Regular Insulin Dosing

Dose (mg)	Regular SC insulin dose	Number of 1-mg Exubera® blister packs/dose	Number of 3-mg Exubera® blister packs/dose
1	3	1	–
2	6	2	–
3	8	–	1
4	11	1	1
5	14	2	1
6	16	–	2

[a]Approximate weight-based dosing.

compressed air disperses the insulin. This could lower the amount of insulin delivered from each blister pack. In addition, it is currently recommended that the device be replaced after 1 year of use.

AERx System

Clinical data is limited, but in a 12 week study in T2DM (52), three AERx insulin inhalations in conjunction with bedtime NPH (baseline A1C 8.5%) produced a similar reduction in A1C ($-0.69 \pm 0.77\%$) as fast-acting human insulin given SC prior to each meal ($-0.77 \pm 0.77\%$). No significant difference between groups was seen for FPG or post-prandial glucose readings. The amount of inhaled insulin versus SC insulin was explored, which is important, as the AERx is dosed in "units" of insulin. The mean meal-related insulin ratio was 0.85 units of SC insulin for each unit of AERx insulin.

AIR System

Clincal data is limited, but in a 12 week study in T1DM subjects who were randomized to receive prandial coverage with SC insulin or AIR system inhaled insulin in conjunction with glargine insulin, the reduction in A1C with AIR system inhaled insulin was equivalent to SC injected insulin (53).

Technosphere/Insulin

Clinical data is limited, but in 227 poorly controlled T2DM subjects given prandial Technosphere/Insulin at doses of 28, 42, and 56 units for 12 weeks in conjunction with insulin glargine, A1C declined by 0.67%, 0.70%, and 0.78%, respectively (54).

SAFETY

Lung Function Parameters

Forced expiratory volume in 1 second (FEV_1) and diffusing capacity of carbon monoxide (DL_{CO}) are parameters utilized to evaluate lung function in studies testing inhaled insulin products. FEV_1 measures the volume of air exhaled during the first second of a forced expiratory maneuver. Values are reported as a percentage of predicted normal values and are used to measure airway obstructions or changes in airway diameter. DL_{CO} reveals oxygen-diffusing capacity influenced by lung surface area, speed of blood flow and thickness of the air–blood barrier. A reduction in DL_{CO} is a hallmark finding in patients with emphysema and interstitial lung disease. Lung function can be affected by various factors. Lung diseases such as COPD (55), interstitial lung disease, or asthma and patient specific factors such as age, smoking, ethnicity, weight or height can have a marked impact on lung function (56,57). Patients with diabetes have an unexplained decrease in lung function compared to those subjects without diabetes (58,59). In one study, the forced vital capacity (FVC) and FEV_1 were approximately 8% lower in diabetic individuals compared to non-diabetics (58). This difference is similar to the difference seen between smokers and a subject who has never smoked.

Inhaled Insulin System Safety

The preponderance of the safety data comes from studies evaluating the Exubera system while there are only a few studies assessing the other systems.

Exubera System
In patients using the Exubera system, cough was reported more frequently (T1DM = 29.5% and T2DM = 21.9%) compared to controls (10%). The cough was generally mild, rarely productive and accounted for 1.2% discontinuation in clinical trials (22). Furthermore, the cough occurred within seconds to minutes of dosing and was not associated with declines in lung function or bronchoconstriction (60).

All reports published regarding pulmonary function with inhaled insulin, regardless of the system, have shown a small decline from baseline and was comparable to SC insulin delivery (39,42,43,47,51–53). In studies investigating Exubera, patients with T1DM who received inhaled insulin experienced a statistically significant decline in DL_{CO} and a non-significant decrease in FEV_1 compared to those patients randomized to SC insulin (Fig. 9) (42,43,60). Although there was a statistically significant decrease in DL_{CO} there were no clinical manifestations and was deemed not clinically significant (42,43). This decline occurred within 2 weeks after insulin initiation and did not progress any further (60). Moreover, patients who did not receive inhaled insulin had a small decline in lung function over time (60). Studies with Exubera, AERx, and AIR systems agree that changes in lung function are not clinically significant and reversible upon treatment discontinuation (42,51–53). Thoracic high resolution computed tomography scans after 2 years of Exubera have shown no significant signals for concern. Additionally, the rate of pulmonary fibrosis, two cases with Exubera (0.40 cases/10,000 subject-months exposure) compared to one case in an oral agent comparator group (0.50 cases/10,000 subject-months exposure), is not increased.

Safety Monitoring

Prescribing information for the Exubera system recommends all patients have baseline spirometry testing, a follow-up 6 months after initiation of

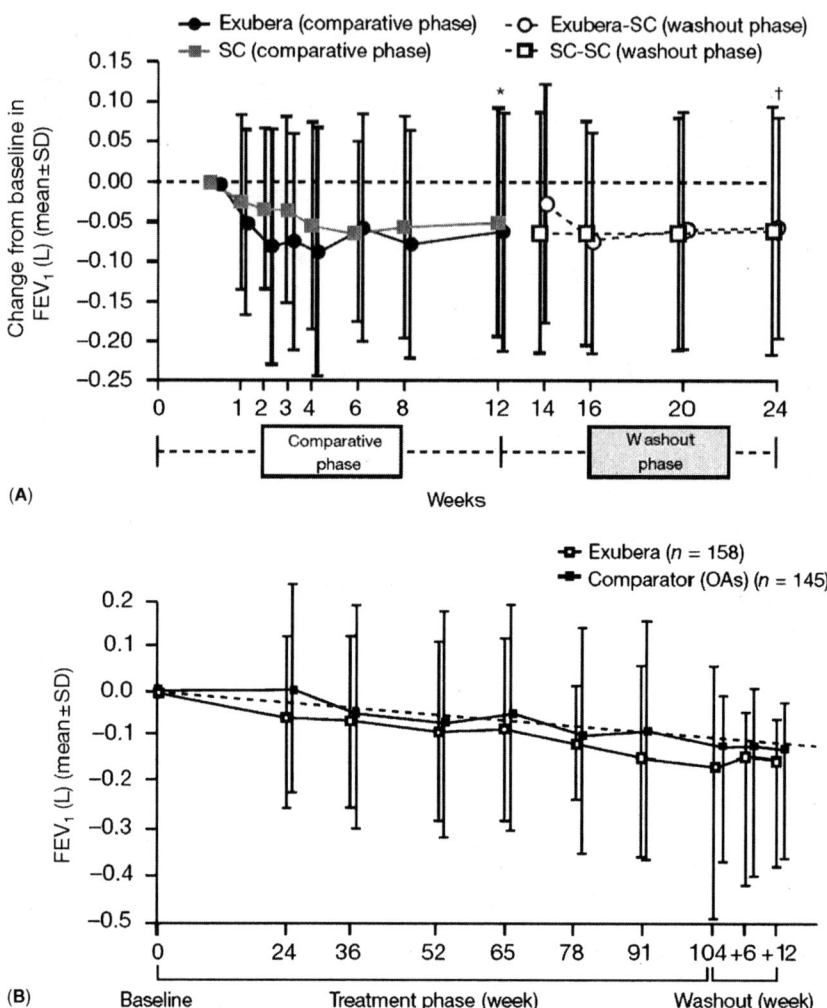

Figure 9 FEV$_1$ with Exubera® inhaled insulin therapy after (**A**) 12 weeks and (**B**) 2 years. *−0.009 (95% CI −0.014 to 0.023); †0.014 (95% CI −0.021 to 0.049). *Abbreviations*: FEV$_1$, forced expiratory volume in 1 s; OA, oral agents.

treatment, and yearly follow-up thereafter. Exubera is not recommended for patients with baseline FEV$_1$ or DL$_{CO}$ < 70% predicted. In patients who experience a ≥20 % decrease in FEV$_1$, tests should be repeated. If this finding is confirmed, then treatment should be discontinued. Treatment should also be stopped in the rare case of acute bronchospasm (22).

Insulin Antibodies

Delivery of insulin to the pulmonary mucosa could induce an immune or allergic reaction which may have safety implications. A large meta-analysis with the Exubera system showed higher insulin antibody levels with inhaled insulin versus comparator therapies (61). The same holds true with the AERx system in one clinical study (52). These increases in insulin antibodies had no significant effect on insulin pharmacokinetics, glucodynamics, glucose control, or safety profile (61). No insulin antibodies were observed with the Technosphere/Insulin technology (62) and no data is available for the AIR system.

Hypoglycemia

As with SC insulin therapy, the inhaled formulations may produce hypoglycemic reactions. In most studies, minor hypoglycemia was defined as: (1) symptoms of hypoglycemia without glucose documented, but with prompt relief from food; or (2) Blood glucose < 60 mg/dl with hypoglycemic symptoms; or (3) < 50 mg/dl with or without hypoglycemic symptoms. "Severe" hypoglycemia was defined as: (1) assistance needed; or (2) neuroglycopenic symptoms with documented blood glucose < 50 mg/dl, or if prompt relief from food.

The nature, frequency and severity (minor and major) of hypoglycemia in patients with T1DM and T2DM receiving inhaled insulin are similar to those patients receiving SC insulin. In patients with T2DM, Exubera was associated with an increased risk of hypoglycemia when compared with comparator metformin or rosiglitazone oral therapy (46,47). Hypoglycemic rates were similar to addition of glibenclamide, but slightly less when compared to a SC regular/NPH insulin regimen (48). Patients receiving the AIR system had increased rates of nocturnal hypoglycemia. Hypoglycemic rates have not been reported in patients receiving the Technosphere formulation.

Weight Change

In T1DM, Exubera resulted in a small amount of weight gain (1–1.5 kg), similar to SQ injected insulin regimens. In most studies in T2DM, Exubera therapy resulted in approximately 2–3 kg of weight gain. This was similar to the weight gain seem with institution of glibenclamide, but more than when metformin (−0.1 kg) was added (47,48). Exubera plus the basal insulin ultralente (0.6 kg) resulted in significantly less weight gain versus regular insulin/NPH therapy (1.4 kg) (51).

PATIENT SATISFACTION AND "INSULIN RESISTANCE"

Insulin therapy continues to be the cornerstone of treatment in T1DM, but is underutilized in the treatment of T2DM. Some investigators believe that multiple daily injections of insulin are disliked by patients and are partly responsible for patients not reaching goal (15,16). This demonstrated "resistance" to the implementation or administration of insulin therapy has been well documented. Often the reasons cited by clinicians to delay insulin therapy include: fear of patient loss from practice, lack of time to implement, and clinical inertia, or the "wait and see" approach. Patient's reasons for delaying insulin therapy include: the frequent asymptomatic course of diabetes, patient refusal to follow directions, and importantly, injections and hypoglycemia (15,16,693). Thus, inhaled insulin may help to overcome some of these concerns.

In a study of theoretical treatment choices in inadequately controlled T2DM individuals, subjects were more likely to choose insulin therapy when the insulin therapy choices included inhaled insulin therapy (64). Rosenstock et al. (65) reported on patient satisfaction in the pooled-analysis of three, 12-week studies in subjects with T1DM and T2DM who were randomized to receive Exubera with ultralente or 2–3 injections of a SC split/mixed insulin regimen, followed by a 1 year open-label extension. Using the Patient Satisfaction with Insulin Therapy questionnaire, overall satisfaction and ease of use were significantly higher with Exubera versus SC insulin, and social comfort tended ($P = 0.11$) to be higher with Exubera. Eighty-five percent of the patients treated with Exubera during the 12-week period chose to continue Exubera during the 1-year extension, whereas only 21% treated with SC insulin during the 12-week study chose to remain on the SC regimen. Most (75%) who were treated with SC insulin during the 12-week study and chose to stay in the 1-year extension were switched to Exubera.

CONCLUSION

Inhalation of insulin is a novel, non-injectable method for delivering insulin into the human body. Many patients do not like injections, and would prefer alternative methods of insulin delivery. Inhaled insulin therapy offers additional options for clinicians to improve glycemic control in patients with diabetes. The dry-power system of Exubera is the most extensively studied and the only Food and Drug Administration approved inhaled insulin. Onset and peak effect are similar to rapid-acting insulins such as lispro, aspart, or glulisine, but Exubera has a duration of action similar to regular SC insulin. Underlying lung disease, smoking, and asthma can all alter the dose of insulin absorbed, and inhaled insulin should be avoided in these populations. Exubera is dosed as milligrams, and this is distinctly different than any other insulin product. It has been shown to be effective in reducing A1C levels in T1DM and T2DM. As with other insulin, a risk of hypoglycemia is present, and Exubera has similar rates of hypoglycemia as SC regular insulin. Safety data is complete through 2 years with Exubera, and small changes in FEV_1 and DL_{co} have been observed, but are reversible upon discontinuation. Other inhalation systems such as AERx, AIR, and Technosphere are in development, and the area of alternative delivery of peptides continues to be an active area of research.

REFERENCES

1. Third Report of the National Cholesterol Education Program (NCEP) Expert Panel on Detection, Evaluation, and Treatment of High Blood Cholesterol in Adults (Adult Treatment Panel III) final report. Circulation 2002; 106(25): 3143–421.
2. Haffner SM, Lehto S, Ronnemaa T, Pyorala K, Laakso M. Mortality from coronary heart disease in subjects with type 2 diabetes and in nondiabetic subjects with and without prior myocardial infarction. N Engl J Med 1998; 339 (4):229–34.
3. Centers for Disease Control and Prevention NCfHS, Division of Health Interview Statistics, data from the National Health Interview Survey. U.S. Bureau of the Census, census of the population and population estimates. Data computed by the Division of Diabetes Translation, National Center for Chronic Disease Prevention and Health Promotion, Centers for Disease Control and Prevention.
4. King H, Aubert RE, Herman WH. Global burden of diabetes, 1995–2025: prevalence, numerical estimates, and projections. Diabetes Care 1998; 21(9): 1414–31.
5. Diagnosis and classification of diabetes mellitus. Diabetes Care 2006; 29 (Suppl 1):S43–8.
6. Foulis AK, McGill M, Farquharson MA. Insulitis in type 1 (insulin-dependent) diabetes mellitus in man–macrophages, lymphocytes, and interferongamma containing cells. J Pathol 1991; 165(2):97–103.
7. DeFronzo RA. Pathogenesis of type 2 diabetes mellitus. Med Clin North Am 2004; 88(4):787-835, ix.
8. The effect of intensive treatment of diabetes on the development and progression of long-term complications in insulin-dependent diabetes mellitus. The Diabetes Control and Complications Trial Research Group. N Engl J Med 1993; 329(14):977–86.
9. Ohkubo Y, Kishikawa H, Araki E, et al. Intensive insulin therapy prevents the progression of diabetic microvascular complications in Japanese patients with non-insulin-dependent diabetes mellitus: a randomized prospective 6-year study. Diabetes Res Clin Pract 1995; 28(2):103–17.
10. Intensive blood-glucose control with sulphonylureas or insulin compared with conventional treatment and risk of complications in patients with type 2 diabetes (UKPDS 33). UK Prospective Diabetes Study (UKPDS) Group. Lancet 1998; 352(9131):837–53.

11. Hanefeld M, Fischer S, Julius U, et al. Risk factors for myocardial infarction and death in newly detected NIDDM: the Diabetes Intervention Study, 11-year follow-up. Diabetologia 996; 39(12):1577–83.
12. Standards of medical care in diabetes—2006. Diabetes Care 2006; 29(Suppl 1): S4–42.
13. Harris MI. Health care and health status and outcomes for patients with type 2 diabetes. Diabetes Care 2000; 23(6):754–8.
14. Saaddine JB, Engelgau MM, Beckles GL, Gregg EW, Thompson TJ, Narayan KM. A diabetes report card for the United States: quality of care in the 1990s. Ann Intern Med 2002; 136(8):565–74.
15. Zambanini A, Newson RB, Maisey M, Feher MD. Injection related anxiety in insulin-treated diabetes. Diabetes Res Clin Pract 1999; 46(3):239–46.
16. Mollema ED, Snoek FJ, Heine RJ, van derPloeg HM. Phobia of self-injecting and self-testing in insulin-treated diabetes patients: opportunities for screening. Diabet Med 2001; 18(8):671–4.
17. Lenzer J. Inhaled insulin is approved in Europe and United States. Br Med J 2006; 332(7537):321.
18. Gansslen M. Uber inhalation von insulin. Klin Wochenschr 1925; 4:71.
19. Wigley FW, Londono JH, Wood SH, Shipp JC, Waldman RH. Insulin across respiratory mucosae by aerosol delivery. Diabetes 1971; 20(8):552–6.
20. Edwards DA, Ben-Jebria A, Langer R. Recent advances in pulmonary drug delivery using large, porous inhaled particles. J Appl Physiol 1998; 85(2): 379–85.
21. Crowder TM, Rosati JA, Schroeter JD, Hickey AJ, Martonen TB. Fundamental effects of particle morphology on lung delivery: predictions of Stokes' law and the particular relevance to dry powder inhaler formulation and development. Pharm Res 2002; 19(3):239–45.
22. Exubera [package insert]. New York, NY: Pfizer/Nektar; 2006May.
23. Heinemann L, Traut T, Heise T. Time-action profile of inhaled insulin. Diabet Med 1997; 14(1):63–72.
24. Rave K, Bott S, Heinemann L, et al. Time-action profile of inhaled insulin in comparison with subcutaneously injected insulin lispro and regular human insulin. Diabetes Care 2005; 28(5):1077–82.
25. Gelfand R, Schwartz S, Horton M. Pharmacological reproducibility of inhaled human insulin pre-meal dosing in patients with type 2 diabetes mellitus (NIDDM). Diabetes 1998; 47(Suppl. 1):A99.
26. Mudaliar S, Henry R, Fryburg D, et al. Within-subject variability of inhaled insulin (Exubera) versus subcutaneous regular insulin in elderly obese patients with type 2 diabetes mellitus (abstract 802). Diabetologia 2003; 52(Suppl 2): A277.
27. Himmelmann A, Jendle J, Mellen A, Petersen AH, Dahl UL, Wollmer P. The impact of smoking on inhaled insulin. Diabetes Care 2003; 26(3):677–82.
28. Farr SJ, Rowe AM, Rubsamen R, Taylor G. Aerosol deposition in the human lung following administration from a microprocessor controlled pressurised metered dose inhaler. Thorax 1995; 50(6):639–44.
29. Farr SJ, McElduff A, Mather LE, et al. Pulmonary insulin administration using the AERx system: physiological and physicochemical factors influencing

insulin effectiveness in healthy fasting subjects. Diabetes Technol Ther 2000; 2(2):185–97.

30. Brunner GA, Balent B, Ellmerer M, et al. Dose-response relation of liquid aerosol inhaled insulin in type I diabetic patients. Diabetologia 2001; 44(3): 305–8.

31. Kapitza C, Hompesch M, Scharling B, Heise T. Intrasubject variability of inhaled insulin in type 1 diabetes: a comparison with subcutaneous insulin. Diabetes Technol Ther 2004; 6(4):466–72.

32. Henry RR, Mudaliar SR, Howland WC, 3rd, et al. Inhaled insulin using the AERx Insulin Diabetes Management System in healthy and asthmatic subjects. Diabetes Care 2003; 26(3):764–9.

33. McElduff A, Mather LE, Kam PC, Clauson P. Influence of acute upper respiratory tract infection on the absorption of inhaled insulin using the AERx insulin Diabetes Management System. Br J Clin Pharmacol 2005; 59(5): 546–51.

34. Valente AX, Langer R, Stone HA, Edwards DA. Recent advances in the development of an inhaled insulin product. BioDrugs 2003; 17(1):9–17.

35. Rave KM, Nosek L, de laPena A, et al. Dose response of inhaled dry-powder insulin and dose equivalence to subcutaneous insulin lispro. Diabetes Care 2005; 28(10):2400–5.

36. Rave K, Hausmann M, De laPena A. Pharmacokinetics (PK) and glucodynamics (GD) of human insulin inhalation powder(HIIP) in subjects with chronic obstructive pulmonary disease (COPD). Diabetes 2006; 55(Suppl 1):A26.

37. Pfutzner A, Mann AE, Steiner SS. Technosphere/Insulin—a new approach for effective delivery of human insulin via the pulmonary route. Diabetes Technol Ther 2002; 4(5):589–94.

38. Steiner S, Pfutzner A, Wilson BR, Harzer O, Heinemann L, Rave K. Technosphere/Insulin—proof of concept study with a new insulin formulation for pulmonary delivery. Exp Clin Endocrinol Diabetes 2002; 110(1):17–21.

39. Rave K, Nosek L, Heinemann L, et al. Inhaled micronized crystalline human insulin using a dry powder inhaler: dose-response and time-action profiles. Diabet Med 2004; 21(7):763–8.

40. Wise S, Chien J, Yeo K, Richardson C. Smoking enhances absorption of insulin but reduces glucodynamic effects in individuals using the Lilly-Dura inhaled insulin system. Diabet Med 2006; 23(5):510–15.

41. Skyler JS, Cefalu WT, Kourides IA, et al. Efficacy of inhaled human insulin in type 1 diabetes mellitus: a randomized proof-of-concept study. Lancet 2001; 357:331–35.

42. Skyler JS, Weinstock RS, Raskin P, et al. Use of inhaled insulin in a basal/bolus insulin regimen in type 1 diabetic subjects: a 6-month, randomized, comparative trial. Diabetes Care 2005; 28(7):1630–5.

43. Quattrin T, Belanger A, Bohannon NJ, Schwartz SL. Efficacy and safety of inhaled insulin (Exubera) compared with subcutaneous insulin therapy in patients with type 1 diabetes: results of a 6-month, randomized, comparative trial. Diabetes Care 2004; 27(11):2622–7.

44. Heise T, Bott S, Tusek C, Stephan J-A, Kawabata T, Finco-Kent D, Liu C, Krasner A. The effect of insulin antibodies on the metabolic action of inhaled

and subcutaneous insulin. A prospective randomized pharmacodynamic study. Diabetes Care 2005; 28:2161–9.

45. Weiss SR, Cheng S-L, Kourides IA, Gelfand RA, Landschculz WH, for the Inhaled Insulin Phase II Study Group: Inhaled Insulin provides improved glycemic control in patients with Type 2 diabetes mellitus inadequately controlled with oral agents: a randomized controlled trial. Arch Int Med 2003; 163:2277–82.

46. DeFronzo RA, Bergenstal RM, Cefalu WT, et al. Efficacy of inhaled insulin in patients with type 2 diabetes not controlled with diet and exercise: a 12-week, randomized, comparative trial. Diabetes Care 2005; 28(8):1922–8.

47. Barnett AH, Dreyer M, Lange P, Serdarevic-Pehar M. An open, randomized, parallel-group study to compare the efficacy and safety profile of inhaled human insulin(Exubera) with metformin as adjunctive therapy in patients with type 2 diabetes poorly controlled on a sulfonylurea. Diabetes Care 2006; 29:1282–87.

48. Barnett AH, Dreyer M, Lange P, Serdarevic-Pehar M. An open, randomized, parallel-group study to compare the efficacy and safety profile of inhaled human insulin(Exubera) with glibenclamide as adjunctive therapy in patients with type 2 diabetes poorly controlled on metformin. Diabetes Care 2006; 29:1818–25.

49. Rosenstock J, Zinman B, Murphy LJ, Clement SC, Moore P, Bowering CK, Hendler R, Lan SP, Cefalu WT. Inhaled insulin improves glycemic control when substituted for or added to oral combination therapy in type 2 diabetes. Ann Int Med 2005; 143:549–58.

50. Cefalu WT, Skyler JS, Kourides IA, et al. Inhaled human insulin treatment in patients with type 2 diabetes mellitus. Ann Int Med 2001; 134:203–7.

51. Hollander PA, Blonde L, Rowe R, et al. Efficacy and safety of inhaled insulin (exubera) compared with subcutaneous insulin therapy in patients with type 2 diabetes: results of a 6-month, randomized, comparative trial. Diabetes Care 2004; 27(10):2356–62.

52. Hermansen K, Ronnemaa T, Petersen AH, Bellaire S, Adamson U. Intensive therapy with inhaled insulin via the AERx insulin diabetes management system: a 12-week proof-of-concept trial in patients with type 2 diabetes. Diabetes Care 2004; 27(1):162–7.

53. Garg S, Rosenstock J, Silverman BL, et al. Efficacy and safety of preprandial human insulin inhalation powder versus injectable insulin in patients with type 1 diabetes. Diabetologia 2006; 49(5):891–9.

54. Tack CJJ, Boss AH, Baughman RA, Ren H, Kramer DA, Diaz M, Richardson P. A randomized, double-blind, placebo-controlled study of the forced titration of prandial Technosphere/Insulin in patients with type 2 diabetes mellitus. Diabetes 2006; 55(Suppl 1):A102.

55. Celli BR, MacNee W. Standards for the diagnosis and treatment of patients with COPD: a summary of the ATS/ERS position paper. Eur Respir J 2004; 23(6):932–46.

56. Mannino DM, Reichert MM, Davis KJ. Lung function decline and outcomes in an adult population. Am J Respir Crit Care Med 2006; 173(9):985–90.

57. Burchfiel CM, Marcus EB, Curb JD, et al. Effects of smoking and smoking cessation on longitudinal decline in pulmonary function. Am J Respir Crit Care Med 1995; 151(6):1778–85.

58. Lange P, Parner J, Schnohr P, Jensen G. Copenhagen City Heart Study: longitudinal analysis of ventilatory capacity in diabetic and nondiabetic adults. Eur Respir J 2002; 20(6):1406–12.
59. Ford ES, Mannino DM. Prospective association between lung function and the incidence of diabetes: findings from the National Health and Nutrition Examination Survey Epidemiologic Follow-up Study. Diabetes Care 2004; 27 (12):2966–70.
60. Brain JD. Unlocking the opportunity of tight glycaemic control. Inhaled insulin: safety. Diabetes Obes Metab 2005; 7(Suppl 1):S14–18.
61. Fineberg SE, Kawabata T, Finco-Kent D, Liu C, Krasner A. Antibody response to inhaled insulin in patients with type 1 or type 2 diabetes. An analysis of initial phase II and III inhaled insulin (Exubera) trials and a two-year extension trial. J Clin Endocrinol Metab 2005; 90(6):3287–94.
62. Rosenstock J, Baughman R, Rivera-Schaub T. A randomized, double-blind, placebo-control study of the efficacy and safety of inhaled Technosphere insulin in patients with type 2 diabetes (T2DM). Diabetes 2005; 54(Suppl 1): A357.
63. Polonsky WH, Fisher L, Guzman S, et al. Psychological insulin resistance in patients with type 2 diabetes. Diabetes Care 2005; 28:2543–5.
64. Freemantle N, Blonde L, Duhot D, et al. Availability of inhaled insulin promotes greater perceived acceptance of insulin therapy in patient with type 2 diabetes. Diabetes Care 2005; 28:427–8.
65. Rosenstock J, Capelleri JC, Bolinder B, Greber RA. Patient satisfaction and glycemic control after 1 year with inhaled insulin (Exubera) in patients with type 2 or type 2 diabetes. Diabetes Care 2004; 27:1318–23.

5

Ophthalmic Infections and Therapeutic Strategies

Ravi S. Talluri, Gyan Prakash Mishra, and Ashim K. Mitra

Division of Pharmaceutical Sciences, School of Pharmacy,
University of Missouri-Kansas City, Kansas City, Missouri, U.S.A.

INTRODUCTION

Viruses, bacteria, and fungi are the three most common pathogens causing various ocular disorders. The infections can be either local, where only the ocular tissues are infected, or systemic where the infection spreads to various other organs.

Immunocompromised patients are more prone to these infections. While some infections are self-limiting and benign, others are associated with serious consequences, such as vision loss. In spite of various anti-infective agents being readily available, delivery of these compounds to the target tissue is still a challenging task. While the anterior chamber infections are treated mostly by topical ophthalmic drops, intravitreal and subconjuntival injections are widely indicated for the treatment of posterior chamber infections.

STRUCTURE OF THE EYE

The ocular globe can be divided into anterior and posterior segments (Fig. 1). The anterior segment consists of external cornea, conjunctiva, aqueous humor, iris-ciliary body and lens. The cornea and the lens consisting of avascular and transparent structures, these structures obtain most of the necessary nutrients from aqueous humor. The cornea also partially depends on the tear fluid for essential nutrients like amino acids

143

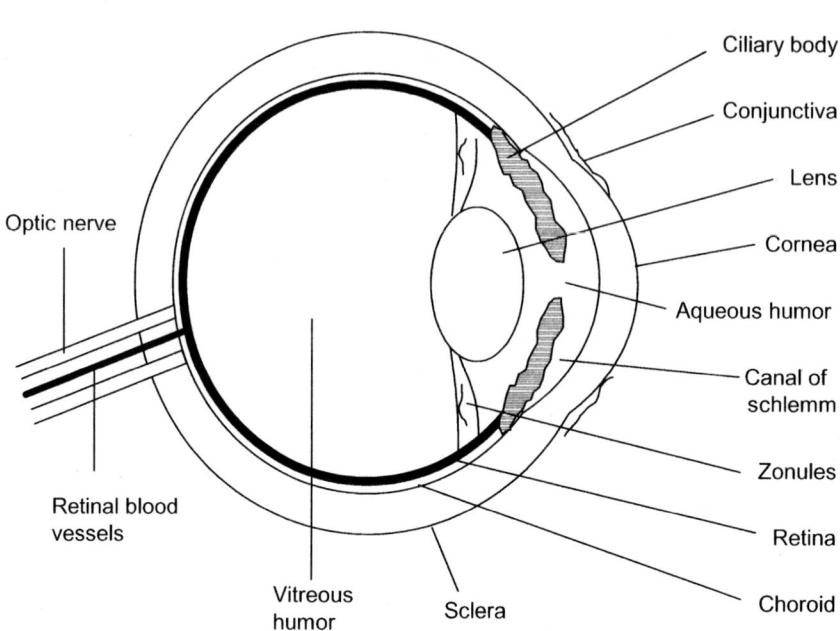

Figure 1 Structure of the human eye.

and vitamins. The iris-ciliary body and conjunctiva are highly vascular tissues. The aqueous humor is a dynamic watery fluid that is continuously secreted by ciliary body and drained out by trabecular meshwork returning to blood stream through schlemm's canal. It maintains a pressure that imparts the convex shape to the front of the globe.

The posterior chamber primarily consists of outer sclera, choroid, retina and the vitreous humor. The sclera is an avascular tissue and acts as outer protective layer. Underneath the sclera is a highly vascular choroid which supplies nutrients to both outer sclera and inner retina. The innermost layer is retina which is primarily responsible for image formation and the thus the vision. The blood–retinal barrier is comprised of retinal pigment epithelium and the endothelium of retinal blood vessels. Unlike the aqueous humor, vitreous humor is a clear watery viscous fluid which is replaced at a very slow rate. The primary purpose of the vitreous humor is to provide a cushioned support for the rest of ocular structures, as well as a clear unobstructed path of light to the retina.

OCULAR DRUG DELIVERY

The ocular tissues most prone to microbial infections are the cornea, the conjunctiva and the iris-ciliary body in the anterior chamber, and the retina in the posterior chamber. To treat ocular infections in the anterior chamber, topical delivery of anti infectives is the most preferred. However, nasolacrimal drainage, tear dilution, conjunctiva absorption, and the outer cornea are the big barriers to topical delivery. Less than 5% of drug reaches the intra ocular tissues after topical administration. The cornea consists of an outer corneal epithelium, the middle stroma and the innermost corneal endothelium. The corneal epithelium is composed of 5–6 layer of columnar epithelial cells with tight junctions presenting a barrier to the hydrophilic compounds, followed by the stroma, which contains more than 90% water and hence acts as a barrier to hydrophobic compounds (1). Hence a drug molecule should posses an optimal hydrophilic–lipophilic balance to permeate across the cornea. The nasolacrimal drianage can be reduced by increasing the residence time of the drug on the cornea which can be achieved by various formulation approaches like liposomes (2), nano-particles (3–5), and other colloidal carriers like microemulsions etc. (5). Oral and systemic administrations have been attempted where the drug molecules need to cross the blood aqueous barrier. This barrier is formed by the non-pigmented layer of the ciliary epithelium and the endothelium of the iridial vessels. It regulates mostly the inward movement of the compounds from blood into the eye (6,7).

Drug delivery to the posterior segment is a significant challenge as the topical delivery may not generate therapeutic concentrations to treat infections. The blood retinal barrier has tight junctions formed by retinal pigment epithelium and endothelium of retinal blood vessels (7). These cells form the tight junctions that can restrict the entry of drug molecules into the posterior chamber after oral or systemic administration. The retina itself is also a big barrier to hydrophilic molecules like ganciclovir and also for macrmolecules (8). Hence high doses are required in the systemic circulation to achieve therapeutic concentrations in retina, which may lead to serious systemic side effects. Intravitreal injection is currently a common mode of drug administration to the posterior chamber. However, this method is associated with retinal detachment, vitreal haemorrahage and cataract (6). Moreover, the frequency of injections depends on the vitreal half life of the drug. Other than systemic and intravitreal routes, periocular routes (subconjunctival, subtenon, and retrobulbular) have also been investigated for drug delivery to the posterior segment. The periocular route is much safer and less invasive than the intravitreal injection. As all these modes involve drug diffusion across the sclera it would be more appropriate to develop transscleral delivery systems. The large surface area of human sclera provides significant avenue for elevated amounts of drug diffusion. Regional differences in the thickness of sclera can further be exploited for designing an optimum delivery system (9). The main mechanism of drug diffusion across sclera is passive diffusion across the aqueous pore pathway. But, diffusion across sclera may not always lead to increased concentrations inside the retina, as molecules still have to cross the highly vascular choroid which can drain a significant fraction of drug into the systemic circulation. Hence the success of this approach depends on whether the choroidal blood supply can limit drug permeation into retina.

VIRAL INFECTIONS

Viruses are capsules of genetic material (DNA or RNA). When a virus enters the body, it invades the cells and takes over the cell machinery, redirecting host cells from their normal function to produce the virus. Viruses may eventually kill their host cells or become part of the cells' genetic material. Ocular manifestations of viral infections vary greatly depending on the type of virus infected and the tissue involved. Some viruses mostly infect only the ocular tissues in the anterior chamber like cornea, conjunctiva, lens and iris-ciliary body, whereas some restrict themselves to the tissues of the posterior segment like retina and choroid. The main problem associated with these viral infections is latency and recurrent infections. Following the primary infection viruses like herpes simplex will be transported into the trigeminal ganglia where the viral particles remain dormant in the latent state (10). Various stimuli reactivate the latent virus that can re-infect the tissues. This process leads to recurrence of the disease which is more vision – threatening and painful. The treatment and the class of antiviral agent indicated depends on the tissue type as well as the type of virus that infects. Most of the infections occurring in the anterior chamber are usually treated with topical antivirals, mostly administered as drops and some times as ointments depending on the physico-chemical properties of the drug molecules. Systemic administration (intravenous and oral) can also achieve therapeutic concentrations in the anterior chamber but may not in posterior chamber, due to restricted movement of drug molecules across the blood–retinal barrier. For the treatment of viral infections in the posterior segment, antivirals are administered either as intravitreal or subconjunctival injections. The following section describes viral infections that frequently occur in the ocular tissues.

Herpes Simplex Virus Infections

Two types of herpes simplex viruses (HSVs) can cause infections, HSV-1 and HSV-2. HSV-1 primarily affects the cornea causing herpes simplex keratitis. A variety of clinical manifestations such as infectious epithelial keratitis, follicular conjunctivitis, neurotrophic keratopathy, necrotizing stromal keratitis, immune stromal keratitis, and endothelitis are associated with HSV keratitis (11). Neurotrophic keratopathy results mainly from decreased corneal innervations and tear secretion as a result of a prior HSV infection of the sensory nerves. The infection may be superficial involving only the outer layers of corneal epithelium which usually do not scar the cornea and can be treated with topical antivirals. Following the primary infection, the virus initially replicates in the corneal epithelium and then eventually enters the axonal ganglia and then translocates to trigeminal ganglia where the virus remain dormant in a latent state. Latency provides a viral reservoir that allows for spontaneous and recurrent reactivation of the disease (10). Various factors including sunlight, trauma, emotional stress, menstruation, and other infections may play a role in reactivation of the virus in the trigeminal nerve. Following the initial episode of HSV keratitis, there is approximately a 50% chance of recurrence within 2 years (12). This reactivation leads to secondary infection that can cause dendritic keratitis which is the most common cause of corneal opacity in developed countries (13). Eventually it also may involve the deeper layers of cornea like stroma causing stromal keratitis. This infection can lead to blindness as it can result in the formation of deep ulcers and visually significant scars on corneal surface. The most common symptoms associated with keratitis are photophobia, blurred vision, tearing, and redness of eye.

Treatment

Various therapeutic agents, such as 5-iodo-2′-deoxyuridine (IDU) (14), phosphano acetic acid (15), trifluorothymidine (TFT) (16), acyclovir (ACV) (14), penciclovir (17), gancoclovir (18), cyclosporine (19), steroids (20), interleukun-10 (21), and trisodium phosphonoformate (foscarnet) (22) were shown to be effective against herpes keratitis. Irrespective of the agent selected to treat the disease the ultimate goal of the medication recommended is to completely eradicate live virus in the cornea. To prevent the recurrence the antiviral agent used should be able to completely eliminate the virus even from the sensory ganglion where the virus remains dormant in a latent state. Keratitis primarily being a corneal disease topical administration of antivirals is the most preferred route of delivery. However, to prevent any recurrences it can also be administered by intravenous and oral routes (23,24).

IDU a thymidine analogue which exerts its action by competitively inhibiting the uptake of thymidine into viral DNA producing a faulty DNA chain which cannot replicate (25). The disadvantages with IDU include its toxicity, poor solubility low corneal penetration, resistance development and failure to effectively treat stromal keratatis (25). IDU has been replaced by more potent drugs like TFT and ACV (26). Other potent antivirals like ganciclovir, foscarnet, bromovinyldeoxyuridine, and cidofovir are also indicated in viral keratitis (26).

TFT acts by inhibiting the thymidylate synthetase and is incorporated into viral DNA, thereby inhibiting the transcription and translation processes. Even though the agent is effective than IDU, long-term treatment with TFT is dose limiting because of its cytotoxicity (27). Currently TFT is available as 1% topical ophthalmic solution.

ACV, previously called acycloguanosine (ACG), is a new, specific, and potent antiviral drug (9-(2-hydroxyethoxymethyl) guanine), which is taken up preferentially by HSV-infected cells (28). The drug acts through a series of phosphorylation reactions, which is initiated by viral specific thymidine kinase followed by two phosphorylation steps by host cellular kinases resulting in formation of ACV-triphosphate. Acyclovir triphosphate competes with 2-deoxyguanosine triphosphate as a substrate for viral DNA polymerase, as well as it acts as a chain terminator. Thus it adversely affects the viral replication. ACV is well tolerated clinically and it appears to be highly potent in inhibiting HSV replication (28). In spite of this high efficacy and low toxicity topical ophthalmic solutions are not available due to poor solubility and low corneal permeability. Several attempts have been made to develop an acyclovir ophthalmic topical formulation. Some of the anatomical constraints including blinking tear turnover and conjunctival absorption cause lower corneal absorption. In Europe, a 3% ACV ophthalmic ointment was reported for treatment of superficial herpes keratitis. It was applied 5 times a day for 14 days (29). However such a formulation is not available in United States probably due to various problems like greasiness, floaters, irritation, variable absorption, and blurred vision associated with ointments. Moreover the ointment formulation is not very effective against stromal keratitis (29). Various formulation approaches including acyclovir encapsulated in liposomes have been investigated for topical delivery. Cationic liposomes have significantly enhanced the corneal absorption of ACV relative to free ACV and anionic liposomes (30). Acyclovir encapsulated in positively charged liposomal formulation has resulted in higher aqueous humor levels of ACV in comparison with the ACV ointment (2). The PEG-coated polyethyl-2-cyanoacrylate nanospheres encapsulated acyclovir showed a significant increase in drug levels (25-fold) in aqueous humor relative to the free drug or a physical mixture of acyclovir with nanospheres (31). A positively charged phospholipid mixture that acts as a permeation enhancer has also improved the corneal permeability and thus ocular bioavailability of acyclovir (32).

A wide variety of prodrugs of acyclovir have been designed and evaluated for antiviral efficacy and corneal permeability. A series of lipophilic acyl esters prodrugs of acyclovir were studied. The study showed that higher the lipophilicity resulted in elevated corneal permeability (33). Inspite of achieving higher corneal permeability there was a significant reduction in the solubility which does not allow the preparation of 1–3% topical ophthalmic formulations (1).

Transporter targeted prodrug delivery has gained immense popularity where the promoiety is attached to the parent molecule such that the nutrient transporters on cell membranes recognize and translocate the conjugates. Various transporters like peptide and aminoacid transporters have been targeted on the eye (27). Peptide transporters (PEPT) have gained immense popularity due to their high capacity and broad substrate specificity. PEPT present on the cornea has been targeted with conjugates like Valine–acyclovir as well as various dipeptide conjugates of acyclovir. These prodrugs have significantly enhanced the corneal permeability as well as the solubility of acyclovir (27). Based on the *in vivo* ocular pharmacokinetics in rabbits, glycine–valine acyclovir and valine–valine acyclovir appear to be promising candidates for further clinical studies in the treatment of herpes simplex infections (34).

In addition to topical route, other routes of administrations have also investigated to treat herpes keratitis. The most promising being the intravenous and oral routes. The highly bioavailable prodrugs like valacyclovir, famciclovir (35) and other dipeptide derivatives of acyclovir (34) provide more favorable pharmacokinetic parameters resulting in reduced dosing frequency and improved therapeutic profiles.

Varicella-Zoster Virus Infections

This infection is caused by varicella-zoster virus (VZV), which is a member of the herpes virus family. It commonly occurs as a childhood infection. The infection frequently causes scars on the eyelids. If left untreated it may lead to serious complications like gangrene of the eyelids, corneal melting, extraocular muscle palsies and optic neuritis (10). Similar to HSV, following primary infection VZV also remains latent in the dorsal root ganglion and then gets reactivated by various stimuli which causes secondary infection (36). The important clinical manifestation of this infection is herpes zoster ophthalmicus. This occurs when the latent VZV is reactivated from ophthalmic trigeminal ganglion. The infection occurs by the secondary perineural and intraneural inflammation of sensory nerves (37,38), and is mainly limited to cornea. Immunocompromised patients, especially those with HIV/AIDS, are more prone to herpes zoster ophthalmicus (39,40). Other ocular complications include: blepharoconjunctivitis, uveitis, sclereitis, retinal necrosis, secondary glaucoma, cataract, and post herpetic neuralgia (38,41).

Treatment

Acyclovir, famciclovir (prodrug of penciclovir) (42) and foscarnet (trisodium phosphonoformate hexahydrate) (43) have been used to treat these infections. These antiviral agents act by inhibiting viral DNA polymerase. Foscarnet exerts its antiviral activity by a selective inhibition at the pyrophosphate binding site on viral-specific DNA polymerases. Unlike acyclovir, foscarnet does not require activation (phosphorylation) by thymidine kinase or other kinases. Due to its poor oral bioavailability foscarnet is administered by the intravenous route. Acyclovir administered as intravenous infusion is a common method of treatment for serious VZV infections in immunocompromised patients (44). Topical corticosteroids may be advantageous in alleviating the inflammations and the immune reactions in the corneal and uveal manifestations (10). The last option available is the corneal transplantation (45).

Epstein-Barr Virus Infections

Epstein-Barr virus infection (EBVs) also belongs to the herpes family. It is transmitted through oral secretions. The virus affects the neuro-ophthalmologic and anterior segment of the eye. It encompasses a wide range of clinical symptoms that include follicular conjunctivitis, dendritic epithelial and stromal keratitis (10,46). Granular or ring shaped lesions may be found distributed through out the cornea and is also associated with corneal vascularization (46). EBV infections of retina are not very common.

Adenovirus Infections

The two most common contagious ocular diseases caused by adenovirus are epidemic keratoconjunctivitis (EKC), and pharyngo conjunctival fever (PCF). The first and most common sign is simple follicular conjunctivitis. EKC is characterized by the presence of corneal lesions, eyelid edema, conjunctival pseudomebrane formation and hemorrhage (10,47). EKC is more serious as it involves both the cornea and conjunctiva and can also have long term affects on vision (48). EKC is followed by immune T cell mediated infiltration of the corneal stroma, leading to the formation of numerous white dots (sub-epithelial filtrates). This symptoms can last for several months that can interfere with normal vision (48). PCF is marked by concurrent pharyngitis and fever (10). Cases may be bilateral where the contralateral eye shows the symptoms in 3–7 days after initial infection in the first eye (10).

Treatment

Cidofovir a phosphonyl acyclic nucleotide is a broad spectrum antiviral agent with significant activity against adenovirus. Topical cidofovir treatment significantly has reduced the adenoviral load in rabbits model (49,50). Excellent activity of cidofovir has been attributed to its high corneal permeability and prolonged intracellular half-life (48,49). However, topical cidofovir has not been approved in USA due to its narrow efficacy/toxicity ratio. Topical cyclosporine (0.05%) also has shown to have some alleviating affects on steroid-resistant atopic keratoconjunctivitis (AKC) (51).

Ganciclovir also posses significant antiviral activity against specific adenovirus types in vitro (48). Other nucleoside analogues with potent anti-adenoviral activity tested in vitro are 2′ 3′–dideoxycitidine (ddC) also known as zalcitibine (52). and cycloferon (10-carboxymethyl-9-acridanone, CMA) (53). Other agents with potent anti adenoviral activity include endogenous microbicide N-chlorotaurine which acts by oxidation of thiols and amines (54,55). doxovir (CTC-96) a cobalt chelate (56).

Cytomegalovirus Infections

Human cytomegalovirus (HCMV) is an opportunistic herpes virus that can cause cytomegalovirus retinitis. It is much more predominant in immuno-compromised patients, especially in AIDS patients with a CD4 cell count of less than 50 cells per cubic millimeter (57). It is an infection, that, if left untreated, can cause partial or complete retinal destruction, ultimately leading to complete vision loss. It develops in about 40% of the patient population with AIDS (58). The disease occurs due to viral replication causing retinal inflammation. The disease initiates at deeper layers of retinal and spread outward towards retinal pigment epithelium (RPE) the site of viral replication and disease progression. The viral replication specifically occurs in RPE and retinal glial cells (59). The spread of disease to RPE from retinal endothelial cells is facilitated by disruption of blood–retinal barrier (60). Histological features of CMV retinitis are thick retinal necrosis and edema, a patch of fluffywhite retinal infiltrate with several areas of retinal hemorrhage (57). Patients with CMV retinitis may have floaters, flashes of light blurred vision, or blind spots (57).

Treatment

Ganciclovir (GCV; a nucleoside analogue), foscarnet (a pyrophosphate analogue), and cidofovir (a nucleotide analogue) are the three antivirals available for treatment of HCMV retinitis. These agents act by inhibiting the activity of viral DNA polymerase thus adversely affecting viral replication (42,43,61). GCV is converted into monophosphate form by viral encoded phosphotransferase in HCMV infected cells. This mono-phosphate is converted into its triphosphate form by cellular kinases. The triphosphate form acts as a competitive inhibitor of natural substrate deoxyguanosine triphosphate. Ganciclovir is a better substrate for viral phosphotransferase than ACV and the cellular half-life of GCV triphos-phate is about 12 hours, which is much longer than 1–2 hours of acyclovir triphosphate. This is the primary reason for higher antiviral efficacy of GCV compared to ACV against HCMV retinitis.

Topical delivery with drops usually does not achieve therapeutic concentrations in the eye. Delivery of hydrophilic molecule like GCV to the posterior segment through sclera and choroid following IV or oral administration is a challenging task. GCV also has lower cellular permeability. In 80–90% of patients i.v. administration of GCV or foscarnet twice daily (induction therapy) for 14–21 days was found to be effective in reducing the viral load (62). After the induction period the maintenance therapy is continued with daily intravenous infusion (62). Despite this pharmacotherapeutic regimen, patients suffer from multiple reactivations causing further damage to retina. Local intravitreal injections of GCV are also effective (63) but short vitreal half-life of GCV requires weekly injections to maintain the therapeutic levels which can cause retinal detachment and vitreous hemorrhage. Various approaches have been investigated to achieve sustained GCV levels for prolonged periods of time.

Intraocular ganciclovir implant: An implant has been developed with 4.5 mg of GCV that can continuously release the drug at the rate of 1.4 µg per hour into vitreous for up to 8 months (62,64,65). This implant can clearly maintain substantially higher levels of GCV for long periods of time. Surgery is needed to insert and remove the implant. Adverse consequences due to this procedure may include retinal detachment, intravitreal bleeding and endophthalmitis (64). Use of ocular implant alone in the absence of any systemic anti CMV treatment may result in any extra ocular infection and contra lateral retinitis (62). Thus concomitant systemic treatment with either oral or intravenous GCV is necessary to prevent these secondary infections.

Application of biocompatible and biodegradable polymers in implants is a viable strategy as the polymers will degrade inside the ocular structures thus eliminating the need for surgical removal of the implant (66).

Oral delivery of ganciclovir has been investigated for the treatment of CMV retinitis. Similar to ACV, ganciclovir has poor oral bioavailability. A dose of 900 mg of ganciclovir results in blood levels similar to those obtained with a dose of 5 mg/kg of body weight given intravenously (67). Orally bioavailable prodrugs like Valganciclovir (Val-Gcv) a monovalyl ester prodrug of ganciclovir have improved he oral bioavailability and patient compliance. Eventually several dipeptide monoesters of GCV were also evaluated for oral bioavailability and anti-CMV activity.

Cidofovir is an antiviral nucleotide analogue with potent activity against CMV (68). Intravenous cidofovir has also shown to be efficacious in delaying the treatment and progression of CMV retinitis (69–72). A single 20 µg intravitreal injection of cidofovir resulted in prolonged arrest of CMV progression (73). The usage of intravenous cidofovir is limited by severe irreversible nephrotoxicity (74) which can be reduced by concomitant administration of probenicid. It also may cause anterior uveitis/iritis and hypotony.

BACTERIAL INFECTIONS

Ocular tissues in both anterior and posterior segments are prone to a wide variety of bacterial infections. Both the gram positive and gram negative species can infect the ocular structures. Most of these infections occur either after a surgery or during trauma. Staphylococcus species are the most common bacterial pathogens responsible for various ocular bacterial infections. These infections produce inflammation, a process generated mostly by the immune system. The symptoms include redness from an increased blood supply, swelling from fluid accumulating in the tissues, pain from nerve irritation and the ultimately vision loss. Even though the infections are caused by different bacterial species, most of them are treated with wide spectrum antibiotics. Antibiotics may be either bactericidal (kills the bacteria) or bacteriostatic (inhibit the growth of bacteria).These agents divided into various classes based on their chemical structure (Table 1). Most of them exert their action is either by inhibiting the cell wall synthesis or by blocking the protein synthesis. Most of the antibiotics have broad spectrum activity where they kill or inhibit a wide range of gram positive or gram negative bacteria.

The delivery of most of these antibiotics to the anterior chamber is also a difficult task as most of them are the substrates for efflux proteins like P-glycoprotein present on cornea (75). These efflux pumps on the cornea restrict the absorption of these antibiotics into cornea, resulting in reduced ocular bioavailability (76). Various strategies have been implemented to improve the corneal permeability of these antibiotics. Both the addition of efflux pump inhibitors and prodrug design have shown to be viable strategies in enhancing the corneal permeability and ocular bioavailability of these antibiotics (77). In a transporter mediated prodrug strategy, the efflux substrate is modified chemically such that it will have diminished affinity towards the efflux transporter but high affinity for an influx transporter. In this way it can circumvent the P-GP mediated efflux. Similar strategies have been employed for delivery of these antibiotics either intravitreally or subconjuntivally for the treatment of posterior chamber infections. In addition, bacterial resistance is also a major obstacle which could lead to treatment failure (78). Antibiotic resistance in bacteria may be an inherent trait of the organism (e.g. a particular type of cell wall structure) that renders it naturally resistant, or it may be acquired by means of mutation in its own DNA or by acquisition of resistance-conferring DNA from another source (79). This problem can be overcome by appropriate use of antibiotic regimens and constant development of novel antibacterial agents which are not substances for efflux proteins.

Blepharitis

Blepharitis is one of the most common bacterial infections of the eyelids and lachrymal apparatus and is caused by *Staphylococcus epidermis* and

Table 1 Classification of Antibiotics

Class	Mechanism of action	Examples	Notes
Aminoglycosides	Inhibit bacterial protein synthesis (translation)	Streptomycin, gentamicin, amikacin	Bactericidal and particularly active against aerobic and gram-negative bacteria.
Beta-lactums	Inhibit bacterial cell wall synthesis by binding penicillin-binding proteins (PBPs) located on the bacterial cell wall.	Penicillins, cephalosporins, carbapenems, monobactums	(1) Contains Betalactum ring (2) Bactericidal and active against gram-positive bacteria
Macrolides	Inhibit bacterial protein synthesis (translation)	Erythromycin, azithromycin, clarithromycin,	Effective against aerobic and anaerobic gram-positive organisms
Fluroquinolines	Inhibit bacterial DNA synthesis by inhibiting bacterial topomerase enzymes	norfloxacin, ciprofloxacin	(1) Broad spectrum antibiotics (2) Relatively less active against staphylococci and streptococci species.
Tetracyclines	Inhibit bacterial protein synthesis (translation)	Tetracycline, doxycycline	(1) Broad spectrum antibiotic (2) Contraindicated with preparations containing vitamins and minerals.
Sulphonamides	Block metabolic processes such as synthesis of folic acid, which is necessary for synthesis of nucleic acids	Co-trimoxizole, trimethaprim	(1) Synthetic antibacterials with sulfonamide group (2) Broad spectrum antibiotic

Lincosamides	Inhibit bacterial protein synthesis (translation)	Clindamycin, lincomycin	(1) Effective against anaerobic gram-positive bacteria (2) Bacteriostatic
Poly Peptides	Inhibit bacterial cell wall synthesis	Bacitracin, polymixin	(1) Bacitracin is active against gram-positive organisms (2) Polymixin is a cationic polypeptide that is active against gram-negative species
Chloramphenicol	Inhibit bacterial protein synthesis (translation)	Chloramphenicol	(1) Bacteriostatic and a broad spectrum antibiotic (2) No longer used due to bone marrow toxicity and resistance
Rifamycins	Inhibit bacterial DNA-dependent RNA polymerase and thus blocks the RNA synthesis	Rifampin, rifabutin	(1) Active against more gram-positive and some gram-negative species (2) Rifampin is rarely used alone due to bacterial resistance
Glycopeptides	Inhibit cell wall synthesis, prevent the transport of cell wall precursors from cytoplasm to the cell wall	Vancomycin	(1) Effective against gram-positive bacteria

S. aureus. If left untreated it can lead to keratitis and chronic conjunctivitis that are described in the subsequent sections.

Bacterial Keratitis

The infection develops in a vascular cornea. It is commonly caused by *S. aureus*, *S. pneumoniae*, and gram negative coliform bacteria. The infection can lead to corneal ulceration. The cornea as well as the eye lids becomes very edematous. The most noticeable feature is rapid progression of the disease that can result in corneal disintegration. Contact lens also serves as exogenous risk factor of ulcerative bacterial keratitis as it can affect the pre-corneal tear film and corneal epithelium (80). Usage of extended wear contact lenses can increase the risk (81). If left untreated it can lead to perforation of cornea that can result in permanent blindness. The common symptoms include severe pain, redness, decreased vision, and photophobia.

Treatment

At the initial stages of infection, broad spectrum antibiotics may be recommended. These antibiotics should be effective against most of the gram positive and gram negative organisms. The treatment should be initiated by the "shot gun" approach which is a combination of cephalosporin and amino glycosides. To achieve the therapeutic concentrations of antibiotics at the site of action "fortified" antibiotics are also recommended (82,83) such form of treatment can assist in achieving higher concentrations in the cornea in a relatively short time period. A loading regimen consisting of 2% tobramaycin and 5% cefamandole may be indicated. This regimen is continued until the specific bacterial strain is identified and then treatment may be gradually shifted to a single drug.

Fluoroquinolones (such as ciprofloxacin, ofloxacin, lomefloxacin and norfloxacin) have shown excellent activity against both gram-negative and gram-positive bacteria and are used frequently for the treatment of bacterial keratitis (84). Ciprofloxacin hydrochloride (0.3%) and ofloxacin (0.3%) have shown to be effective in treatment of bacterial keratitis (85). Subconjunctival delivery of antibiotics might achieve higher concentrations. Collagen shields are produced when the contact lenses, made out of animal collagen, are soaked in antibiotics, this approach can be used to prolong corneal drug levels (86).

Endophthalmitis

This infection involves the inflammation of intra ocular tissues (aqueous humor and vitreous humor). It can be caused by bacteria, fungi and viruses. Most cases of endophthalmitis are caused by bacteria mostly following ocular surgery. The most common causative organisms are *S. aereus*, *S. epidermidis*, and several other streptococcus species. The symptoms depend on the type of endophthalmitis. Post-operative endophthalmitis occurs mostly after the cataract surgery. It can occur by the infection caused either during the surgery or because of the surgical equipmemt that is not sterile (87). Depending on the severity it can also lead to loss of vision. The symptoms include red eye, swollen eyelids and blurred vision. Post-traumatic endophthalmitis is caused by penetrating eye injuries. It is also associated with red eye and swollen eyelids. Hematogenous endophthalmitis occurs when the systemic infection spreads to eye and settles in the ocular tissues. It causes mild decrease in vision and floaters in eye.

Treatment

Since infection is caused by bacteria corticosteroids and flouroquinoline derivatives are usually recommended. The route of administration depends on the type and severity of the disease. The regimen is administered by the intravenous route to treat hematogenous endophthalmitis. Topical antibiotics are given if there is a wound infection on the ocular segment. Corticosteroids are given to decrease inflammation (88). If the vision loss is severe, vitrectomy needs to be performed where a portion of the infected vitreous humor is replaced with sterile saline or other compatible fluid (89).

Conjunctivitis

Conjunctivitis is characterized by irritation, itching and tearing. Bacterial conjunctivitis usually unilateral usually does not spread to the contralateral eye. Staphylococcus species are the most common bacterial pathogens responsible for this disorder. Other species include *Streptococcus pneumoniae*, *Chlamydio trachomatis* and *Haemophilus influenza*. In children it is primarily caused by *H. influenza*, *S. pneumoniae* and *Moraxella catarrhalis* (90). Conjunctivitis can be classified as infectious and non-infectious types. Infectious conjunctivitis also called pinkeye is caused either by bacteria or viruses. It is highly contagious. Non-infectious conjunctivitis can be caused by allergies, pollen and underlying diseases. The common symptoms include red eye, pain, swelling of eyes, blurred vision, itching and gritty feeling. Acute conjunctivitis is self-limiting but it usually takes up to three weeks to treat it completely. Infection is often present in one eye and gradually spreads to the other. No single antibiotic can cover all possible pathogens responsible for conjunctivitis. Since most of the causative organisms are gram positive bacteria it would be appropriate to choose an antibiotic that can cover maximum number of gram positive species (91). Mostly commonly used broad spectrum antibiotics include erythromycin and bacitracin–polymyxin B ointments. Aminoglycosides like gentamicin, neomycin and tobramycin has more of gram negative coverage and are also relatively more toxic to cornea (91). The bacteriostatic agent sulfacetamide (10%) is also commonly prescribed for topical use. Tetracycline and chloramphenicol are also topically used antibiotics. However tetracycline is available only as an ointment. In addition fluroroquinoline derivatives are also commonly prescribed to treat bacterial conjunctivitis.

Hyper Acute Bacterial Conjunctivitis

This is a severe, vision-threatening ocular infection that requires immediate medical attention. The symptoms are characterized by yellow/green purulent discharge that can reaccumulate even after being wiped off (92). The infection is frequently caused by *N. gonorrehoeae* and *N. meningitidis* the former being the most common organisms. If left untreated it may lead to severe perforation and ulceration of cornea leading to partial or total vision loss. It is usually treated with systemic antibiotics. Cephalosporin and spectinimycin are recommended (91).

Chronic bacterial conjunctivitis: This ailment commonly caused by staphylococcus species (93). It often develops in conjunction with blepharitis. The symptoms include itching, burning and morning eyelash crusting (91). Some patients may even have recurrent styes and chalazia. (lipogranulomas) of the eye lid margin which is due to chronic inflammation of meibomian glands (meibomianitis). Treatment is similar to normal conjunctivitis which includes antibiotics such as fluoroquinolines.

Ocular Chlamydial Infections

Ocular Chlamydia trachomatis occur in two forms, that is, trachomatis (chronic keratoconjunctivtis) and inclusive conjunctivitis. Trachoma is the most common cause of ocular morbidity and blindness (91). Inclusive keratoconjunctivitis is a sexually transmitted disease. It is a common cause of conjunctivitis in neonates. Infants are generally exposed to *C. trachomatis* from the infected cervix of mother. It is usually treated with antibiotics and if necessary they are administered systemically (94). Oral azithromycin is also recommended as a cost-effective means of controlling endemic trachoma (95,96). Novel vaccine delivery systems have also been evaluated for trachoma treatment (97).

FUNGAL INFECTIONS

Fungi are organisms that lack chlorophyll but resemble plants. These organisms are saprophytic but can also utilize living matter. Fungi are subdivided into yeasts, which are unicellular and molds that are multicellular with filamentous hyphae. *Fusarium* species like *solani* and *oxysporum* are the most common fungi that cause keratomycosis. It predominantly occurs in tropical and sub-tropical regions, whereas *Candida* infections are observed more in temperate climates. *Aspergillus* species are ubiquitous and commonly seen in a variety of climates. Ocular manifestations of fungal infections include the formation of indolent ulcers which are whitish yellow with raised edges and finger like process extending into the corneal stroma. Moreover satellite lesions and pus can also be observed in the anterior chamber. The diagnosis of these infections can be done by corneal scrapings which are examined for histological changes and also by culture to identify the causative organism. It can be further confirmed by an aqueous specimen or corneal biopsy. The fungal infection if untreated can cause haemorrhage into the anterior chamber (98).

Treatment of ophthalmic fungal infection requires suitable antifungal agent. In 1970s various agents such as clotrimazole, econazole and flucytosine etc were widely employed. The effectiveness of a drug depends upon the concentration reached at the site of action which depends upon several factors such as route of administration, physiochemical properties of drug, solubility and permeation of drug across ocular tissue.

Various antifungals like amphotericin B, natamycin and ketaconazole have molecular weight more than 500 Da. Therefore, these drugs exhibit poor corneal permeability. Collagen shields shaped in the form of contact lens are used to achieve sustained delivery of various antifungal agents (99). It is prepared by rehydrating the solution containing water soluble drug such that it get trapped in the interstices of collagen matrix. Currently the only antifungal agent used as collagen shield is amphotericin B.

Pathogenesis of Fungal Infections

Fungal infections occur due to interaction of host with the fungus. The key factors involved in the pathogenesis involve adhesion, invasiveness, morphogenesis and toxigenicity.

Adhesion: Generally, carbohydrate and protein molecules on fungal surface interact with the host protein in a specific and saturable manner. Various binding proteins like fibronectin and collagen on the cornea interact with fungi to favor adhesion (100).

Invasiveness: Followed by adhesion the fungi invade the anterior chamber and form a lens iris-fungus mass, thereby interfering with the aqueous humor drainage that can lead to rise in the intraocular pressure. The extent of invasiveness depends upon agent factors such as heavy fungal load with deep penetration and host factors such as insufficient inflammatory response (101).

Morphogenesis: Fungi undergo morphological changes to adapt with the changing environment. Features like intrahyphal hyphae and thickened fungal cell walls changes frequently to counteract either host defense mechanism or an antifungal drug (102).

Toxigenicity: It is still unclear whether fungal toxins are responsible for any biological activity but various in vitro experiments suggest this possibility. Studies done on *A. flavus* and *F. solani* suggest that both exhibit metalloproteinase activity (103).

Host Factors Affecting Pathogenesis

Defects in ocular defense mechanism, due to a disease like herpes simplex keratitis, include: impaired tear secretion, defective secretion of immunoglobulin A in tears, and defective positioning of lids. These factors facilitate the progression of diseases. In addition to systemic diseases, immune disorders like HIV/AIDS are also involved in disease progression and aggravation. Polymorphonuclear leukocytes are important in preventing ocular infection by oxygen dependent phagocytic pathway. However, usage of drugs like antibiotics and steroids may lower immune response by inhibiting chemotaxis and ingestion by phagocytes, by blocking degranulation and by reducing the production of phagocytes. These factors further lower the host resistance to fungal infections (104).

Occasionally fungemia with saprophytic fungi will lead to endophthalmitis. It may be of endogenous or exogenous type depending upon the type and nature of infection.

Endogenous Endophthalmitis

Endogenous endophthalmitis occurs due to sustained use of broad spectrum antibiotics, corticosteroids and cytotoxic drugs. Intravenous drug abuse and application of medical device like intravenous catheters may potentiate fungal infections. The main organisms that are responsible for endogenous endophthalmitis are *candida* and *aspergillus*. However sometimes it may occur through dimorphic fungi like *H. capsulatum, Blastomyces dermatitidis* and *C. immitis* (105).

Exogenous Endophthalmitis

Exogenous endopthalmitis occurs by introduction of microorganism into eye from trauma or surgery. These patients are rarely immunocompromised. General procedures such as corneal transplantation may lead to postoperative exogenous endopthalmitis mainly by gram positive bacteria as well as fungi such as *C. glabrata* and *C. famata*.

The mycotic infections involving *fusarium* occur in post keratitis and post traumatic cases, whereas the *C. parapsiloris* occur in lens replacement cases. Fungal contamination of irrigating solution by *Paecilomyces lilacinus* may lead to postoperative infection since the organism mainly grows in sodium bicarbonate medium (106).

Fungal Keratitis

This form of keratitis is frequently caused by filamentous fungi such as species of *Fusarium*, *Aspergillus* and *Curvularia*, and by yeast-like fungi such as *Candida* (107). The symptoms include blurred vision, red and painful eye, increased sensitivity to light, and excessive tearing or discharge. Clinically, it is often difficult to differentiate between fungal and bacterial infections. Diagnosis can be done by corneal culture, tissue biopsy or confocal microscopy (108). The commonly used antifungal agents are fungistatic and require an intact immune system and a prolonged treatment. Without innate immunity, the fungistatic medications are likely to be less effective in suppressing the organism. Drug classes used to treat fungal keratitis include polyene antibiotics (nystatin, amphotericin), pyrimidine analogs (flucytosine), imidazoles (clortrimazole, miconozole, econazole, ketoconazole), triazoles (fluconazole, itraconazole), and silver sulfadiazine (109). The most commonly used antifungal agents are polyenes and azoles. Polyenes bind to ergosterol which is a unique content of fungal cytoplasmic membranes. These compound cause disruption of cytoplasmic membrane integrity resulting into leakage of cellular contents. Natamycin and amphotericin B are the currently recommended polyenes. Natamycin was the first antifungal developed specially for eye which is effective against various fungal species. It is poorly soluble in water therefore available as 5% ophthalmic suspensions which permeate adequately through cornea upon topical application. It is widely recommended for treating fungal keratitis (110). Steroids are contraindicated as they will exacerbate the disease. Alternatives include amphotericin B 0.15% and flucytosine 1%, the same regimen that is recommended to treat yeast infections (108).

Amphotericin B has both fungistatic and fungicidal activity. Generally the antifungal can be administered by intravenous, topical and intravitreal routes. A 0.1 mg/mL dose in a 5% solution of dextrose is used for intravenous infusion. This regimen is recommended for treatment of endophthalmitis due to dimorphic fungi and fungal infections of orbit. A topical application of 0.15–0.3% solutions can be suggested. A 0.15% solution is well tolerated by cornea in fungal keratitis (108). Intravitreal injection of 1–5 µg amphotericin-B has been used for the treatment of fungal endophthalmitis (111).

Azoles bind to cytochrome P-450 which is involved in 14-alpha demethylation of lanosterol. This binding results in the reduction of ergosterol synthesis and accumulation of 14-methyl sterols leading to increased fluidity of fungal cell membrane causing fungal apoptosis. All azoles disrupt host immune response except fluconazole. Miconazole can be used topically or subconjuctivally (10 mg/mL) (112). Miconazole therapy can be effective in *C. albicans*, *A. fumagatis*, and *A. flavons* (113). Ketoconazole is the first orally absorbable antifungal agent but formulations for topical and subconjuctival applications are not available. Oral absorption is highly pH dependent therefore utmost care need to be taken while co-administrating with other agents (114). However several studies have reported the effectiveness of ketoconazole against *fusarium* spp., *aspergillus* spp. and *curvularia* spp. following oral administration. Itraconazole is hydrophobic and very poorly soluble in aqueous solution. The major drawback with itraconazole is its poor permeation inside the ocular tissues particularly cornea, aqueous humor and vitreous. However, itraconazole, either topically or systemically, is effective in treating fungal keratitis, particularly if the infections are due to *aspergillus* spp. or *curvularia* spp. (115). Recent studies have shown that topical econazole 2% appears to be as effective as natamycin 5% for managing fungal keratitis. However, the combined use of these products does not seem to have any advantage over that of natamycin 5% alone (110).

Fluconazole is stable, water soluble with low molecular weight, high oral bioavailability and low toxicity. Oral fluconazole has been recommended in treating patients with fungal keratitis and various other fungal infections (116).

In April 2006, FDA issued warning regarding use of ReNu with MoistureLoc® brand of contact lens as it may cause increase risk for fusarium keratitis.

Candida Infections

Candida can cause both exogenous and endogenous endophthalmitis. Candidal exogenous endophthalmitis is comparatively rare and usually occurs following eye surgery (keratoplasty, cataract extraction), traumatic ulcers, and contaminated ophthalmic irrigation solutions (117). *Candida* spp. commonly causes endogenous endophthalmitis particularly in the retina producing both retinal and vitreal lesions (118). It also causes cotton wool spots and white centered hemorrhages. Sometimes it can affect the contralateral eye structures producing conjunctivitis, episcleritis, iritis or iris abscesses.

Symptoms include decreased vision due to macular involvement or large vitreous lesions, eye discomfort, foreign body sensation, floaters, eye redness, and pain in cases with advanced iritis.

Diagnosis can be done by funduscopic examination with pupillary dilatation. Other methods of diagnosis are not specific. No antifungal agent is considered to be effective; however in case of vitreous infection, vitrectomy is performed. Intravitreal administration of amphotercin B is recommended at the time of performing vitrectomy (119). Fluconazole penetrates freely into the ocular tissues when they are inflamed. However studies in animal model with this drug are not promising. Role of intravenous voriconazole and caspofugin combination is still unclear.

CONCLUSION

Even though the occurance of ocular infections is common, proper diagnosis combined with usage of anti-infectives will reduce the risk. The major problem associated with viral infections is the reoccurrence, which can be prevented only by total elimination of virus particles from the body. No antiviral agent available can achieve this goal. Preventive measures should be taken for contagious infections like bacterial conjunctivitis to prevent the spread of the infection. The most important problem associated with the treatment of bacterial infections is emergence of resistance strains. It would be a major threat if bacteria became resistant to all the available classes of antibiotics. This possibility clearly emphasizes the need for development of a new generation of novel and potent antibiotics. Resistance can also be minimized by appropriate use of antibiotic regimen and by reducing usage of single antibiotics for prolonged periods of time. Most of the symptoms of fungal infections overlap with bacterial infections. Hence proper diagnosis should be done and if necessary histological examination or corneal biopsy should be performed to identify the causative organism to select an appropriate and specific anti-infective.

REFERENCES

1. Anand BS, Mitra AK. Mechanism of corneal permeation of L-valyl ester of acyclovir: targeting the oligopeptide transporter on the rabbit cornea. Pharm Res 2002; 19(8):1194–202.
2. Chetoni P, Rossi S, Burgalassi S, Monti D, Mariotti S, Saettone MF. Comparison of liposome-encapsulated acyclovir with acyclovir ointment: ocular pharmacokinetics in rabbits. J Ocul Pharm Ther 2004; 20(2):169–77.
3. Giannavola C, Bucolo C, Maltese A, et al. Influence of preparation conditions on acyclovir-loaded poly-d,l-lactic acid nanospheres and effect of PEG coating on ocular drug bioavailability. Pharm Res 2003; 20(4):584–90.
4. Vega E, Egea MA, Valls O, Espina M, Garcia ML. Flurbiprofen loaded biodegradable nanoparticles for ophtalmic administration. J Pharm Sci 2006; 95(11):2393–405.
5. Mainardes RM, Urban MC, Cinto PO, et al. Colloidal carriers for ophthalmic drug delivery. Curr Drug Targets 2005; 6(3):363–71.
6. Velez G, Whitcup SM. New developments in sustained release drug delivery for the treatment of intraocular disease. Br J Ophthalmol 1999; 83(11):1225–9.
7. Cunha-Vaz JG. The blood–ocular barriers: past, present, and future. Documenta Ophthalmol 1997; 93(1–2):149–57.
8. Pitkanen L, Ranta VP, Moilanen H, Urtti A. Permeability of retinal pigment epithelium: effects of permeant molecular weight and lipophilicity. Invest Ophthalmol Visual Sci 2005; 46(2):641–6.
9. Ambati J, Adamis AP. Transscleral drug delivery to the retina and choroid. Prog Retinal Eye Res 2002; 21(2):145–51.

10. Ritterband DC, Friedberg DN. Virus infections of the eye. Rev Med Virol 1998; 8(4):187–201.
11. Holland EJ, Schwartz GS. Classification of herpes simplex virus keratitis. Cornea 1999; 18(2):144–54.
12. McGill J, Fraunfelder FT, Jones BR. Current and proposed management of ocular herpes simplex. Surv Ophthalmol 1976; 20(5):358–65.
13. Dawson CR, Togni B. Herpes simplex eye infections: clinical manifestations, pathogenesis and management. Surv Ophthalmol 1976; 21(2):121–35.
14. Klauber A, Ottovay E. Acyclovir and idoxiuridine treatment of herpes simplex keratitis–a double blind clinical study. Acta Ophthalmol 1982; 60(5):838–44.
15. Gordon YJ, Lahav M, Photiou S, Becker Y. Effect of phosphonoacetic acid in the treatment of experimental herpes simplex keratitis. Br J Ophthalmol 1977; 61(8):506–9.
16. Wellings PC, Awdry PN, Bors FH, Jones BR, Brown DC, Kaufman HE. Clinical evaluation of trifluorothymidine in the treatment of herpes simplex corneal ulcers. Am J Ophthalmol 1972; 73(6):932–42.
17. Kaufman HE, Varnell ED, Thompson HW. Trifluridine, cidofovir, and penciclovir in the treatment of experimental herpetic keratitis. Arch Ophthalmol 1998; 116(6):777–80.
18. Colin J, Hoh HB, Easty DL, et al. Ganciclovir ophthalmic gel (Virgan; 0.15%) in the treatment of herpes simplex keratitis. Cornea 1997; 16(4):393–9.
19. Rao SN. Treatment of herpes simplex virus stromal keratitis unresponsive to topical prednisolone 1% with topical cyclosporine 0.05%. Am J Ophthalmol 2006; 141(4):771–2.
20. Wilhelmus KR, Gee L, Hauck WW, et al. Herpetic Eye Disease Study. A controlled trial of topical corticosteroids for herpes simplex stromal keratitis. Ophthalmology 1994; 101(12):1883–95; discussion 95–6.
21. Tumpey TM, Elner VM, Chen SH, Oakes JE, Lausch RN. Interleukin-10 treatment can suppress stromal keratitis induced by herpes simplex virus type 1. J Immunol 1994; 153(5):2258–65.
22. Chatis PA, Miller CH, Schrager LE, Crumpacker CS. Successful treatment with foscarnet of an acyclovir-resistant mucocutaneous infection with herpes simplex virus in a patient with acquired immunodeficiency syndrome. N Engl J Med 1989; 320(5):297–300.
23. Simon AL, Pavan-Langston D. Long-term oral acyclovir therapy. Effect on recurrent infectious herpes simplex keratitis in patients with and without grafts. Ophthalmology 1996; 103 (9):1399–404; discussion 404–5.
24. Fletcher CV, Englund JA, Bean B, Chinnock B, Brundage DM, Balfour HH, Jr. Continuous infusion of high-dose acyclovir for serious herpesvirus infections. Antimicrob Agents Chemother 1989; 33(8):1375–8.
25. Raju VK. Antiviral agents in herpes simplex keratitis. Indian J Ophthalmol 1982; 30(4):209–11.
26. Wilhelmus KR. The treatment of herpes simplex virus epithelial keratitis. Trans Am Ophthalmol Soc 2000; 98:505–32.
27. Anand B, Nashed Y, Mitra A. Novel dipeptide prodrugs of acyclovir for ocular herpes infections: Bioreversion, antiviral activity and transport across rabbit cornea. Curr Eye Res 2003; 26(3–4):151–63.

28. Trousdale MD, Dunkel EC, Nesburn AB. Effect of acyclovir on acute and latent herpes simplex virus infections in the rabbit. Invest Ophthalmol Visual Sci 1980; 19(11):1336–41.

29. Richards DM, Carmine AA, Brogden RN, Heel RC, Speight TM, Avery GS. Acyclovir. A review of its pharmacodynamic properties and therapeutic efficacy. Drugs 1983; 26(5):378–438.

30. Law SL, Huang KJ, Chiang CH. Acyclovir-containing liposomes for potential ocular delivery. Corneal penetration and absorption. J Control Release 2000; 63(1–2):135–40.

31. Fresta M, Fontana G, Bucolo C, Cavallaro G, Giammona G, Puglisi G. Ocular tolerability and in vivo bioavailability of poly(ethylene glycol) (PEG)-coated polyethyl-2-cyanoacrylate nanosphere-encapsulated acyclovir. J Pharmaceut Sci 2001; 90(3):288–97.

32. Montenegro L, Bucolo C, Puglisi G. Enhancer effects on in vitro corneal permeation of timolol and acyclovir. Die Pharmazie 2003; 58(7):497–501.

33. Hughes PM, Mitra AK. Effect of acylation on the ocular disposition of acyclovir. II: Corneal permeability and anti-HSV 1 activity of 2′-esters in rabbit epithelial keratitis. J Ocular Pharmacol 1993; 9(4):299–309.

34. Anand BS, Katragadda S, Mitra AK. Pharmacokinetics of novel dipeptide ester prodrugs of acyclovir after oral administration: intestinal absorption and liver metabolism. J Pharmacol Exp Therapeut 2004; 311(2):659–67.

35. Loutsch JM, Sainz B, Jr., Marquart ME, et al. Effect of famciclovir on herpes simplex virus type 1 corneal disease and establishment of latency in rabbits. Antimicrob Agents Chemother 2001; 45(7):2044–53.

36. Timbury MC, Edmond E. Herpesviruses. J Clin Pathol 1979; 32(9):859–81.

37. Naumann G, Gass JD, Font RL. Histopathology of herpes zoster ophthalmicus. Am J Ophthalmol 1968; 65(4):533–41.

38. Shaikh S, Ta CN. Evaluation and management of herpes zoster ophthalmicus. Am Fam Physician 2002; 66(9):1723–30.

39. Sandor E, Croxson TS, Millman A, Mildvan D. Herpes zoster ophthalmicus in patients at risk for AIDS. N Engl J Med 1984; 310(17):1118–9.

40. Snoeck R, Andrei G, De Clercq E. Current pharmacological approaches to the therapy of varicella zoster virus infections: a guide to treatment. Drugs 1999; 57(2):187–206.

41. Harding SP, Lipton JR, Wells JC. Natural history of herpes zoster ophthalmicus: predictors of postherpetic neuralgia and ocular involvement. Br J Ophthalmol 1987; 71(5):353–8.

42. Crumpacker C. The pharmacological profile of famciclovir. Semin Dermatol 1996; 15(2 Suppl 1):14–26.

43. Wagstaff AJ, Bryson HM. Foscarnet. A reappraisal of its antiviral activity, pharmacokinetic properties and therapeutic use in immunocompromised patients with viral infections. Drugs 1994; 48(2):199–226.

44. Seiff SR, Margolis T, Graham SH, O'Donnell JJ. Use of intravenous acyclovir for treatment of herpes zoster ophthalmicus in patients at risk for AIDS. Ann Ophthalmol 1988; 20(12):480–2.

45. Marsh RJ, Cooper M. Ocular surgery in ophthalmic zoster. Eye (London, England) 1989; 3(Pt 3):313–7.

46. Matoba AY, Wilhelmus KR, Jones DB. Epstein-Barr viral stromal keratitis. Ophthalmology 1986; 93(6):746–51.
47. Rajaiya J, Chodosh J. New paradigms in infectious eye disease: adenoviral keratoconjunctivitis. Arch Soc Espanola Oftalmol 2006; 81(9):493–8.
48. Kinchington PR, Romanowski EG, Jerold Gordon Y. Prospects for adenovirus antivirals. J Antimicrob Chemother 2005; 55(4):424–9.
49. Gordon YJ, Romanowski E, Araullo-Cruz T, De Clercq E. Pretreatment with topical 0.1% (S)-1-(3-hydroxy-2-phosphonylmethoxypropyl)cytosine inhibits adenovirus type 5 replication in the New Zealand rabbit ocular model. Cornea 1992; 11(6):529–33.
50. Romanowski EG, Bartels SP, Gordon YJ. Comparative antiviral efficacies of cidofovir, trifluridine, and acyclovir in the HSV-1 rabbit keratitis model. Invest Ophthalmol Visual Sci 1999; 40(2):378–84.
51. Akpek EK, Dart JK, Watson S, et al. A randomized trial of topical cyclosporin 0.05% in topical steroid-resistant atopic keratoconjunctivitis. Ophthalmology 2004; 111(3):476–82.
52. Mentel R, Kinder M, Wegner U, von Janta-Lipinski M, Matthes E. Inhibitory activity of 3'-fluoro-2' deoxythymidine and related nucleoside analogues against adenoviruses in vitro. Antiviral Res 1997; 34(3):113–9.
53. Zarubaev VV, Slita AV, Krivitskaya VZ, Sirotkin AK, Kovalenko AL, Chatterjee NK. Direct antiviral effect of cycloferon (10-carboxymethyl-9-acridanone) against adenovirus type 6 in vitro. Antiviral Res 2003; 58(2): 131–7.
54. Romanowski EG, Yates KA, Teuchner B, Nagl M, Irschick EU, Gordon YJ. N-chlorotaurine is an effective antiviral agent against adenovirus in vitro and in the Ad5/NZW rabbit ocular model. Invest Ophthalmol Visual Sci 2006; 47 (5):2021–6.
55. Nagl M, Larcher C, Gottardi W. Activity of N-chlorotaurine against herpes simplex- and adenoviruses. Antiviral Res 1998; 38(1):25–30.
56. Epstein SP, Pashinsky YY, Gershon D, et al. Efficacy of topical cobalt chelate CTC-96 against adenovirus in a cell culture model and against adenovirus keratoconjunctivitis in a rabbit model. BMC Ophthalmol 2006; 6:22.
57. Masur H, Whitcup SM, Cartwright C, Polis M, Nussenblatt R. Advances in the management of AIDS-related cytomegalovirus retinitis. Ann Intern Med 1996; 125(2):126–36.
58. Drew WL. Cytomegalovirus infection in patients with AIDS. J Infect Dis 1988; 158(2):449–56.
59. Scholz M, Doerr HW, Cinatl J. Human cytomegalovirus retinitis: pathogenicity, immune evasion and persistence. Trends Microbiol 2003; 11(4):171–8.
60. Bodaghi B, Goureau O, Zipeto D, Laurent L, Virelizier JL, Michelson S. Role of IFN-gamma-induced indoleamine 2,3 dioxygenase and inducible nitric oxide synthase in the replication of human cytomegalovirus in retinal pigment epithelial cells. J Immunol 1999; 162(2):957–64.
61. Lea AP, Bryson HM. Cidofovir. Drugs 1996; 52 (2):225–30; discussion 31.
62. Jacobson MA. Current management of cytomegalovirus retinitis in AIDS update on ganciclovir and foscarnet for CMV infections. Adv Exp Med Biol 1996; 394:85–92.

63. Cantrill HL, Henry K, Melroe NH, Knobloch WH, Ramsay RC, Balfour HH, Jr. Treatment of cytomegalovirus retinitis with intravitreal ganciclovir. Long-term results. Ophthalmology 1989; 96(3):367–74.

64. Martin DF, Parks DJ, Mellow SD, et al. Treatment of cytomegalovirus retinitis with an intraocular sustained-release ganciclovir implant. A random-ized controlled clinical trial. Arch Ophthalmol 1994; 112(12):1 531–9.

65. Sanborn GE, Anand R, Torti RE, et al. Sustained-release ganciclovir therapy for treatment of cytomegalovirus retinitis. Use of an intravitreal device. Arch Ophthalmol 1992; 110(2):188–95.

66. Duvvuri S, Janoria KG, Mitra AK. Development of a novel formulation containing poly(d,l-lactide-co-glycolide) microspheres dispersed in PLGA-PEG-PLGA gel for sustained delivery of ganciclovir. J Control Release 2005; 108(2–3):282–93.

67. Brown F, Banken L, Saywell K, Arum I. Pharmacokinetics of valganciclovir and ganciclovir following multiple oral dosages of valganciclovir in HIV- and CMV-seropositive volunteers. Clin Pharmacokinet 1999; 37(2):167–76.

68. Neyts J, Snoeck R, Schols D, Balzarini J, De Clercq E. Selective inhibition of human cytomegalovirus DNA synthesis by (S)-1-(3-hydroxy-2-phosphonyl-methoxypropyl)cytosine [(S)-HPMPC] and 9-(1,3-dihydroxy-2-propoxymethyl) guanine (DHPG). Virology 1990; 179(1):41–50.

69. Rahhal FM, Arevalo JF, Munguia D, et al. Intravitreal cidofovir for the maintenance treatment of cytomegalovirus retinitis. Ophthalmology 1996; 103 (7):1078–83.

70. Lalezari JP. Cidofovir: a new therapy for cytomegalovirus retinitis. J Acquir Immune Defic Syndr Hum Retrovirol 1997; 14(Suppl 1):S22–6.

71. Lalezari JP, Kuppermann BD. Clinical experience with cidofovir in the treatment of cytomegalovirus retinitis. J Acquir Immune Defic Syndr Hum Retrovirol 1997; 14(Suppl 1):S27–31.

72. Berenguer J, Mallolas J. Intravenous cidofovir for compassionate use in AIDS patients with cytomegalovirus retinitis. Spanish Cidofovir Study Group. Clin Infect Dis 2000; 30(1):182–4.

73. Kirsch LS, Arevalo JF, De Clercq E, et al. Phase I/II study of intravitreal cidofovir for the treatment of cytomegalovirus retinitis in patients with the acquired immunodeficiency syndrome. Am J Ophthalmol 1995; 119(4):466–76.

74. Plosker GL, Noble S. Cidofovir: a review of its use in cytomegalovirus retinitis in patients with AIDS. Drugs 1999; 58(2):325–45.

75. Dey S, Patel J, Anand BS, et al. Molecular evidence and functional expression of P-glycoprotein (MDR1) in human and rabbit cornea and corneal epithelial cell lines. Invest Ophthalmol Visual Sci 2003; 44(7):2909–18.

76. Dey S, Gunda S, Mitra AK. Pharmacokinetics of erythromycin in rabbit corneas after single-dose infusion: role of P-glycoprotein as a barrier to in vivo ocular drug absorption. J Pharmacol Exp Therapeut 2004; 311(1):246–55.

77. Katragadda S, Talluri RS, Mitra AK. Modulation of P-glycoprotein-mediated efflux by prodrug derivatization: an approach involving peptide transporter-mediated influx across rabbit cornea. J Ocul Pharmacol Ther 2006; 22(2):110–20.

78. Livermore DM. Bacterial resistance: origins, epidemiology, and impact. Clin Infect Dis 2003; 36(Suppl 1):S11–23.

79. Barker KF. Antibiotic resistance: a current perspective. Br J Clin Pharmacol 1999; 48(2):109–24.
80. Vallas V, Stapleton F, Willcox MD. Bacterial invasion of corneal epithelial cells. Aust New Zeal J Ophthalmol 1999; 27(3–4):228–30.
81. Tabbara KF, El-Sheikh HF, Aabed B. Extended wear contact lens related bacterial keratitis. Br J Ophthalmol 2000; 84(3):327–8.
82. Lauffenburger MD, Cohen KL. Topical ciprofloxacin versus topical fortified antibiotics in rabbit models of Staphylococcus and Pseudomonas keratitis. Cornea 1993; 12(6):517–21.
83. Lin CP, Boehnke M. Effect of fortified antibiotic solutions on corneal epithelial wound healing. Cornea 2000; 19(2):204–6.
84. Mah FS. Fourth-generation fluoroquinolones: new topical agents in the war on ocular bacterial infections. Curr Opin Ophthalmol 2004; 15(4):316–20.
85. Prajna NV, George C, Selvaraj S, Lu KL, McDonnell PJ, Srinivasan M. Bacteriologic and clinical efficacy of ofloxacin 0.3% versus ciprofloxacin 0.3% ophthalmic solutions in the treatment of patients with culture-positive bacterial keratitis. Cornea 2001; 20(2):175–8.
86. Callegan MC, Engel LS, Clinch TE, Hill JM, Kaufman HE, O'Callaghan RJ. Efficacy of tobramycin drops applied to collagen shields for experimental staphylococcal keratitis. Curr Eye Res 1994; 13(12):875–8.
87. Field D, Merrick E. Postoperative endophthalmitis: caution is the watchword. J Perioperat Pract 2006; 16(1):16–20.
88. Elston RA, Chattopadhyay B. Postoperative endophthalmitis. J Hospit Infect 1991; 17(4):243–53.
89. Das T, Sharma S. Current management strategies of acute post-operative endophthalmitis. Semin Ophthalmol 2003; 18(3):109–15.
90. Weiss A, Brinser JH, Nazar-Stewart V. Acute conjunctivitis in childhood. J Pediat 1993; 122(1):10–4.
91. Morrow GL, Abbott RL. Conjunctivitis. Am Fam Physician 1998; 57(4):735–46.
92. Mannis MJ, Sugar J. Syphilis, serologic testing, and the setting of standards for eye banks. Am J Ophthalmol 1995; 119(1):93–5.
93. Boustcha E, Nicolle LE. Conjunctivitis in a long-term care facility. Infect Control Hosp Epidemiol 1995; 16(4):210–6.
94. Sommer A. Systemic antibiotics for communitywide trachoma control. Arch Ophthalmol 2005; 123(5):687–8.
95. Tabbara KF, Abu-el-Asrar A, al-Omar O, Choudhury AH, al-Faisal Z. Single-dose azithromycin in the treatment of trachoma. A randomized, controlled study. Ophthalmology 1996; 103(5):842–6.
96. Schachter J, West SK, Mabey D, et al. Azithromycin in control of trachoma. Lancet 1999; 354(9179):630–5.
97. Igietseme J, Eko F, He Q, et al. Delivery of Chlamydia vaccines. Expert Opin Drug Delivery 2005; 2(3):549–62.
98. Jones DB, Green MT, Osato MS, Broberg PH, Gentry LO. Endogenous Candida albicans endophthalmitis in the rabbit. Chemotherapy for systemic effect. Arch Ophthalmol 1981; 99(12):2182–7.

99. Schwartz SD, Harrison SA, Engstrom RE, Jr., Bawdon RE, Lee DA, Mondino BJ. Collagen shield delivery of amphotericin B. Am J Ophthalmol 1990; 109(6):701–4.

100. Bouchara JP, Bouali A, Tronchin G, Robert R, Chabasse D, Senet JM. Binding of fibrinogen to the pathogenic Aspergillus species. J Med Vet Mycol 1988; 26(6):327–34.

101. Kuriakose T, Thomas PA. Keratomycotic malignant glaucoma. Indian J Ophthalmol 1991; 39(3):118–21.

102. Thomas MA, Kaplan HJ. Surgical removal of subfoveal neovascularization in the presumed ocular histoplasmosis syndrome. Am J Ophthalmol 1991; 111(1):1–7.

103. Zhu WS, Wojdyla K, Donlon K, Thomas PA, Eberle HI. Extracellular proteases of Aspergillus flavus. Fungal keratitis, proteases, and pathogenesis. Diagn Microbiol Infect Dis 1990; 13(6):491–7.

104. Stern GA, Buttross M. Use of corticosteroids in combination with antimicrobial drugs in the treatment of infectious corneal disease. Ophthalmology 1991; 98(6):847–53.

105. Klotz SA, Penn CC, Negvesky GJ, Butrus SI. Fungal and parasitic infections of the eye. Clin Microbiol Rev 2000; 13(4):662–85.

106. Pflugfelder SC, Flynn HW, Jr., Zwickey TA, et al. Exogenous fungal endophthalmitis. Ophthalmology 1988; 95(1):19–30.

107. Iyer SA, Tuli SS, Wagoner RC. Fungal keratitis: emerging trends and treatment outcomes. Eye Contact Lens 2006; 32(6):267–71.

108. Thomas PA. Fungal infections of the cornea. Eye (London, England) 2003; 17 (8):852–62.

109. Ganegoda N, Rao SK. Antifungal therapy for keratomycoses. Expert Opin Pharmacother 2004; 5(4):865–74.

110. Prajna NV, Nirmalan PK, Mahalakshmi R, Lalitha P, Srinivasan M. Concurrent use of 5% natamycin and 2% econazole for the management of fungal keratitis. Cornea 2004; 23(8):793–6.

111. Shah CV, Jones DB, Holz ER. Microsphaeropsis olivacea keratitis and consecutive endophthalmitis. Am J Ophthalmol 2001; 131(1):142–3.

112. Garcia deLomas J, Fons MA, Nogueira JM, Rustom F, Borras R, Buesa FJ. Chemotherapy of Aspergillus fumigatus keratitis: an experimental study. Mycopathologia 1985; 89(3):135–8.

113. Fitzsimons R, Peters AL. Miconazole and ketoconazole as a satisfactory first-line treatment for keratomycosis. Am J Ophthalmol 1986; 101(5):605–8.

114. Thomas PA. Current perspectives on ophthalmic mycoses. Clin Microbiol Rev 2003; 16(4):730–97.

115. Rajasekaran J, Thomas PA, Kalavathy CM, Joseph PC, Abraham DJ. Itraconazole therapy for fungal keratitis. Indian J Ophthalmol 1987; 35(5–6): 157–60.

116. Thakar M. Oral fluconazole therapy for keratomycosis. Acta Ophthalmol 1994; 72(6):765–7.

117. Aguilar GL, Blumenkrantz MS, Egbert PR, McCulley JP. Candida endophthalmitis after intravenous drug abuse. Arch Ophthalmol 1979; 97(1): 96–100.

118. Rodriguez-Adrian LJ, King RT, Tamayo-Derat LG, Miller JW, Garcia CA, Rex JH. Retinal lesions as clues to disseminated bacterial and candidal infections: frequency, natural history, and etiology. Medicine 2003; 82(3):187–202.
119. Barza M. Treatment options for candidal endophthalmitis [editorial; comment]. Clin Infect Dis 1998; 27(5):1134–6.

6

Strategies for Delivery of Cancer Chemotherapy

David Taft and Xudong Yuan

*Division of Pharmaceutical Sciences, Long Island University,
Brooklyn, New York, U.S.A.*

INTRODUCTION

Cancer is a class of disorders characterized by abnormal growth of cells that proliferate in an uncontrolled way. Cancer cells are able to spread in the body by two processes: invasion and metastasis. Invasion involves direct growth of cancer into adjacent tissue. Metastasis is a process where cancer cells are transported through the bloodstream or lymphatic system, resulting in implantation into distant sites (organs, tissues).

Cancer is the second leading cause of death worldwide. Globally an estimated 11 million new cancer cases occur each year (1). Skin cancer is the most common type of malignancy (Table 1). Prostate cancer and breast cancer are the second most common types of cancer in men and women, respectively. Once diagnosed, cancer is usually treated with a combination of surgery, chemotherapy, and radiotherapy.

Cancer research is aimed at the discovery of magic bullets, strategies that specifically target cancer cells with no effects on healthy organs and tissues. Over the past decade, significant progress has been made in the development of new chemical approaches for treating cancer. These approaches include not only identifying new agents that are effective against cancer, but also crafting new ways to deliver both old and new therapeutic agents with the goal of targeted drug therapy. Therapeutic outcomes—enhancing quality of life and extending life expectancy—are improved by this targeting ability of treatment.

Table 1   List of Most Common Cancer Types[a]

Cancer type	Estimated new cases (yearly)	Estimated deaths (yearly)
Skin (non-melanoma)	>1,000,000	<2,000
Prostate	218,890	27,050
Lung	213,380	160,390
Breast	178,480	40,460
Colon and rectal	153,760	52,180
Bladder	67,160	13,750
Non-Hodgkin's lymphoma	63,190	18,660
Melanoma	59,940	8,110
Leukemia (all)	44,240	21,790
Kidney (renal dell)	43,512	10,957
Endometrial	39,080	7,400
Pancreatic	37,170	33,370
Thyroid	33,550	1,530

[a] Data are reported for most common cancer types in the United States.
Source: From Ref. 2.

This chapter provides an overview of drug delivery in cancer chemotherapy. The chapter begins with an overview of existing cancer drug therapy and emerging therapeutic approaches to treat cancer. This is followed by a review of the various barriers (physiologic, physicochemical, and pharmacokinetic) that must be breached in order to deliver medications to tumor cells. Next, drug delivery strategies that have been devised to overcome these barriers are presented together with relevant examples. While much of the research in this area remains at preclinical stages, formulation scientists have successfully applied various technologies to develop delivery systems that have optimized therapeutic outcomes across several types of cancers. These examples are provided later in the chapter. The chapter concludes with a look to the future of cancer drug therapy, the challenges that remain and the opportunities that exist for expanded development of clinically effective formulations for treating patients with malignancy.

OVERVIEW OF CANCER DRUG THERAPY

Drug treatment for cancer is traditionally called "chemotherapy," but this term refers to the use of cytotoxic compounds to destroy cancer cells. While these medications are still widely used to treat cancer, recently developed therapies cause cancer cell death (apotosis) through other biological pathways. Currently, cancer drug therapy can be classified as chemotherapy, hormonal therapy, and biological therapy. These are briefly summarized below.

Chemotherapy

Since cancer cells divide more rapidly than normal cells, traditional chemotherapy involves treating patients with cytotoxic compounds designed to kill growing cells. The general approach of all chemotherapy is to decrease the growth rate (cell division) of the cancer cell. These medications target various phases of the cell cycle (Fig. 1), resulting in one of the following effects: damaging cellular DNA, inhibiting DNA synthesis, or stopping mitosis (splitting of the original cell into two new cells).

A classification of cancer chemotherapeutic agents is presented in Table 2. Overall, there are more than 50 different chemotherapic drugs. While these medications may be administered alone, often a combination of two, three, or more drugs is used in cancer therapy.

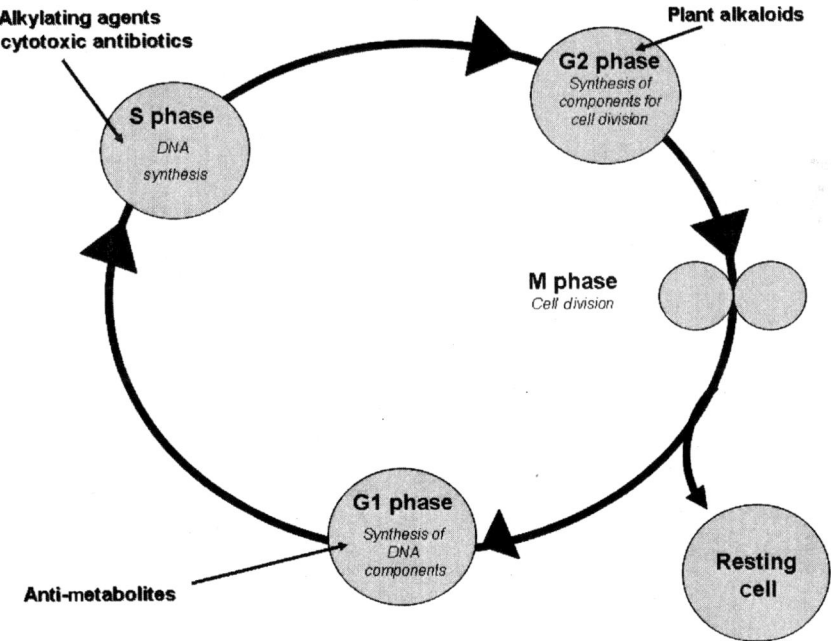

Figure 1 Schematic representation of the cell cycle showing phases targeted by various classes of chemotherapeutic agents.

Table 2 Classification of Anti-Cancer Agents

Classification	Mechanism	Examples
Alkylating agents	Stop tumor growth by cross-linking guanine nucleobases in DNA double-helix strands, directly attacking DNA	Nitrogen mustards: Cyclophosphamide, chlorambucil Nitrosureas: carmustine Platinum: cisplatin, carboplatin, oxalplatin Others: bsulfan, temozolomide
Antimetabolites	Interfere with the production of nucleic acids by preventing synthesis of normal nucleoside triphosphates, resulting in decreased DNA or RNA synthesis	Folic acid: methotrexate Purine: clofarabine, fludarabine, cladribine, pentostatin, mercaptopurine Pyrimidine: cytarabine, fluorouracil, gemcitabine
Cytotoxic antibiotics	Inhibit DNA and RNA synthesis by intercalating between base pairs of the DNA/RNA strand	Anthracyclines: doxorubicin, daunorubicin, mitoxantrone Others: bleomycin, hydroxyurea
Plant alkaloids	Bind to tubulin, inhibiting microtubule and mitotic spindle formation. Cell division is stopped	Taxanes: docetaxel, paclitaxel Vinca: vinblastine, vincristine,
Topoisomerase inhibitors	Inhibit enzymes blocks topoisomerase enzymes (I, II), which are involved in DNA structure and cell growth	Topotecan, irinotecan, etoposide, teniposide

.The majority of drugs currently on the market are not specific, which leads to the many common side effects associated with cancer chemotherapy. Because the common approach of all chemotherapy is to decrease the growth rate (cell division) of the cancer cells, adverse reactions typically involve organs and tissues that have a rapid turnover of cells such as blood cells forming in the bone marrow and cells in the digestive tract, reproductive system, and hair follicles.

Hormonal Therapy

Hormones are naturally occurring substances produced by endocrine glands (e.g. testes, ovaries) that act as chemical messengers to influence the growth and activity of cells. Hormonal therapies act by altering the production or activity of particular hormones in the body. Hormonal therapy is most commonly used to treat breast and prostate cancer.

Drugs used in hormone therapy for prostate cancer inhibit the action or block the production of testosterone and other male hormones, thereby slowing tumor growth (3). Medications include gonadotropin-releasing hormone agonists (block testosterone production) such as leuprolide acetate (Lupron®) goserelin acetate (Zolade®) an histrelin acetate (Vantas® Implant). Androgen antagonists include flutamide (Eulexin®), bicalutamide (Casodex®) and nilutamide (Nilandron®). Alternatively, estrogen therapy involves the use of the female hormone diethylstilbestrol to suppress the production of testosterone.

For patients with breast cancer, there are two types of hormone therapy. Selective estrogen receptor modulators are medications that inhibit estrogen and progesterone from promoting breast cancer cell growth (4,5) Examples include tamoxifen and toremifine (Fareston®). Additionally, there are several compounds that turn off the production of estrogen by ihibiting the enzyme aromatase. Medications in this category include anastrozole (Armidex®), letrozole (Femara®) and exemestane (Aromasin®).

Biological Therapy

Over the past few years, a number of novel new cancer medicines have been marketed that have collectively been termed biological therapy. Biological therapy encompasses those strategies that employ substances that occur naturally in the body to destroy cancer cells. Compared to chemotherapy, which exert cytotoxic effects on both cancer and normal cells, biological therapy is regarded as targeted therapy because its effects are directed toward the cancer cell itself. Medicines in this category include monoclonal antibodies, protein kinase inhibitors, and proteosome inhibitors (Table 3).

Table 3 Biological Cancer Therapy

Classification	Examples	Cancer indication
Monoclonal antibodies	Alemtuzumab (Campath®)	Leukemia
	Bevacizumab (Avastin®)	Lung
	Cetuximab (Erbitux®)	Head/neck
	Gemtuzumab (Mylotarg®)	Leukemia
	Rituximab (Rituxan®)	Lymphoma
	Trastuzumab (Herceptin®)	Breast
	Panitumumab (Vectibix®)	Colorectal
Tyrosine kinase inhibitors	Imatinib (Gleevec®)	Leukemia
	Dasatinib (Sprycel®)	Leukemia
	Sunitinib (Sutent®)	GI, renal
EGFR kinase inhibitors	Gefitinib (Iressa®)	Lung
	Erlotinib (Tarceva®)	Pancreatic, lung
Raf kinase/VEGF kinase/ angiogenesis inhibitor	Sorafenib (Nexavar®)	Renal
Proteosome inhibitor	Bortezomib (Velcade®)	Lymphoma, myeloma
Histone deacetylase inhibitor	Vorinostat (Zolinza®)	Lymphoma

Monoclonal Antibodies

Humans have the ability to make antibodies able to recognize and bind to an antigen (a molecule that stimulates an immune response). While antibodies help the body fight infection, antibodies also represent a useful approach target other types of molecules found in the body, including cancer cells.

For cancer therapy, monoclonal antibodies (produced from a single cell) work by recognizing the protein on the surface of the cancer cell and then locking onto it. Anticancer antibodies work by one of the following mechanisms: (1) binding to tumor cells either specifically or selectively, causing their death; (2) interfering with signals transmitted from receptors for growth factors; (3) preventing angiogenesis, growth of new blood vessels from pre-existing vessels; (4) enhancing the cellular immune response against cancer by binding to molecules on the surface of immune system cells (6).

Protein Kinase Inhibitors

Protein kinases play important roles in regulating cellular function in normal and neoplastic cells. These enzymes function as components of signal transduction pathways, playing a role in the control of cell growth, metabolism, differentiation, and apoptosis. Numerous kinases are deregulated in human cancers including Bcr-Abl, HER (epidermal growth factor-receptor), and vascular endothelial growth factor-receptor (VEGF) kinases.

Consequently, these enzymes are prime targets for the development of selective inhibitors (7,8).

A number of kinase inhibitors have already been developed and approved for cancer treatment (Table 3). Many other drugs are currently being tested in cancer clinical trials for their ability to target kinases. Compared to chemotherapy, kinase inhibitors display lower levels of toxicity in patients.

Other Biological Therapies

Proteosomes are a group of enzymes found in all cells in the body. They play a critical role in the degradation of many proteins involved in cell cycle regulation, apoptosis and angiogenesis. Since these pathways are fundamental for cancer cell survival, inhibition of proteasome is an attractive therapeutic target for cancer (9).

Among the other biologic therapies used to treat cancer are intererons and interleukins. These substances occur naturally and are used to stimulate the body's own immune system to attack the cancer cells. Examples include interferon alfa-2b (Intron® A) and aldesleukin (Proleukin®). These medications appear to be useful in a limited number of cancer patients (10).

EMERGING CANCER THERAPIES

Gene Therapy

Cancer is a multigenic disorder involving mutations of both tumor suppressor genes and oncogenes. A large body of preclinical data, however, suggest that cancer growth can be arrested or reversed by delivering a single growth inhibitory or pro-apoptotic gene, or a gene that can elicit immune responses against the tumor.

Gene therapy is a technique that is currently being developed to treat a number of different cancers. A main requirement for gene therapy is the development of efficient, non-toxic gene carriers that can encapsulate and deliver foreign genetic materials into specific cell types such as cancer cells (11). Both viral and non-viral vectors have been developed.

Many gene transfer vectors are modified viruses that retain the capability of the virus for efficient gene delivery but are safer than the native virus due to modifications that eliminate or alter one or more essential viral functions. The field of viral-based gene transfer vectors for the treatment of cancer has now entered the final stage of clinical testing prior to possible product approvals (12). Three viral vectors are currently undergoing this Phase III or Phase II/III clinical testing for cancer treatment. All three of these vectors are based on adenovirus, a common human virus that in its native state can cause cold or flu-like symptoms.

Although non-viral vectors are generally less efficient delivery systems than viral vectors, they have the advantages of safety, simplicity of preparation and high gene encapsulation capability. Non-viral vectors are generally cationic in nature (13). They include cationic polymers such as poly(ethylenimine) (PEI) and poly(L-lysine), cationic peptides, and cationic liposomes.

Gene-directed enzyme prodrug therapy (GDEPT), also known as suicide gene therapy, is an approach where a gene encoding an enzyme is delivered to tumor cells. A prodrug is then administered, which is converted to a cytotoxin by the enzyme at the tumor site. Clinical trials with GDEPT have produced promising results (14).

The utilization of combined gene therapy is another important area for future research. In terms of cancer treatment, the identification of tumor rejection antigens and the defective immune response observed in cancer patients are important topics for future studies. Additionally, new approaches for formulating and delivering plasmid DNA and alphaviral replicon vectors have resulted in increased potency of gene-based vaccines (15).

Clinical utilization of gene therapy is a long way off and will require the efforts of investigators in the basic and clinical sciences. Despite substantial progress, a number of key technical issues need to be resolved before gene therapy can be effectively applied in the clinic (16).

Antisense Oligonucleotides

Antisense oligonucleotides inhibit gene expression by binding in a sequence-specific manner to an RNA target. Modern nucleotide chemistry has enabled the synthesis of chemically modified oligonucleotides that are highly resistant to nuclease degradation. With progress made in chemical modifications, target selection and drug delivery systems, antisense oligonucleotides are emerging as a novel approach to cancer therapy used alone or in combination with conventional treatments such as chemotherapy and radiation therapy (17). Further improvements in antisense chemistry and nanoparticles are promising avenues in antisense therapy of cancer (18).

Several unsolved problems remain with antisense therapeutics. Poor efficiency of delivery to cells, tissue toxicity and antisense-independent biological effects of oligonucleotides currently limit the widespread application of antisense oligonucleotides to human disease (19). Potential nonspecific effects of antisense oligonucleotides should be carefully considered in studies in which antisense agents are used to define biological functions of specific genes. At the clinical level, the systemic effects of antisense oligonucleotides, the dosage required, the timing of administration compared with mechanical intervention, and the toxicity of breakdown products all need to be established (20).

RNAi

One of the most dramatic events of the past few years in the field of molecular biology has been the discovery of RNA interference (RNAi). Although RNAi is an evolutionarily conserved phenomenon for sequence-specific gene silencing in mammalian cells, exogenous small interfering RNA (siRNA) and vector-based short hairpin RNA (shRNA) can also invoke RNAi responses. Both are now not only experimental tools for analyzing gene function but are also expected to be excellent avenues for drug target discovery and represent an emerging class of gene medicine for targeting incurable diseases such as cancer. The success of cancer therapeutic use of RNAi relies on the development of safe and efficacious delivery systems that introduce siRNA and shRNA expression vectors into target tumor cells (21).

RNAi has rapidly become a powerful tool for drug target discovery and validation in cell culture, and now has largely displaced efforts with antisense and ribozymes. Consequently, interest is rapidly growing for extension of its application to in vivo systems, such as animal disease models and human therapeutics. Studies on RNAi have resulted in two basic methods for its use for gene selective inhibition: (1) cytoplasmic delivery of siRNA, which mimics an active intermediate of an endogenous RNAi mechanism and (2) nuclear delivery of gene expression cassettes that express a shRNA, which mimics the micro interfering RNA (miRNA) active intermediate of a different endogenous RNAi mechanism.

RNAi has the potential to knock down oncogenes in cancer, including brain cancer. RNAi-based gene therapy can be coupled with gene therapy that replaces mutated tumor suppressor genes to build a polygenic approach to the gene therapy of cancer (22). However, the therapeutic potential of RNAi will not be realized until targeted delivery systems are available, particular for central nervous system (CNS) delivery.

Tumor Vaccines

The development of cancer vaccines, aimed to enhance the immune response against a tumor, is a promising area of research. A better understanding of both the molecular mechanisms that govern the generation of an effective immune response and the biology of a tumor has contributed to substantial progress in the field.

Areas of intense investigation in cancer immunotherapy include the following: (1) discovery and characterization of novel tumor antigens to be used as targets for vaccination; (2) investigation of different vaccine-delivery modalities such as cellular-based vaccines, protein- and peptide-based vaccines, and vector-based vaccines; (3) characterization of biological adjuvants to further improve the immunogenicity of a vaccine; and (4) investigation of multimodal therapies where vaccines are being combined with other oncological treatments such as radiation and chemotherapy (23).

Various strategies have been used to design peptide-based vaccines for cancer (24). DNA vaccines are typically comprised of plasmid DNA molecules that encode an antigen(s) derived from a pathogen or tumor cell. DNA vaccine technology, even though still in its infancy, has the potential to be used in humans in future (25). Liposomes can be used to incorporate immunomodulatory factors and targeting molecules, and hence can serve as potent vaccine carrier system (26).

Photodynamic Drug Therapy

Photodynamic therapy (PDT) is a cancer treatment modality that is based on the administration of a photosensitizer, which is preferentially retained in tumors compared to normal tissues. This is followed by illumination of the tumor with visible light in a wavelength range matching the absorption spectrum of the photosensitizer. The photosensitizer absorbs light energy and induces production of reactive oxygen species in the tumor environment, generating a cascade of events that kills the tumor cells.

The first generation photosensitizer, porfirmer sodium (Photofrin®), has been approved for treatment esophageal and lung cancer, and is being studied as a potential therapy for other malignant and non-malignant diseases. Sub-optimal light penetration at the treatment absorption peak and prolonged skin photosensitivity in patients are drawbacks to Photofrin. Several new photosensitizers have been developed, and these agents allow for deeper penetration of light into the tumor and more rapid faster clearance from normal tissue.

The use of PDT in oncology has been restricted to certain cancer indications and has not yet become an integral part of cancer treatment in general. The main advantage of PDT is that the treatment can be repeated multiple times safely, without producing immunosuppressive and myelo-suppressive effects. PDT can be administered even after surgery, chemotherapy or radiotherapy (27–30). Recently, PDT has been proposed for use in combination with anticancer chemotherapy with a view to exploiting any additive antitumor effect to reduce the effective doses of cytoctoxic drugs, thus lowering their toxic effects on normal host tissues (31).

BARRIERS AND OBSTACLES TO DRUG DELIVERY TO CANCER CELLS

Despite the progress that has been made in the development of novel cancer medicines, a number of barriers exist in terms of systemic delivery of drugs to cancer cells. These include both physiologic and physicochemical barriers.

Formulation scientists are faced with the challenge of devising novel delivery systems to overcome these obstacles, in order to provide therapies that target and kill tumor cells while minimizing adverse effects.

Physiologic Barriers

For effective drug delivery, physiologic barriers exist at both the tumor-level and cellular level. Additionally, systemic drug delivery is limited by restrictive mechanisms in organs such as the GI tract and CNS. These barriers are listed in Table 4, and are summarized below.

Physiologic Barriers at the Tumor Level

Physiologic factors affecting drug delivery to solid tumors have been reviewed in the literature (32,33). The malformed vasculature of the tumor region and an impaired lymphatic system creates a microenvironment that acts as a barrier to cancer chemotherapy. Hallmarks of the tumor microenvironment include the following:

- *High interstitial fluid pressure (IFP)*. The leaky tumor vasculature combined with the lack of lymphatic draining drainage results in an enhanced permeability and retention (EPR) effect, which applies to large molecular weight compounds (33). However, this includes increased retention of plasma proteins in the interstitium (space between vasculature and tumor cells) that in turn increases osmotic pressure or IFP. Increased IFP creates a net outward convective flow from the center of the tumor, thereby limiting drug transport into the tumor.
- *Hypoxia*. Impaired blood flow to a tumor results in reduced oxygen delivery, which is unable to meet the oxygen demands of the cancer cells. These creates hypoxic regions within the tumor. In addition to limiting response to radiotherapy (oxygen is a potent radiosenstizer), hypoxia is a contributing factor to decreased efficacy of chemotherapeutic agents by decreasing the rate of cell division (32).
- *Low extracellular pH*. Compared to surrounding tissues, the pH of tumor microenvironment is acidic. This is due, in part, to impaired

Table 4 Barriers to Cancer Drug Delivery

Physiologic barriers (tumors)		
Tumor level	Cellular level	Other barriers
High IFP	Mulitdrug resistant (ABC transporters)	Physicochemical (drug solubility, lipophilicity)
Hypoxia	Stress-mediated resistance	Blood-brain barrier
Low extracellular pH	Apoptosis-resistance	Intestinal absorption

drainage of metabolic end products from the tumor interstitium. The acidic environment can limit the permeability of chemotherapeutic agents that are organic cations. While these molecules are generally lipophilic, they will undergo ionization at the lower extracellular pH, reducing drug uptake into the tumor cell.

Physiologic Barriers at the Cellular Level

At the cellular level, resistance of tumors to drug therapy can develop through alterations in the biochemistry of cancer cells (34). Specific pathways leading to drug resistance include the following:

- *Multidrug resistance by ABC transporters.* Cancer cells possess membrane proteins that catalyze drug transport and enable cells to overcome toxicity by limiting intracellular drug concentrations. Multidrug resistance (MDR), a major challenge for cancer chemotherapy, is defined as when cells to develop resistance to a broad range of structurally and functionally unrelated compounds. This resistance is generally produced upon exposure to a single substance (35). In 1976, Juliano and Ling identified a plasma membrane protein that was overexpressed in colchicine-resistant tumor cells (36). This protein was called P-glycoprotein (Pgp). Pgp belongs to a family of membrane transporters known as the ATP-binding cassette (ABC) superfamily (37). Comprehensive reviews of the ABC superfamily are provided in the literature (38). Several members of the ABC transporter family have been associated with MDR in human cancer cells: Pgp, multidrug resistance associated proteins, and breast cancer resistance protein (BCRP) (39). Reversal of MDR via transporter inhibition is an approach that has been studied to improve efficacy of cancer chemotherapy.
- *Stress-mediated resistance.* Physiologic barriers in the tumor micro-environment (hypoxia, low pH) commonly cause glucose-regulated stress response of cancer cells. This leads to induction of resistance to multiple chemotherapeutic agents drugs. Specific mechanisms, such as the decreased expression of DNA topoisomerase, results in resistance of cancer cells to topoisomerase inhibitors such as etoposide and topotecan. This type of drug resistance is reversible and dissipates when stress conditions are removed (34,40).
- *Apoptosis resistance.* The goal of cancer drug therapy is to initiate tumor-selective cell death, or apoptosis. Since chemotherapeutic agents induce apoptosis in tumor cells, disruption of the apoptotic machinery is another important determinant of resistance to anticancer drugs (41). Expression of the enzyme glycoxalase I is known to be elevated in apoptosis-resistance cells, making the enzyme a potential target for reversing this phenomenon (34).

Other Physiologic Barriers

There are other physiologic barriers that must be overcome for effective drug delivery. For example, the blood-brain barrier (BBB) presents a formidable obstacle to effective delivery of cancer drugs to the CNS (42). As described in Chapter 10, active efflux of chemotherapeutic agents from the CNS results in reduced drug uptake in the brain. Efflux mechanisms in the CNS include members of the ABC family of transporters such as Pgp and BCRP. Inhibition of these transporters may improve CNS penetration of anti-cancer agents.

The GI tract represents a challenge for systemic drug delivery of orally administered medications. By understanding the barriers involved in oral drug absorption, strategies can be developed to enhance drug delivery of poorly bioavailable cancer medications. For example, peptides and proteins are unstable in GI fluids. Delivery systems that protect these compounds from GI degradation may allow for oral drug delivery of these compounds. Membrane transport systems are present in the intestine to facilitate the absorption of essential nutrients, systems that may also targeted for oral drug deliver of cancer agents. Conversely, efflux transporters in the intestinal cell are known to limit bioavailability of some chemotherapeutic compounds, and inhibition of these transport systems may improve bioavailability.

Physicochemical Barriers

In addition to the physiologic barriers discussed above, physicochemical properties of the compound can also limit the effectiveness of cancer medications. Among these, molecular weight, chemical structure, and pKa are important determinants of drug delivery to tumor cells. Additionally, compounds that are substrates for Pgp and other transporters are susceptible to multi-drug resistance as well as limited transport across the GI tract (limited bioavailability) and the BBB.

Aqueous solubility is an important physicochemical parameter for effective drug therapy. Chemotherapeutic agents, including taxanes, are poorly soluble. Consequently, formulations of these medications include non-ionic surfactants to solubilize the drug. Among the most commonly used surfactants are Cremophor EL® (polyoxyethyleneglycerol triricinoleate 35) and Tween® 80 (polyoxyethylene-sorbitan-20-monooleate). These solubilizers are pharmacologically active, and their use in drug formulations is associated with acute hypersensitivity reactions, peripheral neurotoxicty, and dyslipidemia (43). Additionally, both compounds can influence the pharmacokinetic profile of cancer drugs (44). These drawbacks have led to the development of alternative delivery systems for poorly soluble cancer drugs (discussed later).

STRATEGIES TO ENHANCE CANCER DRUG DELIVERY

Chemotherapy is a major therapeutic approach for the treatment of both localized and metastasized cancers. Since chemotherapeutic agents are neither specific nor targeted to the cancer cells, improved delivery of anticancer drugs to tumor tissues in humans appears to be a reasonable and achievable challenge (45). Current cancer drug delivery is no longer limited to traditional methods and dosage forms. It utilizes extensively some state-of-the-art technologies, such as nanotechnology, polymer chemistry, and electronic engineering (46). Our expanding knowledge of the molecular biology of cancer and the pathways involved in malignant transformation of cells have revolutionized cancer treatment with a focus on targeted cancer therapy. New approaches to cancer treatment not only supplement conventional chemotherapy and radiotherapy, but also aim to prevent damage to the normal tissues and overcome drug resistance. Innovative methods of cancer treatment require new concepts of drug delivery in cancer (47). The combination of diagnostics with therapeutics, an important feature of personalized cancer therapy, is facilitated by the use of monoclonal antibodies and nanobiotechnology (48).

Development of alternative delivery systems dosage existing anticancer drugs may enhance their therapeutic value. Among the delivery systems being developed for anticancer agents are colloidal systems (liposomes, emulsions, nanoparticles, and micelles), polymer conjugates, and carrier systems targeting the tumor microenvironment. These delivery systems are capable of providing enhanced therapeutic activity and reduced toxicity of anticancer agents, mainly by altering their pharmacokinetics and biodistribution. Furthermore, the identification of cell-specific receptor/antigens on cancer cells has led to the development of ligand- or antibody-bearing delivery systems which can be targeted to cancer cells by specific binding to receptors or antigens.

Besides conventional small molecule anticancer agents, research efforts have led to the development of new cancer therapeutics including as oligonucleotides, genes, peptides, and proteins. Because of to the macromolecular properties of these agents, new strategies of delivery for them are required to achieve therapeutic efficacy in clinical setting. The blending of new technologies and multidisciplinary expertise has resulted in the development of advanced drug delivery systems that are applicable to a wide range of anticancer agents. While the majority of these approaches remain at the preclinical stages of development, they hold promise as effective cancer therapies in the future (49). These delivery systems are detailed below.

Prodrugs

A prodrug is a pharmacological substance that is administered in an inactive (or significantly less active) form. Once administered, the prodrug is metabolized in vivo into an active compound. The rationale behind the use of a prodrug is generally to optimize absorption, distribution, metabolism, and excretion properties. For example, prodrugs are usually intended to improve oral bioavailability of poorly absorbed compounds.

In addition to improving pharmacokinetic properties, a prodrug strategy can also be used to increase the selectivity of the drug for its intended target. This is particularly important for cancer chemotherapy, where the reduction of adverse effects is always of paramount importance. Prodrugs used to target hypoxic cancer cells, through the use of redox-activation, utilize the large quantities of reductase enzyme present in the hypoxic cell to convert the drug into its cytotoxic form, essentially activating it. As the prodrug has low cytotoxicity prior to this activation, there is a markedly lower chance of it "attacking" healthy, non-cancerous cells which reduces the side-effects associated with these chemotherapeutic agents (50).

Over the past decades, numerous prodrugs have been developed that are designed to be enzymatically activated by tumor cells into anti-cancer agents. The most important enzymes involved in prodrug activation in terms of tissue distribution, up-regulation in tumor cells and turnover rates include the following: aldehyde oxidase, amino acid oxidase, cytochrome P450 reductase, DT-diaphorase, cytochrome P450, tyrosinase, thymidylate synthase, thymidine phosphorylase, glutathione S-transferase, deoxycytidine kinase, carboxylesterase, alkaline phosphatase, beta-glucuronidase and cysteine conjugate beta-lyase. Accordingly, a number of prodrugs have been synthesized that are designed for tumor-selective activity of various chemotherapeutic medications including alkylating agents (cyclophosphamide, ifosfamide, cisplatin), antimetabolites (mercaptopurine, thioguanine), antibiotics (doxorubicin, daunorubicin, epirubicin), alkaloids (paclitaxel) and topoisomerase inhibitors (etoposide, ironotecan).

Besides utilizing unique aspects of tumor physiology such as selective enzyme expression, hypoxia, and low extracellular pH, others prodrug approaches are based on tumor-specific delivery techniques, including activation of prodrugs by exogenous enzymes delivered to tumor cells via monoclonal antibodies (ADEPT), or generated in tumor cells from DNA constructs containing the corresponding gene (GDEPT).

Because only a small proportion of the tumor cells may be able to activate the prodrug, regardless of which activating mechanism is targeted, tumor-activated prodrugs (TAP) must be also capable of killing activation-incompetent cells through a "bystander effect". A wide variety of chemistries have been explored for the selective activation of TAP. These include reduction of quinones, N-oxides, nitroaromatics and metal complexes by endogenous enzymes or radiation, amide cleavage by endogenous peptidases, and metabolism by a variety of exogenous enzymes, including phosphatases, kinases, amidases, and glycosidases (51,52).

The development of prodrugs has been relatively successful; however, all prodrugs lack complete selectivity. Therefore, more work is needed to explore the differences between tumor and non-tumor cells and to develop optimal substrates in terms of substrate affinity and enzyme turnover rates for prodrug-activating enzymes resulting in more rapid and selective cleavage of the prodrug inside the tumor cells (53).

Liposomes and Lipoplexes

Liposomes are the most widely studied carrier system for anticancer drug delivery. A liposome is a spherical vesicle ranging from 50 nm to several microns in size, with a membrane composed of a phospholipid and cholesterol bilayer, usually containing a hydrophilic core. The main advantages of liposomal formulations include the following: (1) improved pharmacokinetics and drug release, (2) enhanced intracellular penetration, (3) tumor targeting and preventing adverse side effects and (4) ability to combine several medications into a single drug delivery system (54).

First developed in the 1960s, therapeutic liposomes became clinically available over a decade ago. First-generation liposome systems (conventional liposomes) include liposomes containing the antifungal amphotericin B (Ambisome®) and a doxorubicin-containing liposome (Myocet®). Second-generation liposomes (pure lipid approach) are long-circulating liposomes, such as Daunoxome®, a daunorubicin-containing liposome to treat AIDS-related Kaposi's sarcoma. More recently, liposomes have been surface-modified with gangliosides or sialic acid (third-generation liposomes), in order to evade elimination by the immune system. Fourth-generation liposomes, included pegylated liposomal doxorubicin (described later in this chapter), are called "stealth liposomes" because of their ability to evade interception by the immune system.

Cell targeted liposomes can be produced by attaching an antigen-directed monoclonal antibody or small protein (e.g. folate, epidermal growth factor, transferrin) to the distal end of polyethylene glycol in pegylated liposomal formulation. Once the liposomes localize in the tumor interstitial space, the cytotoxic drug is slowly released within the tumor. Liposomes can act as sustained release delivery systems that can significantly impact the therapeutic outcomes of medications. By manipulating properties such as liposome diameter and drug release rate, drug bioavailability can be improved and dosing frequency can be extended (55).

Liposome formulations can be designed to serve as a multifunctional carrier system. These delivery systems are designed to simultaneously target three molecular targets: (1) extracellular receptors or antigen expressed on the surface of plasma membrane of cancer cells in order to direct the liposomal system specifically to the tumor, preventing adverse side effects on healthy tissues; (2) inhibition of drug efflux pumps (e.g. Pgp) to enhance drug retention by cancer cells, increasing intracellular drug accumulation and thereby limiting the need for prescribed high drug doses that cause adverse drug side effects; and (3) intracellular controlling mechanisms of apoptosis in order to suppress cellular antiapoptotic defense (54).

Perhaps the most promising therapeutic application of liposomes is as non-viral vectors for gene therapy. Lipoplexes, cationic liposome-DNA complexes, have been used extensively for gene transfer into cells, including tumor cells. By understanding the mechanisms involved in the formation of the complexes and their intracellular delivery, the design of non-viral vectors for gene therapy applications can be optimized. (56). The effectiveness of cancer genotherapy will depend on the ability to target lipoplexes into tumor sites with minimal gene dosage to normal tissues (57). However, plasma instability of lipoplexes remains a major obstacle that needs to be addressed with these systems (58).

Polymer Conjugates

Macromolecular drug carriers are an attractive cancer drug delivery method because they appear to target tumors and have limited toxicity in normal tissues. The polymer-based medicines are the most important macromolecular drug carriers for the diagnosis and treatment of cancer, and they have been extensively investigated for this purpose. Polymer drug conjugate (PDC) delivery relies on the EPR effect for targeting tumors. EPR allows for passive targeting of large molecular weight compounds, and passive targeting due to EPR has become one of the breakthrough areas for targeting solid tumors (59).

At least 11 PDCs have entered clinical trials. Polymer–drug conjugation promotes tumor targeting through the EPR effect and, at the cellular level following endocytic capture, relies on lysosomotropic drug delivery. Various polymeric carriers have been utilized that contain a hydrolyzable linker arm for conjugation with a bioactive moiety. These hydrolyzable linkages are broken down by acid lysosomal hydrolysis in the cancer cell, thereby releasing the chemotherapeutic agent in high concentrations near the cell nucleus, the target site.

PDCs are water-soluble hybrid constructs designed for intravenous administration. PDCs fall into two main categories: polymer–protein conjugates or polymer–drug conjugates. Polymer conjugation to proteins reduces immunogenicity, prolongs plasma half-life and enhances protein stability. The successful clinical application of polymer–protein conjugates (PEGylated enzymes and cytokines) and promising results arising from clinical trials with polymer-bound chemotherapy, has provided a firm foundation for more sophisticated second-generation constructs that deliver newly emerging target-directed anticancer agents (e.g. modulators of the cell cycle, signal transduction inhibitors and antiangiogenic drugs) as well as traditional cytotoxic drugs (60,61). Studies show that increasing the molecular weight of PDCs reduces vascular permeability but increase plasma half-life, which promotes greater extravasation (i.e. to enter tumor tissue from the vasculature) (62).

Conceptually, polymer conjugates share many features with other macromolecular drugs, but they have the added advantage of the versatility of synthetic chemistry that allows for tailoring of molecular mass and addition of biomimetic features. Formulation characteristics must be carefully optimized to ensure that the polymeric carrier is biocompatible and that the polymer molecular mass enables tumor-selective targeting followed by endocytic internalization. Additionally, the polymer-drug linker must be stable in transit, but undergo lysosomal degradation within the tumor cell at an optimal rate to liberate the active moiety. Recent evidence suggests that inappropriate trafficking and/or malfunction of enzymatic activation can lead to new mechanisms of clinical resistance (63).

Polymer conjugates can also be used to increase the solubility of poorly water soluble compounds. For example, 1,5-diazaanthraquinones are promising anticancer drugs, however, their clinical potential is limited due to poor solubility. Conjugation of anticancer agents to hydrophilic water-soluble N-(2-hydroxypropyl) methacrylamide (HPMA) polymers can overcome this problem, and this approach has already been used to generate conjugates with demonstrated clinical benefit (64).

HPMA conjugates are among the most extensively studied polymer conjugates. An HPMA conjugate containing doxorubicin became the first synthetic polymer-drug conjugate to undergo clinical testing (phase I/II). Compared to free drug, the HPMA-doxorubicin conjugate showed enhanced antitumor activity in vivo in a T-cell lymphoma mouse model (65). The conjugate has also been shown to induce apoptosis in ovarian cancer cells, and showed greater potency than free doxorubicin in the treatment of ovarian cancer in vivo and in vitro (66).

Research has also shown that a conjugate that combines immunoglo-bulin therapy and doxorubicin is markedly more active than individual conjugates carrying a single drug. At the same time, most of the systemic toxicity of the doxorubicin is avoided (67).

In addition to doxorubicin, other PDCs have been developed that utilize HPMA. An HPMA-mitoxantrone conjugate increased drug accumulation in solid tumors, therefore achieving the targeting to the tumor tissue (68). A water soluble HPMA-9-aminocamptothecin (9-AC) conjugate was designed for colon-specific drug delivery to treat of colon cancer following oral administration. As a consequence of the colon-specific release of unmodified 9-AC from the polymer conjugate, antitumor efficacy is anticipated to be enhanced due to prolonged colon tumor exposure to higher and more localized drug concentrations (69). HPMA copolymer-based anticancer drug delivery systems have also been investigated for intratumoral chemotherapy as an alternative route of drug administration (70).

Water soluble polymer anticancer conjugates can improve the pharmacokinetics of covalently bound drugs by limiting cellular uptake to the endocytic route, thus prolonging plasma circulation time and consequently facilitating tumor targeting by the EPR effect. Polyethylene glycol (PEG) drug conjugates have been studied for over twenty years. However, the limited success of these drug formulations was disappointing, and has been attributed in part to the use of low molecular weight PEG (<20000). Recent research has looked toward higher molecular weight conjugates, resulting in a renaissance in the field of PEG-anticancer drug conjugates. Higher molecular weight PEGs, such as PEG 40000, have longer plasma half-lives and have shown promise following testing using established in vivo tumor models. Additionally, high molecular weight PEG prodrug strategies have also been applied to amino containing drugs and proteins (71).

The low water solubility of the drug camptothecin (CPT) and its unique pharmacodynamics and reactivity in vivo limit its delivery to cancer cells. To increase the anticancer efficacy of CPT, a special drug delivery system is needed. CPT–PEG and CPT–PEG-biotin conjugates were synthesized and studied in vitro in A2780 sensitive and A2780/AD multidrug-resistant human ovarian carcinoma cells. Conjugating CPT to PEG/PEG-biotin polymers increased drug cytotoxicity as well as the ability of CPT to induce apoptosis by activation of caspase-dependent cell death signaling pathway and simultaneous suppression of antiapoptotic cellular defense (72).

Microparticles, Nanoparticles, Nanofibers and Nanotubes

The use of microparticles (>1 μm) and nanoparticles (<1 μm) as delivery vehicles for anticancer therapeutics has great potential to revolutionize the future of cancer therapy. Microparticles have been used a carrier system for anticancer drugs using different types of polymers, such as poly(d,l-lactic-co-glycolic acid) (PLGA). For example, sustained-release anastrozole-loaded PLGA microparticles have been developed for the long-term treatment of breast cancer (73).

Nanoparticles have been extensively studied in recent years as more promising delivery system. Since the tumor architecture causes nanoparticles to preferentially accumulate at the tumor site, their use as drug delivery vectors results in the localization of a greater amount of the drug load at the tumor site. As a result, this passive targeting of anticancer drug loaded nanoparticles improves cancer therapy and reduces the harmful nonspecific side effects of chemotherapeutics. In addition, formulation of these nanoparticles with imaging contrast agents provides a very efficient system for cancer diagnostics. Given the exhaustive possibilities available to polymeric nanoparticle chemistry, research has quickly been directed at multi-functional nanoparticles, combining tumor targeting, tumor therapy, and tumor imaging in an all-in-one system, providing a useful multi-modal approach in the battle against cancer (74).

A number of anticancer compounds have been successfully loaded into nanoparticles for tumor delivery. Among these compound is paclitaxel. Paclitaxel was loaded into biodegradable polymeric particles by electro-hydrodynamic atomization to provide sustained drug delivery following local application to treat malignant glioma (75). Nanoparticles of poly (lactide)-Vitamin E TPGS (PLA-TPGS) copolymers are a novel paclitaxel formulation for oral chemotherapy (76). Nanoparticles can also be coated with PEG (a "stealth" approach to avoid mononuclear phagocytic uptake in plasma) to deliver the highly potent hydrophobic anticancer drug docetaxel to solid tumors (77).

Nanoparticles can also be prepared as magnetic nanovectors to achieve targeted anticancer drug delivery. PLGA nanoparticles loaded with magnetite/maghemite have been used as magnetically-controlled drug delivery systems. Electrophoretic properties (zeta potential) were found to be comparable for both composite and ferrite-free PLGA sub-micron particles, thus indicating that the polymeric coating masks the surface of ferrite nanoparticles buried inside (78). Oleic acid-coated magnetite was encapsulated in biocompatible magnetic nanoparticles by a simple emulsion evaporation method. By including an anticancer drug in the formulation, these magnetic nanoparticles represent a promising mean for simultaneous tumor imaging, drug delivery and real time monitoring of therapeutic effect (79).

Micro- and nanoparticles can be characterized in terms of encapsulation efficiency, particle size distribution, surface morphology, and drug release profile. Surface morphology can be studied by scanning electron microscopy, field emission electron microscope, and X-ray photoelectron spectroscopy. Particle size is usually measured by laser diffraction particle sizer and charge can also be measured by Zetasizer. Analytical methods, such as HPLC, GC-MS, and LC-MS may be used for the quantitation of the drug. Preparative variables such as concentrations of stabilizer, drug-polymer ratio, polymer viscosity, stirring rate, and ratio of internal to external phases are important factors for the preparation of both micro-particles and nanoparticles. Fourier transform infrared with attenuated total reflectance analysis and differential scanning calorimetry may be employed to determine any interactions between drug and polymer. The data can be fit to various dissolution kinetics models for multiparticulate systems, including the zero order, first order, square root of time kinetics, and biphasic models (73).

Microfibers, nanofibers, and nanotubes are alternative carrier systems for anticancer drugs. Electrospun paclitaxel-loaded biodegradable micro- and nanofibers show promise for the treatment of brain tumors by providing sustained delivery of paclitaxel (75). Oxidized single-wall carbon nanohorns, a type of single-wall nanotube, were developed to entrap the anticancer agent cisplatin. Drug was house inside nanotubes and was slowly released from the nanotubes in aqueous environments. The released cisplatin was found to be effective in terminating the growth of human lung-cancer cells, while the nanotubes themselves had no such effect (80).

Microemulsions

Microemulsions (oil-in-water) can be used as vehicles for targeted delivery of chemotherapeutic or diagnostic agents to neoplastic cells. By designing microemulsions with a chemical composition similar to the lipid component of low density lipoprotein, these artificial microemulsion particles can incorporate plasma apolipoproteins on to their surface following intra-venous administration. The microemulsions can then target LDL receptors on tumor cells and deliver drug molecules that are incorporated in the core or at the surface of the microemulsion.

A number of microemulsion formulations have been developed for anticancer therapy including delivery systems for arsenic oxide (81) and paclitaxel (82). An injectable microemulsion of vincristine (M-VCR) was prepared containing a vitamin E solution of oleic acid,, PEG-lipid and cholesterol (surfactants) and an aqueous drug solution (83). Additionally, folate-linked microemulsions were synthesized as a tumor-targeted drug carrier for lipophilic antitumor antibiotics. Folate modification with a sufficiently long PEG chain was found to be an effective way of targeting emulsion to tumor cells (84).

Microemulsion technology can also be used as a template to prepare other delivery systems. For example, microemulsions have been used to engineer stable emulsifying wax and Brij 72 (polyoxyl 2 stearyl ether) nanoparticles. The technique is simple, reproducible, and amenable to large-scale production of stable nanoparticles having diameters below 100 nm. Investigation of the process variables showed that the amount of surfactant used in the preparation of microemulsion templates had the greatest influence on the microemulsion window, as well as the properties and stability of the cured nanoparticles (85). Long-circulating liospheres containing 6-mercaptopurine (6-MP) were prepared by solidification of warm microemulsion at low temperature. Palmitoyl PEG was incorporated in the system to confer stealth-type properties (86). A water-in-oil microemulsion approach has been used to synthesize the superparamagnetic core and the polymeric microsphere in one continuous step. The synthesis allows for control of the magnetic nanoparticle size and the thickness of the hydrogel, ranging from 80 to 320 nm (87).

Micelles

Micelles are self-assembling nanosized colloidal particles with a hydrophobic core and hydrophilic shell (Fig. 2). Micelles have been successfully used to deliver poorly soluble pharmaceuticals, and they are attractive drug carriers for cancer therapy. Polymeric micelles, micelles formed from amphiphilic block co-polymers, possess high stability both in vitro and in vivo and good biocompatibility. Lipid-core micelles are micelles that are formed by conjugating soluble copolymers with lipids. Of these, polyethylene glycol-phosphatidyl ethanolamine conjugate (PEG-PE) lipid-core micelles are of special interest. These micelles can effectively solubilize a broad variety of poorly soluble drugs (anticancer drugs in particular) and diagnostic agents.

Drug-loaded lipid-core micelles can spontaneously target body areas with compromised vasculature (tumors, infarcts) via EPR effect. Lipid-core mixed micelles containing certain specific components (such as positively charged lipids) are capable of escaping endosomes and delivering incorporated drugs directly into the cell cytoplasm. Various specific targeting ligand molecules (such as antibodies) can be attached to the surface of the lipid-core micelles and bring drug-loaded micelles to and into target cells. Lipid-core micelles carrying various reporter (contrast) groups may become the imaging agents of choice in different imaging modalities (89).

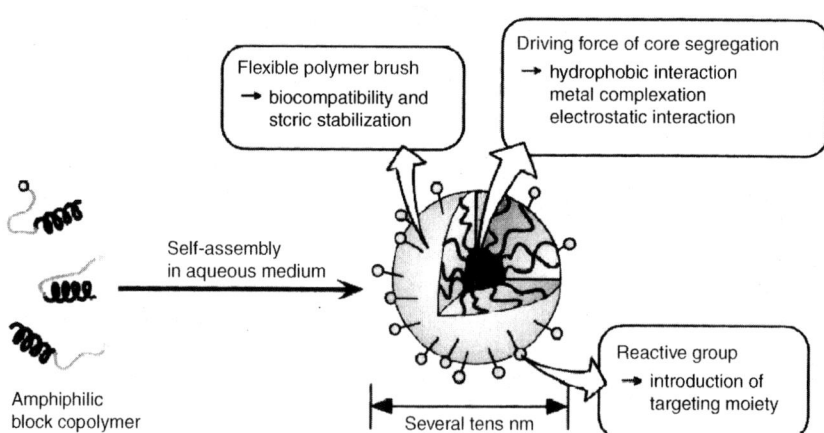

Figure 2 Depiction of polymeric micelles as carriers for drug delivery. *Source:* Reprinted from Ref. 88 with permission from Elsevier.

Micelles prepared from amphiphilic PEG-PE represent a particulate drug delivery system capable of accumulation in tumors via EPR and can be used as a tumor-specific delivery system for poorly soluble anticancer drugs (90). Micelles were prepared by mixing PEG-PE and d-alpha-tocopheryl polyetheyene glycol 1000 succinate (TPGS), and loaded with the poorly soluble anticancer drug camptothecin (CPT). The cytotoxicity of the CPT-loaded mixed micelles against various cancer cells in vitro was remarkably higher than that of the free drug (91). Paclitaxel-loaded mixed polymeric micelles consisting of PEG-PE, solid triglycerides (ST), and cationic lipofectin lipids (LL) were able to escape from endosomal degradation and deliver drug to the cytoplasm of BT-20 cancer cells, increasing anticancer efficiency (92).

Environmentally sensitive micelles prepared from poly(2-ethyl-2-oxazoline)-b-poly(L-lactide) diblock copolymers (PEOz-PLLA) make a promising carrier to transport anticancer drugs specifically to tumor cells and release the drug molecules inside the cell cytosol for improved chemotherapy (93,94). Polymer micelles can also be prepared with cross-linked ionic cores that display high stability. Block ionomer complexes of poly(ethylene oxide)-b-poly(methacrylic acid) copolymer and divalent metal cations have been utilized as micellar templates for the synthesis of the cross-linked micelles. Such micelles represent hydrophilic nanospheres of core-shell morphology. Cisplatin, a potent chemotherapeutic agent, was incorporated into the ionic core of the micelles with remarkably high efficiency (22% w/w) (95).

Nanogels

Another potential approach for cancer drug delivery are nanogels. Nanogels are hydrophilic nanosized particles consisting of a cross-linked cationic polymer network. Nanogels (particle size 100–300 nm) can be synthesized by a micellar approach or prepared by an emulsification/evaporation method.

Nanogels composed of amphiphilic polymers (Pluronic(R) F68 and P123) were synthesized and cationic PEI was used for encapsulation and cellular delivery of cytotoxic nucleoside analogs 5'-triphosphates. Formulations of nucleoside analogs with these nanogels improved the delivery of these cytotoxic drugs to cancer cells, increasing their therapeutic potential as anticancer chemotherapy (96). The drug-nanogel formulation showed significantly enhanced cytotoxicity in cultured cancer cells compared to drug alone.

Cancer cell-targeting molecules, such as folate, can easily be attached to nanogels, and this modification has resulted in enhanced internalization in human breast carcinoma MCF-7 cells. Moreover, transcellular transport of folate-nanogel polyplexes was found to be more effective compared to the drug alone using Caco-2 cell monolayers as an in vitro intestinal model (97). A nanogel formulation containing 3′-azido-2′,3′-dideoxythymidine 5′-triphosphate (AZTTP) showed enhanced cytotoxicity in two breast cancer cell lines, MCF-7 and MDA-MB-231, demonstrating IC50 values 130–200 times lower than those values for AZT alone. A substantial release of encapsulated drug was observed following interactions of drug-loaded nanogels with cellular membranes (98).

Photocrosslinked nanogels, containing a hydrophobic core and hydrophilic shell, can be prepared with amphiphilic triblock copolymers, poly(D,L-lactic acid)/poly(ethylene glycol)/poly(D,L-lactic acid) (PLA–PEG–PLA) and acrylated groups at the end of the PLA segments (99). Copolymers are synthesized by ring-opening polymerization and possess a low CMC, which easily helps to form micelles by self-assembly. The acrylated end groups allow the micelles to be photocrosslinked by ultraviolet irradiation, which converts the micelles into nanogels. These nanogels exhibit excellent stability as a suspension in aqueous media at ambient temperature as compared to the micelles. Moreover, the size of the nanogels is easily manipulated in a range of 150–250 nm by changing the concentration of crosslinkers (e.g. ethylene glycol dimethacrylate) and ultraviolet light irradiation time. These nanogels possess high encapsulation efficiency and offer a steady and long-term release mechanism for the hydrophobic anticancer drug CPT.

Dendrimers

Dendrimers are versatile, derivatizable, well-defined, compartmentalized chemical polymers with sizes and physicochemical properties similar to biomolecules such as proteins (100). Dendrimers evolve from the cross-linked and branched polymers (Fig. 3). There are numerous applications of dendrimers as tools for efficient multivalent presentation of biological ligands in biospecific recognition, inhibition and targeting. Dendrimers can be used for antibacterial and antiviral therapy, and have shown potential as drug or gene delivery devices in anticancer therapy. Additionally, dendrimers can serve as "glycocarriers" for the controlled multimeric delivery of biologically relevant carbohydrate moieties, agents that are useful for targeting modified tissue in malignant diseases for diagnostic and therapeutic purposes (102).

A star polymer composed of amphiphilic block copolymer arms has been synthesized and characterized. The core of the star polymer is polyamidoamine (PAMAM) dendrimer, the inner block in the arm is lipophilic poly(epsilon-caprolactone) (PCL), and the outer block in the arm is hydrophilic PEG. A loading capacity of up to 22% (w/w) into the star-PCL-PEG polymer was achieved with etoposide, a hydrophobic

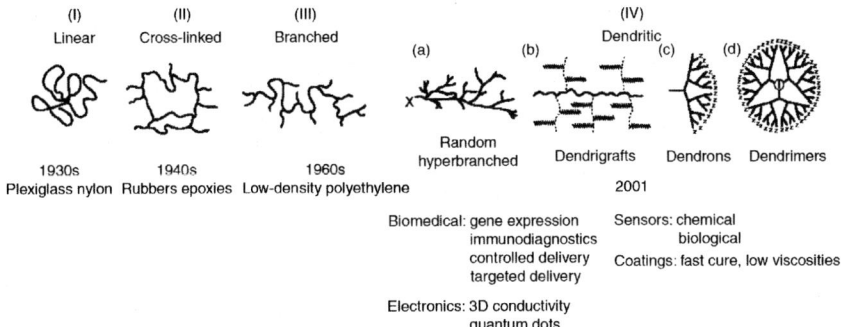

Figure 3 Evolution of dendrimers from synthetic polymers: (I) linear, (II) cross-linked (bridged), (III) branched, (IV) structure-controlled dendritic polymers, and finally, dendrimers. *Source*: Reprinted from Ref. 101 with permission from Elsevier.

anticancer drug. A cytotoxicity assay demonstrated that the star-PCL–PEG copolymer is nontoxic in cell culture. This type of block copolymer can be used as a drug delivery carrier (103). Paclitaxel, a poorly soluble anticancer drug, was covalently conjugated with PAMAM G4 hydroxyl-terminated dendrimer and bis(PEG) polymer for the potential enhancement of drug solubility and cytotoxicity. Cytotoxicity increased by 10-fold with PAMAM dendrimer-succinic acid-paclitaxel conjugate when compared with free nonconjugated drug (104). Dendrimers based on melamine can reduce the organ toxicity of solubilized cancer drugs administered by intraperitoneal injection (105).

Peptide dendrimers are radial or wedge-like branched macromolecules consisting of a peptidyl branching core and/or covalently attached surface functional units. The multimeric nature of these constructs, the unambiguous composition and ease of production make this type of dendrimer well suited to various biotechnological and biochemical applications. Applications include use as biomedical diagnostic reagents, protein mimetics, anticancer and antiviral agents, vaccines and drug and gene delivery vehicles (106).

pH Responsive Carrier Systems

As discussed earlier, drug–polymer conjugates are potential candidates for the selective delivery of anticancer agents to tumor tissue. Incorporating acid-sensitive bonds between the drug and the polymer is an attractive approach because it ensures effective release of the polymer-bound drug at the tumor site. This release is either extracellular, resulting from the slightly acidic pH of the tumor microenvironment, or intracellular, via acidic endosomes or lysosomes following cellular uptake of the drug–polymer conjugate (107).

Non-targeted or antibody-targeted pH-sensitive polymer-doxorubicin conjugates were designed to facilitate site-specific chemotherapy. Doxorubicin is attached to the polymer carrier via a simple hydrolytically labile spacer containing either a hydrazone bond or cis-aconitic acid residue (108). In vitro incubation of the conjugates in various buffers demonstrated fast drug release from the polymer at pH 5 (tumor environment) with minimal release at pH 7.4 (physiologic plasma). Thus, doxorubicin conjugates were stable and inactive during transport in the body, but activate inside target cells as a result of regional differences in pH. Cytotoxicity of the conjugate depends on the detailed structure of the polymer and of the spacer between the drug and polymer carrier. In vivo antitumor activity of the pH-sensitive conjugates containing doxorubicin was significantly enhanced compared with free drug or conjugate alone (109).

pH-sensitive materials can also be incorporated into nanovectors, such as nanoparticles and liposomes. Cisplatin-loaded nanoparticles with pH-responsive poly[2-(N,N-diethylamino)ethyl methacrylate] (PDEA) cores were synthesized from PDEA-block-poly(ethylene glycol) (PDEA-PEG) copolymer. Studies showed that PDEA-PEG nanoparticles dissolved at pH below 6, and the particles were rapidly internalized and transferred to lysosomes. In vitro experiments found that the relative proportion of viable cells was diminished to a greater extent following exposure to fast-releasing nanoparticles compared to slow-releasing nanoparticles or an equivalent dose of free cisplatin (110).

The potential of pH-sensitive, serum-stable and long-circulating liposomes as drug delivery systems has also been investigated. Liposomes were prepared with dioleoylphosphatidylethanolamine (DOPE) and oleic acid (DOPE/oleic acid liposome) or DOPE and 1,2-dipalmitoylsuccinylglycerol (DOPE/DPSG liposome) (111). The inclusion of polyethylene glycol-derived phosphatidylethanolamine (DSPE-PEG) enhanced the serum stability of both liposomes, but also shifted the pH-response curve of pH-sensitive liposomes to more acidic regions and reduced the maximum leakage percentage. In tumor tissue homogenates, where the pH is lower than normal healthy tissues, the pH-sensitive DOPE/DPSG liposomes released the entrapped markers rapidly compared to pH-insensitive dipalmitoylphosphatidylcholine/cholesterol/DSPE-PEG liposomes. The blood circulation time of methotrexate was significantly prolonged following incorporation in DOPE/UDPSG liposomes with increasing content of DSPE-PEG. The complete destabilization of the liposomes at tumor tissues suggests that the liposomes might be useful for the targeted delivery of drugs such as anticancer agents.

Thermosensitive Carrier Systems

Thermoresponsive materials have been used to develop into thermosensitive carrier systems, designed to release drug when a certain phase transition temperature is achieved at or near the target site (e.g. tumor).

Temperature-sensitive liposomes (TS-liposomes) have been studied for chemotherapeutic purposes to enhance the release of anticancer drugs at tumor sites. TS-liposomes can be prepared by thin film hydration and subsequent sonication and characterized for size, phase transition temperature, in vitro drug release and stability. The phase transition temperature of the liposomes can be determined by differential scanning calorimetry (112). PEG-modified TS-liposomes (PETS-liposomes) were found to be highly efficacious carriers for the in vivo delivery of anticancer drugs, and to have potential anticancer applications in combination with hyperthermia (113). In addition to synthetic lipids, TS-liposomes can also be prepared from natural lipids such as dipalmitoyl phosphatidylcholine, distearoyl phosphatidyl choline and cholesterol for local drug release in response to hyperthermia for achieving tumor drug targeting. These liposomes are biodegradable, non-toxic and more cost effective in comparison with liposomes prepared from synthetic lipids for use in multimodality cancer therapy (114,115).

Targeting Transport Mechanisms

Receptor-Mediated Delivery

Receptor mediated transport is a potential pathway for delivery of large molecule peptides or proteins. This is particularly important for the drug delivery in the brain, in which the molecule is shuttled across the BBB into brain interstitial fluid by specific receptors (discussed in Chapter 10). The endothelial transferrin receptor (TfR) is a bidirectional transporter, and is the major route of cellular iron uptake. This efficient cellular uptake pathway has been exploited for the site-specific delivery not only of anticancer drugs and proteins, but also of therapeutic genes into proliferating malignant cells that overexpress the transferrin receptors. This is achieved either chemically by conjugation of transferrin with therapeutic drugs, proteins, or genetically by infusion of therapeutic peptides or proteins into the structure of transferrin. The resulting conjugates significantly improve the cytotoxicity and selectivity of the drugs. The coupling of DNA to transferrin via a polycation or liposome serves as a potential alternative to viral vector for gene therapy. Moreover, the OX26 monoclonal antibody against the rat transferrin receptor offers great promise in the delivery of therapeutic agents across the BBB to the brain (116).

Carrier-Mediated Delivery

Membrane transporters perform a central function in drug disposition and activity. Significant advances in experimental methodology have resulted in the identification of transporters in the liver, gastrointestinal (GI) tract, kidney, and CNS. These transporters have been extensively reviewed in the literature (117–119). Modulation of these transport systems can elicit changes in distribution, clearance, and bioavailability and, consequently, drug activity.

For anti-cancer delivery, two organs that can be targeted in terms of carrier-mediated transport systems are the brain and the intestine. For treatment of CNS tumors, drug delivery can be enhanced by developing prodrugs that are substrates for transporters in the BBB. An alternative strategy involves inhibition of efflux mechanisms in the BBB that restrict access of drugs to the brain. The ABC transporters such as Pgp and BCRP are potential targets, since many chemotherapeutic agents are substrates for these transport systems. For example, Pgp inhibitors have been shown to increase the accumulation of docetaxel in the brain without significant effects on systemic exposure (39,120).

Oral drug delivery of anticancer agents can be improved by targeting or modulating transporters in the GI tract. An understanding of the molecular and functional characteristics of the intestinal membrane transporters will be helpful in developing tools to enhance oral delivery of poorly absorbed drugs. For example, the intestinal peptide transport system could be exploited to improve oral bioavailability through a prodrug approach. Since drug efflux by Pgp has been shown to limit oral absorption of several cancer agents (etoposide, paclitaxel, topotecan), bioavailability can be improved by co-administration with Pgp inhibitors (39,121). Furthermore, there appears to be a cooperative function of this efflux system and intestinal metabolism in limiting drug absorption.

Direct Intratumoral Delivery and Convection-Enhanced Delivery

In an effort to improve survival from malignant gliomas, investigators have used intratumoral chemotherapy protocols to deliver high doses of tumoricidal agents directly to the brain. Theoretically, these infusions bypass the BBB, minimize systemic drug levels and the side effects of chemotherapy, and achieve prolonged elevations of intracerebral chemotherapeutic agents relative to those obtainable by systemic administration. Almost all major classes of chemotherapeutic agents have been examined as possible intratumoral therapies via delivery approaches ranging from simple intratumoral injections to implantable computer-driven constant infusion pumps and biodegradable polymer matrices (122).

Solvent facilitated perfusion (SFP) has been proposed as a technique to increase the delivery of chemotherapeutic agents to tumors. SFP entails direct injection of the agent into the tumor in a water-miscible organic solvent, and because the solvent moves easily through both aqueous solutions and cellular membranes it drives the penetration of the solubilized anticancer agent throughout the tumor (123).

A polymeric paste formulation of paclitaxel has been developed. The formulation is designed to be injected through a narrow gauge needle at room temperature and set to a solid implant in vivo for the intratumoral treatment of localized cancer. Pastes were manufactured from a triblock copolymer composed of poly(D,L-lactide-co-caprolactone)-block-polyethylene glycol-block-poly(DL-lactide-co-caprolactone) (PLC-PEG-PLC) or triblock blended with a low molecular weight polymer methoxypolyethylene glycol (MePEG). These pellets released paclitaxel in a controlled manner over 7 weeks. Pastes composed of 40:60 triblock:MePEG blends containing 10% paclitaxel are proposed as suitable injectable formulations of the drug for intratumoral therapy (124).

Convection-enhanced delivery (CED) is a technique for local delivery of agents to a large area of tissue in the CNS. CED of nanoliposomal CPT-11 greatly prolonged tissue residence while also substantially reducing toxicity, resulting in a highly effective treatment strategy in tumor (125).

Polymer Implants

Cancer chemotherapy is not always effective. Difficulties in drug delivery to the tumor, drug toxicity to normal tissues, and drug stability in the body contribute to this problem. Polymeric materials provide an alternate means for delivering chemotherapeutic agents. When anticancer drugs are encapsulated in polymers, they can be protected from degradation. Implanted polymeric pellets or injected microspheres localize therapy to specific anatomic sites, providing a continuous sustained release of anticancer drugs while minimizing systemic exposure. In certain cases, polymeric microspheres delivered intravascularly can be targeted to specific organs or tumors (126).

A unique preparation composed of 5-fluorouracil (5-FU) and biodegradable polymer, poly L-lactic acid was formed into plastic needles designed for the topical application in tumor tissue. When the needles were implanted in the liver, 5-FU concentrations in the treated tissues were sustained at high levels, whereas levels in the peripheral blood were minimized. This, the 5-FU-PLA needle can be potentially useful as a drug delivery system in cancer chemotherapy (127).

Gliadel® Wafer (polifeprosan 20 with carmustine implant) is indicated for newly diagnosed patients with high-grade malignant glioma as an adjunct to surgery and radiation. Gliadel is a white, dime-sized wafer made up of a biocompatible polymer that contains the cancer chemotherapeutic drug, carmustine (BCNU). Gliadel provides localized delivery of chemotherapy directly to the site of the tumor and is the only FDA approved brain cancer treatment capable of doing so (128).

The DepoFoam® drug delivery system consists of microscopic, spherical particles composed of hundreds of non-concentric aqueous chambers encapsulating the drug to be delivered. The individual chambers are separated by bilayer lipid membranes made up of synthetic analogues of naturally occurring lipids, resulting in a delivery vehicle that is both biocompatible and biodegradable. The rate of drug release from a DepoFoam formulation is controlled by modification of the lipid components, aqueous excipients and manufacturing parameters used in production of the formulation. The DepoFoam system provides either local site or systemic delivery, and can be administered by a number of routes; these include under the skin, within muscle tissue, and into spinal fluid, joints and the abdominal cavity (129).

Controlled release of doxorubicin was obtained using thermosensitive poly(organophosphazene) hydrogels. An aqueous solution of poly(organophosphazenes) containing doxorubicin was transformed at physiologic temperature into a hydrogel with good gel strength. The efficacy of doxorubicin was observed to be maintained over a prolonged time period without any initial burst release, making the delivery system was an excellent candidate as a locally injectable gel-depot system (130).

Electrochemotherapy

Electrochemotherapy (ECT) is a novel cancer treatment in which electric pulses (EPs) inducing cell membrane pores (electroporation) are used as a means of delivering antitumor drugs to the cytoplasm of cancer cells. ECT has demonstrated inhibitory effects on cells in vitro and on solid tumors in clinical trials. More than 10 different antitumor drugs have been delivered using ECT. Of these, the most promising results were seen with bleomycin and cisplatin (131).

Factors influencing ECT effects include the electric parameters, diameter of electrode, distribution of electric field lines, size of tumor, model of drugs injection and kinds of drugs. Some questions of ECT still remain, such as the required dosages and kinds of drugs for clinical trials, model of drug injection, influence on normal tissues, and therapeutic mechanism (131). Electrochemotherapy, a treatment relevant to some precise clinical situations, might be widely instrumental in the treatment of cancer (132,133). It was found that, during electrochemotherapy, only minimal local side-effects were observed, whereas no systemic side-effects of the treatment were noted (134).

Reversal of Multidrug Resistance

Agents (modulators) that reverse the in vitro resistance of tumor cells to anticancer drugs that are substrates for Pgp and other ABC transporters have been studied as an approach to improve therapeutic outcomes in cancer patients. There have been a number of clinical studies conducted to determine whether modulation of Pgp activity improves the efficacy of cancer chemotherapy (135). Studies have shown that Pgp modulators such as verapamil and cyclosporin indeed reverse resistance in a small number of patients, but significant side effects are observed with these therapeutically active medications. Additionally, it is likely that the principal effect of these modulators may be through altered pharmacokinetics and not MDR reversal in cancer cells (136).

Presently, a number of second and third generation Pgp-specific modulators that are in various stages of development. These agents are specific inhibitors of Pgp; that is, they that demonstrate inhibitory activity but are devoid of pharmacologic effects. Examples include PSC-833 (valspodar), GF120918 (elacridar), VX-710 (biricodar), LY335979 (zosu-quidar), and XR9576 (39, 137). Not only might these compounds prove useful in the treatment of cancer, but they may also be able to enhance oral absorption or CNS distribution of drug products.

So far, none of these Pgp inhibitors have been approved for clinical use. There are two reasons for the limited success towards reversing MDR. First, multidrug resistance is a complex phenomenon involving several different biochemical mechanisms, and transporter inhibition alone may be insufficient to reverse it. Second, the physiological role of Pgp and other ABC transporters requires more potent modulators with proper selectivity in order to avoid unwanted side effects (138).

The vast majority of investigations into these drugs indicate that P-gp modulators decrease the systemic clearance of anticancer drugs, thus potentially nonselectively increasing exposure to normal and malignant cells and thereby potentially increasing the severity and/or incidence of adverse effects associated with the anticancer therapy. Mechanisms by which P-gp modulators could alter the pharmacokinetics of the anticancer agent include competition for cytochrome P450 intestinal or liver metabolism, inhibition of P-gp mediated biliary excretion or intestinal transport, or inhibition of renal elimination. Administration of P-gp modulators is unlikely to improve the therapeutic index for anticancer drugs unless agents that lack significant pharmacokinetic interactions are found (136).

OPTIMIZING THERAPEUTIC OUTCOMES IN CANCER

As described in the previous section, numerous approaches are being studied to improve drug delivery to cancer cells. As our understanding of the barriers to effective cancer therapy continues to increase, new strategies are being investigated to target drug delivery to the tumor cells. Through this research, scientists continue to devise new drug formulations and delivery systems to improve therapeutic outcomes for cancer patients disorders. While the advent of new biological therapies (e.g. monoclonal antibodies) has dramatically impacted cancer therapy, novel formulations have recently been developed for conventional chemotherapeutic drugs (doxorubicin, pacltixel, cytarabine, carmustine) and hormones (histrelin), resulting in improved therapeutic outcomes (Table 5). These formulation strategies for specific cancer medications are highlighted below.

Doxorubicin

Doxorubicin is an anthracycline antibiotic used to treat various cancers including breast, ovarian and lung cancer. While anthracyclines are among the most effective chemotherapeutic agents in the treatment of numerous

Table 5 Advanced Formulation Strategies: Examples of Drug Products Used to Treat Cancer

Drug	Product(s)	Description
Pegylated liposomal doxorubicin	Doxil® Caelyx®	Surfactant-free formulation that prolongs circulation time, lowers drug distribution to healthy tissues, promoting tumor uptake
Nanoparticle doxorucin	Transdrug	Doxorubicin polyisohexylcyanoacrylate nanoparticles designed to overcome multidrug resistance caused by ABC transporters. Provides controlled release for prolonged activity
Albumin-bound paclitaxel	Abraxane	Nanoparticle formulation that reduces incidence of adverse reactions and avoids need for adjuvant therapy (steroids, antihistamines) associated with surfactant formulations
Liposomal cytarabine	Depocyte®	Depot liposomal formulation of cytarabine used as an intrathecal treatment of neoplastic and lymphomatous meningitis
Carmustine wafer	Gliadel	Wafer formulation containing carmustine that provides localized delivery of chemotherapy directly to tumor site. Wafers are implanted for removal of malignant glioma, releasing high concentrations of drug targeting microscopic tumor cells that sometimes remain after surgery

malignancies, their use is limited by a dose-dependent cardiotoxicity. Over recent years, novel doxorubicin formulations have been marketed to improve therapeutic outcomes in cancer patients. These nanoscale drug delivery systems allow for optimal control over the pharmacokinetic and pharmacodynamic profiles of doxorubicin.

Liposomal Doxorubicin (Doxil®, Caelx)

Liposomal doxorubicin represents an advanced generation of chemotherapy delivery systems with distinct pharmacokinetic advantages and improved control of drug biodistribution. These formulations rely on Stealth® technology, where doxorubicin is encapsulated in liposomes with surface-bound PEG (i.e. pegylated liposomes, Fig. 4). As a result, the liposomes are protected from detection by the mononuclear phagocyte system. These features produce a pharmacokinetic profile of doxorubicin characterized by an extended circulation time and a reduced volume of distribution, thereby promoting tumor uptake (13,140).

Pegylated liposomes have long circulating half-lives, and are able to retain more that 90% of doxorubicin liposome-encapsulated in the plasma. This allows more time for passive targeting of drug across the leaky vasculature of the tumor. Clinical studies demonstrate a significantly lower rate of both heart failure in patients treated with liposomal-encapsulated

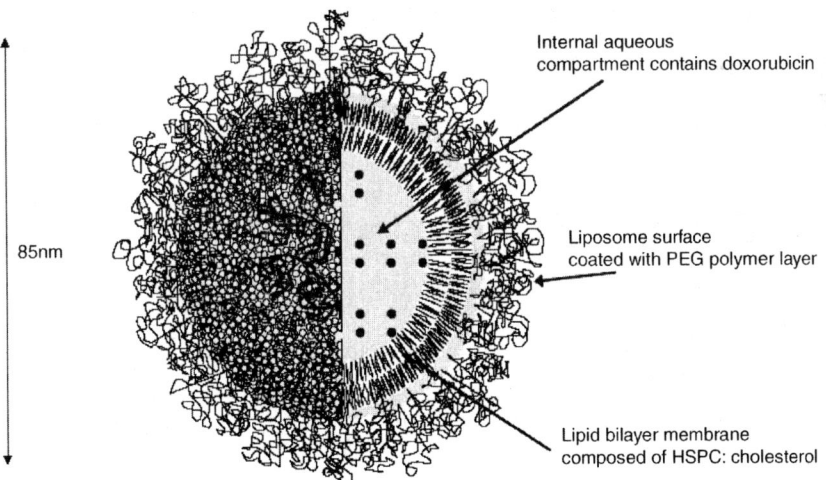

Figure 4 Cross-sectional view of a pegylated doxorubicin liposome. A single lipid bilayer membrane separates an internal aqueous compartment (containing drug) from the external medium. PEG molecules are engrafted onto the liposome surface to form a protective hydrophilic layer. The liposome has a mean diameter of ~85 nm. *Source*: From Ref. 137.

doxorubicin compared to doxorubicin alone. This reduced cardiotoxicity allows for administration of a larger cumulative dose of doxorubicin than what is tolerable for free doxorubicin drug (141).

Nanoparticulate Doxorubicin

Doxorubicin Transdrug® is a novel formulation of polyisohexylcyanoacrylate (PIHCA) nanoparticles containing doxorubicin. Compared to other available polymers, PIHCA shows a high efficacy/tolerance ratio in preclinical model, is produced from highly purified and characterized monomers, and the subsequent polymerization procedure is well defined and controlled. These nanoparticles are able bypass MDR mechanisms in the cancer cell, making the formulation capable of restoring the sensitivity of cancer cells to chemotherapy, overcoming resistance in cancer therapy, and thereby enhancing therapeutic outcomes in cancer treatment (142).

The mechanism through which Transdrug nanoparticles reverse MDR may involve saturation of efflux capacity of MDR transporters. This is presumably the result of the concentration gradient of drug across the cancer cell membrane that is generated by strong adsorption of the nanoparticles onto the cell surface (143–145). Doxorubicin Transdrug is undergoing clinical trials in patients hepatocellular carcinoma. These tumors are known to develop resistance to chemotherapy, so this novel formulation aims to provide preferential distribution of doxorubicin to liver and with increased efficacy by circumventing MDR (142).

Paclitaxel

Taxanes (paclitaxel, docetaxel) are chemotherapeutic agents used to treat a variety of malignancies including breast cancer, ovarian cancer, non–small cell lung cancer and prostate cancer. As discussed previously in this chapter, physicochemical properties of taxanes (poor solubility) necessitates the inclusion of surfactant vehicles (Cremophor®, Tween 80) in commercial formulations containing these medications. In addition to influencing the pharmacokinetic profile of cancer drugs, these solubilizers are pharmacologically active, and their use in drug formulations is associated with acute hypersensitivity reactions, peripheral neurotoxicty, and dyslipidemia (43,44). Patients generally require corticosteroid and antihistamine premedication to reduce the risk of these solvent-mediated hypersensitivity reactions.

Scientists are investigating various approaches in order to improve therapeutic outcomes in patients receiving taxane therapy. A number of formulations are presently undergoing clinical trials. These novel formulations have been reviewed in the literature (146), and are summarized next.

Albumin-Bound Paclitaxel (Abraxane®)

Abraxane is a nanoparticle colloidal suspension of paclitaxel and human serum albumin used to treat patients with metastatic breast cancer (147). The size of the nanoparticles (~150 nm) in the formulation is much smaller than an erythrocyte (one-hundreth the size), thus circumventing the need for any solvent.

The formulation technology relies on the natural properties of albumin to reversibly bind paclitaxel, transport it across the endothelial cell and concentrate it in areas of tumor. Clinical studies have shown that superior clinical results with Abraxane compared to the conventional surfactant formulation, with greater response rate, increased time to disease progression and increased survival. The lower side-effect profile allows albumin-bound paclitaxel to be administered at higher doses. Overall, the albumin-bound formulation has increases the therapeutic index of paclitaxel, and improves therapeutic outcomes in cancer patients (148).

Paclitaxel Prodrugs

DHA-paclitaxel (Taxoprexin®) is a prodrug comprised of the naturally occurring omega-3 fatty acid docosahexaenoic acid (DHA) covalently conjugated to paclitaxel. Since tumor cells take up DHA, DHA-paclitaxel is designed to improve the safety and effectiveness of taxane chemotherapy by delivering more therapeutic agent to the tumor with reduced exposure to healthy tissues, thereby reducing side effects. Studies indicate that DHA-paclitaxel is stable in plasma ($t_{1/2}$ ~2 days), with minimal release of free paclitaxel. Thus, DHA-paclitaxel demonstrates a dramatically different pharmacokinetic profile compared to standard paclitaxel, with reduced toxicity in preclinical models. The formulation is currently undergoing clinical trials (146,149,150).

Paclitaxel poliglumex (Xyotax®) is a macromolecular taxane conjugate of paclitaxel and poly-L-glutamic acid. The design of the formulation is intended to promote retention in the tumor intersitium by EPR and limited lymphatic drainage. Preclinical studies demonstrate reduced toxicity of Xyotax in animal tumor models. Likewise, early clinical trials indicated that paclitaxel poliglumex was associated with improved outcomes (reduced alopecia and neutropenia) compared to standard taxanes. Xyotax is presently undergoing clinical evaluation in patients with advanced non-small cell lung carcinoma (151,152).

Other Strategies

Additional strategies are being tested to circumvent the toxicity associated with conventional taxane formulations. For example, several taxane analogs have been developed including BMS-184476, DJ-927, and RPR 109881A

(146). These compounds are intended to provide improved bioavailability, comparable efficacy, and a reduced side effect profile.

Additionally, a number of formulations for paclitaxel are being tested in clinical trials such as polymeric miceller paclitaxel, liposomal encapsulated paclitaxel, paclitaxel vitamin E emulsion, and paclitaxel microspheres. While it is presently unknown whether any of these formulations will improve survival in cancer patients, their value may lie in improved quality of life for cancer patients compared to standard therapy.

Cytarabine

Neoplastic meningitis is a feared complication in cancer patients, with median survival ranging from weeks to a few months. Management is palliative and aims to provide symptoms relief while delaying neurological deterioration (153). A depot formulation of cytarabine (DepoCyte®) has proven to be useful as intrathecal treatment of neoplastic and lymphomatous meningitis. Clinical studies indicate that, compared to conventional treatment, liposomal cytarabine provides superior response rates, improved patient quality of life, and a prolongation of the time to neurological progression. Additionally, side effects of the formulation were mild and manageable in all patients (154,155).

Carmustine

Gliadel is a white, dime-sized wafer made up of a biocompatible polymer that contains the cancer chemotherapeutic drug, carmustine (BCNU). It is a new adjuvant chemotherapy for glioblastoma that provides localized delivery of chemotherapy directly to the site of the tumor. After a neurosurgeon removes a high-grade malignant glioma, up to eight wafers can be implanted in the cavity where the tumor resided. Once implanted, Gliadel slowly dissolves, releasing high concentrations of BCNU at the tumor site, targeting microscopic tumor cells that sometimes remain after surgery. The specificity of Gliadel minimizes drug exposure to other areas of the body (156,157). Clinical studies indicate that both the clinical progression and radiological progression of cancer were significantly delayed in patients treated with Gliadel, resulting in an increased survival time and improved therapeutic outcomes (158).

CONCLUSIONS: PERSPECTIVES AND FUTURE DIRECTIONS

Although cancer remains one of the leading causes of morbidity and mortality across the globe, research efforts over the past 20 years have led to improved patient survival and quality of life for cancer sufferers. During this time, cancer treatment has progressively moved away from therapies that exert nondiscriminating cytotoxicity on the body, and toward approaches that specifically target cancer cells. The identification of new therapeutic targets, such as blood vessels fueling tumor growth and other biological agents that are more specific in their activity, has resulted in the development of new classes of medications that act specifically at the tumor site. The goal of these new chemotherapeutics is to increase therapeutic outcomes for cancer patients, and achieving this goal is directly related to the treatment's ability to target and to kill the cancer cells while affecting as few healthy cells as possible (159).

Correspondingly, scientists have implemented various formulation approaches to improve therapeutic outcomes with conventional chemotherapeutic agents. As highlighted in this chapter, a number of drug delivery strategies have been devised to overcome the physiologic and physicochemical barriers that must be breached in order to deliver medications to tumor cells. Reformulated delivery systems for several cytotoxic drugs such as doxorubicin and paclitaxel have increased the safety profile of these medications and enhanced their activity by delivery the drug to the tumor microenvironment and reversing MDR. Ongoing research efforts are likely to result in other "alternative" formulations becoming commercially available in the coming years.

Cancer researchers are close to unlocking the complete genetic makeup of many common cancers. As our understanding of how tumor cells differ from normal cells continues to evolve, treatments will be designed to work on specifically the cancer cells. Encompassing a range of disciplines including medicine, biology, chemistry, physics, and engineering, nanotechnology will undoubtedly provide the foundation for developing these "magic bullets," with the ultimate goal of finding a cure for cancer and eradicating the disease worldwide.

REFERENCES

1. The Nation's Investment in Cancer Research: A Plan and Budget Proposal for Fiscal Year 2008. National Cancer Institute, http://plan.cancer.gov/science.shtml (accessed March 2007).
2. National Cancer Institute, http://www.cancer.gov/cancertopics/commoncancers (accessed March 2007).
3. Sowery RD, So AI, Gleave ME. Therapeutic options in advanced prostate cancer: present and future. Curr Urol Rep 2007; 8:53–9.
4. Howell A. Future use of selective estrogen receptor modulators and aromatase inhibitors. Clin Cancer Res 2001; 7 (12 Suppl):4402s–10s; discussion 4411s–12s.
5. Berry J. Are all aromatase inhibitors the same? A review of controlled clinical trials in breast cancer. Clin Ther 2005; 27(11):1671–84.

6. Murillo O, Arina A, Tirapu I, et al. Potentiation of therapeutic immune responses against malignancies with monoclonal antibodies. Clin Cancer Res 2003; 9:5454–64.

7. Fabbro D, Ruetz S, Buchdunger E, et al. Protein kinases as targets for anticancer agents: from inhibitors to useful drugs. Pharmacol Ther 2002; 93: 79–98.

8. Shchemelinin I, Sefc L, Necas E. Protein kinase inhibitors. Folia Biol (Praha) 2006; 52:137–48.

9. Montagut C, Rovira A, Albanell J. The proteasome: a novel target for anticancer therapy. Clin Transl Oncol 2006; 8(5):313–7.

10. Pouessel D, Culine S. Targeted therapies in metastatic renal cell carcinoma: the light at the end of the tunnel. Expert Rev Anticancer Ther 2006; 6(12): 1761–7.

11. El-Aneed A. An overview of current delivery systems in cancer gene therapy. J Control Release 2004; 94:1–14.

12. Wilson DR. Viral-mediated gene transfer for cancer treatment. Curr Pharm Biotechnol 2002; 3:151–64.

13. Schatzlein AG. Non-viral vectors in cancer gene therapy: principles and progress. Anticancer Drugs 2001; 12:275–304.

14. Dachs GU, Tupper J, Tozer GM. From bench to bedside for gene-directed enzyme prodrug therapy of cancer. Anticancer Drugs 2005; 16:349–59.

15. Dubensky TW, Jr., Liu MA, Ulmer JB. Delivery systems for gene-based vaccines. Mol Med 2000; 6:723–32.

16. Kouraklis G. Progress in cancer gene therapy. Acta Oncol 1999; 38:675–83.

17. Rayburn ER, Wang H, Zhang R. Antisense-based cancer therapeutics: are we there yet? Expert Opin Emerg Drugs 2006; 11:337–52.

18. Zhang C, Pei J, Kumar Det al. Antisense oligonucleotides: target validation and development of systemically delivered therapeutic nanoparticles. Methods Mol Biol 2007; 361:163–85.

19. Heidenreich O, Kang SH, Xu X, Nerenberg M. Application of antisense technology to therapeutics. Mol Med Today 1995; 1:128–33.

20. Bennett MR, Schwartz SM. Antisense therapy for angioplasty restenosis. Some critical considerations. Circulation 1995; 92:1981–93.

21. Takeshita F, Ochiya T. Therapeutic potential of RNA interference against cancer. Cancer Sci 2006; 97:689–96.

22. Pardridge WM. Intravenous, non-viral RNAi gene therapy of brain cancer. Expert Opin Biol Ther 2004; 4:1103–13.

23. Palena C, Abrams SI, Schlom J, Hodge JW. Cancer vaccines: preclinical studies and novel strategies. Adv Cancer Res 2006; 95:115–45.

24. Pietersz GA, Pouniotis DS, Apostolopoulos V. Design of peptide-based vaccines for cancer. Curr Med Chem 2006; 13:1591–607.

25. Pachuk CJ, McCallus DE, Weiner DB, Satishchandran C. DNA vaccines – challenges in delivery. Curr Opin Mol Ther 2000; 2:188–98.

26. Altin JG, Parish CR. Liposomal vaccines – targeting the delivery of antigen. Methods 2006; 40:39–52.

27. Dalla Via L, Marciani Magno S. Photochemotherapy in the treatment of cancer. Curr Med Chem 2001; 8:1405–18.

28. Sibata CH, Colussi VC, Oleinick NL, Kinsella TJ. Photodynamic therapy in oncology. Expert Opin Pharmacother 2001; 2:917–27.

29. Chen WR, Huang Z, Korbelik M, et al. Photoimmunotherapy for cancer treatment. J Environ Pathol Toxicol Oncol 2006; 25:281–91.

30. Wilson BC. Photodynamic therapy for cancer: principles. Can J Gastroenterol 2002; 16:393–6.

31. Canti G, Nicolin A, Cubeddu R, et al. Antitumor efficacy of the combination of photodynamic therapy and chemotherapy in murine tumors. Cancer Lett 1998; 125:39–44.

32. Cairns R, Papandreou I, Denko N. Overcoming physiologic barriers to cancer treatment by molecularly targeting the tumor microenvironment. Mol Cancer Res 2006; 4:61–70.

33. Jang SH, Wientjes MG, Lu D, Au JL-S. Drug delivery and transport to solid tumors. Pharm Res 2003; 20:1337–50.

34. Tsuruo T, Naito M, Tomida A, et al. Molecular targeting therapy of cancer: drug resistance, apoptosis and survival signal. Cancer Sci 2003; 94:15–21.

35. Fardel O, Lecureur V, Guillouzo A. The P-glycoprotein multidrug transporter. Gen Pharmac 1996; 27(8):1283–91.

36. Juliano RL, Ling V. A surface glycoprotein modulating drug permeability in chinese hamster ovary cell mutants. Biochim Biophys Acta 1976; 455:152–62.

37. van Veen HW, Konings WN. The ABC family of multidrug transporters in microorganisms. Biochim Biophys Acta 1998; 1365:31–36.

38. Borst P, Elferink, RO. Mammalian ABC transporters in health and disease. Ann Rev Biochem 2002; 71:537–592.

39. Breedveld P, Beijen JH, andSchellens JHM. Use of p-glycoprotein and BCRP inhibitors to improve oral bioavailability and CNS penetration of anticancer drugs. Trends Pharm Sci 2006; 27:17–24.

40. Tomida A and Tsuruo T. Drug resistance mediated by cellular stress response to the microenvironment of solid tumors. Anti-Cancer Drug Design 1999; 14: 169–77.

41. Zhivotovsky B, Orrenius S. Defects in the apoptotic machinery of cancer cells: role in drug resistance. Semin Cancer Biol 2003; 13(2):125–34.

42. Begley DJ. Understanding and circumventing the blood-brain barrier. Acta Paediatr Suppl 2003; 92(443):83–91.

43. ten Tije AJ, Verweji J, Loos WJ and Sparreboom A. Pharmacologicial effects of formulation vehicles: Implications for cancer chemotherapy. Clin Pharmaokinet 2003; 42:665–85.

44. Hennenfent KL and Govindan R. Novel formulations of taxanes: a review. Old wine in a new bottle? Ann Oncol 2006; 17:735–49.

45. Zee-Cheng RK, Cheng CC. Delivery of anticancer drugs. Methods Find Exp Clin Pharmacol 1989; 11:439–529.

46. Jain KK. Editorial: targeted drug delivery for cancer. Technol Cancer Res Treat 2005; 4:311–3.

47. Jain KK. Nanotechnology-based drug delivery for cancer. Technol Cancer Res Treat 2005; 4:407–16.

48. Jain KK. Personalised medicine for cancer: from drug development into clinical practice. Expert Opin Pharmacother 2005; 6:1463–76.

49. Kim CK, Lim SJ. Recent progress in drug delivery systems for anticancer agents. Arch Pharm Res 2002; 25:229–39.

50. Yoon KJ, Potter PM, Danks MK. Development of prodrugs for enzyme-mediated, tumor-selective therapy. Curr Med Chem Anticancer Agents 2005; 5:107–13.

51. Denny WA. Prodrug strategies in cancer therapy. Eur J Med Chem 2001; 36:577–95.

52. Denny WA. Tumor-activated prodrugs – a new approach to cancer therapy. Cancer Invest 2004; 22:604–19.

53. Rooseboom M, Commandeur JN, Vermeulen NP. Enzyme-catalyzed activation of anticancer prodrugs. Pharmacol Rev 2004; 56:53–102.

54. Minko T, Pakunlu RI, Wang Y, et al. New generation of liposomal drugs for cancer. Anticancer Agents Med Chem 2006; 6:537–52.

55. Allen TM, Cheng WW, Hare JI, Laginha KM. Pharmacokinetics and pharmacodynamics of lipidic nano-particles in cancer. Anticancer Agents Med Chem 2006; 6:513–23.

56. Wasungu L, Hoekstra D. Cationic lipids, lipoplexes and intracellular delivery of genes. J Control Release 2006; 116:255–64.

57. Dass CR, Burton MA. Lipoplexes and tumours. A review. J Pharm Pharmacol 1999; 51:755–70.

58. Ross PC, Hui SW. Lipoplex size is a major determinant of in vitro lipofection efficiency. Gene Ther 1999; 6:651–9.

59. Modi S, Prakash Jain J, Domb AJ, Kumar N. Exploiting EPR in polymer drug conjugate delivery for tumor targeting. Curr Pharm Dis 2006; 12:4785–96.

60. Vicent MJ, Duncan R. Polymer conjugates: nanosized medicines for treating cancer. Trends Biotechnol 2006; 24:39–47.

61. Greco F, Vicent MJ, Gee S, et al. Investigating the mechanism of enhanced cytotoxicity of HPMA copolymer-Dox-AGM in breast cancer cells. J Control Release 2007; 117:28–39.

62. Dreher MR, Liu W, Michelich CR, et al. Tumor vascular permeability, accumulation, and penetration of macromolecular drug carriers. J Natl Cancer Inst 2006; 98:335–44.

63. Duncan R. Designing polymer conjugates as lysosomotropic nanomedicines. Biochem Soc Trans 2007; 35:56–60.

64. Vicent MJ, Manzanaro S, de la Fuente JA, Duncan R. HPMA copolymer-1, 5-diazaanthraquinone conjugates as novel anticancer therapeutics. J Drug Target 2004; 12:503–15.

65. Chytil P, Etrych T, Konak C, et al. Properties of HPMA copolymer-doxorubicin conjugates with pH-controlled activation: effect of polymer chain modification. J Control Release 2006; 115:26–36.

66. Malugin A, Kopeckova P, Kopecek J. HPMA copolymer-bound doxorubicin induces apoptosis in ovarian carcinoma cells by the disruption of mitochondrial function. Mol Pharm 2006; 3:351–61.

67. Sirova M, Strohalm J, Subr V, et al. Treatment with HPMA copolymer-based doxorubicin conjugate containing human immunoglobulin induces

long-lasting systemic anti-tumour immunity in mice. Cancer Immunol Immunother 2007; 56:35–47.

68. Huang Y, Zhang ZR. [Selective tumor-accumulation of HPMA copolymer-mitoxantrone conjugates]. Yao Xue Xue Bao 2004; 39:374–9.

69. Gao SQ, Lu ZR, Kopeckova P, Kopecek J. Biodistribution and pharmaco-kinetics of colon-specific HPMA copolymer-9-aminocamptothecin conjugate in mice. J Control Release 2007; 117:179–85.

70. Lammers T, Peschke P, Kuhnlein R, et al. Effect of intratumoral injection on the biodistribution and the therapeutic potential of HPMA copolymer-based drug delivery systems. Neoplasia 2006; 8:788–95.

71. Greenwald RB. PEG drugs: an overview. J Control Release 2001;74:159.

72. Minko T, Paranjpe PV, Qiu B, et al. Enhancing the anticancer efficacy of camptothecin using biotinylated poly(ethylene glycol) conjugates in sensitive and multidrug-resistant human ovarian carcinoma cells. Cancer Chemother Pharmacol 2002; 50:143–50.

73. Zidan AS, Sammour OA, Hammad MA, et al. Formulation of anastrozole microparticles as biodegradable anticancer drug carriers. AAPS Pharm Sci Tech 2006; 7:61.

74. van Vlerken LE, Amiji MM. Multi-functional polymeric nanoparticles for tumour-targeted drug delivery. Expert Opin Drug Deliv 2006; 3:205–16.

75. Xie J, Marijnissen JC, Wang CH. Microparticles developed by electro-hydrodynamic atomization for the local delivery of anticancer drug to treat C6 glioma in vitro. Biomaterials 2006; 27:3321–32.

76. Zhang Z, Feng SS. Self-assembled nanoparticles of poly(lactide) – Vitamin E TPGS copolymers for oral chemotherapy. Int J Pharm 2006; 324:191–8.

77. Khalid MN, Simard P, Hoarau D, et al. Long circulating poly(ethylene glycol)-decorated lipid nanocapsules deliver docetaxel to solid tumors. Pharm Res 2006; 23:752–8.

78. Okassa LN, Marchais H, Douziech-Eyrolles L, et al. Optimization of iron oxide nanoparticles encapsulation within poly(d,l-lactide-co-glycolide) sub-micron particles. Eur J Pharm Biopharm 2007; 67:31–38.

79. Hamoudeh M, Faraj AA, Canet-Soulas E, et al. Elaboration of PLLA-based superparamagnetic nanoparticles: Characterization, magnetic behaviour study and in vitro relaxivity evaluation. Int J Pharm 2007; 338:248–257.

80. Ajima K, Yudasaka M, Murakami T, et al. Carbon nanohorns as anticancer drug carriers. Mol Pharm 2005; 2:475–80.

81. Karasulu HY, Karabulut B, Kantarci G, et al. Preparation of arsenic trioxide-loaded microemulsion and its enhanced cytotoxicity on MCF-7 breast carcinoma cell line. Drug Deliv 2004; 11:345–50.

82. Kang BK, Chon SK, Kim SH, et al. Controlled release of paclitaxel from microemulsion containing PLGA and evaluation of anti-tumor activity in vitro and in vivo. Int J Pharm 2004; 286:147–56.

83. Junping W, Takayama K, Nagai T, Maitani Y. Pharmacokinetics and antitumor effects of vincristine carried by microemulsions composed of PEG-lipid, oleic acid, vitamin E and cholesterol. Int J Pharm 2003; 251: 13–21.

84. Shiokawa T, Hattori Y, Kawano K, et al. Effect of polyethylene glycol linker chain length of folate-linked microemulsions loading aclacinomycin A on targeting ability and antitumor effect in vitro and in vivo. Clin Cancer Res 2005; 11:2018–25.

85. Oyewumi MO, Mumper RJ. Gadolinium-loaded nanoparticles engineered from microemulsion templates. Drug Dev Ind Pharm 2002; 28:317–28.

86. Khopade AJ, Shelly C, Pandit NK, Banakar UV. Liposphere based lipoprotein-mimetic delivery system for 6-mercaptopurine. J Biomater Appl 2000; 14:389–98.

87. Song L, Liu T, Liang D, et al. Coupling of optical characterization with particle and network synthesis for biomedical applications. J Biomed Opt 2002; 7:498–506.

88. Kataoka K, Harada A, Nagasaki Y. Block copolymer micelles for drug delivery: design, characterization and biological significance. Adv Drug Deliv Rev 2001; 47:113–31.

89. Torchilin VP. Lipid-core micelles for targeted drug delivery. Curr Drug Deliv 2005; 2:319–27.

90. Lukyanov AN, Gao Z, Mazzola L, Torchilin VP. Polyethylene glycol-diacyllipid micelles demonstrate increased acculumation in subcutaneous tumors in mice. Pharm Res 2002; 19:1424–9.

91. Mu L, Elbayoumi TA, Torchilin VP. Mixed micelles made of poly(ethylene glycol)-phosphatidylethanolamine conjugate and d-alpha-tocopheryl poly-ethylene glycol 1000 succinate as pharmaceutical nanocarriers for camptothe-cin. Int J Pharm 2005; 306:142–9.

92. Wang J, Mongayt D, Torchilin VP. Polymeric micelles for delivery of poorly soluble drugs: preparation and anticancer activity in vitro of paclitaxel incorporated into mixed micelles based on poly(ethylene glycol)-lipid conjugate and positively charged lipids. J Drug Target 2005; 13: 73–80.

93. Liu SQ, Wiradharma N, Gao SJ, et al. Bio-functional micelles self-assembled from a folate-conjugated block copolymer for targeted intracellular delivery of anticancer drugs. Biomaterials 2007; 28:1423–33.

94. Hsiue GH, Wang CH, Lo CL, et al. Environmental-sensitive micelles based on poly(2-ethyl-2-oxazoline)-b-poly(L-lactide) diblock copolymer for application in drug delivery. Int J Pharm 2006; 317:69–75.

95. Bontha S, Kabanov AV, Bronich TK. Polymer micelles with cross-linked ionic cores for delivery of anticancer drugs. J Control Release 2006; 114: 163–74.

96. Vinogradov SV, Kohli E, Zeman AD. Comparison of Nanogel Drug Carriers and their Formulations with Nucleoside 5'-Triphosphates. Pharm Res 2006; 23: 920–930.

97. Vinogradov SV, Zeman AD, Batrakova EV, Kabanov AV. Polyplex Nanogel formulations for drug delivery of cytotoxic nucleoside analogs. J Control Release 2005; 107:143–57.

98. Vinogradov SV, Kohli E, Zeman AD. Cross-linked polymeric nanogel formulations of 5'-triphosphates of nucleoside analogues: role of the cellular membrane in drug release. Mol Pharm 2005; 2:449–61.

99. Lee WC, Li YC, Chu IM. Amphiphilic poly(D,L-lactic acid)/poly(ethylene glycol)/poly(D,L-lactic acid) nanogels for controlled release of hydrophobic drugs. Macromol Biosci 2006; 6:846–54.

100. Boas U, Heegaard PM. Dendrimers in drug research. Chem Soc Rev 2004; 33: 43–63.

101. Esfand R, Tomalia DA. Poly(amidoamine) (PAMAM) dendrimers: from biomimicry to drug delivery and biomedical applications. Drug Discovery Today 2001; 6(8):427–36.

102. Hong S, Leroueil PR, Majoros IJ, et al. The binding avidity of a nanoparticle-based multivalent targeted drug delivery platform. Chem Biol 2007; 14: 107–15.

103. Wang F, Bronich TK, Kabanov AV, et al. Synthesis and evaluation of a star amphiphilic block copolymer from poly(epsilon-caprolactone) and poly (ethylene glycol) as a potential drug delivery carrier. Bioconjug Chem 2005; 16:397–405.

104. Khandare JJ, Chandna P, Wang Y, et al. Novel polymeric prodrug with multivalent components for cancer therapy. J Pharmacol Exp Ther 2006; 317: 929–37.

105. Neerman MF, Chen HT, Parrish AR, Simanek EE. Reduction of drug toxicity using dendrimers based on melamine. Mol Pharm 2004; 1:390–3.

106. Sadler K, Tam JP. Peptide dendrimers: applications and synthesis. J Biotechnol 2002; 90:195–229.

107. Kratz F, Beyer U, Schutte MT. Drug-polymer conjugates containing acid-cleavable bonds. Crit Rev Ther Drug Carrier Syst 1999; 16:245–88.

108. Ulbrich K, Etrych T, Chytil P, et al. HPMA copolymers with pH-controlled release of doxorubicin: in vitro cytotoxicity and in vivo antitumor activity. J Control Release 2003; 87:33–47.

109. Ulbrich K, Etrych T, Chytil P et al. Polymeric anticancer drugs with pH-controlled activation. Int J Pharm 2004; 277:63–72.

110. Xu P, Van Kirk EA, Murdoch WJ, et al. Anticancer efficacies of cisplatin-releasing pH-responsive nanoparticles. Biomacromolecules 2006; 7:829–35.

111. Hong MS, Lim SJ, Oh YK, Kim CK. pH-sensitive, serum-stable and long-circulating liposomes as a new drug delivery system. J Pharm Pharmacol 2002; 54:51–8.

112. Tiwari SB, Pai RM, Udupa N. Temperature sensitive liposomes of plumbagin: characterization and in vivo evaluation in mice bearing melanoma B16F1. J Drug Target 2002; 10:585–91.

113. Han HD, Choi MS, Hwang T, et al. Hyperthermia-induced antitumor activity of thermosensitive polymer modified temperature-sensitive liposomes. J Pharm Sci 2006; 95:1909–17.

114. Chelvi TP, Jain SK, Ralhan R. Hyperthermia-mediated targeted delivery of thermosensitive liposome-encapsulated melphalan in murine tumors. Oncol Res 1995; 7:393–8.

115. Chelvi TP, Ralhan R. Designing of thermosensitive liposomes from natural lipids for multimodality cancer therapy. Int J Hyperthermia 1995; 11:685–95.

116. Qian ZM, Li H, Sun H, Ho K. Targeted drug delivery via the transferrin receptor-mediated endocytosis pathway. Pharmacol Rev 2002; 54:561–87.

117. Giacomini KM. Membrane transporters in drug disposition. J Pharmaco Biopharm 1997; 25(6):731–41.

118. Ayrton A, Morgan P. Role of transport proteins in drug absorption, distribution and excretion. Xenobiotica 2001; 31:469–497.

119. Kushuhara H, Sugiyama Y. Role of transporters in the tissue-selective distribution and elimination of drugs: transporters in the liver, small intestine, brain and kidney. J Control Rel 2002; 78:43–54, 2002.

120. Kemper EM, Verheij M, Boogerd W, et al. Improved penetration of docetaxel into the brain by co-administration of inhibitors of P-glycoprotein. Eur J Cancer 2004; 40:1269–74.

121. Leu BL, and Huang JD. Inhibition of intestinal P-glycoprotein and effects on etoposide absorption. Cancer Chemother Pharmacol 1995; 35:432–6.

122. Walter KA, Tamargo RJ, Olivi A, et al. Intratumoral chemotherapy. Neurosurgery 1995; 37:1128–45.

123. Hamstra DA, Moffat BA, Hall DE, et al. Intratumoral injection of BCNU in ethanol (DTI-015) results in enhanced delivery to tumor – a pharmacokinetic study. J Neurooncol 2005; 73:225–38.

124. Jackson JK, Zhang X, Llewellen S, et al. The characterization of novel polymeric paste formulations for intratumoral delivery. Int J Pharm 2004; 270: 185–98.

125. Noble CO, Krauze MT, Drummond DC, et al. Novel nanoliposomal CPT-11 infused by convection-enhanced delivery in intracranial tumors: pharmacology and efficacy. Cancer Res 2006; 66:2801–6.

126. Saltzman WM, Fung LK. Polymeric implants for cancer chemotherapy. Adv Drug Deliv Rev 1997; 26:209–30.

127. Yamashita R. [Experimental study of an anticancer drug delivery system using polylactic acid]. Nippon Geka Gakkai Zasshi 1987; 88:401–12.

128. http://www.mgipharma.com/wt/page/gliadel (accessed March2007).

129. Mantripragada S. A lipid based depot (DepoFoam technology) for sustained release drug delivery. Prog Lipid Res 2002; 41(5):392–406.

130. Kang GD, Cheon SH, Song SC. Controlled release of doxorubicin from thermo-sensitive poly(organophosphazene) hydrogels. Int J Pharm 2006; 319:29–36.

131. Yang K, Yue B, Wang Z. [Progress in electrochemotherapy]. Sheng Wu Yi Xue Gong Cheng Xue Za Zhi 2004; 21:1043–6.

132. Mir LM. [Antitumor electro-chemotherapy]. Bull Cancer 1994; 81:740–8.

133. Domenge C, Orlowski S, Luboinski B, et al. Antitumor electrochemotherapy: new advances in the clinical protocol. Cancer 1996; 77:956–63.

134. Rebersek M, Cufer T, Cemazar M, et al. Electrochemotherapy with cisplatin of cutaneous tumor lesions in breast cancer. Anticancer Drugs 2004; 15:593–7.

135. M Raderer and W Scheitauer. Clinical trials of agents that reverse multidrug resistance. A literature review. Cancer 1993; 72:3353–63.

136. Relling MV. Are the major effects of P-glycoprotein modulators due to altered pharmacokinetics of anticancer drugs? Ther Drug Monit 1996; 18:350–6.

137. Perez-Tomas R. Multidrug resistance: retrospect and prospects in anti-cancer drug treatment. Curr Med Chem 2006; 13(16):1859–76.

138. Teodori E, Dei S, Martelli C, Scapecchi S, Gualtieri F. The functions and structure of ABC transporters: implications for the design of new inhibitors of

Pgp and MRP1 to control multidrug resistance (MDR). Curr Drug Targets 2006; 7(7):893–909.

139. Gabizon A, Shmeeda H and Barenholz Y. Pharmacokinetics of pegylated liposomal doxorubicin: review of animal and human studies. Clin Pharmacokinet 2003; 42:419–36.

140. http://www.doxil.com/pdf/DOXIL_PI_Booklet.pdf (accessed March 2007).

141. van Dalen EC, Michiels EM, Caron HN, Kremer LC. Different anthracycline derivates for reducing cardiotoxicity in cancer patients. Cochrane Database Syst Rev 2006; (4):CD005006.

142. http://www.bioalliancepharma.com/products_transdrug.asp (accessed March 2007).

143. Couvreur P and Vauthier C. Nanotechnology: intelligent design to treat complex disease. Pharm Res 2006; 23:1414–9.

144. deVerdiere AC, Dubernet C, Nemati F, et al. Reversion of multidrug resistance withpolyalkylcyanoacrylate nanoparticles: towards amechanism of action. Brit J Cancer 1997; 76:198–205.

145. Soma CE, Dubernet C, Barratt G, et al. Ability of doxorubicin-loaded nanoparticles to overcome multidrug resistance of tumor cells after their capture by macrophages. Pharm Res 1999; 16:1710–16.

146. Hennenfent KL and Govindan R. Novel formulations of taxanes. Old wine in a new bottle? Ann Oncol 2006; 17:735–49.

147. http://www.abraxane.com/resources/Abraxane_PI.pdf (accessed March 2007).

148. Gradishar WJ. Albumin-bound paclitaxel: a next-generation taxane. Expert Opin Pharmacother 2006; 7(8):1041–53.

149. National Cancer Institute, http://www.cancer.gov/Templates/db_alpha.aspx?CdrID = 45043 (accessed March 2007).

150. Taxoprexin DHA-paclitaxel allows higher taxane doses with fewer side effects. http://www.scienceblog.com/community/older/2000/C/200002582.html (accessed March 2007).

151. Singer JW. Paclitaxel poliglumex (XYOTAX, CT-2103): a macromolecular taxane. J Control Release 2005; 109:120–6.

152. Singer JW, Shaffer S, Baker B, et al. Paclitaxel poliglumex (XYOTAX; CT-2103): an intracellularly targeted taxane. Anticancer Drugs 2005; 16:243–54.

153. Rueda Dominguez A, Olmos Hidalgo D, Viciana Garrido R, Torres Sanchez E. Liposomal cytarabine (DepoCyte) for the treatment of neoplastic meningitis. Clin Transl Oncol 2005; 7(6):232–8.

154. Brem H and Gabikian P. Biodegradable polymer implants to treat brain tumors. J Control Release 2001; 74:63–67.

155. Sancho JM, Ribera JM, Romero MJ, Martin-Reina V, Giraldo P, Ruiz E. Compassionate use of intrathecal depot liposomal cytarabine as treatment of central nervous system involvement in acute leukemia: report of 6 cases. Haematologica 2006; 91(3):ECR02.

156. Ducray F, Honnorat J. New adjuvant chemotherapy for glioblastoma. Presse Med 2007; (Epub ahead of print).

157. http://www.mgipharma.com/wt/page/gliadel (accessed February 2007).

158. Giese A, Kucinski T, Knopp U, Goldbrunner R, Hamel W, Mehdorn HM, Tonn JC, Hilt D, Westphal M. Pattern of recurrence following local

chemotherapy with biodegradable carmustine (BCNU) implants in patients with glioblastoma. J Neurooncol 2004; 66(3):351–60.

159. Brannon-Peppas, L and Blanchette, JO. Nanoparticle and targeted systems for cancer therapy. Adv Drug Deliv Rev 2004; 56:1649–59.

7

Targeting Infections within the Gastrointestinal Tract

Hannah Batchelor and Barbara Conway

Medicines Research Unit, School of Life and Health Sciences, Aston University, Birmingham, U.K.

INTRODUCTION

Pharmaceuticals remain the least invasive and most cost-effective means of patient treatment, and represent a substantially important factor in the potential reduction of long-term, overall healthcare costs. The pharmaceutical industry continues to evolve within a transitional environment as most of the elements affecting its core business have changed. Drug companies are confronting the consequences of managed care, increased pricing pressures, generic competition, and overall margin shrinkage. In response, they are vigorously seeking new ways to increase profitability, differentiate their products, and add depth and diversity to their product lines. One of these alternatives is to exploit improvements upon the method of drug delivery.

Advanced drug delivery entails the use of one or more technologies to create a system or vehicle that will facilitate the entry of a therapeutic substance into the body in a manner that demonstrates better performance, fewer side effects, greater efficacy, and less wasted product than a prior system. In the current financial climate the pharmaceutical industry is capitalizing on improved drug delivery technology as a means of improving profitability. Newer advanced drug delivery methods can differentiate products from the competition by improved performance and increased compliance over standard delivery methods, which is increasingly important due to generic competition. This chapter focuses on advanced drug delivery systems that target the GI tract using infectious diseases as examples of the benefits of these delivery systems where appropriate.

Structure of the GI Tract

The gastrointestinal tract can be divided into five regions in terms of drug targeting: the oral cavity, esophagus, stomach, intestine, and colon, each of these regions may be further subdivided for specific targeted drug delivery. Each of these regions is discussed specifically in terms of drug delivery within this chapter. Along the GI tract there are many differences in terms of epithelial topography, luminal contents, pH, enzymatic profile and obviously time from ingestion, each of these factors can be used as a trigger for drug delivery. Along the GI tract there are many differences in terms of epithelial topography, luminal contents, pH, enzymatic profile and obviously time from ingestion; each of these factors can be used as a trigger for drug delivery. Figure 1 illustrates the regions of the GI tract alongside drug delivery formulations that can be used to target these regions, providing an overview of drug delivery along the GI tract.

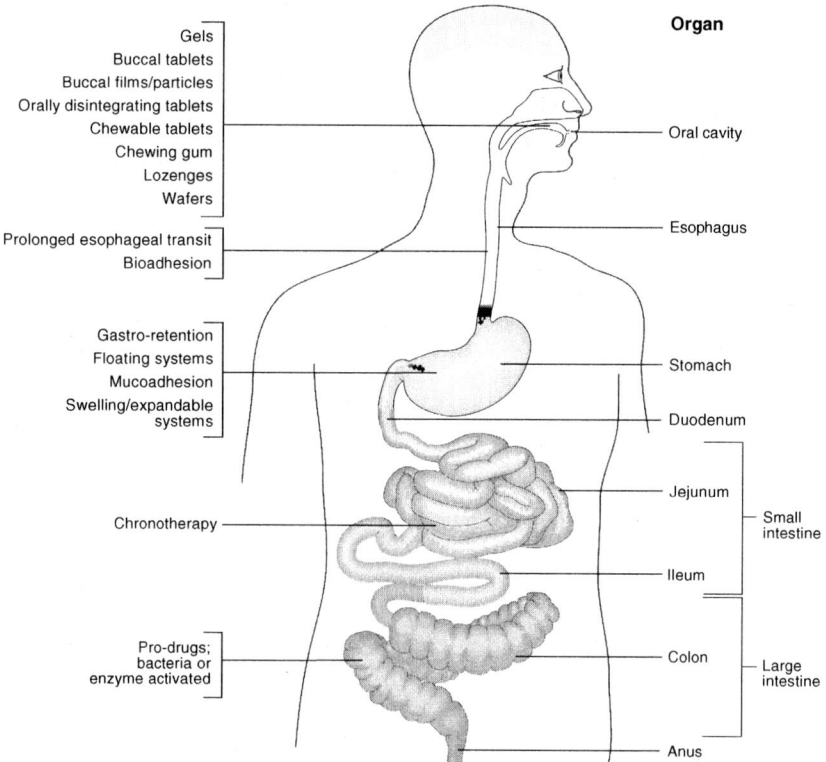

Figure 1   The GI tract can be divided in terms of regions (righthand side) and in terms of methods used for drug delivery within those regions (lefthand side).

Infectious Diseases of the GI Tract

Diarrhea is the most frequent manifestation of gastrointestinal infection. Diarrhea is common in both the developed and developing world. Poor water quality and sewage disposal are the main factors for intestinal infections in the developing world, whereas the widespread use of broad-spectrum antibiotics and impaired host community owing to greater numbers of immunocompromised individuals are factors in the developed world. Symptoms of enteric infections are not limited to the gastrointestinal system, the respiratory system, nervous system eyes and skin may also be affected by microorganisms multiplying within the GI tract. For example infection by *E. coli* can result in kidney damage; even so called simple GI infections can result in severe vomiting and diarrhea which in turn can lead to dehydration. Other examples of gastrointestinal infections include: esophagitis from candidiasis; gastritis from anisakiasis; intestinal obstruction from tuberculosis, and proctitis from *Chlamydia* infections.

Current and Future Agents Used in Infections

Table 1 lists drugs currently used in the treatment of infections within the GI tract. Standard therapy is often an oral tablet or capsule of the drug administered for systemic uptake from the intestine via a conventional drug delivery formulation as described below.

Conventional Drug Delivery

The primary aim of the management of many gastrointestinal infections is appropriate therapy for self-limiting disorders. However, in immunocompromised patients often the infections are not self-limiting and the pathogen needs to be identified to enable administration of an effective, safe antimicrobial chemotherapeutic agent. In other gastro-intestinal infections specific therapy is required.

In this chapter conventional drug delivery refers to current usual dosage forms for oral drug delivery including: oral liquids (solutions, suspensions, and emulsions) and solid dosage forms such as tablets, capsules, and granules. In each of these cases the drug is contained within the dosage form for absorption within the GI tract for systemic action. If a rapid onset of action is required, an oral liquid is chosen with solutions being most rapid-acting as they bypass the need for disintegration and drug dissolution. Oral liquids are also preferred in patients that have difficulty in swallowing; the elderly or very young. Emulsions are a useful means of delivering hydrophibic drugs as the bioavailability of such compounds may be improved via an appropriate formulation. The most common dosage form is the tablet, where the most popular form of tablet is one that is

Table 1   Drug Therapy Used in Infections Within the GI Tract

Disease	Therapy	Formulation
Dental infection	Phenoxymethylpenicillin	Tablets or oral solution
	Amoxicillin	Capsules, oral suspension, sachets, injection (powder for reconstitution)
	Erythromycin	Capsules, tablets, suspension
	Metronidazole	Tablets, suspension, intravenous infusion
	Amoxicillin + Clavulanic acid	Tablets, oral suspension, dispersible tablets, injection (powder for reconstitution)
Oral herpes	Tetracycline	Capsule, mouthwash
	Chlorhexidine	Mouthwash, oral gel, oral spray
Throat infection	Phenoxymethylpenicillin	Tablets or oral solution
	Amoxicillin	Capsules, oral suspension, sachets, injection (powder for reconstitution)
	Oral cephalosporin (e.g. cefalexin, cefradine, cefadroxil, cefalcor, cefprozil)	Capsules, tablets, syrup, suspension
Oropharyngeal candidiasis	Amphotericin	Tablets, lozenges, suspension, injection (powder for reconstitution)
	Nystatin	Suspension, pastilles
	Fluconazole	Capsules, suspension, injection (powder for reconstitution)
	Itraconazole	Capsules, oral liquid
	Ketaconazole	Tablets
	Miconazole	Oral gel
H. pylori infection	Amoxicillin	Capsules, oral suspension, sachets, injection (powder for reconstitution)
	Clarithromycin	Tablets, granules, pediatric suspension, intravenous infusion
	Metronidazole	Tablets, suspension, intravenous infusion
Gastroenteritis	Antibacterial not indicated, usually self-limiting	

(*Continued*)

Table 1 Drug Therapy Used in Infections Within the GI Tract (*Continued*)

Disease	Therapy	Formulation
Campylobacter enteritis	Ciprofloxacin	Tablets, suspension, intravenous infusion
	Erythromycin	Capsules, tablets, suspension
Invasive salmonellosis	Ciprofloxacin	Tablets, suspension, intravenous infusion
	Trimethoprim	Tablets, suspension, injection
Shingellosis	Ciprofloxacin	Tablets, suspension, intravenous infusion
	Trimethoprim	Tablets, suspension, injection
Typhoid Fever	Ciprofloxacin	Tablets, suspension, intravenous infusion
	Cefotaxime	Injection (powder for reconstitution)
	Chloramphenicol	Capsules, injection (powder for reconstitution)
Intestinal candidiasis	Amphotericin	Tablets, suspension, injection (powder for reconstitution)
Antibiotic-associated colitits	Metronidazole	Tablets, suspension, intravenous infusion
	Vancomycin	Capsules, injection (powder for reconstitution)
Peritonitis	Cephalosporin (e.g. cefalexin, cefradine, cefadroxil, cefalcor, cefprozil)	Capsules, tablets, syrup, suspension
	Gentamycin with Metronidazole	Injection Tablets, suspension, intravenous infusion
	Clindamycin	Injection, capsules

swallowed whole and releases the drug within the GI tract for systemic action. Capsules have increased in popularity in recent times with new liquid capsules offering the convenience of a solid dosage form with the rapid onset of an oral liquid. Buccal and sublingual tablets as well as coated tablets are described in more detail within the advanced drug delivery section.

Limitations of Current Therapy

One example of the limitations of current therapy is the standard local treatment for oropharyngeal infections: nystatin suspension (1–3 million units orally four times daily) or clotrimazole (10 mg orally five times daily) are treatment regimes that have similar efficacy, but such frequent dosing leads to poor patient compliance. Systemic therapy is used in severe cases and in immunocompromised patients with the treatment of choice being fluconazole 100–200 mg daily. Fluconazole is used in preference to ketoconazole and itraconazole. However, in fluconazole-resistant patients, combination therapy of itraconazole (100–200 mg daily) and flucocytosine (100 mg/kg daily) or intravenous amphotericin B (AmB; 3–5 mg/kg daily) are equally as effective options. More advanced drug combinations are always being investigated due to the development of resistance that can occur in immunocompromised individuals with long-term therapy where prophylactic therapy is required.

AmB is a poorly soluble anti-fungal agent, and therefore absorption within the GI tract is limited. To overcome this AmB is used parenterally as a liposomal (AmBisome®) or as a colloidal dispersion (Fungizone®, Abelcet®) for the treatment of systemic infection. Parenteral formulations are disadvantageous due to the high costs and low compliance associated with treatment, particularly in resource-low countries. Advanced drug delivery that provides a means by which such drugs can be administered orally offers many advantages for such low solubility drugs.

Novel agents to treat infectious diseases are constantly required as new resistance mechanisms and resistance strains continually emerge. However, as with all medicines there is a need to design appropriate delivery systems for the agents to maximize their efficacy. Genomics was anticipated to be the solution for the discovery of novel acting antibacterials following the sequencing of the first bacterial genome in 1995. However, no novel agent has progressed beyond Phase I testing, despite much industrial research and development activity.

Caspofungin was approved by the FDA in 2004 for the treatment of fungal infections. Caspofungin belongs to the echinocandin class of anti-fungal agents along with micafungin and anidulafungin. Currently, these drugs are not available orally and are limited to indications requiring parenteral therapy. An ultimate goal in drug delivery is to formulate these agents so that they can be self-administered by the patient, with the oral route being most popular, for efficacy superior to current azole therapy.

Augmentin® was the first β-lactum/β-lactamase inhibitor combination to be marketed in the 1980s, and was a best-selling antibacterial with sales of $2 billion per annum. However, despite this success and the launch of other β-lactum/β-lactamase inhibitor drugs (Timentin®, Unasyn®, Zosyn® and Sulperazone®) before 1990, there have been no further advances over the past 20 years.

Novexel is a pharmaceutical company that researches and develops novel antimicrobial compounds with activity against multi-resistant organisms, and with a low propensity to generate resistance (www.novexel.com). The company currently has four compounds in development; the oral Streptogramin compound NXL103 entered a Phase I multi-dose study in October 2005. In addition to NXL103, the

Aminocandin anti-fungal, NXL201, is also in Phase I for the treatment of severe fungal infections. NXL104, a novel non-β-lactum/β-lactamase inhibitor for the treatment of nosocomial infections in also in pre-clinical development. NXL101 is a bacterial topoisomerase/gyrase inhibitor and the first drug candidate in a new class of antibacterials to treat hospital-based infections caused by gram positive pathogens. NXL101 is being developed in both intravenous and oral dosage formulations, and this compound has recently progressed into Phase I clinical development.

Tigecycline (Wyeth Research) received FDA approval in 2005 as the first marketed glycylcycline antibacterial agent for intra-abdominal infections. Glycylcyclines are an advanced form of tetracyclines with the benefit of defeating the defined mechanisms of tetracycline resistance.

There have also been advances in antimycotic drugs due to the emergence of strains resistant to currently used antifungal drugs. Antimicrobial peptides that are often cationic and amphipathic including defensins, protegrins, histatins, and lactoferrin derived peptides have been investigated for their antifungal efficacy in the treatment of candidiasis (1).

There is clearly a clinical need for newer, more potent, broader spectrum anti-infective agents as increasing resistance makes currently marketed therapies less effective. Effective delivery of these compounds is also crucial for clinical success.

THE ORAL CAVITY

Infections of the Oral Cavity

Infections of the oral cavity are often caused by overactivity of endogenous flora (e.g. dental caries) and also by primary pathogens (e.g. candidiasis). Typical disorders include viral infections such as *Herpes simplex* (cold sores); fungal infections such as oral thrush caused by the *Candida albicans* and other *Candida* species; bacterial-induced infection such as gingivitis leading to periodontitis in some cases.

Standard treatments for oral infections are systemic administration of antifungal or antibacterial agents as required. Local therapies exist for many conditions and these include mouthwashes containing antibacterial agents such as chlorhexidine for bacterial infection (Corsodyl®, GSK Consumer Healthcare, U.K.), and antibacterial sprays. Mechanical protection for ulcers within the oral cavity is provided by a thick carmellose gelatine paste that prevents further physical or chemical damage at the ulcer site (Orabase®, ConvaTec, U.K.). Further alternatives include steroid tablets that are allowed to dissolve next to an ulcer for specific targeted therapy (Corlan®, Celltech, U.K.). Oral gels that incorporate drugs such as miconazole (Daktarin®, Janssen-Cilag) are used for topical fungal infections within the oral cavity. Lozenges are also available including nystatin pastilles (Nystan®, Squibb), amphotericin lozenges (Fungilin®, Squibb). In addition, suspensions of both nystatin and amphotericin that are intended to be held in the mouth are also available. Conventional oral formulations including mouthwashes, rinses and oral gels are limited as they are incapable of maintaining salivary concentrations of drugs for prolonged periods. Therefore a system that provides sustained release of antifungal agents within the oral cavity is desirable.

Advanced Formulations Targeting the Oral Cavity

Advantages of targeting the oral cavity include the relatively large surface area for drug absorption [the total surface area is approximately $170\,cm^2$ with the buccal membranes being approximately $50\,cm^2$ of this total area (2)] and the ready accessibility of this site. The rich blood supply of the oral cavity is also advantageous for systemic absorption of drugs as well as the low metabolic activity within the mouth compared to the remainder of the GI tract. In the treatment of local infections the biggest advantage in targeting the oral cavity is retention of the dosage form, allowing prolonged exposure of the infection to the drug thereby enhancing the efficacy. This can also be coupled with zero-order release profiles that have been demonstrated with certain dosage forms. However, there are also disadvantages associated with drug delivery targeted within the oral cavity. These relate to the size of the dose that can be administered and this is of particular relevance for systemic absorption. Variability in bioavailability may also be an issue, due to variable saliva flow rates and clearance. Additionally, accidental swallowing of delivery devices designed for oral retention could be problematic. The two most important factors in formulation design for orally delivered formulations are taste masking and patient comfort, as these will dictate the compliance of such systems.

In a previous study, a comparison of release from chewing gum, sublingual tablets and lozenges was performed using ^{99m}Tc E-HIDA as a model hydrophilic agent. Gamma scintigraphy was used to visualize the distribution. No differences in distribution were observed among the three formulations in the oral cavity, esophagus, and glottis, although the duration within the oral cavity was greatest for the sublingual tablet and shortest for the lozenge (3).

Solutions, Suspensions, and Emulsions

There are many formulations readily available for the treatment of infections within the oral cavity. Reports suggest that fluconazole suspension may be better able to treat oropharyngeal candidiasis via a "swish and swallow" technique (4) compared to the tablet formulation. The application of a suspension intended for systemic absorption has not been fully exploited yet, and this indicates that local action plays a part in therapy.

Gels

Daktarin gel is a popular therapy for mild oral infections, and this has been used successfully for many years although the high doses and relatively short retention time within the oral cavity means that it is less effective compared to more sophisticated delivery systems. Chitosan has previously demonstrated anti-fungal action, and this combined with chlorhexidine was examined as buccal gels or films for anti-fungal action by prevention of *Candida* binding to the mucosal surface. In vitro studies suggested that these formulations would provide prolonged levels of drug compared to oral rinses (5). Incorporation of amphotericin into Orabase prior to administration to the buccal mucosa resulted in sufficiently high salivary drug concentrations overnight to maintain therapy (6).

Buccal Tablets, Patches, and Films

Buccal tablets are small tablets designed for local administration to the buccal cavity. These can be subdivided into two categories: bioadhesive tablets that release drug in one direction into the buccal mucosa for systemic uptake and erodible tablets that erode within the oral cavity releasing drug into the saliva for both local action and systemic uptake from the oral cavity and the GI tract once swallowed. Buccal patches and films are similar to tablets in mechanism of release yet manufactured in different ways. Buccal patches refer to unidirectional release systems where an impermeable backing exists on one side of the formulation. Buccal films are flexible and have a larger surface area compared to tablets. Buccal tablets and films are generally manufactured with polymeric excipients that control the rate of drug release; these polymers may also have bioadhesive properties to retain the tablet/film within the oral cavity.

An orally erodible buccal formulation releases dissolved drug within the saliva providing an ideal means to treat local infections including mucositis. The rate at which an erodible device dissolves is dependant upon the formulation, with high molecular weight polymers providing sustained release.

A number of studies have been performed measuring bioadhesive strength and in vitro drug release from buccal tablets, patches and films for infection control. Many of the studies have been performed using miconazole as an antifungal agent (7,8); other studies have investigated clotrimazole (9); some chlorhexidine (10–12).

A comparison of a 10 mg once-daily mucoadhesive buccal tablet of miconazole with 400 mg once daily ketoconazole administered for systemic uptake was performed in HIV-positive patients with oropharyngeal candidiasis. The results demonstrated that the low dose treatment via the buccal tablet was not statistically inferior to the systemic treatment. Local therapy was associated with a lower incidence of gastrointestinal disorders and drug-related adverse events (13), demonstrating the advantages of advanced drug delivery over conventional systemic delivery.

Several studies have compared the salivary concentrations of miconazole from a buccal tablet to an oral gel, in all cases the dose within the tablet was much lower than the gel although both are designed for release into the oral cavity and local action (14,15). Buccal tablets, by providing sustained release, are more cost effective as lower drug loading is required. In one study, salivary miconazole pharmacokinetics of a once-daily 50 mg bioadhesive eroding tablet (Lauriad®, Bioalliance Pharma) was compared to a gel directed to be applied three times-daily with a total dose of 375 mg. The results demonstrated that salivary concentrations following the administration of the tablet were higher with prolonged duration above the MIC of some *Candida* species (16). Phase III studies have been carried out by Bioalliance Pharma evaluating the company's miconazole bioadhesive buccal tablet (Lauriad). The tablet has proven efficacy in cancer patients suffering from oropharyngeal candidiasis following radiotherapy, and this product is expected to launch in Europe in the very near future. In addition, this company is developing a follow-on product, acyclovir Lauriad for the treatment of oral herpes, utilizing the same delivery platform; they have completed a Phase I clinical study and a phase 2/3 study is expected in Europe in 2006 (http://www.bioalliancepharma.com/products.asp).

Orally Disintegrating Tablets

Orally disintegrating tablets (ODTs) are those that disintegrate upon contact with saliva within the oral cavity releasing the drug into the saliva for rapid uptake or local action. The demand for fast-dissolving/ disintegrating tablets or fast-melting tablets that can dissolve or disintegrate in the mouth has been growing particularly for those with difficulty swallowing tablets such as elderly and children. They are referred to using a range of terminologies: fast-dissolving, orodispersible, and fast-melting and the FDA has adopted the term ODTs.

ODTs disintegrate and/or dissolve rapidly in the saliva without the need for water, within seconds to minutes. Some tablets are designed to dissolve rapidly in saliva, within a few seconds, and are true fast-dissolving tablets. Others contain agents to enhance the rate of tablet disintegration in the oral cavity, and are more appropriately termed fast-disintegrating tablets, as they may take up to a minute to completely disintegrate. Increased bioavailability using such formulations is sometimes possible if there is sufficient absorption via the oral cavity prior to swallowing (17). However, if the amount of swallowed drug varies, there is the potential for inconsistent bioavailability. Patented ODTs technologies include OraSolv®, DuraSolv®, Zydis®, FlashTab®, WOWTAB®, and others.

A fluconazole orally dispersible tablet (100 mg once-daily) was prepared via microencapsulation of the drug to allow rapid dispersion of the drug. This was found to be effective for the treatment of oropharyngeal candidiasis (18). The ODT fluconazole tablet was bioequivalent to the solid dosage form of fluconazole (capsule) yet the salivary concentration of the drug was 63 times higher with the ODT compared to the capsule (18).

Chewable Tablets

Chewable tablets are designed to be mechanically disintegrated in the mouth. Potential advantages of chewable tablets mainly concern patient convenience and acceptance although enhanced bioavailability is also claimed. This can be due to a rapid onset of action as disintegration is more rapid and complete compared to standard formulations that must disintegrate in the GI tract. The dosage form is an appealing alternative for pediatric and geriatric consumers. Chewable tablets are also desirable because they offer convenience for consumers, avoiding the necessity of co-administration with water, and creation of palatable formulations could increase compliance. A limitation with this system is that many pharmaceutical actives have an unpleasant bitter taste that can reduce compliance among patients.

Antacid and pediatric vitamins are often formulated as chewable tablets but other formulations include anti-histamines (Zyrtec®), antimotility agents (Imodium® Plus), antiepileptic agents (Epanutin Infatabs®), antibiotics (Augmentin Chewable), asthma treatments (Singulair®), and analgesics (Motrin®).

Augmentin chewable tablets show statistically similar pharmacokinetic profiles to the equivalent dose of Augmentin suspension (Augmentin prescribing information GSK, http://www.gsk.com/products/prescription_-medicines/us/augmentin_us.htm) in the treatment of severe dental infections, although a chewable tablet is more convenient and often associated with greater stability profile compared to liquids.

Chewing Gums

Medicated chewing gum is a drug delivery system containing gum base with a pharmacologically active ingredient that can be used for local delivery within the oral cavity or for systemic absorption. Chewing gum offers advantages in that it can be taken without water (similar to oral liquids) yet possesses stability and shelf life associated with solid dosage forms. In addition, the discrete nature and unit dose capabilities could also improve patient acceptability. Drugs that are intended for local action within the oral cavity may have low saliva solubility and chewing gum can assist in improving solubility and retention of the drug within the gum in the oral cavity. Disadvantages include the need for taste-masking of the active agent, and this has been reported to be problematic with chlorhexidine which also stains the teeth and tongue (19).

Chewing gums are formulated from water insoluble bases that incorporate elastomer to control the gummy texture and plasticizers to regulate the cohesion of the formulation. Fillers and water soluble sweeteners are also added as required. By altering the composition of the gum the release of the active ingredient can be manipulated. Factors affecting the release of medicament from chewing gum can be divided into three groups: the physicochemical properties of the drug, the gum properties, and chew-related factors, including rate and frequency. A special apparatus has been developed to measure drug release from chewing gums that accounts for these variables (Fig. 2). Drugs can be incorporated into gums as solids or liquids. For most pharmaceuticals, aqueous solubility of the drug will be a major factor affecting the release rate. In order for drugs to be released, the gum would need to become hydrated; the drugs can then dissolve and diffuse through the gum base under the action of chewing.

The promotion of sugar-free gums to counteract dental caries by stimulation of saliva secretion, which increases plaque pH aiding in the prevention of caries, has led to a more widespread use and acceptance of gums. Medicated gums for delivery of dental products to the oral cavity are marketed in a number of countries, for example, fluoride-containing gums as an alternative to mouthwashes and tablets or chlorhexidine gum for treatment of gingivitis. The potential use of medicated chewing gums in the treatment of oral infections has also been reported. Gums have been prepared containing antifungal agents such as nystatin (20) and miconazole (21) or antibiotics, such as penicillin and metronidazole for the treatment of oral gingivitis (22).

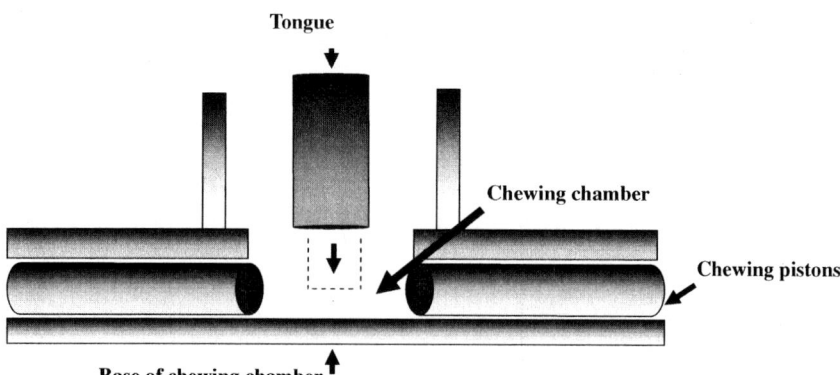

Figure 2 Schematic diagram of the chewing chamber of an in vitro chewing apparatus.

Clinical trials that compared miconazole oral gel with miconazole chewing gum demonstrated that the gum was at least as efficient as oral gel in the treatment of fungal infections within the mouth, although very low doses of the drug were released from the chewing gum indicating greater efficacy, likely to be due to the retention of the formulation within the oral cavity (21,23). Analgesic-antibiotic chewing troches and chewing gum have been used in post-operative care of tonsillectomized patients, however, this is now inappropriate considering sensitization and toxicology (24).

Lozenges

Lozenges are tablets that dissolve or disintegrate slowly in the mouth to release drug into the saliva. They are easy to administer to pediatric and geriatric patients and are useful for extending drug form retention within the oral cavity. They usually contain one or more ingredients in a sweetened flavored base. Drug delivery can be either for local administration in the mouth, such as anesthetics, antiseptics, and antimicrobials or for systemic effects if the drug is well absorbed through the buccal lining or is swallowed. More traditional drugs used in this dosage form include phenol, sodium phenolate, benzocaine, and cetylpyridinium chloride. Antifungal lozenges of AmB, and nystatin are also currently commercially available. These provide local therapy for oral infections and are used in early stage therapy prior to systemic delivery of azoles.

Lozenges containing both miconazole and chlorhexidine have been developed and initial studies demonstrated that, in comparison to a proprietary oral gel formulation, the bioadhesive lozenges produced much more uniform and effective salivary levels of miconazole over a prolonged period (25).

Wafers

A novel wafer drug-delivery system intended for the treatment of microbial infections associated with periodontis is inserted into the periodontal pocket providing controlled drug release over four weeks (26). This formulation consists of poly(lactic-co-glycolic acid) (PLGA) as a primary bioerodible polymeric component, poly(ethylene glycol) as a plasticizer and encapsulation aid, and silver nitrate to provide the bioactive silver. Such sustained release is highly advantageous in terms of patient compliance. Existing products such as PerioChip®, a biodegradable, cross-linked gelatin matrix that is capable of maintaining an efficacious chlorhexidine concentration in gingival crevicular fluid for up to 7 days, are already on the market (www.periochip.com).

Modulation of Candidal Adherence

Adhesion of micro-organisms, such as *C. albicans*, is the first step in the establishment of infection; therefore, prevention of this step is a strategy in reducing infection incidence. The adhesion properties of antimicrobial agents has been examined at very low (sub-MIC) concentrations (27–30). More recent studies have examined bioadhesive polymers coating a surface as a means of modulating *Candida* adherence; Barembaum et al. (29) examined sodium alginate and chitosan as agents to prevent the first stage adhesion and thus combat infection and found that these agents had in vitro MIC values of 0.1% and 0.25% w/v, respectively (29). The use of non-drug loaded polymeric nanoparticles to disrupt microbial adherence was measured in vitro using buccal epithelial cells with a reduction in adherence observed, and this may be a future direction for prophylaxis of candidiasis of the oral cavity (30).

THE ESOPHAGUS

Infections of the Esophagus

Esophageal infections are usually, although not always, associated with immunocompromised hosts, as the increase in cancer chemotherapy and emergence of HIV has created greater numbers of immunocompromised patients. *Candida* and *H. simplex* virus are the most common infections although bacterial (e.g. *Mycobacterium tuberculosis*) and viral (e.g. *Varicella-zoster* virus) infections have also been cited.

Advanced Formulations Targeting the Esophagus

Infections of the esophagus are primarily associated with immunocompromised hosts, including patients undergoing cancer chemotherapy or those with HIV. Available antifungals for the treatment of esophageal candidiasis include polyenes, azoles, nystatin, and AmB. Therapy with these agents is not always successful due to the poor permeation of the esophagus and their low systemic availability. The treatment of choice is fluconazole 100–200 mg daily which will achieve endoscopic clearance in more than 90% of patients (31). In patients who are susceptible to recurrent esophageal infections, prophylaxis with azoles reduces the risk of relapse. Recently, however, fluconazole-resistant *Candida* infections have been reported (32) thus necessitating alternative therapeutic strategies.

Topical therapy designed to target the infection from the luminal aspect of the esophagus bypasses the need for high doses of systemic azoles. Although topical therapy has been used in the oral cavity with, for example, miconazole oral gel (Daktarin, Janssen-Cilag), topical therapy within the esophagus has had limited success. Itraconazole oral solution (intended for systemic uptake) has been used in the treatment of oropharyngeal candidiasis, although the rationale for the solution was ease of administration rather than topical therapy. However, the benefits of topical action were recognized (33). Likewise, fluconazole suspension led to a more rapid clinical cure compared to capsules and the topical effect was considered to be important and warrants further investigation (34).

Delivery devices targeting the esophagus have generated recent interest with predominant research focusing on bioadhesive, viscous liquids. A viscous liquid will demonstrate a longer transit time through the esophagus, thereby increasing the contact time between the drug and the site of action. In addition, there is no need for disintegration or dissolution as with solid dosage forms. Many of the strategies employed for targeting the oral cavity can be translated into targeting the esophagus with drug release into saliva over prolonged periods leading to a constant swallowing of antifungal agents. In addition, concomitant release of viscosity-enhancing polymers from dosage forms such as erodible lozenges would lead to even longer transit times of the drug containing saliva through the esophagus.

THE STOMACH

Infections of the Stomach

Helicobacter pylori is an infection that occurs within the stomach and is interesting in terms of advanced drug delivery. *H. pylori* is a gram-negative, microaerophilic, spiral and flagellated bacterium, with a unipolar-sheathed flagella that provides motility. Its spiral shape and high motility allow it to penetrate mucus, resist gastric emptying and remain in the host gastric mucosa. Since its isolation in the early 1980s, it has now been firmly established that infection with this bacterium is the cause of chronic active gastritis and an aetiological factor in peptic ulcer disease, gastric mucosal-associated lymphoid tissue lymphoma and gastric carcinoma.

Although *H. pylori* is sensitive to many antibiotics in vitro, no single agent is effective alone in vivo and the treatment for eradication of *H. pylori* is complicated, requiring a minimum of two antibiotics in combination with gastric acid inhibitors. *Helicobacter* infections are currently treated with a first-line triple therapy treatment, consisting of a proton pump inhibitor and two antibiotics. None of the antibiotics used achieved sufficient eradication, therefore adjuvant therapy is required (35).

However, increasing resistance to current antibiotics is driving research to produce alternatives to the commonly used therapies. In addition to increasing levels of antibiotic resistance, the hostile environment of the stomach, reducing antibiotic bioavailability at the site of action, contributes to failures in treatment. The bacterium resides in the surface layers of the stomach (Fig. 3), both in the mucus and the mucosa. With conventional oral therapy, the majority of antibiotic present at the site of *Helicobacter* infection (the gastric mucus) must be exsorbed from the blood. Delivery of amoxicillin across the gastric mucosa is poor compared with other drugs such as metronidazole and clarithromycin (36,37). Instability of antibiotics such as

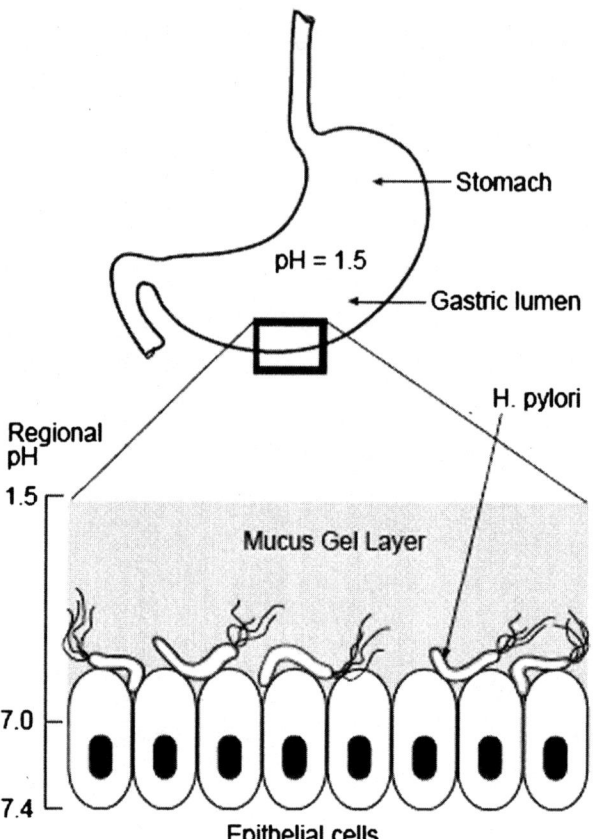

Figure 3 Schematic representation of *H. pylori* location within the stomach.

amoxicillin and clarithromycin in gastric acid may also contribute to the failure of such agents to completely eradicate the infection (38).

Major obstacles in the eradication of *H. pylori* are therefore the presence of antibiotic-resistant bacteria, a therapy requiring multiple drugs with complicated dosing schedules and bacterial residence in an environment where high drug concentrations are difficult to achieve. In order to ensure that the therapy is adequately delivered to the unique niche of the gastric mucosa, it will be necessary to employ a drug delivery system.

Advanced Formulations Targeting the Stomach

Local delivery could increase drug levels in the gastric mucus and mucosa to effective bactericidal levels against *H. pylori*. As gastric residence time is generally short in fasted conditions and variable in the fed state, even high doses may not reach desired concentrations at the site of action. The development of oral dosage forms with prolonged gastric residence is therefore desirable. Examples of strategies for gastric drug delivery are outlined below.

Gastro-Retentive Drug Delivery

The general requirements for gastro-retentive delivery systems are that the vehicle exhibits prolonged residence time in the stomach whilst maintaining a controlled release of the drug. Overcoming the physiological barriers of the human GI tract is a major challenge facing successful development of gastro-retentive systems, and inter- and intra-patient variability in retention are major impediments in product development. In addition to the thick protective mucus layer, gastrointestinal motility patterns need to be considered. In the fasted state, the interdigestive myoelectric motor complex (IMMC) is a 2-h cycle of peristaltic activity that regulates motility patterns. Therefore, gastric residence time will depend on which phase of IMMC is active when the dosage form is administered. The gastric residence time of dosage forms is also influenced by posture, age, gender, disease status, and concomitant medication. A number of different techniques have been explored to increase gastric retention including high density and magnetic systems, but the three main systems are: floating systems, bio/mucoadhesive systems, and swelling systems.

Floating systems: Various approaches have been made since the late 1970s to utilize floating behavior in order to prolong residence including hydrodynamically balanced systems (HBS), gas-generating systems, raft-forming systems, and hollow microspheres. HBS have a bulk density lower than gastric fluids and contain one or more colloids formulated into a single unit with the drug and other additives which swell on contact with water and facilitate floating. A density of less than 1.0 g/ml is required. Such systems can be successful in vitro although it is difficult to reproduce the results in vivo.

Mucoadhesive systems: Mucoadhesive delivery systems can adhere to mucus coated epithelial surfaces and potentially prolong residence time and provide a more intimate contact with the mucosal barrier. Mucoadhesive drug delivery systems have potential for the treatment of infections of the GI tract, including *H. pylori*, if they can adhere to the mucus layer and provide a controlled release of the drug. Thus, antibiotics could be delivered locally, resulting in increased retention and concentration at the site of infection. Chitosan is a cationic polymer, and so protonation at low pH enables electrostatic interaction and it has been shown to adhere to mucus (39) and porcine stomach (40). A range of dosage forms containing chitosan has been proposed for sustained antibiotic release in the stomach (41).

Swelling and expandable systems: The pyloric sphincter, situated between the antrum of the stomach and the duodenum has a diameter of about 12.8 ± 7 mm in humans. The use of expandable dosage forms exploits the mechanical restrictions placed on the exit of large particles from the stomach. If gastric retention is to be achieved merely by physical restriction of emptying, then the size required may be longer than 5 cm with a diameter greater than 3 cm (42). The device must be sufficiently strong to prevent breaking down under gastric contractile forces while not interfering with the normal functions of the stomach. The dosage form is designed to expand either by swelling or unfolding to prevent passage through the pyloric sphincter, and then it must degrade or disintegrate in the stomach to a smaller form to allow elimination once the drug is released.

Vaccination as Therapy for Stomach Infections

Although effective antimicrobial delivery strategies are a major component in the treatment of *H. pylori* infections, there is also a need for a preventative strategy as many long-term infections are asymptomatic. The underlying chronic gastritis is a risk factor for gastric cancer and unless the host develops gastrointestinal disease (e.g. peptic ulcer), it is likely to remain untreated.

A critical question is whether it is necessary or possible to induce local mucosal immune responses in the stomach. Synthetic, recombinant or highly purified sub-unit antigens are considered safe but are inherently of low immunogenicity. They often require effective delivery systems to reach the mucosal lymphoid tissues and mucosal adjuvants to enhance and direct the immune response (43). Mucosal delivery, in addition to being less invasive than parenteral administration, enables the antigen to be delivered to the major site of pathogen entry. Oral administration of protein antigens also requires the protein to be protected from the harsh acidic environment. Delivery systems for vaccines include adjuvant emulsions, liposomes and microspheres.

A number of vaccine candidates for *Helicobacter* immunization have been identified, including urease, neutrophil activating protein (NAP), bacterial lysates, vacuolating cytotoxin (VacA), cytotoxin-associated antigen (CagA), and inactivated whole cell preparations. Following infection with *H. pylori*, significant antibody responses are raised in the host, both in serum and in the stomach, and a strong mucosal T-cell response is present. Mucosal immunization can result in stimulation of both the systemic and mucosal immune responses and induction of mucosal immunity leads to production of secretory IgA. In order to gain protection against subsequent bacterial challenges, it has been proposed that mice need to have an enhanced Th2 response indicated by balanced IgG2a:IgG1 response. This can be achieved using mucosal adjuvants. The toxicity of currently available mucosal adjuvants such as cholera toxin, its B-subunit and *E. coli* heat-labile entertoxin (LT) has been a major limiting factor in the successful development of a mucosally administered vaccine.

Microspheres as Vaccine Vectors

Biodegradable microspheres prepared using synthetic and natural polymers have been investigated as vaccine adjuvant systems. An advantage of microspheres as a delivery system for vaccines is protection of the antigen from mucosal degradation. Additionally, the release rate from the microspheres is controlled by degradation and the antigen can be released into an environment rich in immunocompetent cells with induction of secretory immunity, resulting in a local and disseminated mucosal response. Delivery of the antigen in a particulate form can lead to more efficient phagocytosis and sustained release of antigen over extended time periods. Using antigen delivery systems applied at mucosal surfaces, it is therefore possible to stimulate immune responses in systemic and mucosal compartments. Due to their long, safe history, and biodegradation, microspheres consisting of poly (D, L-lactide-co-glycolide) (PLGA) have been the major system studied. Antigen release can be controlled by molecular weight, polymer composition (co-polymer ratio), particle size and loading and formulation processes. Following oral delivery, particles of less than 5 μm can be taken up intact by the mucosal associated lymphoid tissue (Peyers' patches, PP). Thus microspheres can act as an adjuvant, enhancing systemic and mucosal immune responses following uptake into the PP or lymphoid tissue at other mucosal sites.

Mucosal Adjuvants for Vaccination Against *H. pylori*

A vaccine that is administered via the oral route and induces local and systemic responses would be ideal for *H. pylori* eradication. Mucosal delivery eliminates the need for injections with associated risk of infection. However, the harsh and variable environment within the GI tract means that it is difficult to achieve consistent, reproducible levels of uptake, even when using particulate dosage forms. Despite many successful oral immunization experiments in animal studies, human trials have not met with the same degree of success. Coadministration with mucosal adjuvants may be one way to promote a protective response and modification of adjuvants has been carried out to reduce toxicity. These modified adjuvants have been assessed with *H. pylori* vaccines. A genetically modified non-toxic mutant of LT (LTK63) was found to be a potent oral adjuvant when coadministered with VacA and total bacterial lysates to mice, improving protection rates from 23.1% to 79.7% for VacA and 37% to 90% for bacterial lysates following intragastric administration (44). Systemic immunization with NAP and CagA in MF59 (an oil-in-water emulsion) was compared with nasal and oral immunisation with the same antigens combined with two genetically modified mutants, LTK63 and LTR72, respectively (45). A mucosal priming route followed by systemic boosting was shown to induce the most potent antigen-specific mucosal and systemic humoral responses in mice, selectively inducing a Th2-type cytokine responses.

In addition to development of candidate immunogens and adjuvants, there is also a need to address the issue of induction of suitable local immune responses in the host stomach. It is suggested that MHC class II-restricted CD4+ cells are important but the relative importance of Th1 or Th2 response is more controversial. Evidence from past studies suggests that pro-inflammatory Th1 responses are associated with infection while Th2 responses are required for vaccine-mediated protection (46). However, recent evidence suggests that both Th1 and Th2 responses may be protective against *Helicobacter* infection (47).

THE SMALL INTESTINE

Infections of the Small Intestine

Small bowel infections often result in diarrhea as intestinal fluid secretions prevent absorption of water and electrolytes. These infections are intriguing due to the difficulty in identifying the cause; they can be viral, bacterial, protozoan or helminth.

Amoebiasis is an infection caused by *Entamoeba histolytica*, a single celled protozoan parasite. *E. histolytica* is often found in food or water contaminated with human faeces, which once ingested settles in the large intestine. Amoebiasis can develop when something contaminated with *E. histolytica* is eaten or swallowed. The parasites then move through the digestive system and settle in the small and large intestine. Amoebiasis is the second leading cause of death from parasitic disease worldwide (48). Conventional drug therapy for amoebiasis is metronidazole and tinidazole given systemically, however, the pharmacokinetic profiles suggest that these drugs are absorbed in the upper gastro-intestinal tract and little drug is available for local action within the large intestine.

Advanced Formulations Targeting the Small Intestine

Transit time through the upper small intestine is relatively short and a prolonged residence in the stomach can be used to target drugs to this area. Time-controlled oral delivery systems which delay the release of the drug until they reach the target tissue can also be used. Delayed delivery systems can be categorized into reservoir, capsular, and osmotic devices. Reservoirs comprise single or multiple units and, according to their functional coating characteristics, can be rupturable or erodible. Rupturable reservoirs are subject to a volume increase resulting from water uptake, causing a rise in internal pressure that can alter the rate controlling membrane exposing the core contents to the external fluids and the drug is immediately available.

Reservoir systems based on erodible coating layers are mainly prepared with swellable hydrophilic polymeric materials, such as hydroxypropyl methylcellulose, hydroxyethylcellulose, and hydroxypropylcellulose. Upon contact with aqueous media, they typically undergo combined swelling, dissolution and mechanical erosion phenomena, which contribute to delay the onset of release. Hydrophilic swellable coating agents can be applied by press-coating or, more recently, through spray-coating and powder-layering techniques.

Pulsincap® is an example of a capsular system and comprises a water-insoluble capsule enclosing the drug reservoir with a swellable hydrogel plug that seals the drug contents into the capsule body. Upon contact with dissolution medium or gastro-intestinal fluids, the plug swells, pushing itself out of the capsule after a lag time, rapidly releasing the drug. The dimensions of the plug and its point of insertion into the capsule are varied to control the lag time.

A gelatin capsule coated with a semi-permeable membrane can be used to osmotically control expulsion of a plug. The drug is contained within the capsule along with an osmotic agent to attract water into the formulations, increasing pressure and expelling a water-insoluble plug.

Site specific delivery into the upper intestine has been achieved for many years by the use of pH-sensitive coatings. Enteric coated products are intended to remain intact in the stomach and then to release the active substance in the upper intestine. Reasons for using enteric coated preparations may be to delay release of drugs inactivated by the stomach, to delay release of drugs inducing nausea or bleeding or to give a repeat or pulsed action or to deliver the drug intact to its site of action. Coating materials are usually weak acids that remain undissociated at low pH but ionise at pH > 5 such as anionic polymethacrylates [copolymerisate of methacrylic acid and either methylmethacrylate or ethyl acrylate (Eudragit®)], cellulose based polymers [e.g. cellulose acetate phthalate (Aquateric®)] or polyvinyl derivatives [e.g. polyvinyl acetate phthalate (Coateric®)].

Chronotherapy

A chronotherapeutic agent represents a pharmaceutical product that contains a dynamic element such as a delivery system to deliver the drug at the time when it is needed. A pulsatile drug delivery system is characterized by a lag time that is an interval of no drug release followed by rapid drug release. The interest in pulsatile delivery has been developing in close connection with emerging chronotherapeutic views. In this respect, it is by now well established that the symptoms of many pathologies are subject to circadian variation patterns. Hence, the possibility of accomplishing effective drug levels in accordance with the specific temporal requirements of an illness state holds considerable appeal in that it could improve therapeutic outcome.

Bacteria exposed to antibiotics in short repetitive bursts, or pulses, may be eliminated more efficiently and tend not to develop more resistant forms (49). Delivery systems combining a range of targeting techniques including sustained release and pH-coating have been developed to result in a pulsed release profile within the GI tract.

Targeting to Intestinal M-Cells

Orally administered micro- and nanoparticles can be absorbed, albeit in small quantities, not only by way of the membranous epithelial cells (M-cells) of the Peyer's patches in the gut-associated lymphoid tissue but also by the much more numerous gut enterocytes. It has yet to be established whether there can be sufficient absorption of nanoparticulate drug carriers by way of normal enterocytes and by M-cells to allow therapeutic quantities of drugs to be absorbed and whether experimental data derived from animal studies be extrapolated to humans.

Successful mucosal absorption of drugs requires formulations capable of extending residence time at the site of absorption thus providing intimate contact to the absorptive tissue. The concentration gradient of the drug between the lumen and the enterocytes is, therefore, increased and absorption of drugs can be facilitated by passive diffusion. Buparvaquone is used in the treatment of *Cryptosporidium parvum*, a protozoan parasite that persists in the entire gastrointestinal tract. The infection with *C. parvum* causes watery diarrhea, cramps and nausea, especially in immunocompromised patients and symptoms from colonization with this parasite over months or even years. A strategy to enhance solubility of the poorly soluble buparvaquone is to reduce the particle size which can lead to prolonged retention in the GI tract. Studies have been carried out to further increase the duration of contact of nanosuspensions by combination with mucoadhesive polymers such as polymers like Carbopols® (50) and chitosan (51). Mucoadhesive chitosan nanosuspensions of bupravaquone have been demonstrated to show anticryptidosporidial activity in animal studies (52).

In contrast to classical mucoadhesion, which relies on non-specific interpenetration of polymer chains and mucus, lectin–sugar interactions may be exploited to extend the duration of delivery to the intestine. Provided that the lectin-grafted drug delivery system can penetrate the mucus layer, the carbohydrate layer surrounding each mammalian cell represents a second target. The glycocalyx is built up of oligosaccharide moieties of proteoglycans, glycoproteins, and glycolipids anchored in the lipid bilayer of the cell membrane. Both the glycocalyx of intestinal epithelial cells and those of the oral cavity could be targeted.

In terms of drug delivery, the carbohydrate-mediated biorecognition of lectins, resulting in mucoadhesion, cytoadhesion, and/or cytoinvasion, might be advantageous for drug delivery to the small intestine. One approach is the preparation of prodrugs consisting of the lectin as the glycotargeting moiety, the drug as the active ingredient, and the spacer as a link. The second approach is the development of lectin-grafted carrier systems. A carrier such as a microparticle, nanoparticle or liposome would contain the drug and lectins would be immobilised on the surface of the particles, affording interaction with the target site.

THE COLON

Infections of the Colon

The human colon maintains a homeostatic relationship with the huge numbers of organisms inhabiting its lumen, allowing them to grow and reproduce but not invade. This unique setting of the GI tract means that there are few agents that can outwit the colon and lead to disease. Shigellae are gram-negative bacilli associated with dysentery which can lead to an infected colon. The infection and host response lead to crypt abscesses, ulceration, vascular congestion, oedema, hemorrhage, and epithelial regeneration. Standard treatment for Shigella is antibiotic therapy including ampicillin, sulfonamides, quniolones, aminoglycosides, and cephalosporins.

Advanced Formulations Targeting the Large Intestine

The treatment of diseases such as inflammatory bowel disease (including ulcerative colitis and Crohn's disease), has lead to the need for colon-targeted drug delivery. In addition systemic absorption from the colon offers advantages over absorption from the intestines including:

■ a long transit time for extending the absorption window;
■ relatively low enzyme levels, particularly important for enzymatically labile drugs.

Modified release solid dosage forms including coated, osmotic pumps, and chronotherapeutic systems as described previously may be used to target the colon via oral administration. Suppositories or enemas may be used for rectal administration to the colon.

Bacteria/Enzyme Activated Drug Delivery

The concentration of microflora along the GI tract varies from 10^3 to 10^4 colony forming units per milliliter (CFU/mL) in the stomach and small intestine with 10^{11}–10^{12} CFU/mL in the colon. These bacteria survive via fermentation of a range of substrates left undigested in the small intestine (e.g. oligosaccharides, polysaccharides).

The use of a prodrug whereby the active is released via a reaction with enzymes found only in the colon is a relatively new idea for colon specific drug delivery. Glucoside and glucuronide prodrugs of dexamethasone are examples where the prodrugs are poorly absorbed in the upper gastro-intestinal tract but rapid hydrolysis by beta-D-glucuronidase occurs in the colon to release dexamehtasone and glucuronic acid.

Azo-bonded prodrugs, for example, sulfasalazine (sulfapyridine azo-linked to a salicylate radical) was introduced for the treatment of inflammation; the colonic bacteria split the azo link releasing the sulfapyridine. However, it was later thought that sulfapyridine was responsible for the side effects of sulfasalazine and newer treatments have been introduced. Alternative azo-bonded prodrugs have included; p-aminohippurate (4-amino benzoyl glycine) in ipsalazine, 4-amino benzoyl-b-alanine in balsalazine, p-aminobenzoate in HB-313 or a nonabsorbable sulphanilamide ethylene polymer in poly-ASA. Another azo-linked prodrug is olsalazine which is a dimer representing two molecules of 5-ASA, so for every molecule of olsalazine administered two molecules of 5-ASA are released. Despite the early promise ipsalazine, this has not progressed into the market.

Hydrogels of N-(2-hydroxypropyl)methacrylamide (HPMA) containing azo-bond cleaving bricks have also been used in polymer-linked drug delivery to the colon.

In addition to azo-reduction, the colonic microflora also offers the opportunity for hydrolysis of saccharide substrates. Furthermore, the colonic microflora is responsible for numerous metabolic reactions via a wide spectrum of enzymes, from which reduction and hydrolysis are the predominate processes relevant to colonic delivery. This has been exploited using saccharide and polysaccharide delivery systems; however, the major drawback of polymer linked drug delivery to the colon is associated with premature release for the drug due to swelling of the polymer in the upper GI tract.

Coated tablets that protect against pH or that provide a time delay for drug release have been described previously, for colonic drug delivery polymers that are specifically degraded by the microflora environment within the colon can be used to control drug release targeted to the colon. Pectin (a natural polysaccharide) and insoluble salts of pectin as well as guar gum have been widely investigated as polymers that specifically degrade within the colon.

Recent work has examined pressure controlled drug delivery systems. The viscosity of the luminal contents within the colon is greater than at other sites within the GI tract due to the reabsorption of water from the large intestine. This change in viscosity leads to an increase in pressure resulting from the peristaltic forces. Moreover, this pressure change can be used to trigger drug release. A pressure-controlled colon delivery capsule has been investigated in both animals and humans.

CONCLUSIONS

Conventional drug delivery is still widely used in disorders within the GI tract and this is often associated with recognized treatment patterns, manufacturing processes, and benefits from the associated knowledge. Advanced drug delivery offers the opportunity to not only improve efficacy of treatment to individuals but also can extend the patent life of existing drugs.

This chapter has shown that many studies have revealed that advanced formulation offers benefits over conventional treatments and that this is an area of both scientific and commercial interest.

REFERENCES

1. Lupetti A, Danesi R, van't Wout JW, van Dissel JT, Senesi S, Nibbering PH. Antimicrobial peptides: therapeutic potential for the treatment of Candida infections. Expert Opin Invest Drugs 2002; 11:309–18.
2. Salamat-Miller N, Chittchang M, Johnston TP. The use of mucoadhesive polymers in buccal drug delivery. Adv Drug Deliv Rev 2005; 57:1666–91.
3. Christrup LL, David SS, Frier M, et al. Deposition of a model substance, 99mTc E-HIDA, in the oral cavity after administration of lozenges, chewing gum and sublingual tablets. Int J Pharm 1990; 66(1–3):169–174.
4. Martins MD, Rex JH. Fluconazole suspension for oropharyngeal candidiasis unresponsive to tablets. Ann Internal Med 1997; 126(4):332–3(Abstract).
5. Senel S, Ikinci G, Kas S, Yousefi-Rad A, Sargon MF, Hincal AA. Chitosan films and hydrogels of chlorhexidine gluconate for oral mucosal delivery. Int J Pharmaceut 2000; 193:197–203.
6. deVries-Hospers HG, vander WD. Amphotericin B concentrations in saliva after application of 2% amphotericin B in orabase. Infection 1978; 6:16–20.
7. Nafee NA, Ismail FA, Boraie NA, Mortada LM. Mucoadhesive buccal patches of miconazole nitrate: in vitro/in vivo performance and effect of ageing. Int J Pharm 2003; 264:1–14.
8. Bouckaert S, Remon JP. In-vitro bioadhesion of a buccal, miconazole slow-release tablet. J Pharm Pharmacol 1993; 45:504–7.
9. Khanna R, Agarwal SP, Ahuja A. Preparation and evaluation of bioerodible buccal tablets containing clotrimazole. Int J Pharm 1996; 138:67–73.

10. Senel S, Ikinci G, Kas S, Yousefi-Rad A, Sargon MF, Hincal AA. Chitosan films and hydrogels of chlorhexidine gluconate for oral mucosal delivery. Int J Pharm 2000; 193:197–203.

11. Giunchedi P, Juliano C, Gavini E, Cossu M, Sorrenti M. Formulation and in vivo evaluation of chlorhexidine buccal tablets prepared using drug-loaded chitosan microspheres. Eur J Pharm Biopharm 2002; 53:233–9.

12. Carlo CG, Bergamante V, Calabrese V, Biserni S, Ronchi C, Fini A. Design and evaluation in vitro of controlled release mucoadhesive tablets containing chlorhexidine. Drug Dev Ind Pharm 2006; 32:53–61.

13. Van RJ, Haxaire M, Kamya M, Lwanga I, Katabira E. Comparative efficacy of topical therapy with a slow-release mucoadhesive buccal tablet containing miconazole nitrate versus systemic therapy with ketoconazole in HIV-positive patients with oropharyngeal candidiasis. J Acquir Immune Defic Syndr 2004; 35:144–50.

14. Bouckaert S, Schautteet H, Lefebvre RA, Remon JP, van CR. Comparison of salivary miconazole concentrations after administration of a bioadhesive slow-release buccal tablet and an oral gel. Eur J Clin Pharmacol 1992; 43: 137–40.

15. Mohammed FA, Khedr H. Preparation and in vitro/in vivo evaluation of the buccal bioadhesive properties of slow-release tablets containing miconazole nitrate. Drug Dev Ind Pharm 2003; 29:321–37.

16. Cardot JM, Chaumont C, Dubray C, Costantini D, Aiache JM. Comparison of the pharmacokinetics of miconazole after administration via a bioadhesive slow release tablet and an oral gel to healthy male and female subjects. Br J Clin Pharm 2004; 58:345–51.

17. Habib W, Khankari R, Hontz J. Fast-dissolve drug delivery systems. Crit Rev Ther Drug Carrier Syst 2000; 17:61–72.

18. Vandercam B, Gibbs D, Valtonen M, JSger H, Armignacco O. Fluconazole orally dispersible tablets for the treatment of patients with oropharyngeal candidiasis. J Int Med Res 1998; 26(4):209–18 (Abstract).

19. Addy M, Roberts WR. Comparison of the bisbiguanide antiseptics alexidine and chlorhexidine. II. Clinical and in vitro staining properties. J Clin Periodontol 1981; 8:220–30.

20. Andersen T, Gram-Hansen M, Pedersen M, Rassing MR. Chewing gum as a drug delivery system for nystatin. Influence of solubilising agents on the release of water-soluble drugs. Drug Dev Ind Pharm 1990; 16: 1985–94.

21. Bastian HL, Rindum J, Lindeberg H. A double-dummy, double-blind, placebo-controlled phase III study comparing the efficacy and efficiency of miconazole chewing gum with a known drug (Brentan gel) and a placebo in patients with oral candidosis. Oral Surg Oral Med Oral Pathol Oral Radiol Endod 2004; 98:423–8.

22. Emslie RD. Treatment of acute ulcerative gingivitis. A clinical trial using chewing gums containing metronidazole or penicillin. Br Dent J 1967; 122: 307–8.

23. Rindum JL, Holmstrup P, Pedersen M, Rassing MR, Stoltze K. Miconazole chewing gum for treatment of chronic oral candidosis. Scand J Dent Res 1993; 101:386–90.

24. Pavelic RA. Use of an analgesic-antibiotic chewing troche (orabitic) in post-tonsillectomy and adenoidectomy complications. Eye Ear Nose Throat Mon 1960; 39:644–5.

25. Codd JE, Deasy PB. Formulation development and in vivo evaluation of a novel bioadhesive lozenge containing a synergistic combination of antifungal agents. Int J Pharm 1998; 173:13 24.

26. Bromberg LE, Buxton DK, Friden PM. Novel periodontal drug delivery system for treatment of periodontitis. J Control Release 2001; 71:251–9.

27. Gorman SP, McCafferty DF, Woolfson AD, Jones DS. A comparative study of the microbial antiadherence capacities of three antimicrobial agents. J Clin Pharm Ther 1987; 12:393–9.

28. Tobgi RS, Samaranayake LP, MacFarlane TW. Adhesion of Candida albicans to buccal epithelial cells exposed to chlorhexidine gluconate. J Med Vet Mycol 1987; 25:335–8.

29. Barembaum S, Virga C, Bojanich A, et al. Effect of Chitosan and Sodium Alginate on the adherence of autochthonous C. Albicans to oral epithelial cells (in vitro). Med Oral 2003; 8:188–96.

30. McCarron PA, Donnelly RF, Canning PE, McGovern JG, Jones DS. Bioadhesive, non-drug-loaded nanoparticles as modulators of candidal adherence to buccal epithelial cells: a potentially novel prophylaxis for candidosis. Biomaterials 2004; 25:2399–407.

31. Laine L, Dretler RH, Conteas CN, et al. Fluconazole compared with ketoconazole for the treatment of Candida esophagitis in AIDS. A randomized trial. Ann Intern Med 1992; 117:655–60.

32. Rex JH, Rinaldi MG, Pfaller MA. Resistance of Candida species to fluconazole. Antimicrobial Agents Chemother 1995; 39(1):1–8.

33. Wilcox CM, Darouiche RO, Laine L, Moskovitz BL, Mallegol I, Wu J. A randomized, double-blind comparison of itraconazole oral solution and fluconazole tablets in the treatment of esophageal candidiasis. J Infect Dis 1997; 176:227–32.

34. Laine L, Rabeneck L. Prospective study of fluconazole suspension for the treatment of oesophageal candidiasis in patients with AIDS. Aliment. Pharmacol Ther 1995; 9:553–6.

35. Megraud F. Strategies to treat patients with antibiotic resistant Helicobacter pylori. Int J Antimicrob Agents 2000; 16:507–9.

36. Goddard AF, Erah PO, Barrett DA, Shaw PN, Spiller RC. The effect of protein binding and lipophilicity of penicillins on their *in-vitro* flux across gastric mucosa. J Antimicrobial Chemother 1998; 41:231–6.

37. Nakamura M, Spiller RC, Barrett DA, et al. Gastric juice, gastric tissue and blood antibiotic concentrations following omeprazole, amoxicillin and claithromycin triple therapy. Helicobacter 2003; 6:294–9.

38. Erah PO, Goddard AF, Barrett DA, Shaw PN, Spiller RC. The stability of amoxicillin, clarithromycin and metronidazole in gastric juice: relevance to the treatment of *Helicobacter pylori* infection. J Antimicrob Chemother 1997; 39: 5–12.

39. Lehr C-M, Bouwstra JA, Schacht EH, Junginger HE. *In vitro* evaluation of mucoadhesive properties of chitosan and some other natural polymers. Int J Pharm 2006; 78:43–8.

40. Gaserød O, Joliffe IJ, Hampson FC, Dettmar PW, Skjåk-Bræk G. The enhancement of the bioadhesive properties of alginate gel beads by coating with chitosan. Int J Pharmaceut 1998; 175:237–46.

41. Chandy T, Sharma CP. Chitosan matrix for oral sustained delivery of ampicillin. Biomaterials, 1993; 14:939–44.

42. Klausner EA, Lavy E, Friedman M, Hoffman A. Expandable gastroretentive dosage forms. J Control Release 2003; 90:143–62.

43. Chen J, Blevins WE, Park H, Park K. Gastric retention properties of superporous hydrogel composites. J Control Release 2000; 64:39–51.

44. Marchetti M, Rossi M, Gianelli V, et al. Protection against *Helicobacter pylori* infection in mice by intragastric vaccination with *H. pylori* antigens is achieved using a non-toxic mutant of *E. coli* heat-labile enterotoxin (LT) as adjuvant. Vaccine 1998; 16:33–7.

45. Vajdy M, Singh M, Ugozzoli M, et al. Enhanced mucosal and systemic immune responses to *Helicobacter pylori* antigens through mucosal priming followed by systemic boosting immunizations. Immunology 2003; 100:86–94.

46. Rugiero P, Peppoloni S, Rappuoli R, Del Guidice G. The quest for a vaccine against *Helicobacter pylori*: how to move from mouse to man? Microbial Infect 2003; 5:749–56.

47. Sutton P, Doidge C. *Helicobacter* vaccines spiral into the new millennium. Digestive Liver Dis 2003; 25:675–87.

48. Stanley J. Amoebiasis. The Lancet 2003; 361:1025–34.

49. Rudnic EM, Isbister JD, Treacy DJ, Wassink SE. Antibiotic product, use and formulation thereof US Patent No. 6,723,341, 2004 http://www.patentstorm.us/patents/6723341.html.

50. Mortazavi SA, Smart JD. An investigation of some factors influencing the in vitro assessment of mucoadhesion. Int J Pharm 1995; 116:223–30.

51. Henrisksen I, Green KL, Smart JD, Smistad G, Karlsen J. Bioadhesion of hydrated chitosans: An in vitro and in vivo study. Int J Pharm 1996; 145: 231–40.

52. Kayser O. A new approach for targeting to *Cryptosporidium parvum* using mucoadhesive nanosuspensions: research and applications. Int J Pharm 2001; 214:83–5.

8

Local Controlled Drug Delivery to the Brain

Juergen Siepmann and Florence Siepmann
College of Pharmacy, University of Lille, Lille, France

INTRODUCTION

The treatment of brain diseases is particularly challenging because of the blood-brain barrier (BBB) (1–3). Generally, only low molecular weight, lipid-soluble molecules and a few peptides and nutrients can cross the BBB to a notable extent, either by passive diffusion or via specific transport mechanisms. Thus, for most drugs it is very difficult to achieve therapeutic levels within the brain tissue via the common administration routes (e.g. intravenous, intramuscular, subcutaneous, or oral). In addition, highly potent drugs (e.g. anticancer drugs and neurotrophic factors) that may be necessary to be delivered to the central nervous system (CNS) often cause serious toxic side effects when administered systemically. To overcome these restrictions, three major approaches have been proposed (2):

(1) The drug is chemically modified in order to enhance its BBB permeability or is linked to a carrier which is capable to cross this barrier. For example, two more lipophilic derivatives of the anticancer drug carmustine (lomustine and semustine) have been synthesized to increase the resulting brain tissue concentration and, thus, the efficiency of CNS tumor treatments. Unfortunately, clinical trials with these more lipophilic derivatives did not show any significant advantage compared to carmustine administration (4). A comprehensive overview on *vector-mediated* drug delivery to the brain has been given by Pardridge (5). For example, certain monoclonal antibodies undergoing receptor-mediated transcytosis through the BBB can successfully be used to delivery drugs to the target tissue.

(2) The BBB is temporarily disrupted to allow drug transport from the blood into the brain. For instance, when infusing hyperosmolar mannitol solutions intra-arterially, the endothelial cells dehydrate and shrink, resulting in a widening of the tight junctions. Also the bradykinin agonist RMP-7 is able to disrupt the BBB (6). Elliot et al. (7) have shown that the uptake of carboplatin into the brain can effectively be increased upon i.v. RMP-7 co-administration.

(3) The drug is directly administered into the brain tissue (intra-cranially) (3). In this case, the BBB is crossed by mechanical disruption. This approach offers the major advantage to provide high drug concentrations locally at the site of action, while minimizing the drug levels in the rest of the human body, resulting in reduced side effects. However, the intracranial administration route cannot frequently be used due to the considerable risk of CNS infections. As most drugs are rapidly cleared from the brain tissue (exhibiting only short half-lives), this is a serious restriction: Therapeutic drug concentrations at the target site are only achieved during short time periods and the efficiency of the treatments is limited. Importantly, delivery systems that are able to control the rate at which the drug is released during several weeks or months can overcome this restriction (3,8–15). Ideally, one single intracranial administration is sufficient to achieve therapeutic drug levels at the site of action for the entire treatment period. If the system is composed of biodegradable compounds, it is not necessary to remove empty remnants after drug exhaust.

MASS TRANSPORT MECHANISM

Transport Mechanisms within Controlled Drug Delivery Systems

Often, the active agent is embedded within a polymeric matrix that hinders instantaneous drug dissolution/release (16,17). Different types of macromolecules and device geometries (shapes and sizes) can be used. Practical examples include poly(anhydride)-based, flat cylinders and poly (lactic-co-glycolic acid) (PLGA)-based, spherical microparticles. Depending on the composition, dimension, and sometimes even prepa-ration method of the system, different physical and chemical phenomena can be involved in the control of the resulting drug release kinetics, including: (*i*) water penetration into the system; (*ii*) drug dissolution; (*iii*) dissolution/degradation of the polymer chains; (*iv*) precipitation and re-dissolution of polymer degradation products; (*v*) structural changes within the system occurring during drug release, such as creation/closure of water-filled pores; (*vi*) changes in micro-pH (e.g. generation of acidic microclimates in PLGA-based delivery systems); (*vii*) diffusion of drug and/or degradation products of the polymer out of the device with constant or time/position-dependent diffusion coefficients; (*viii*) osmotic effects; (ix) convectional processes; as well as (*x*) adsorption/desorption phenomena. In contrast to oral controlled drug delivery systems, significant swelling of the polymer (a phenomenon that can also be effectively used to control drug release) must be avoided due to the limited space in the crane: Intracranially administered devices that remarkably swell would generate mechanical pressure on the surrounding (highly sensitive) brain tissue and, thus, lead to serious side effects.

The physical and chemical processes involved in the control of drug release from a particular delivery system can affect each other in a rather complex way. For example, PLGA-based devices can show autocatalytic effects: Water penetration into PLGA-based microparticles and implants is much faster than the subsequent ester bond cleavage (18). Thus, the entire drug delivery system is rapidly wetted and polymer degradation occurs throughout the device, generating shorter chain *acids* (and alcohols). Due to concentration gradients these species diffuse out of the dosage form. In addition, bases from the surrounding environment diffuse into the drug delivery system, neutralizing the generated acids. However, diffusional processes are generally slow and the rate at which the acids are produced within the dosage forms can be higher than the rate at which they are neutralized. Consequently, the micro-pH within the system can significantly drop (19,20). As ester bond cleavage is catalyzed by protons, this leads to accelerated polymer degradation and drug release (21). It has to be pointed out that device characteristics, for example, porosity and size, can significantly affect the relative diffusion rates of the involved acids and bases and, thus, alter the underlying drug release mechanisms (22). Consequently, systems with identical chemical composition that are prepared using different methods (e.g. water-in-oil-in-water vs. oil-in-water solvent extraction/evaporation techniques for microparticles) resulting in porous versus non-porous devices can exhibit very different drug release patterns due to altered drug release mechanisms.

To get deeper insight into the physical and chemical processes that govern drug release within a particular controlled delivery system, adequate, *mechanistic* mathematical theories can be used. Obviously, *empirical* models are not suitable for this purpose. The application of *semi-empirical* theories should be viewed with great caution as the conclusions can be misleading. Examples for appropriate mechanistic theories describing drug transport in biodegradable controlled delivery systems include those developed by Goepferich et al. (21–30). These models combine Monte Carlo simulations (quantifying polymer degradation) with Fick's second law (describing drug diffusion). Importantly, the theories are applicable to both, surface as well as bulk eroding polymeric systems [31]. Furthermore, they have been extended to describe the erosion of composite matrices made of bulk and surface eroding polymers [e.g. poly(D,L-lactic acid) and poly[bis (p-carboxyphenoxy) propane – sebacic acid]] (28).

Siepmann et al. (32) proposed a mathematical theory considering: (*i*) polymer degradation (based on Monte Carlo simulations); (*ii*) drug diffusion (based on Fick's second law); and – optionally – (*iii*) limited drug solubilities within spherical microparticles. Figure 1 shows a schematic presentation of such a system for mathematical analysis. To minimize computation time, it is assumed that there are no concentration gradients in the direction of the angle θ. Thus, a two-dimensional grid (Fig. 1B) can be defined, which upon rotation around the z-axis describes the three-dimensional structure of the microparticle. Considering symmetry planes at $z = 0$ and $r = 0$, the mathematical analysis can be further reduced to only one quarter of the two-dimensional circle (Fig. 2A). Before exposure to the release medium, each pixel represents either non-degraded polymer or drug. Knowing the initial drug loading of the microparticles as well as the initial drug distribution within the system, direct Monte Carlo techniques can be used to define which pixel represents non-degraded polymer and which pixel represents drug. Figure 2A shows an example for a homogeneous initial drug distribution. Importantly, all pixels are defined in such a way that they have the same height, but different widths. The coordinates are chosen to assure that the volumes of the rings, which are described by the rectangular pixels upon rotation around the z-axis, are all identical. This results in about equal numbers of cleavable ester bonds within each ring. Thus, the probability with which the polymer pixels erode within a certain time period after contact with water can be assumed to be similar (being essentially a function of the number of cleavable polymer bonds). As polymer degradation is a random process, not all pixels degrade exactly at the same time point. Each pixel is characterized by an individual, randomly distributed "lifetime" ($t_{lifetime}$), which can be calculated as in Equation 1, as a function of the random variable ε (integer between 0 and 99):

$$t_{lifetime} = t_{average} + \frac{(-1)^{\varepsilon}}{\lambda} \cdot \ln\left(1 - \frac{\varepsilon}{100}\right) \tag{1}$$

where $t_{average}$ is the average "lifetime" of the pixels, and λ is a constant (being characteristic for the type and physical state of the polymer). As soon as a pixel comes into contact with water, its "lifetime" starts to decrease. After the latter has expired, the pixel is assumed to erode instantaneously and to be converted into a water-filled pore. Once the initial condition (Fig. 2A) and the specific "life times" of all polymer pixels are defined, it is possible to determine the status of each pixel (representing drug, non-degraded polymer or a water-filled pore) at any time point. Figure 2B shows an example for the composition and structure of a microparticle at a specific time point *during* drug release. This structural information is of major

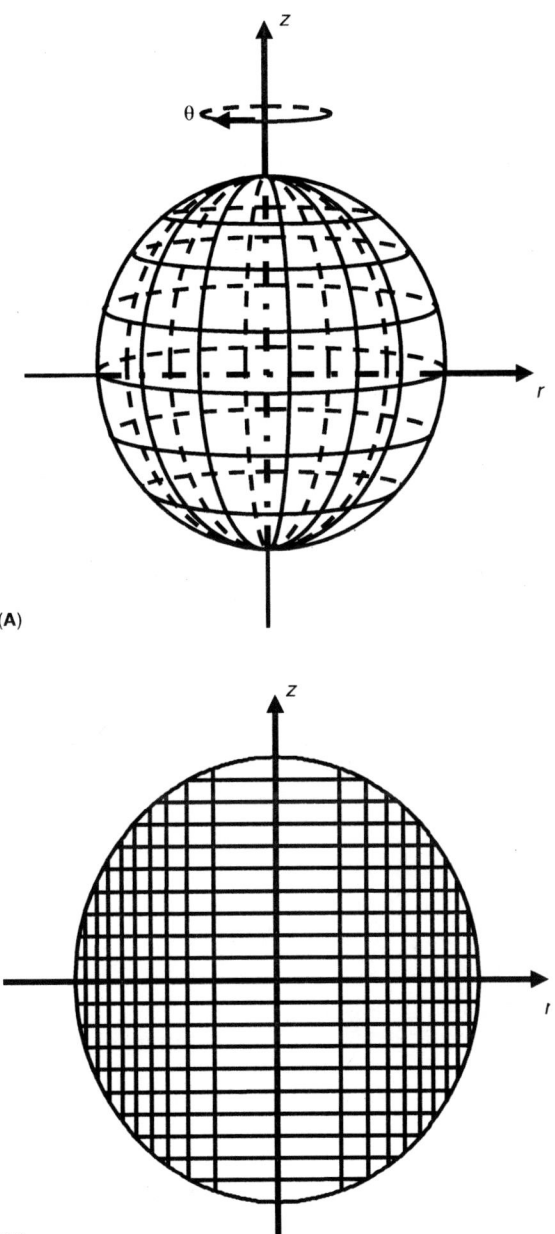

(A)

(B)

Figure 1 Schematic presentation of a single bioerodible microparticle for mathematical analysis: **(A)** Three-dimensional geometry; **(B)** two-dimensional cross-section with two-dimensional pixel grid used for numerical analysis. *Source*: From Ref. 32.

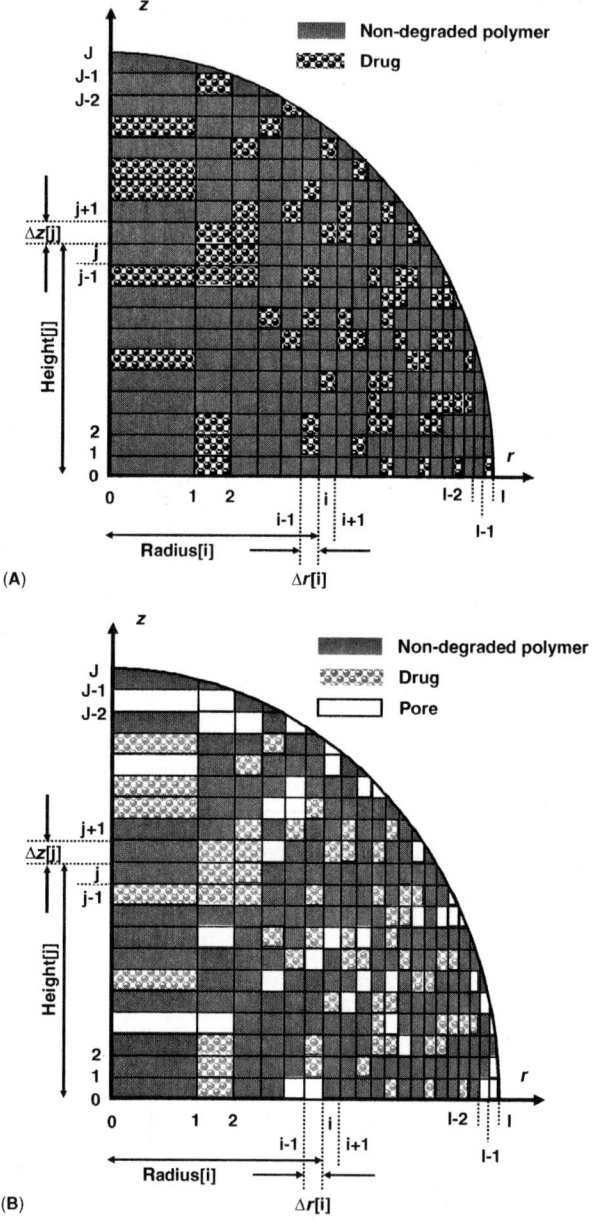

Figure 2 Principle of the Monte Carlo-based approach to simulate polymer degradation and diffusional drug release; schematic structure of the system: (**A**) At time $t = 0$ (before exposure to the release medium); and (**B**) during drug release. Gray, dotted, and white pixels represent non-degraded polymer, drug, and pores, respectively. *Source*: From Ref. 32.

importance, because it allows to calculate the porosity of the microparticles at any time point in radial and axial direction [$\varepsilon(z,t)$ and $\varepsilon(r,t)$] (Eq. 2, 3):

$$\varepsilon(r,t) = 1 - \frac{1}{n_z} \cdot \sum_{j=1}^{j=n_z} s(i(r),j,t) \tag{2}$$

$$\varepsilon(z,t) = 1 - \frac{1}{n_r} \cdot \sum_{i=1}^{i=n_r} s(i,j(z),t) \tag{3}$$

with s being the "status function" of the pixel $x_{i,j}$ at time t, defined as Equations 4 and 5:

$$s(i,j,t) = 1 \quad \text{for non-eroded polymer} \tag{4}$$

$$s(i,j,t) = 0 \quad \text{for pores} \tag{5}$$

where, n_z and n_r represent the number of pixels in the axial and radial direction at r and z, respectively. Using Equations 2–5, the time- and direction-dependent porosities within the microparticles can be calculated at any gridpoint. This is essential information for the accurate calculation of the time-, position-, and direction-dependent diffusivities (Eq. 6, 7):

$$D(r,t) = D_{cirt} \cdot \varepsilon(r,t) \tag{6}$$

$$D(z,t) = D_{crit} \cdot \varepsilon(z,t) \tag{7}$$

where D_{crit} represents a critical diffusion coefficient, being characteristic for a specific drug-polymer combination. These equations are combined with Fick's second law of diffusion and the resulting set of Partial Differential Equations is solved numerically. Fitting the model to sets of experimentally measured drug release kinetics from 5-fluorouracil-loaded, PLGA-based microparticles (which are used for the treatment of brain tumors), good agreement between experiment and theory was obtained (Fig. 3). Based on these calculations important information on the underlying drug release mechanisms could be gained: It was shown that the initial high drug release rate from these systems (also called "burst effect") can primarily be attributed to pure drug diffusion (at early time points the diffusion pathways are short, thus, drug release is rapid). The second release phase (with an about constant release rate) can be attributed to a combination of two processes: drug diffusion and polymer erosion. The increase in the length of the diffusion pathways with time (which should result in a decrease in the drug release rate) is compensated by the increase in drug mobility within the polymeric matrix (due to the decreasing chain length of the macromolecules upon ester hydrolysis and, thus, increased macromolecular mobility). The final, again more rapid drug release phase can be attributed to the disintegration of the microparticles: As soon as a critical minimal polymer

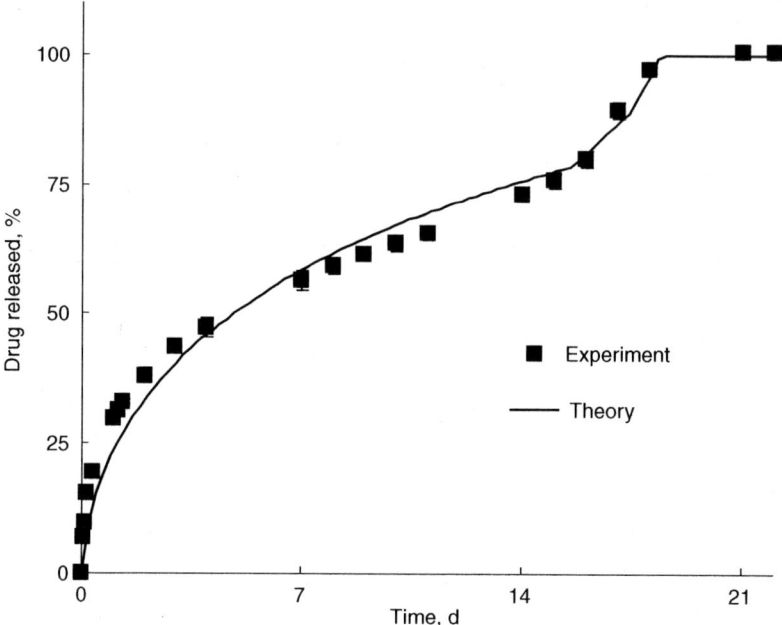

Figure 3 Experiment (symbols) and theory (curve): Fitting of a Monte Carlo-based mathematical model to experimentally determined 5-fluorouracil release from PLGA-based microparticles. *Source*: From Ref. 32.

molecular weight is reached, the mechanical stability of the system is no more guaranteed. Consequently, the relative surface area of the system increases and the diffusion pathway lengths decrease.

Transport Mechanisms within Living Brain Tissue

The physical and chemical phenomena which can affect the transport of a drug (once it is released from its dosage form) within the living brain tissue are very complex and yet not fully understood. Excellent overviews on the importance of diffusion and related processes have been given by Tao and Nicholson (33,34). Nicholson and his co-workers as well as the group of Saltzman made major contributions towards a better understanding of these phenomena (35–41). A large variety of processes can be involved in the transport of a drug in the brain, including: (1) diffusion within the extracellular space; (2) reversible and irreversible binding to the extracellular matrix (which is built of long-chain macromolecules); (3) degradation/metabolism (e.g. by enzymes or hydrolysis); (4) different types of passive and active uptake into CNS cells (e.g. by "simple" diffusion or receptor-mediated internalization); (5) release from endolysosomes into the cytosol; (6) diffusion and convection within the cytosol of the cells; (7) uptake into the cell nuclei; (8) elimination into the blood stream; (9) bulk flow within the extracellular space; (10) direction-dependent drug transport (anisotropy), because the brain is not one homogeneous mass. Figure 4 shows a schematic presentation of some of these processes.

The extracellular space represents approximately 20% of the total human brain volume. Its geometry can be compared to that of the water phase in an aqueous foam. Drug transport in this region can often (surprisingly well) be described based on Fick's second law of diffusion. Important aspects to be taken into account include the volume fraction in which diffusion can take place and the tortuosity of the diffusion pathways. Recently, the group of Charles Nicholson studied the effects of the geometry of CNS cells on the tortuosity of the extracellular space (42–44). Considering uniformly spaced convex cells, they found that the presence of dead-space micro-domains can help to better understand the difference between the experimentally measured tortuosity and the theoretically calculated one. It has to be pointed out that many brain diseases can significantly affect the conditions for drug transport within the CNS (45,46). For example, cellular swelling can lead to a significant shrinkage of the extracellular space, because the total brain volume is

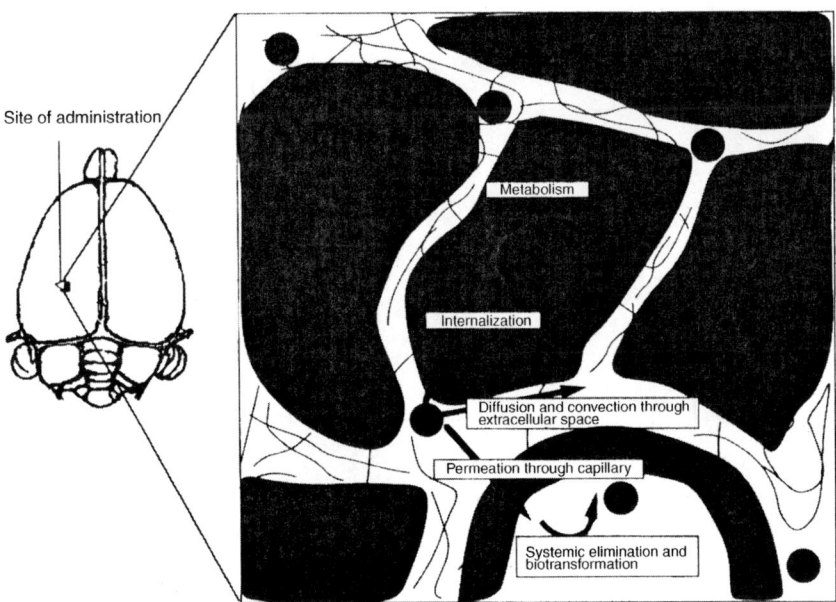

Figure 4 Schematic presentation of some of the processes that can be involved in drug transport through the living brain tissue (indicated in the figure). The black circles represent drug molecules in the interstitial space. *Source*: From Ref. 40.

restricted by the rigid crane. Also the tortuosity can strongly be altered. In some cases, the non-physiological mass transport conditions are not the *consequence* of the disease, but its *cause*: The appropriate transport of oxygen, glucose, neurotransmitters, and many other substances is vital for a normal functioning of the brain. In addition, the conditions for drug transport within the brain can be significantly age-dependent: Lehmenkuhler et al. (47) showed that the volume fraction of the extracellular space of rats decrease from about 0.4 inches for 2–3 day old animals to only 0.2 inches 21 day old animals. Obviously, these changes can strongly affect the transport of intracranially administered drugs.

Interestingly, so far only a very few mechanistic mathematical models have been reported in the literature that quantitatively describe drug transport in living brain tissue (Fig. 5). In particular, Nicholson and co-workers as well as the group of Saltzman made major contributions to this field. For example, Saltzman and Radomsky (36) proposed an interesting theory, considering drug diffusion from intracranially administered cylindrical delivery systems. The model is based on the following assumptions: (1) The drug concentration at the surface of the dosage forms is constant. (2) The elimination of the drug from the brain tissue follows first order kinetics (the elimination rate of the drug is proportional to its concentration). (3) Diffusion is isotropic (does not dependent on a spatial direction). (4) Convectional processes are negligible. The model is based on Fick's second law of diffusion (considering one dimension), which is coupled with a first order elimination term (Eq. 8):

$$\frac{\partial c}{\partial t} = D \left[\frac{\partial^2 c}{\partial x^2} \right] - k \cdot c \tag{8}$$

where c is the concentration of the drug within the brain tissue; t is time ($t = 0$ at the time of device administration); D represents the apparent diffusion coefficient of the drug within the brain; x is the spatial coordinate; and k is the first order elimination rate constant of the drug. The initial and boundary conditions were considered as shown in Equations 9, 10, and 11:

$$c = 0 \quad \text{for } t = 0; \quad x \geq a \tag{9}$$

$$c = c_0 \quad \text{for } t > 0; \quad x = a \tag{10}$$

$$c = 0 \quad \text{for } t > 0; \quad x \to \infty \tag{11}$$

where, a is the half-thickness of the cylindrical dosage form and c_0 the (constant) drug concentration at the surface of the device. Equation (9) indicates that the brain tissue is free of drug prior to the administration of the dosage form. Equation (10) states that the constant drug concentration at the interface "dosage form – brain tissue" (c_0) is time-independent, and

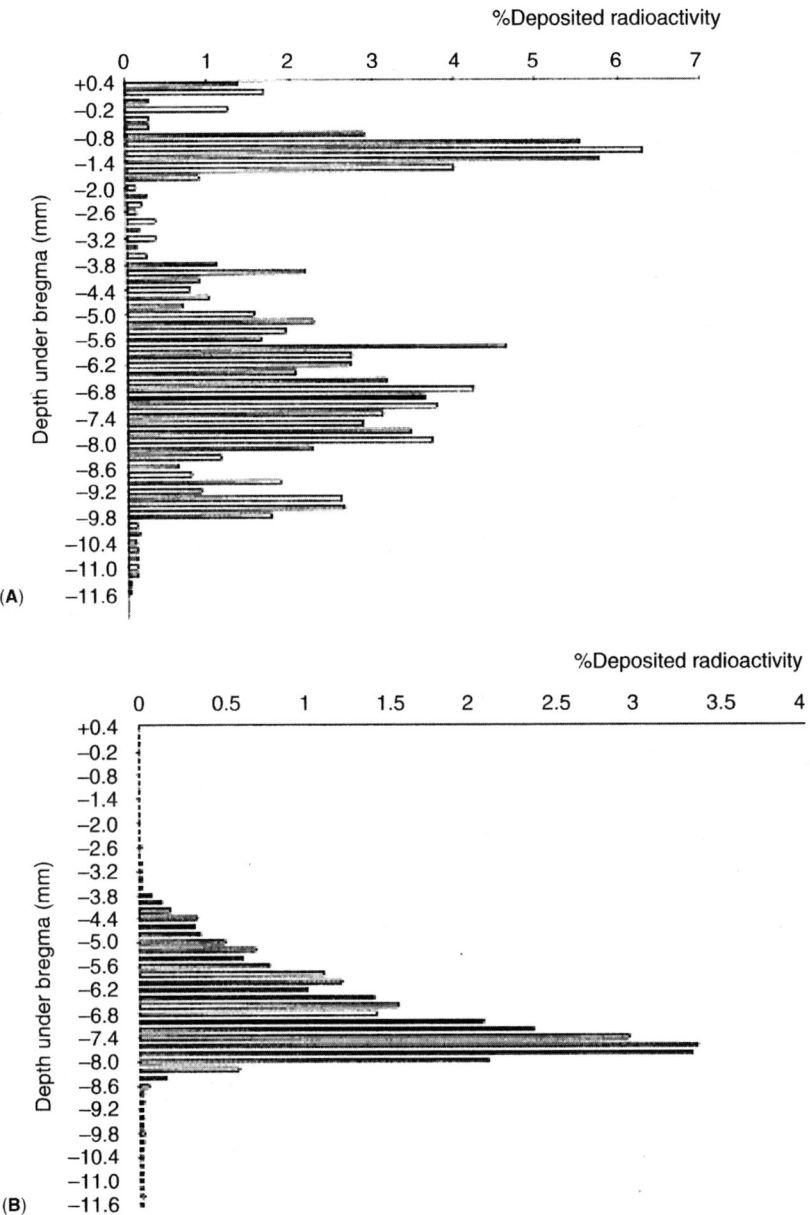

Figure 5 Monitoring of the drug distribution within: (**A**) healthy, and (**B**) tumor-bearing rat brain using autoradiography: Relative radioactivity measured as a function of the distance from the bregma; 168 hours after intracranial administration of [^{3}H]5-fluorouracil-loaded microspheres. The black bars indicate the administration site (7 mm under the bregma). *Source*: From Ref. 87.

Equation 11 expresses the fact that the drug concentration vanishes to zero at large distances from the cylinder. Assuming steady state conditions (the concentration of the drug within the brain tissue does not vary with *time*, only with *position*), this set of equations can be solved to give Equation 12:

$$c(x) = c_0 \cdot \exp\left(-a\sqrt{\frac{k}{D}} \cdot \left[\frac{x}{a} - 1\right]\right) \tag{12}$$

On the other hand, considering non-steady state conditions (the drug concentration varies with time and position), the following solution can be derived (Eq. 13) (47):

$$c(x,t) = \frac{c_0}{2} \cdot \left[\exp\left(-x\sqrt{\frac{k}{D}}\right) \cdot \mathrm{erfo}\left[\frac{x}{\sqrt{4 \cdot D \cdot t}} - \sqrt{k \cdot t}\right]\right.$$
$$\left. + \exp\left[x\sqrt{\frac{k}{D}}\right] \cdot \mathrm{efro}\left[\frac{x}{\sqrt{4 \cdot D \cdot t}} + \sqrt{k \cdot k}\right]\right] \tag{13}$$

where, x is the distance from the interface "delivery system–brain tissue." Both Equations (12) and (13) allow us to calculate the drug concentration at any distance from the axial surface of the cylindrical dosage form. Figure 6 shows examples of fittings of these models to sets of experimentally

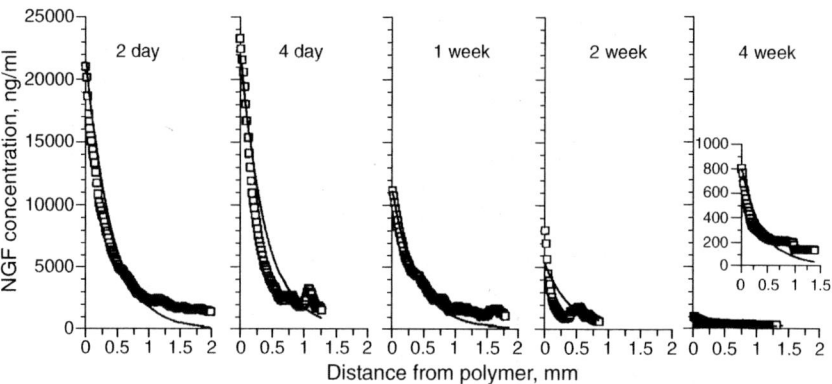

Figure 6 Experiment (symbols) and theory (curve): Concentration profiles of radioactively labeled NGF within rat brain upon intracranial administration of a cylindrical controlled drug delivery system. The distance from the surface of the device is plotted on the *x*-axis; the times elapsed after implantation are indicated in the diagrams. The non-steady state and steady state model of Saltzman and co-workers [Equations (12) and (13)] were fitted to the experimental results obtained after 2, 4 days and 1, 2, 4 weeks, respectively. *Source*: From Ref. 48.

measured concentration profiles of radioactively labeled nerve growth factor (NGF). The drug was incorporated within cylindrical poly(ethylene-co-vinyl acetate) (EVAc)-based discs and intracranially administered into rats. After pre-determined time intervals (indicated in the figures), the animals were sacrificed, the brains sliced and the radioactivity measured. The non-steady state model [Equation (13)] was fitted to the experimentally determined NGF concentration profiles at days 2 and 4, the steady state model [Equation (12)] was fitted to the concentration profiles measured after 1, 2, and 4 weeks. Clearly, good agreement between theory and experiment was obtained in all cases. Thus, NGF transport through the living brain tissue seems to be dominated by diffusion and first order elimination. The partially observed deviations between theory and experiment might be attributable to experimental errors or violation of model assumptions (e.g. time-dependent drug concentrations at the surface of the delivery systems). Importantly, the distance that NGF can penetrate into the brain tissue is rather limited: After 2–3 mm its concentration decreases to only 10% of the maximal value (at the interface "dosage form–brain tissue"). Furthermore, Saltzman and coworkers compared the transport of NGF in rat brain upon its release from three different types of intracranial, controlled drug delivery systems: (1) slowly releasing EVAc discs; (2) fast releasing PLGA-based microparticles; and (3) PLGA-based microparticles with an intermediate release rate (48). In all cases, good agreement between theory and experiment was obtained. An apparent diffusion coefficient of about 8×10^{-7} cm^2/s could be determined for NGF in rat brain.

BRAIN CANCER TREATMENT

Controlled Release Implants

The first pharmaceutical product available on the market being based on the principle of intracranial controlled drug delivery was Gliadel® (41,50–57). It is a disc-shaped wafer (flat cylinder), containing 3.85% of the anticancer drug BCNU [1,3-bis(2-chloroethyl)-1-nitrosourea; carmustine]. BCNU alkylates the nitrogen bases of DNA. Although it is a low molecular weight, lipophilic drug (and can, thus, cross the BBB to a certain extent), a systemic treatment is not efficient due to the severe, dose-limiting side effects (in particular bone marrow suppression and pulmonary fibrosis) and relatively short half-life (<15 min) (58,59). In Gliadel, BCNU is embedded within a biodegradable matrix based on poly[bis(p-carboxyphenoxy) propane–sebacic acid] [p(CPP:SA)], a polyanhydride that is hydrolytically cleaved into shorter chain acids upon contact with water. This advanced drug delivery system has been developed by the group of Brem, presenting a major breakthrough in the field of controlled local brain delivery. It got approved by the Food and Drug Administration (FDA) in 1996 for the

treatment of recurrent glioblastoma multiforme. The basic principle of this
treatment method is illustrated in Figure 7: Schematic cross-sections of
a human brain are shown. The tumor is represented by the black circle
(Fig. 7A), the surrounding tissue is infiltrated by tumor cells. If operable, the
surgeon removes the tumor (Fig. 7B). Unfortunately, large quantities of
surrounding tissue cannot be removed at the same time, because the risk to
affect vital brain functions is considerable. Thus, the probability that tumor
cells remain within the brain is important (infiltrated neighboring tissue).
Consequently, many patients die due to local tumor recurrence in the direct
vicinity of the primary tumor. To reduce this risk one or more (maximal 8)
disc-shaped, BCNU-loaded wafers are placed into the resection cavity of the
tumor (during the same operation, not requiring an extra opening of the
crane) (Figs. 7C and 8). The anticancer drug is then released in a time-
controlled manner into the resection cavity and penetrates into the
surrounding tissue.

In the early nineties, a phases I–II clinical trial demonstrated the safety
of using this type of p(CPP:SA)-based wafers loaded with BCNU for
intracranial tumor treatment (50): Twenty one patients with recurrent
malignant glioma (that had previously undergone a craniotomy for
debulking and in whom standard therapy had failed) were treated with p
(CPP:SA)-based discs containing 1.93%, 3.85%, and 6.35% (w/w) BCNU.
Each wafer had a mass of 200 mg and in most patients the maximal number
of discs (eight) was implanted. Importantly, there was no evidence for
systemic toxicity (including bone marrow suppression). Based on this phases
I–II clinical trial, the efficiency of the 3.85% BNCU-loaded, polyanhydride-
based wafers were studied in a phase III clinical trial (51). The latter was
multicentered, randomized, double-blinded and placebo-controlled.

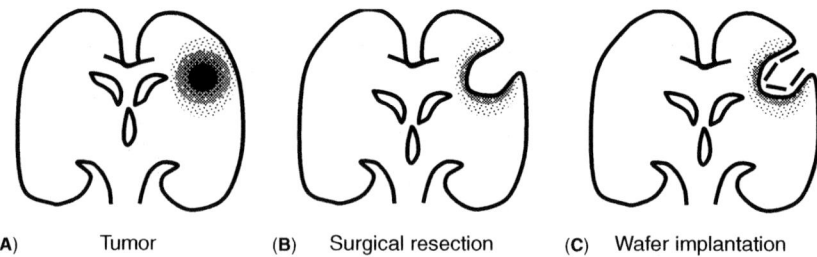

(A)   Tumor   (B)   Surgical resection   (C)   Wafer implantation

Figure 7 Principle of the treatment of operable brain tumors with BCNU-loaded,
p(CPP:SA)-based wafers (Gliadel). Schematic cross-sections through a human brain:
(A) The tumor is illustrated as a black circle; the surrounding tissue is infiltrated by
tumor cells. (B) The tumor has been removed surgically. (C) To minimize the risk of
local tumor recurrence, drug-loaded wafers (up to 8) are placed into the resection
cavity of the tumor (see Fig. 8).

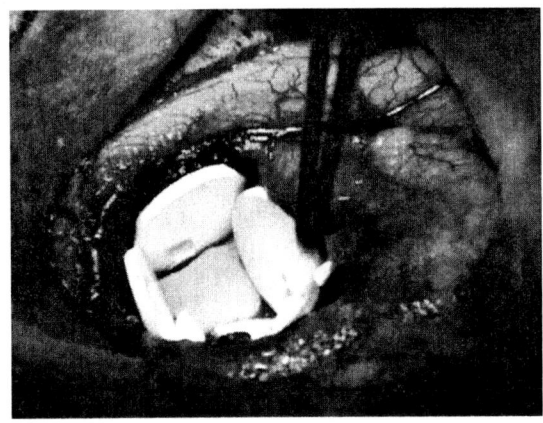

Figure 8 Principle of the treatment of recurrent glioblastoma multiforme with Gliadel®. Upon tumor resection, up to 8 anticancer drug-loaded, polymeric wafers are placed into the resection cavity. The drug is then released in a time-controlled manner from the wafers into the cavity. *Source*: From Ref. 57.

Two hundred and twenty two patients with recurrent malignant brain tumors requiring re-operation were randomly assigned to receive intracranially implanted p(CPP:SA)-based discs containing or not BCNU. Randomization balanced the treatment groups for all examined prognostic factors. Importantly, the median survival time of the 110 patients who were treated with drug-loaded wafers was 31 weeks compared to only 23 weeks in the case of the 112 patients who received placebo discs. The 6-month survival rate was 60% in the treatment population vs. 47% in the placebo group. Figure 9A illustrates the overall survival of the patients (after adjustment for the examined prognostic factors). Furthermore, there were no clinically important side effects caused by the drug-loaded, biodegradable wafers: neither locally within the brain, nor systemically. Based on these results the FDA approved this treatment method for recurrent glioblastoma multiforme. It was the first time in 23 years that the FDA approved a novel treatment method for malignant gliomas.

Afterwards, Gliadel was also used for the initial treatment of malignant gliomas (52). Generally, it can be expected that an anticancer treatment that has proven to be effective against recurrent tumors is likely to be more effective when used as initial therapy. A phases I–II clinical trial with 21 patients receiving postoperative external beam radiation therapy (5000 rad) showed a median survival of 44 weeks from the time of implantation, with 4 patients surviving >18 months. Thus, this trial demonstrated the safety of Gliadel in combination with radiation therapy also for patients with newly diagnosed malignant gliomas. Based on these results, a phase III clinical trial (multicentered, randomized, double-blinded and placebo-controlled) was planned and started for 100 patients (56). Unfortunately, this study had to be prematurely terminated because of a temporary unavailability of the anticancer drug BCNU. At the end, only 32 patients could be included in the trial. The obtained results were nevertheless encouraging: The median

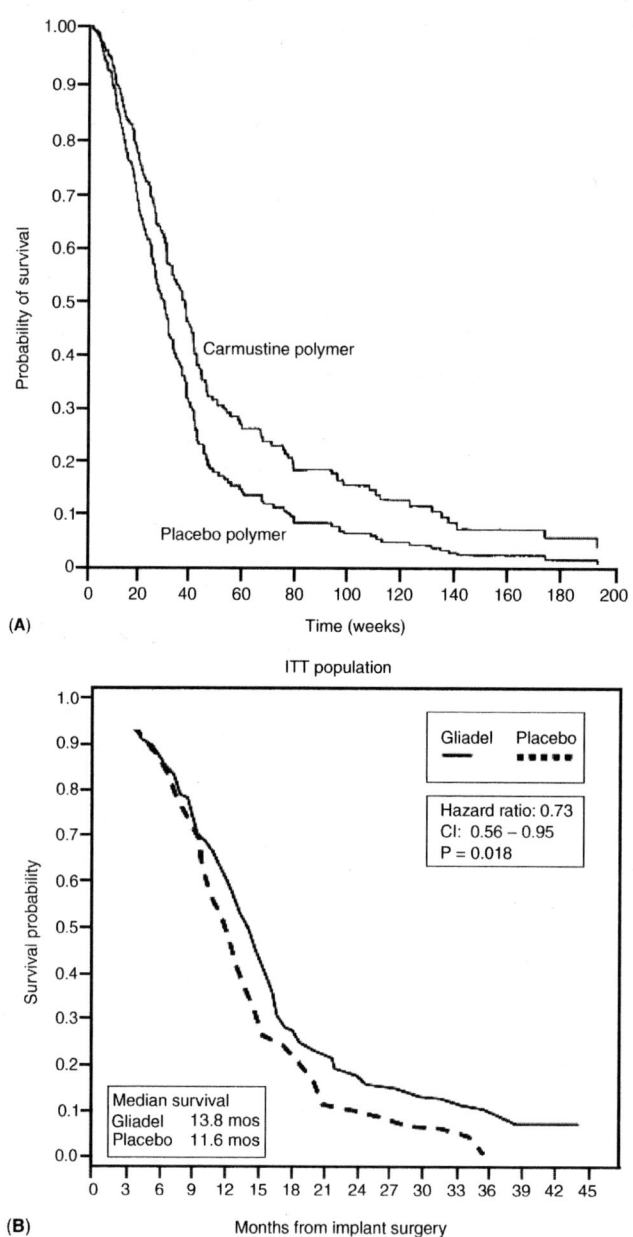

(A)

(B)

Figure 9 Clinical efficiency of Gliadel® for the treatment of *recurrent* and *newly diagnosed* malignant gliomas. Overall survival of the **(A)** 222 patients with recurrent brain tumors participating in a phase III clinical trial after adjustment for prognostic factors; **(B)** 240 patients with newly diagnosed brain tumors participating in a phase III clinical trial, including results from the long-term follow-up. *Source*: From Refs. 51 and 60.

survival time was 58 versus 40 weeks in the treatment versus placebo group. Due to the limited number of patients a further, larger (also randomized and placebo-controlled) phase III study was performed, including 240 patients with primary malignant gliomas (60). Upon surgical tumor resection the patients received either BCNU-loaded or drug-free wafers. Both groups were postoperatively treated with external beam radiation. The median survival time in the intent-to-treat group was 13.9 months for the Gliadel treated patients and 11.6 months for the placebo group. The 1-year survival rates were 59.2% and 49.6%, respectively. Recently, a long-term follow-up of this phase III clinical trial was reported (61). Of the 59 patients available for this follow-up, 11 were alive at 56 months: 9 of them had been treated with Gliadel, and only 2 with placebo wafers. The extended Kaplan-Meier curves for all 240 patients are shown in Figure 9B. As it can be seen, the survival advantage of the Gliadel treated group was clearly maintained at 1, 2, and 3 years.

The effects of the BCNU loading and composition of the biodegradable matrix former p(CPP:SA) on the efficiency of this type of disc-shaped implants to treat intracranial 9L-gliosarcomas bearing rats were studied by Sipos et al. (55). The anticancer drug loading was varied from 4% to 32% and the monomer ratio CPP:SA of the co-polymer was either 20:80 or 50:50. In the latter case, the degradation rate of the matrix is reduced, resulting in decreased drug release rates and prolonged release periods. Figure 10 (A and B) shows the observed survival curves in the respective animal groups. Clearly, the BCNU loading had a significant effect on the efficiency of the treatment: A drug content of 20 and 32% led to the longest survival rates in the case of p (CPP:SA) 20:80 and 50:50-based discs, respectively. In contrast, the pharmacodynamic effects of the wafers were similar for the two types of co-polymers differing in the monomer ratio (at equal drug contents) (Fig. 10A vs. 10B).

Controlled Release Microparticles

Different types of anticancer drugs have been incorporated into biodegradable microparticles (62–69). Compared to implants these systems offer the advantage that they can be injected directly into the *wall* of the resection cavity (the neighboring tissue of the debulked tumor) at multiple positions. As drug transport within the brain tissue is generally restricted to a few millimeters, this type of devices allows to more easily cover larger brain regions. For instance, Emerich et al.(67) compared the efficiency of carboplatin- and BCNU-loaded, PLGA-based microparticles that were injected either into the resection *cavity* or into the *tissue* surrounding the resection cavity of tumor bearing rats that underwent surgical removal. The microparticles that were injected into the tissue were administered at four

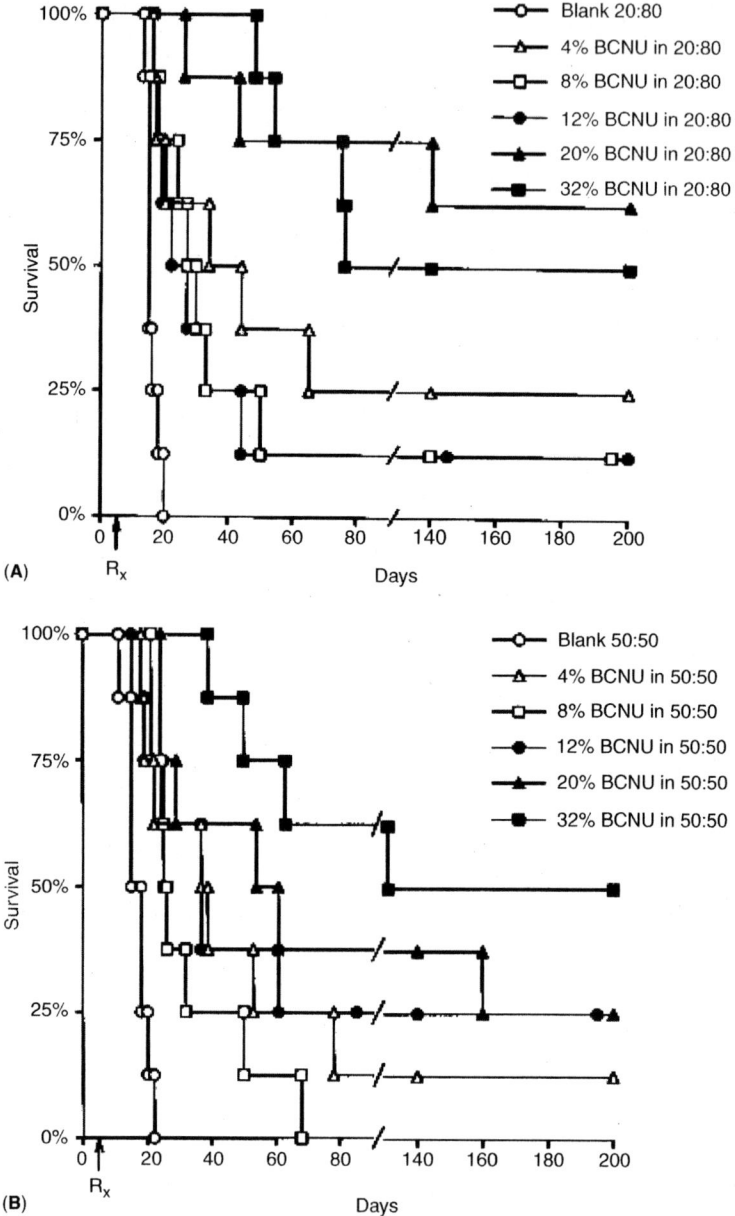

(A) R$_x$

(B) R$_x$

Figure 10 Effects of the drug loading and co-polymer composition of BCNU-loaded p(CPP:SA)-based wafers on the survival of 9L-gliosarcomas-bearing rats treated 5 days after tumor implantation with discs based on: **(A)** 20:80 p(CPP:SA); and **(B)** 50:50 p(CPP:SA). The drug contents are indicated in the figures. Blank wafers (free of drug) were studied for reasons of comparison (n = 8 for each group). *Source*: From Ref. 55.

separate sites, approximately 5 mm from the edge of the resection cavity in a diamond-shaped configuration. In addition, control groups that: (1) underwent no resection; (2) underwent resection only; or (3) received a bolus drug injection were studied. The drug amount administered with the microparticles was either 10, 50, or 100 μg/rat, the bolus control drug injection contained 100 μg drug/rat. As an example, Figure 11(A and B) shows the observed survival rates of the respective animal groups treated with carboplatin (please note the different scaling of the *x*-axes). Clearly, the efficiency of the treatments increased in the following order: no resection < resection only < bolus injection < sustained release microparticles injection (drug amount: 10 μg < 50 μg < 100 μg). This was true also for BCNU (data not shown). Importantly, carboplatin-loaded microparticles were more effective when injected into the surrounding tissue compared to an administration into the resection cavity (Fig. 11B vs. 11A). This can at least partially be attributed to the fact that carboplatin diffusion into the brain tissue is limited: Atomic absorption spectrophotometry measurements showed that the drug was distributed within an area of approximately 0.5 mm from the implantation site.

Another interesting advantage of controlled release *microparticles* compared to larger implants is the possibility to inject them directly into *inoperable* tumors. Using standard needles anticancer drug-loaded systems can be administered by stereotaxy without causing major damage to the brain tissue. For instance, the group of Benoit and Menei developed 5-fluorouracil-loaded, PLGA-based microparticles that can be used for both: the treatment of *operable* as well as *inoperable* malignant gliomas (62–66,70). Due to the significant progress in neurosurgery it is nowadays possible to precisely inject small volumes of microparticle suspensions at any position in the brain. Using a stereotaxic frame (or even frameless) computer-assisted neurosurgery (neuronavigation) can guide the implantation of electrodes, catheters, injections or the realization of biopsies (71). The morbidity of stereotaxic procedures is low (<1%). Due to their small size (often <100 μm), anticancer drug-loaded microparticles can be easily administered using these techniques in discrete, precise regions of the brain. For example, Figure 12 shows: (1) the stereotaxic intracranial implantation of microparticles into an anaesthetized rat using a stereotaxic frame; (2) successfully administered microparticles in the striatum of a rat brain; (3) a navigation software neurosurgeons can use to guide their injections; and (4) the stereotaxic implantation of anticancer drug-loaded microparticles into a human brain. Importantly, PLGA-based microparticles are completely biodegradable and are biocompatible with brain tissue (72,73). Figure 13A and Figure 13B shows optical micrographs of such particles 24 h and 3 weeks after injection into rat striatum, respectively. As it can be seen in Figure 13B, some microparticles were vacuolized.

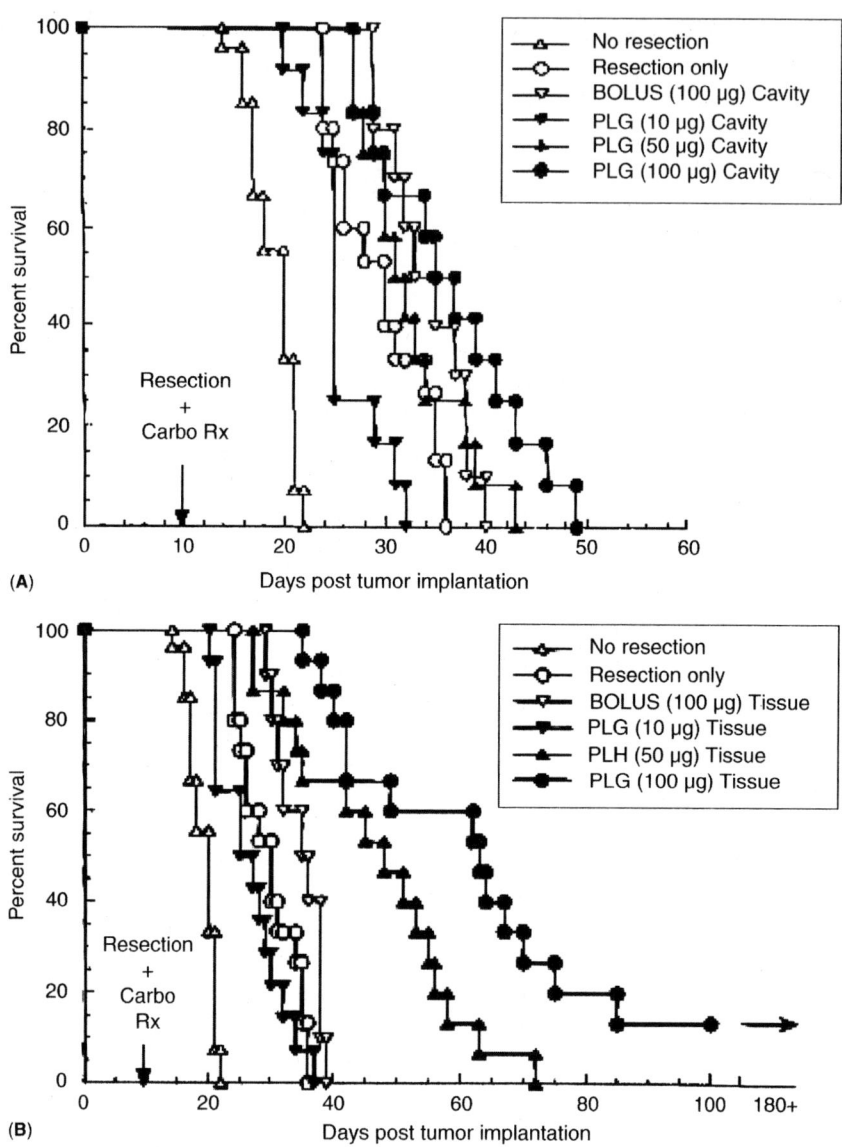

Figure 11 Survival of rats upon surgical brain tumor removal and subsequent carboplatin administration into (**A**) the resection cavity; or (**B**) the surrounding tissue. The drug was administered either as a bolus injection (drug amount = 100 μg/rat), or in the form of controlled release microparticles (drug amount = 10, 50 or 100 μg/rat, as indicated). In the case of microparticle injection into the surrounding tissue, the total dose was divided into 4 equal parts that were administered at 4 separate sites. "No resection" and "resection only" controls were studied for reasons of comparison. *Source*: From Ref. 67.

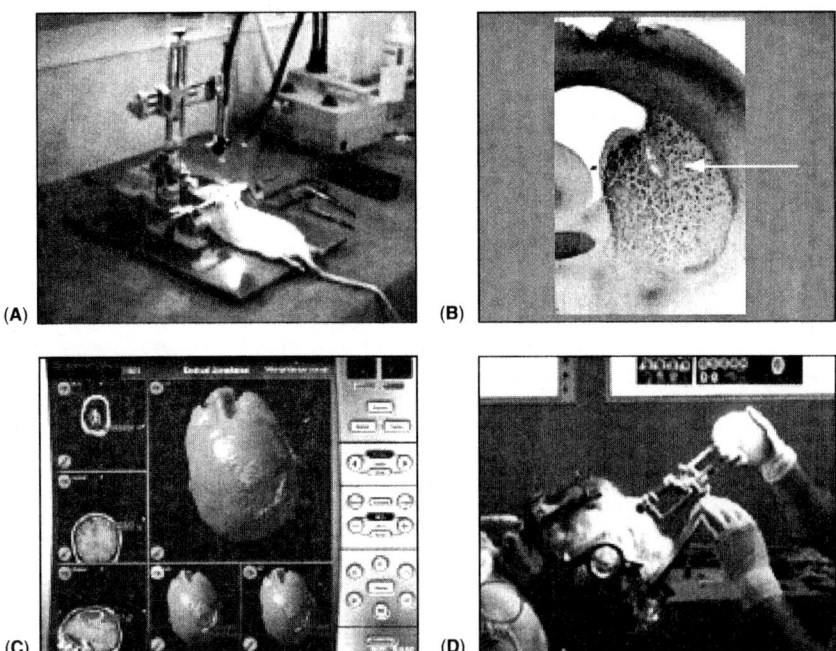

Figure 12 Stereotaxic administration of biodegradable microparticles into the brain: (**A**) In an anaesthetized rat using a stereotaxic frame; (**B**) imaging of microparticles that have successfully been injected into the striatum of a rat brain (indicated by the arrow); (**C**) programming of a stereotaxic intratumoral administration of microparticles using computer-assisted neurosurgery; (**D**) stereotaxic administration of 5-fluorouracil-loaded microparticles into a human brain. *Source*: From Ref. 65.

A phase I clinical trial with 5-fluorouracil-loaded, PLGA-based microparticles for the treatment of *operable* malignant gliomas was reported in 1999 (63). The principle of this treatment method is illustrated in Figure 14. The tumor (marked as a black circle) is removed by surgical resection; the surrounding infiltrated tissue remains in the brain. To reduce the resulting risk of local tumor recurrence, the anticancer drug-loaded microparticles are injected into the wall of the resection cavity. Eight patients with newly diagnosed glioblastoma were included in this trial, who received in addition external beam radiation. They were followed by clinical examination, magnetic resonance imaging as well as 5-fluorouracil assays in the blood and cerebrospinal fluid (CSF). The anticancer drug was detectable during at least 1 month upon administration in the CSF, whereas the concentrations in the blood were lower and transitory. The systemic

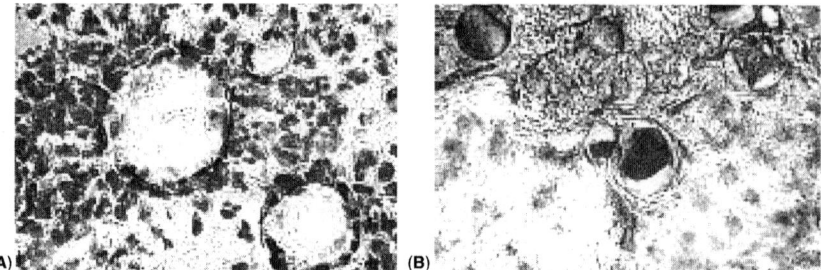

Figure 13 Optical microscopy pictures of PLGA-based microparticles upon implantation into the rat striatum, after (**A**) 24 hours, and (**B**) 3 weeks (some microparticles are vacuolized). *Source*: From Ref. 72.

tolerance to the treatment was good. Importantly, the median survival time was 98 weeks from the time of implantation and 2 patients achieved complete remission at 139 and 153 weeks, respectively. Based on these encouraging results, a randomized, multicenter phase II clinical trial with these 5-fluorouracil-loaded, PLGA-based microparticles was conducted (64,65). All patients that were included suffered from high-grade gliomas, underwent tumor resection and received external beam radiation. One group of patients also received the anticancer drug-loaded microparticles (130 mg 5-fluorouracil, injected at multiple sites into the surrounding tissue) (Arm A), the other group of patients did not (Arm B). Ninety five patients were randomized, 75 were treated and analyzed in intention to treat for efficacy

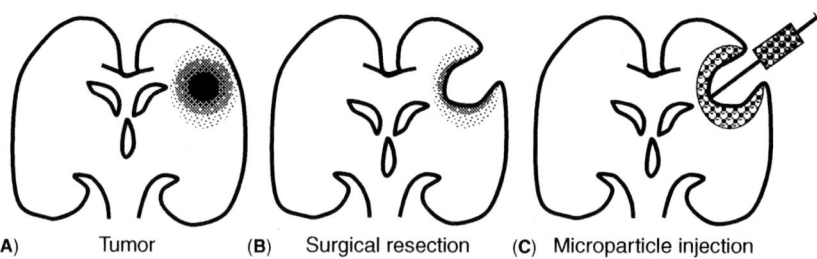

(**A**)   Tumor   (**B**)   Surgical resection   (**C**)   Microparticle injection

Figure 14 Principle of the treatment of *operable* brain tumors with 5-fluouracil-loaded, PLGA-based microparticles. Schematic cross-sections through a human brain: (**A**) The tumor is illustrated as a black circle; the surrounding tissue is infiltrated by tumor cells. (**B**) The tumor has been removed surgically. (**C**) To minimize the risk of local tumor recurrence, drug-loaded microparticles are injected into the wall of the resection cavity at multiple locations.

and safety. The overall survival in Arm A was 15.2 months versus 13.5 months in Arm B. In the subpopulation of patients with complete resection, the overall survival was 15.2 months in Arm A versus 12.3 months in Arm B. Thus, the treatment of the patients with this type of local controlled drug delivery system increased the overall survival. However, the differences were not statistically significant in this study (that was not designed and sufficiently powered to demonstrate this).

A phase I clinical trial with this type of 5-fluorouracil-loaded, PLGA-based microparticles to treat *inoperable* brain tumors was reported in 2004 (74). Ten patients with newly diagnosed, inoperable malignant gliomas were included. The microparticles (containing 132 mg anticancer drug) were administered by stereotaxy directly into the tumor in one or several trajectories with 1–7 deposits per trajectory. The patients also received external beam radiation and were followed by clinical examination, computed tomography scanning, magnetic resonance imaging and 5-fluorouracil assays in the blood and CSF. Importantly, the microparticle implantation was well tolerated. No acute intracranial hypertension was observed despite of the intracranial injection of 2.5 mL suspension. This can be attributed to the fast resorption of the liquid vehicle. However, the four patients who received only one single trajectory (with 1–5 deposits) experienced a transitory worsening of their pre-existing neurological symptoms. Thus, it seems to be preferable to divide the total suspension volume into several parts and to deposit them in separate trajectories. Importantly, there were no episodes of edema or hematological complications, the anticancer drug was detectable in the CSF and the median overall survival was 40 weeks (two patients survived 71 and 89 weeks, respectively).

Importantly, tumor cells express proteins that are foreign to the host because of their genetic mutations. Thus, they are vulnerable to an immune response of the human body. That is why efforts have been made to delivery immune response stimulating substances locally and in a controlled manner to brain tumors. For instance, interleukin-2 (IL-2)-loaded, gelatin- and chondroitin-6-sulfate-based microparticles have been proposed by Hanes et al. (71). Bioactive IL-2 was found to be released over at least 2 weeks in vitro; in vivo significant concentrations could be detected up to 3 weeks. The efficiency of these microparticles to protect mice challenged intracranially with B16-F10 melanoma cells is illustrated in Figure 15A (the melanoma cell challenge and microparticle injection were simultaneous). Clearly, the IL-2-loaded microparticles were able to protect the mice, whereas placebo systems were not. Interestingly, even autologous B16-F10 cells engineered to secrete IL-2 were not as effective as the controlled release microparticles. In Figure 15B, the survival curves of *rats* challenged intracranially with a lethal dose of 9L gliosarcoma cells are illustrated. The animals received a simultaneous injection of IL-2-loaded or drug-free microparticles. Clearly, drug-containing, controlled release microparticles were able to prolong the

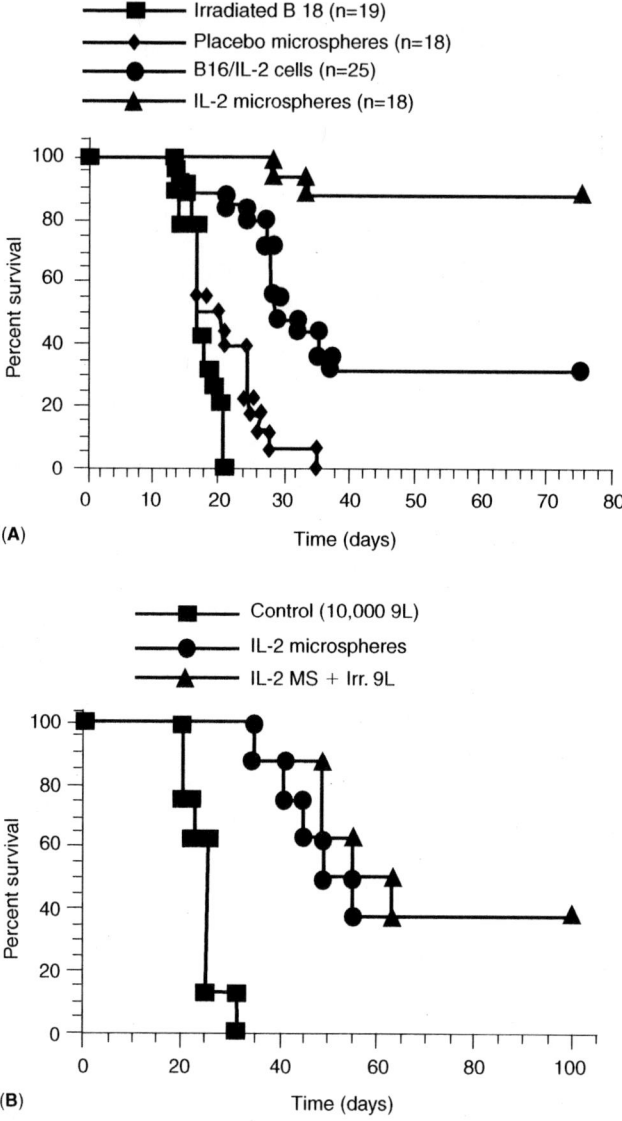

Figure 15 Efficiency of IL-2-loaded, gelatin-chondroitin-6-sulfate-based micropar-ticles to protect: (**A**) Mice challenged intracranially with B16-F10 melanoma. For reasons of comparison also placebo microparticles, autologous B16-F10 cells engineered to secrete IL-2, and B16-F10 cells (antigen control) were administered. (**B**) Rats challenged intracranially with wild-type 9L gliosarcoma cells. For reasons of comparison also placebo microparticles (control) and IL-2-loaded microparticles plus extra tumor antigen in the form of irradiated 9L tumor cells (Irr. 9L) were administered. *Source*: From Ref. 69.

survival of the rats. Interestingly, the addition of extra tumor antigen in the form of irradiated 9L tumor cells (Irr. 9L) did not have any significant effect.

Controlled Release Microchip Technology

Microchip-based controlled drug delivery systems offer an interesting potential as future alternatives for the above described polyanhydride- or polyester-based implants and microparticles (3,74–78). The idea is to use silicon microchips with multiple microreservoirs and to fill the latter with one or more drugs (solids, liquids, or semi-solids). The microreservoirs are then sealed with polymeric membranes (that degrade after pre-determined time periods) or gold membranes (that dissolve upon electrochemical stimulation). Thus, in theory this type of microchip-based device can provide any kind of release kinetics for one or more drugs, representing a very attractive tool to optimize the treatment of various brain diseases. Figure 16A and Figure 16B shows an example for such as device, a "microelectromechanical system (MEMS)". The drug(s) is (are) filled into the microreservoirs located in the middle plates. This type of system was recently filled with C^{14}-labeled BCNU and studied in a rat flank model (77). Interestingly, only 40% of the drug content was recovered from the activated devices in this first study. This could be attributed to the fact that BCNU penetrated into the neoprene gasket used in the stainless steel frame package (78). Importantly, this restriction could be overcome by a new packaging method (using Pyrex plates) and adding poly(ethylene glycol) (PEG) into the microreservoirs. Figure 16C shows the in vitro drug release kinetics obtained in phosphate buffer saline from such MEMS upon activation with a potentiostat at pre-determined time points (indicated by the dashed lines). Clearly, the rate and extent of BCNU release significantly increased upon co-formulation with PEG. The total radioactivity recovery was close to 100%.

TREATMENT OF NEURODEGENERATIVE DISEASES

The local controlled delivery of neurotrophic factors (e.g. NGF, and glial cell-derived neurotrophic factor, GDNF) can be very advantages for the treatment of neurodegenerative diseases, such as Parkinson's, Huntington's and Alzheimer's Disease. For this purpose, biodegradable controlled release microparticles have been proposed (71,63,79–85). For instance, Pean et al. (82) prepared PLGA-based microparticles loaded with NGF and evaluated their in vivo performance in rat brain. Drug-loaded as well as placebo microparticles were injected near the septal cholinergic neurons, axotomized by an unilateral transection of the fornix-fimbria (Fig. 17A and 17B). The histological analysis 2 and 4 weeks after administration revealed a non-

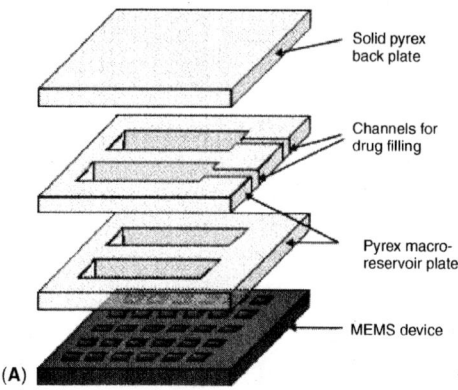

Solid pyrex
back plate

Channels for
drug filling

Pyrex macro-
reservoir plate

MEMS device

(A)

(B)

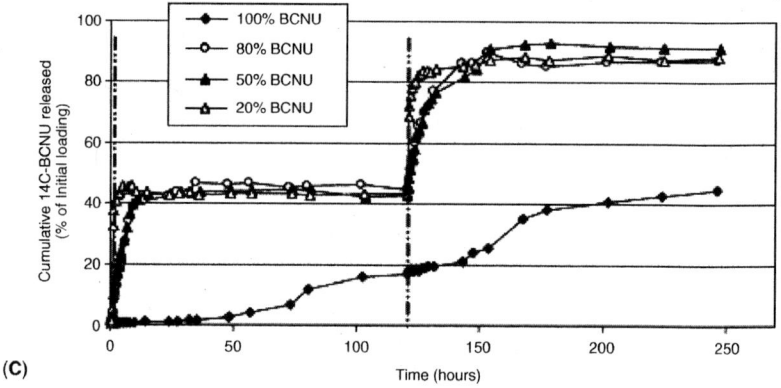

(C)

Figure 16 Novel MEMS for controlled local drug delivery within the brain.
(A) Schematic presentation of the different layers the devices are composed of,
(B) photographs showing the top and bottom sides of an assembled device, (C) in
vitro BCNU release into phosphate buffer saline upon activation using a potentiostat
(at the time points indicated by the dashed lines). The microreservoirs were filled
either with pure drug or with a BCNU:PEG blend (the drug percentage is indicated
in the figure). *Source*: From Ref. 20.

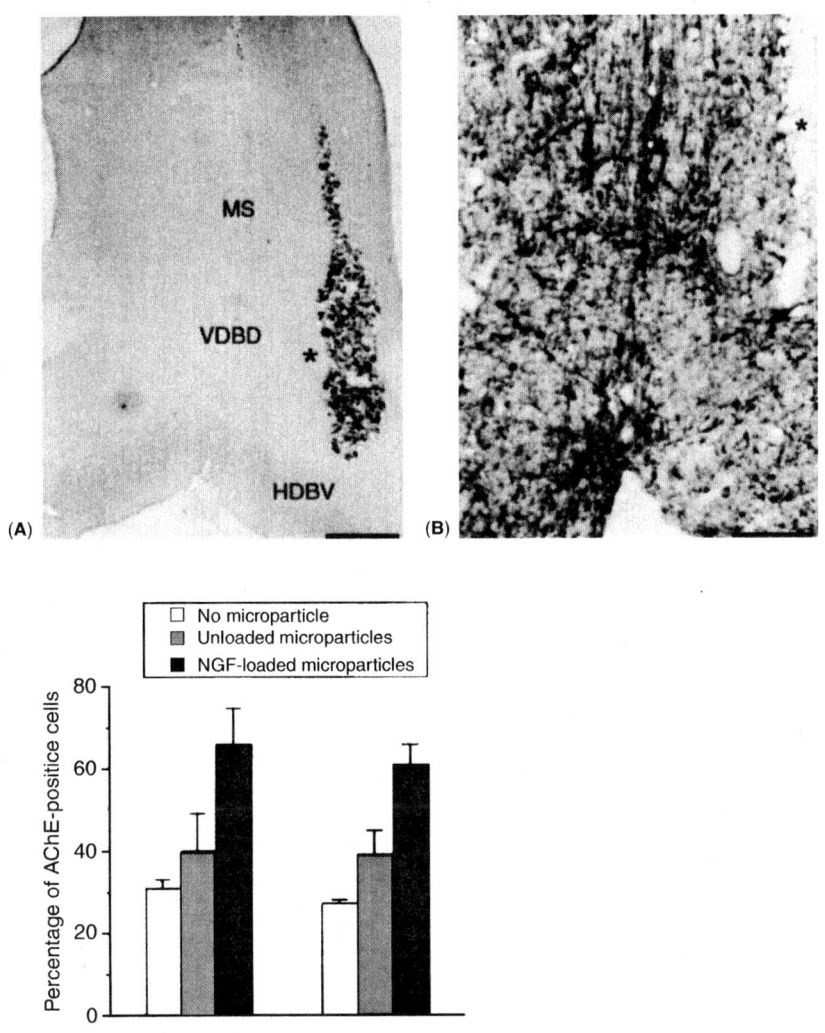

Figure 17 Intracranial administration of NGF-loaded, PLGA-based microparticles in rats: (**A**) Microscopic picture of the septal area 2 weeks upon microparticle injection, NGF immunostaining (scale bar = 500 μm); (**B**) adjacent section stained with AchE histochemistry (* = microparticles). Note the apparently healthy neurons in contact with the microspheres (scale bar = 200 μm); (**C**) percentage of surviving cholinergic neurons stained by AchE histochemistry (compared to the contralateral, intact side) after 2 and 6 weeks upon microparticle administration (drug-free or NGF-loaded systems as indicated in the figure; a control group without microparticle injection is included for reasons of comparison) *Abbreviations*: HDBV, horizontal limb of the diagonal band of Broca; MS, medial septal nucleus; VDBD, vertical limb of the diagonal band of Broca. *Source:* From Ref. 79.

specific astro- and micro-glial reaction around the microspheres, which was similar for drug-loaded and drug-free devices. Importantly, no neuronal toxicity was observed and healthy appearing neurons were visible that were in direct contact with the microparticles. In the non-treated control group, the percentage of axotomized surviving neurons (compared to the contralateral, intact side) was $31(\pm2)\%$ and $27(\pm1)\%$ at $t = 2$ and 6 weeks, respectively (Fig. 17C, white bars). Drug-free microparticles exhibited neither protective nor toxic effects for the neurons: The percentage of surviving neurons after 2 and 6 weeks was equal to $40(\pm9)\%$ and $39(\pm6)\%$, respectively (gray bars in Fig. 17C). Importantly, the NGF-loaded, controlled release microparticles could significantly increase the survival rate of the neurons: up to $66(\pm9)\%$ and $61(\pm5)\%$ after 2 and 6 weeks, respectively (Fig. 17C, black bars). Thus, this type of biodegradable microparticles is able to release sufficient amounts of bioactive NGF in a time-controlled manner into the brain tissue to limit the lesion-induced disappearance of cholinergic neurons.

Recently, Jollivet et al. (83,84) proposed GDNF-loaded, PLGA-based microparticles for the treatment of Parkinson's Disease. The particles release the neurotrophic factor during at least 2 months in vivo and are well tolerated upon intracranial administration into rat brain. Importantly, they were able to stimulate the axonal regeneration of mesencephalic dopaminergic neurons in "Parkinsonian rats" (the animals received two injections of 10 μg 6-hydroxydopamine inducing a partial progressive and retrograde lesion of the nigrostriatal system). The administration of the GDNF-loaded microparticles led to sprouting of the preserved doperminergic fibers with synaptogenesis. As it can be seen in Figure 18, this neural regeneration was accompanied by a functional improvements of the rats. The amphetamine-induced rotational behavior was measured 1, 4, 6, 8, and 10 weeks after the lesion, the microparticles were administered 2 weeks after the lesion. For reasons of comparison, also placebo microparticles were administered and an untreated animal group included as control. Clearly, all animals in this study were sucessfully lesioned (results obtained after 1 week, Fig. 18). Importantly, the number of ipsiversive turns per minute (tpm) increased in the non-treated and placebo group: from $10.7(\pm2.2)$ to $15.4(\pm2.3)$ tpm and from $11.6(\pm2)$ to $20.4(\pm2.2)$ tpm, respectively (white and gray bars). In contrast, the number of ipsiversive tpm significantly decreased in the rats that received GDNF-loaded microparticles: from $15.4(\pm0.9)$ to $8.8(\pm2.1)$ tpm (Fig. 18, black bars).

A further promising approach to treat neurodegenerative diseases is to transplant living cells into the human brain which continuously produce biologically active agents, e.g. missing neurotransmitters (such as dopamine in the case of Parkinson's Disease) ("cell therapy"). Unfortunately, so far the success of this type of advanced treatment method is limited due to the: (1) low-survival rate of the transplanted cells within the brain tissue; and (2)

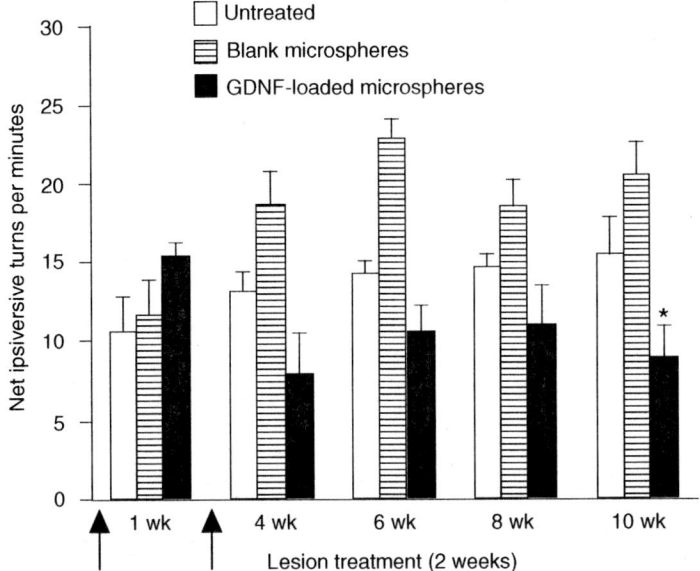

Figure 18 Effects of GDNF-loaded, PLGA-based microparticles on the behavior of "Parkinsonian rats" (the animals received two injections of 10 μg 6-hydroxydopamine inducing a partial progressive and retrograde lesion of the nigrostriatal system): Number of ipsiversive turns per minute (tpm) in the amphetamine-induced rotation test at 1, 4, 6, 8, and 10 weeks after the lesion. GDNF-loaded microparticles as well as placebo microparticles were injected 2 weeks after the lesion. Untreated animals were included for reasons of comparison. *Source*: From Ref. 83.

poor integration of the cells in their new environment. Generally, about 90% of the transplanted cells die within the first 2 weeks after administration. A promising approach aiming to overcome these restrictions is the combination of controlled release microparticles with cell transplantation (65). The microparticles can for instance release growth factors and/or cytokines in a pre-determined manner, helping to reduce cell death and to improve cell integration into the brain tissue. An even more sophisticated approach is to use the microparticles not only as controlled drug delivery systems, but also as *microcarriers* for the transplanted cells (65,86). These systems are also called "pharmacologically active microcarriers" (PAM). Figure 19A shows a schematic illustration of this concept. The microparticles are coated with cell adhesion or extracellular matrix molecules and release the biologically active agents at pre-determined rates. Figure 19B and C show optical and scanning electron micrographs of such devices carrying PC12 cells on their surfaces. The microparticles may also release drugs that are able to modify the microenvironment, for example,

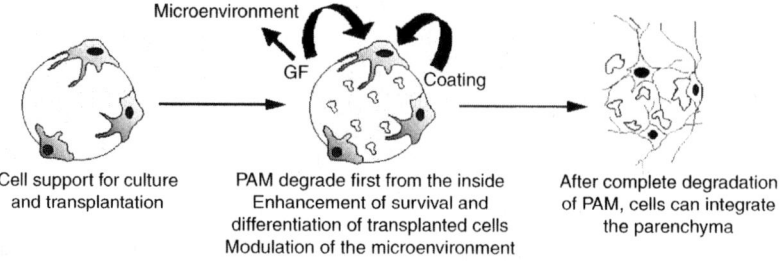

| Cell support for culture and transplantation | PAM degrade first from the inside Enhancement of survival and differentiation of transplanted cells Modulation of the microenvironment | After complete degradation of PAM, cells can integrate the parenchyma |

(A) GF : growth factor and/or cytokine

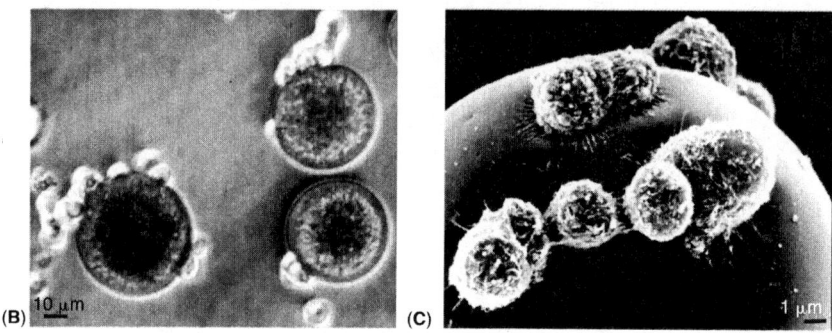

Figure 19 PAM used to improve the efficiency of cell therapies: (A) Schematic illustration of concept of the approach; (B and C) optical and scanning electron micrographs of cells adhering onto PAM. *Source*: From Ref. 86.

favor angiogenesis or local immuno-depression. After complete micro-particle degradation, the cells can integrate the parenchyma. Recently, NGF-releasing PAM conveying PC12 cells were transplanted into "Parkinsonian rats" (65). When PC12 cells, which express TH, are exposed to NGF, they stop cell division, extend long neuritis, become excitable and after depolarization they can release significant amounts of dopamine (the neurotransmitter missing in Parkinson's Disease). First results showed that these NGF-releasing PAM can reduce cell death and improve the amphet-amine-induced rotational behavior of the rats.

CONCLUSIONS AND FUTURE PERSPECTIVES

The potential of intracranial controlled drug delivery systems to allow new therapeutic strategies and/or to optimize existing ones for various brain diseases is considerable. However, the mass transport mechanisms that are involved in the control of drug transport within the pharmaceutical dosage forms as well as within the living brain tissue are complex and not yet fully understood. Thus, the design of this type of advanced drug delivery systems is not straightforward. Recent clinical trials aiming to improve the treatment of brain tumors have shown promising results. Novel approaches to improve the treatment of neurodegenerative diseases (e.g. Parkinson's or Huntington's Disease) are currently in the pre-clinical state. For the future it can be expected that based on a more and more comprehensive understanding of the involved phenomena and on the progress in technology (e.g. microchip-based delivery systems, computer-assisted neurosurgery) this type of advanced drug delivery system will gain in practical importance.

REFERENCES

1. Grieg NH. Optimizing drug delivery to brain tumors. Cancer Treat Rev 1987; 14:1–28.
2. Abott NJ, Romero IA. Transporting therapeutics across the blood-brain barrier. Mol Med Today 1996; 2:106–13.
3. Wang PP, Frazier J, Brem H. Local drug delivery to the brain. Adv Drug Deliver Rev 2002; 54:987–1013.
4. Kornblith PL, Walker M. Chemotherapy for malignant gliomas. J Neurosurg 1988; 68:1–17.
5. Pardridge WM. Vector–mediated drug delivery to the brain. Adv Drug Deliver Rev 1999; 36:299–321.
6. Sanovich E, Bartus RT, Friden PM, Dean RL, Le HQ, Brightman MW. Pathway across blood-brain barrier opended by the bradykinin RMP–7. Brain Res 1995; 705:125–35.
7. Elliot PJ, Hayward NJ, Dean RL, Blunt DG, Bartus RT. Intravenous RMP–7 selectively increases uptake of carboplatin in experimental brain tumors. Cancer Res 1996; 56:3998–4005.
8. Langer R, Folkman J. Polymers for the sustained release of proteins and other macromolecules. Nature 1976; 263:797–800.
9. Langer RS, Wise DL. Medical Applications of Controlled Release. Boca Raton FL: CRC Press; 1984.
10. Leong KW, Brott BC, Langer R. Bioerodible polyanhydrides as drug–carrier matrices. I: Characterization, degradation, and release characteristics. J Biomed Mater Res 1985; 19:941–55.
11. Chasin M, Langer R. Biodegradable Polymers as Drug Delivery Systems. New York: Marcel Dekker1990.
12. Tamargo RJ, Sills AKJ, Reinhard CS, Pinn ML, Long DM, Brem H. Interstitial delivery of dexamethasone in the brain for the reduction of peritumoral edema. J Neurosurg 1991; 74:956–61.
13. Tamargo RJ, Myseros JS, Epstein JI, Yang MB, Chasin M, Brem H. Interstitial chemotherapy of the 9L gliosarcoma: controlled release polymers for drug delivery in the brain. Cancer Res 1993; 53:329–33.
14. Brem H, Walter K, Langer R. Polymers as controlled drug delivery devices for the treatment of malignant brain tumors. Eur J Pharm Biopharm 1993; 27:2–7.

15. Brem H, Langer, R. Polymer–based drug delivery to the brain. Sci Med 1996; 52–61.
16. Siepmann J, Goepferich A. Mathematical modeling of bioerodible; polymeric drug delivery systems. Adv Drug Deliver Rev 2001; 48:229–47.
17. Siepmann J, Siepmann F, Florence AT. Local controlled drug delivery to the brain: Mathematical modeling of the underlying mass transport mechanisms. Int J Pharm 2006; 314:101–19.
18. Burkersroda F. von, Schedl L, Goepferich A. Why degradable polymers undergo surface erosion or bulk erosion. Biomaterials 2002; 23: 4221–31.
19. Brunner A, Maeder K, Goepferich A. pH and osmotic pressure inside biodegradable microspheres during erosion. Pharm Res 1999; 16:847–53.
20. Li L, Schwendeman SP. Mapping neutral microclimate pH in PLGA microspheres. J Control Release 2005; 101:163–73.
21. Goepferich A, Langer R. Modeling monomer release from bioerodible polymers. J Control Release 1995; 33:55–69.
22. Goepferich A, Langer R. Modeling of polymer erosion in three dimensions— rotationally symmetric devices. AIChE J1995; 41: 2292–9.
23. Siepmann J, Elkharraz K, Siepmann F, Klose D. How autocatalysis accelerates drug release from PLGA-based microparticles: A quantitative treatment. Biomacromolecules 2005; 6:2312–19.
24. Klose D, Siepmann F, Elkharraz K, Krenzlin S, Siepmann J. How porosity and size affect the drug release mechanisms from PLGA–based microparticles. Int J Pharm 2006; 314:198–206.
25. Goepferich A, Shieh L, Langer R. Aspects of polymer erosion. Mat Res Soc Symp Proc 1995; 394:155–60.
26. Goepferich A. Mechanisms of polymer degradation and erosion. Biomaterials 1996; 17:103–14.
27. Goepferich A. Polymer degradation and erosion: Mechanisms and applications. Eur J Pharm Biopharm 1996; 42:1–11.
28. Goepferich A. Erosion of composite polymer matrices. Biomaterials 1997; 18:397–403.
29. Goepferich A. Bioerodible implants with programmable drug release. J Control Release 1997; 44:271–81.
30. Goepferich A. Mechanisms of polymer degradation and elimination. In Domb A, Kost, J, Wiseman D, eds. Handbook of Biodegradable Polymers. Amsterdam: Harwood Academic Publishers; 1997.
31. Goepferich A. Polymer bulk Erosion. Macromolecules.1997; 30: 2598–604.
32. Siepmann J, Faisant N, Benoit JP. A new mathematical model quantifying drug release from bioerodible microparticles using Monte Carlo simulations. Pharm Res 2002; 19:1885–93.
33. Tao L, Nicholson C. Diffusion of albumins in rat cortical slices and relevance to volume transmission. Neuroscience 1996; 75:839–47.
34. Nicholson C. Diffusion and related transport mechanisms in brain tissue. Rep Prog Phys 2001; 64:815–84.
35. Reinhard CS, Radomsky ML, Saltzman WM, Hilton J, Brem H. Polymeric controlled release of dexamethasone in normal rat brain. J Control Release 1991; 16:331–9.

36. Saltzman WM, Radomsky; ML. Drugs released from polymers: Diffusion and elimination in brain tissue. Chem Eng Sci 1991; 46:2429–44.

37. Dang W, Colvin OM, Brem H, Saltzman WM. Covalent coupling of methotrexate to dextran enhances the penetration of cytotoxicity into a tissue-like matrix. Cancer Res 1994; 54:1729–35.

38. Haller MF, Saltzman WM. Localized delivery of proteins in the brain: Can transport be customized? Pharm Res1998; 15: 377–85.

39. Haller MF, Saltzman; WM. Nerve growth factor delivery systems. J Controlled Release 1998; 53:1–6.

40. Fung LK, Shin M, Tyler B, Brem H, Saltzman WM. Chemotherapeutic drugs released from polymers: Distribution of 1,3-bis(2-chloroethyl)-1-nitrosourea in the rat brain. Pharm Res 1996; 13:671–82.

41. Fung LK, Ewend MG, Sills A, et al. Pharmacokinetics of interstitial delivery of carmustine, 4–hydroperoxycyclophosphamide, and paclitaxel from a biodegradable polymer implant in the monkey brain. Cancer Res 1998; 58: 672–84.

42. Tao L, Nicholson C. Maximum geometrical hindrance to diffusion in brain extracellular space surrounding uniformly spaced convex cells. J Theor Biol 2004; 229:59–68.

43. Tao A, Tao L, Nicholson C. Cell cavities increase tortuosity in brain extracellular space. J Theor Biol 2005; 234:525–36.

44. Hrabetova S, Nicholson C. Contribution of dead–space microdomains to tortuosity of brain extracellular space. Neurochem Int 2004; 45:467–77.

45. Sykova E. The extracellular space in the CNS: Its regulation; volume and geometry in normal and pathological neuronal function. Neuroscientist 1997; 3:28–41.

46. Sykova E. Diffusion properties of the brain in health and disease. Neurochem Int 2004; 45:453–66.

47. Lehmenkuhler A, Sykova E, Svoboda J, Zilles K, Nicholson C. Extracellular space parameters in the rat neocortex and subcortical white matter during postnatal development determined by diffusion analysis. Neuroscience 1993; 55:339–51.

48. Krewson CE, Saltzman WM. Transport and elimination of recombinant human NGF during long–term delivery to the brain. Brain Res 1996; 727:169–81.

49. Saltzman WM, Mak MW, Mahoney MJ, Duenas ET, Cleland JL. Intracranial delivery of recombinant nerve growth factor: Release kinetics and protein distribution for three delivery systems. Pharm Res 1999; 16:232–40.

50. Brem H, Mahaley MS, Vick NA, et al. Interstitial chemotherapy with drug polymer implants for the treatment of recurrent gliomas. J Neurosurg 1991; 74:441–6.

51. Brem H, Piantadosi S, Burger PC, et al. Placebo-controlled trial of safety and efficacy of intraoperative controlled delivery by biodegradable polymers of chemotherapy for recurrent gliomas. The Polymer-brain Tumor Treatment Group. Lancet 1995; 345:1008–12.

52. Brem H, Ewend MG, Piantadosi S, Greenhoot J, Burger PC, Sisti M. The safety of interstitial chemotherapy with BCNU–loaded polymer followed by radiation therapy in the treatment of newly diagnosed malignant gliomas: Phase I trial. J Neurooncol 1995; 26:111–23.

53. Grossman SA, Reinhard C, Colvin OM, Chasin M, Brundrett R, Tamargo RJ, Brem, H. The intracerebral distribution of BCNU delivered by surgically implanted biodegradable polymers. J Neurosurg 1992; 76:640–7.

54. Ewend MG, Williams JA, Tabassi K, et al. Local delivery of chemotherapy and concurrent external beam radiotherapy prolongs survival in metastatic brain tumor models. Cancer Res 1996; 56:5217–23.

55. Sipos EP, Tyler B, Piantadosi S, Burger PC, Brem H. Optimizing interstitial delivery of BCNU from controlled release polymers for the treatment of brain tumors. Cancer Chemother Pharmacol 1997; 39:383–9.

56. Valtonen S, Timonen U, Toivanen P, et al. Interstitial chemotherapy with carmustine–loaded polymers for high–grade gliomas: A randomized double-blind study. Neurosurgery 1997; 41:44–8.

57. Moses MA, Brem H, Langer R. Advancing the field of drug delivery Taking aim at cancer. Cancer Cell 2003; 4: 337–41.

58. Walker MD, Green SB, Byar DP, et al. Randomized comparisons of radiotherapy and nitrosoureas for the treatment of malignant glioma after surgery. N Engl J Med 1980; 303:1323–9.

59. Green SB, Byar DP, Walker MD, et al. Comparisons of carmustine; procarbazine; and high-dose methylprednisolone as additions to surgery and radiotherapy for the treatment of malignant glioma. Cancer Treat Rep 1983; 67:121–32.

60. Westphal M, Hilt DC, Bortey E, et al. A phase 3 trial of local chemotherapy with biodegradable carmustine (BCNU) wafers (Glidel wafers) in patients with primary malignant glioma. Neuro-Oncology (serial online). 2003; 5: Doc. 02–023; http://neuro–oncology.mc.duke.edu.

61. Westphal M, Ram Z, Riddle V, Hilt DC, Bortey E, and On behalf of the Executive Committee of the Gliadel Study Group. Gliadel wafer in initial surgery for malignant glioma: long-term follow-up of a multicenter controlled trial. Acta Neurochir 2006; 148:269–75.

62. Menei P, Benoit JP, Boisdron-Celle M, Fournier D, Mercier P, Guy G. Drug targeting into the central nervous system by stereotactic implantation of biodegradable microspheres. Neurosurgery 1994; 34:1058–64.

63. Menei P, Venier MC, Gamelin E, et al. Local and sustained delivery of 5-fluorouracil from biodegradable microspheres for the radiosensitization of glioblastoma: A pilot study. Cancer 1999; 86:325–30.

64. Menei P, Capelle L, Guyotat J, et al. Local and sustained delivery of 5-Fluorouracil from biodegradable microspheres for the radiosensitization of malignant glioma: A randomized phase II trial. Neurosurgery 2005; 56: 242–8.

65. Menei P, Montero-Menei C, Venier MC, Benoit JP. Drug delivery into the brain using poly(lactide-co-glycolide) microspheres. Expert Opin 2005; 2:363–76.

66. Benoit JP, Faisant N, Venier–Julienne MC, Menei P. Development of microspheres for neurological disorders: From basics to clinical applications. J Control Release 2000; 65:285–96.

67. Emerich DF, Winn SR, Hu Y, et al. Injectable chemotherapeutic microspheres and glioma I: Enhanced survival following implantation into the cavity wall of debulked tumors. Pharm Res 2000; 17:767–75.

68. Emerich DF, Winn SR, Snodgrass P, et al. Injectable chemotherapeutic microspheres and glioma II: Enhanced survival following implantation into deep inoperable tumors. Pharm Res 2000; 17:767–75.

69. Hanes J, Sills A, Zhao Z, et al. Controlled local delivery of interleukin–2 by biodegradable polymers protects animals from experimental brain tumors and liver tumors. Pharm Res 2001; 18:899–906.

70. Menei P, Pean JM, Nerriere-Daguin V, Jollivet C, Brachet P, Benoit JP. Intracerebral implantation of NGF-releasing biodegradable microspheres protects striatum against excitotoxic damage. Exp Neurology 2000; 161:259–72.

71. Ohye C. The idea of stereotaxy toward minimally invasive surgery. Sterotact Funct Neurosurg 2000; 74:185–93.

72. Menei P, Daniel V, Montero-Menei C, Brouillard M, Pouplard-Barthelaix A, Benoit JP. Biodegradation and brain tissue reaction to poly(D-L lactide-co-glycolide) microspheres. Biomaterials 1993; 14:470–8.

73. Menei P, Jadaud E, Faisant N, et al. Stereotaxic implantation of 5 fluorouracil–releasing microspheres in malignant glioma. Cancer 2004; 100: 405–10.

74. Veziers J, Lesourd M, Jollivet C, Montero-Menei C, Benoit JP, Menei P. Analysis of brain biocompatibility of dug-releasing biodegradable microspheres by scanning and transmission electron microscopy. J Neurosurg 2001; 95:489–94.

75. Santini JT Jr, Cima MJ, Langer R. A controlled-release microchip. Nature 1999; 397:335–8.

76. Santini JT Jr, Richards AC, Scheidt R, Cima MJ, Langer R. Microchips as controlled drug-delivery devices. Angewandte Chimie 2000; 39:2396–407.

77. Li Y, Shawgo RS, Tyler B, et al. In vivo release from a drug delivery MEMSs device. J Control Release2004; 100: 211–19.

78. Li Y, Duc HLH, Tyler B, et al. In vivo delivery of BCNU from a MEMS devices to a tumor model. J Control Release 2005; 106:138–45.

79. Pean JM, Venier-Julienne MC, Boury F, Menei P, Denizot B, Benoit JP. NGF release from poly(DL(lactide-co-glycolide) microspheres. Effect of some formulation parameters on encapsulated NGF stability. J Control Release 1998; 56:175–87.

80. Pean JM, Venier-Julienne MC, Filmon R, Sergent M, Phan-Tan-Luu R, Benoit JP. Optimization of HAS and NGF encapsulation yields in PLGA microparticles. Int J Pharm 1998; 166:105–15.

81. Pean JM, Boury F, Venier-Julienne MC, Menei P, Proust J, Benoit JP. Why does PEG 400 co–encapsulation improve NGF stability and release from PLGA biodegradable microspheres? Pharm Res 1999; 16: 1294–9.

82. Pean JM, Menei P, Morel O, Montero-Menei C, Benoit JP. Intraseptal implantation of NGF-releasing microspheres promote the survival of axotomized cholinergic neurons. Biomaterials 2000; 21:2097–101.

83. Jollivet C, Aubert-Pouessel, A, Clavreul A, et al. Striatal implantation of GDNF releasing biodegradable microspheres promotes recovery of motor function in a partial model of Parkinson's Disease. Biomaterials 2004; 25: 933–42.

84. Jollivet C, Aubert-Pouessel A, Clavreul A, et al. Long-term effect of intra-striatal glial cell line-derived neurotrophic factor-releasing microparticles in a partial rat model of Parkinson's Disease. Neurosci Lett 2004; 356:207–10.

85. Clavreul A, Sindji L, Aubert–Pouessel A, Benoit J.P, Menei P, Montero-Menei C. Effect of GDNF–releasing biodegradable microspheres on the function and the survival of intrastriatal fetal ventral mesencephalic cell grafts. Eur J Pharm Bioppharm 2006; 63:221–8.

86. Tatard VM, Venier-Julienne MC, Benoit JP, Menei P, Montero-Menei CN. In vivo evaluation of pharmacologically active microcarriers releasing nerve growth factor and conveying PC12 cells. Cell Transplantation 2004; 13(5): 573–83.

87. Roullin VG, Deverre JR, Lemaire L, et al. Anti–cancer drug diffusion within living rat brain tissue: an experimental study using [^{3}H](6)-5-fluorouracil-loaded PLGA microspheres. Eur J Pharm Biopharm 2002; 53:293–9.

9

Enhanced Nasal Delivery with Lyophilized Inserts

Fiona McInnes

Strathclyde Institute of Pharmacy and Biomedical Sciences, University of Strathclyde, Glasgow, U.K.

Panna Thapa

Department of Pharmacy, Kathmandu University, Dhulikhel, Kavre, Nepal

Howard N.E. Stevens

Strathclyde Institute of Pharmacy and Biomedical Sciences, University of Strathclyde, Glasgow, U.K.

INTRODUCTION

Nasal delivery is increasingly considered to be an alternative route for drugs that currently require parenteral administration to achieve good efficacy, or where circumstances make oral delivery difficult. As a site for systemic absorption the nasal route provides a means of avoiding first pass metabolism, and the thin epithelia (1) with a large surface area ($150\,cm^2$) (2), combined with a high perfusion of arterial blood mean that it is ideally suited to drug absorption (3). The presence of many commercial nasal preparations already available on the market (4) confirms the patient acceptability of this route of administration. The majority of commercially available formulations however, are designed for topical treatment of conditions, such as allergic rhinitis, colds, and nasal congestion, with a few products available for systemic delivery of small peptide molecules such as desmopressin and calcitonin. Investigation of nasal absorption of larger peptide molecules, primarily insulin, has steadily continued over many years, with mixed success. Initial investigations were carried out using commercially available insulin preparations intended for subcutaneous (SC) administration, given intranasally to healthy and diabetic subjects (5). Recent research has focused on more specialized delivery vehicles (6–9). The opportunity for utilizing nasal drug delivery for systemic effect has increasingly become the focus of research for many other therapeutic areas, with applications such as delivery of heparins (10), melatonin (11), apomorphine (12), human growth hormone (13), and combined systemic and mucosal immunization against anthrax (14,15) and influenza (16). For many therapeutic indications, an attractive advantage of the avoidance of first pass metabolism through nasal delivery is a reduction in dosage requirements, with a corresponding theoretical decrease in adverse side effects. Achieving systemic levels of nasally absorbed therapeutic compounds is dependant on a number of complex factors, and in order to understand the obstacles to nasal delivery, it is necessary to consider the physiology of the human nose.

NASAL ANATOMY AND PHYSIOLOGY

Anatomy

The human nose consists of two nasal cavities separated by the septal wall, the narrowest section being the nasal valve near the front of the nose. Each nasal cavity opens onto the face through the nostril and extends posteriorly to the nasopharanyx with a length of approximately 10 cm. The combined surface area of both cavities is approximately $150 \, cm^2$ and the volume approximately 15 mL. The vestibular area serves as a baffle system whose surface is covered by a common pseudostratified epithelium and long hairs, to provide filtering of airborne particles. Found in the posterior region of the nose are the inferior, middle, and superior turbinates (Fig. 1), which combined with the nasal blood vessels play an important role in the humidification and temperature regulation of inspired air (17), a prime function of the human nose. Breathing through the mouth can sustain life, but without the air conditioning effect the air flow is both unpleasant and potentially harmful. In one inspiration via the nose, room temperature air (23°C) at 40% humidity is conditioned to 32°C at 98% humidity (18), suited to conditions required by the lungs. Under normal conditions, respiration predominantly occurs via one side of the nose, with the other side becoming congested, a cycle which alternates every 3–7 h (9,19). There is also a diurnal variation, with reduced nocturnal secretion rate (18) when the clearance of secretion is also markedly reduced (20).

The olfactory region, the area responsible for the detection of smell is situated in the posterior of the nose, near the top of the nasal cavity. It has been suggested that absorption directly to the central nervous system (CNS) is provided by formulations which can be administered directly to the olfactory region (21,22). This is of interest for the administration of drugs for conditions such as Parkinson's disease (23), which tend to have a high incidence of side effects that may be reduced by the lower doses

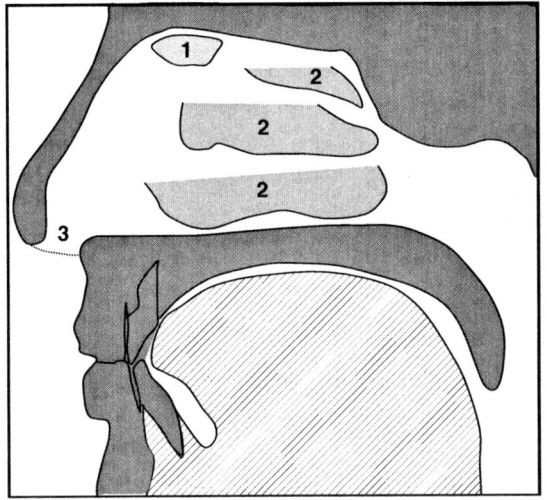

Figure 1 The nasal cavity.

required for direct CNS delivery, or centrally acting analgesics such as morphine (24).

The majority of the nasal mucosa is composed of ciliated columnar cells, which are further covered in microvilli, providing an ideal large surface area for absorption. Secretions from the mucosal and submucosal glands form the mucus layer of the nose, which covers the mucosa and functions in conjunction with the cilia to trap foreign particles and remove them from the nose.

The Mucus Layer

Water, containing proteins (including albumin, immunoglobulins, lysozyme and lactoferin) and electrolytes amongst other compounds, makes up the largest part of the mucous layer (90–95%) (25,26). Mucins or glycoproteins of varying molecular weights (100–10,000,000 Da) constitute 3% of the mucus layer, with carbohydrates making up 80% of the weight of these mucins (26).

The mucus layer is actually thought to be a double layer about 5–10 µm thick (12,27), with an upper slow moving viscous layer (gel layer) that is capable of trapping particles, and a lower layer (sol layer), less viscous and faster moving (12,27), through which the cilia can beat with little resistance. The tips of the cilia just reach the bottom of the viscous layer and thereby propel this layer towards the nasopharynx. After a co-ordinated beat and full extension, the cilia can then move back to their original position easily through the watery layer below, in preparation for the next beat.

The nose is an important site of particle deposition although the efficacy of the nasal filter depends on the diameter of the inhaled particles. Almost all particles larger than 10 μm (e.g. pollen grains) are retained in the nose during breathing at rest, while most particles smaller than 2 μm (e.g. mould spores) pass through the nose (27,28). The rate of mucocilliary clearance varies between individuals and can be altered by factors such as cigarette smoking, medicinal products and disease states such as allergic rhinitis, nasal polyps, cystic fibrosis or the common cold (9,12). The mucus layer and any foreign particles are swept towards the back of the nasal cavity where they can be swallowed and subsequently destroyed in the gastro-intestinal tract. This forms part of the body's defence system, preventing substances such as bacteria or allergens from reaching the lungs (12). It is therefore important that any nasal drug formulations do not have an irreversible effect on this process. Cilia beat about 1000 times per minute in the backward direction and thereby convey the mucus, with its trapped inhaled particles, to the throat where it is swallowed. Transport rates of 3–25 mm/min (average 6 mm/min) have been reported in normal subjects (29). Thus, the nasal mucus layer is replaced every 10–15 min and under normal conditions, inhaled substances or a drug delivery system are cleared from the nose within 15–20 min (30).

Factors Affecting Nasal Absorption

A combination of the structure and function of the nasal mucosa as a barrier to absorption, the properties of the active drug and the formulation within which it is contained will affect the extent to which a compound is absorbed nasally. The following factors are commonly reported as significantly influencing the extent of nasal absorption.

Hydrophilicity/Lipophilicity

Generally, drugs which are in solution will be rapidly absorbed nasally (31), although it appears that hydrophilic compounds are not readily absorbed above around 1000 Da (32,33), as they are thought to be absorbed via aqueous channels in the nasal mucosa (34), and so absorption decreases as molecular weight increases.

The absorption of lipophilic drugs is thought to occur via an alternative pathway to aqueous channels, namely the transcellular route. The extent of absorption is linked to lipophilicity, and the partition coefficient between the nasal mucosa and any buffer solution used (4,35), allowing the drug to partition into the lipid cell membrane (36).

The Effect of pH

The pH of the nasal surface is 7.39 (37) and nasal secretions in the adult have a pH in the range of 5.5–6.5 (38). The effect of pH can be variable, but can generally be linked to how the pH would affect the ionization or structure of the molecule (4,6) , as drugs are generally more likely to be absorbed in the unionized state. For example, midazolam (pKa 6.1) absorption in rats was found to be dependant on a pH greater than 4, when it existed in at least 1% of the unionized form (39). Given the relative sensitivity of the nasal mucosa however, nasal formulations should be adjusted to a pH in a range close to that of the nasal surface, which may lead to difficulty where drug physicochemistry is not amenable to this range. For example, some other increases of absorption at low pH are thought to be a result of damage to the nasal mucosa (40,41), rather than physicochemical effects.

Osmolarity

Osmolarity has been reported to play a role in nasal absorption, although accounts are conflicting. A group of researchers (42) found that hypertonic solutions promoted absorption of secretin in rats, thought to be a result of the observed shrinkage of cells in the mucosa, allowing more drug to permeate. It was found elsewhere that addition of osmotic agents to adhesive gels containing insulin resulted in further decreases in plasma glucose concentrations in rats (43). The finding that hypertonic solutions promote absorption would appear to agree with the theory that some polymers promote absorption via uptake of water and subsequent shrinkage of epithelial cells, resulting in a widening of tight junctions between the cells as shown by (44). However, contrary to these results it has been reported that in rats only very hypoosmotic solutions led to significant absorption of midazolam (39). The authors suggested that a resultant swelling of the mucosa aided paracellular diffusion of midazolam, however it was also indicated that pH effects may also have influenced the results, showing that the complex number of determining factors involved in nasal absorption can result in difficulty in discerning the precise mode of action.

Disease States

Drug absorption can be affected by the condition of nasal mucosa. Disease states such as allergic rhinitis, sinusitis, the common cold or nasal infection can result in increased nasal secretions. Such conditions may also result in increased or decreased viscosity of the mucus layer, and the resultant outcome for any of the above occurrences will be reduced absorption, either due to rapid clearance from the nasal cavity or the increased physical barrier between drug and mucosa. Physical abnormalities such as a deviated septum or nasal polyps may also affect the dynamics of mucociliary clearance and therefore drug absorption.

Administration Technique

Individual administration devices and technique can result in different sites of deposition within the nasal cavity, influencing absorption and clearance. Harris et al. (45), found that administration of a nasal pump spray resulted in deposition on the non-absorptive anterior region of the nasal cavity, and Soane et al. (46) reported that nasal formulations were deposited in either the anterior or turbinate region, depending on the administration technique of the volunteer. Ideally, the drug would be deposited in the turbinate site for predictable absorption (12).

Other Factors

When a drug is administered as a solution the volume applied at one time is of consequence, and is also inherently restricted due to the size of the nasal cavity (12). Optimal volumes appear to be between 50 and 100 μL, with improved absorption obtained by halving the dose and administering twice if the volume is too large (6). This effect was demonstrated in a study by Harris et al. (47) who reported that administration of desmopressin as a $2 \times 50\,\mu L$ dose gave significantly higher peak plasma and area under the curve (AUC) values than either a 1×50 or $1 \times 100\,\mu L$ dose.

Cytochrome P450 enzymes are thought to be responsible for the metabolism of many therapeutic compounds in the nasal mucosa. A further barrier to the absorption of peptide drugs such as insulin nasally may be degradation of the peptide by the various enzymes present in the nasal cavity. For example, research by Hirai et al. (48) has shown a high level of insulin degradation on exposure to rat nasal enzyme homogenates.

METHODS FOR ENHANCING NASAL ABSORPTION

Absorption Enhancers

The use of absorption enhancers in the formulation is a commonly employed method for increasing absorption of substances across the nasal mucosa. Many reports on the use of absorption enhancers have shown positive results, although it is often unclear whether or not long term use will result in harmful side effects to the nasal mucosa.

Bile salts have often been investigated as absorption enhancers and are possibly the most widely used means of aiding absorption via use of a surfactant (49). Typically used bile salts include cholate, sodium taurocholate, sodium glycocholate, and sodium deoxycholate (49–52). The mechanisms by which bile salts are thought to promote absorption include increasing the permeability of the membrane, inhibiting proteolytic enzymes, formation of aqueous pore pathways in the membrane or solubilization of the drug in aqueous solution as a result of their surfactant properties (49).

Another commonly investigated group of absorption enhancers are the fusidate derivatives, such as sodium fusidate and sodium dihydrotaurofusidate, which has been shown to enhance nasal absorption of insulin in a study by Shao and Mitra (51). These absorption enhancers are generally sodium salts of fusidic acid, an antibiotic compound. Fusidate derivatives show similar physical and chemical properties to the bile salts, and so have been suggested to enhance absorption in a similar manner (49).

Chitosan polymers are often employed for their bioadhesive properties, however they also appear to show absorption enhancing effects that are the result of a separate mechanism. It is thought that chitosans cause a transient opening in the tight junctions between the epithelial cells of the mucosa, thereby increasing the permeability of the mucosa to the drug in question (53,54). This may be a result of the cationic Ca^{2+}-binding properties of chitosans, 'trapping' the Ca^{2+} in the mucosa, as a correlation has been observed between Ca^{2+}-binding properties and octreotide absorption in rats (55).

Cyclodextrins are cyclical oligomers of glucose, and are capable of forming "inclusion complexes" with drugs. The properties of the drug in question can be altered by this non-covalent 'inclusion', where the drug sits in lipophilic cavities in the cyclodextrin molecules (49), effectively disguising the drugs' profile. Dimethyl-β-cyclodextrin (DMβCD) has been shown to significantly improve nasal insulin absorption in rabbits at a 5% concentration (56), thought to be a result of interaction of DMβCD with lipids in the membrane.

Phospholipids investigated as absorption promoters include phosphatidylcholines such as lysophosphatidlycholine and didecanoyl-L-phosphatidylcholine (49). Lysophosphatidylcholine has been shown to have a significant enhancing effect on nasal absorption of a model drug in rats at 0.5% (57), and in combination with degradable starch microspheres resulted in an increase in absorption of human growth hormone in sheep (58).

Other absorption enhancers that have been investigated with varying degrees of success include substances such as menthol (59), ammonium glycyrrhizinate and glycyrrhetinic acid (60), aminated gelatin (61), enzyme inhibitors (e.g. aminoboronic acid derivatives, amastatin) (62), microparticle resins (63), alkyl maltosides and alkyl sucrose esters (64), glycofurol (65), polyacrylic acid gel (66) and particulate carrier systems (e.g. alginic acid, microcrystalline cellulose) (55). A summary of the effects of various absorption enhancers on the nasal absorption of insulin in pre-clinical models is shown in Table 1. Aside from the mechanisms described above, other proposed mechanisms of action include pore formation in the membrane, lowering the membrane potential and increasing blood flow to the nose (67).

Table 1 Agents Used to Enhance Nasal Absorption of Insulin in Animals

Species	Insulin dose	Absorption enhancer	Insulin C_{max} (μIU/mL)	Bioavailability as a % relative to (X)	Reference
Dog	50 IU	Sodium glycocholate or polyoxyethylene-9-lauryl ether	75%[a]	–	37
Sheep	100 IU	Chitosan/ chitosan nanoparticles	743 ± 259/ 106 ± 99	17.0 ± 7 (SC) 1.3 ± 0.8(SC)	7
Sheep	2 IU/kg	Lysophosphatidyl choline	380 ± 58	25.3 (SC)	102
		Glycodeoxyxholate	776 ± 155	31.9 (SC)	
		Sodium taurodihydrofusidate	409 ± 59	16.5 (SC)	
Sheep	1.39 IU/kg	Sodium taurodihydrofusidate	1250	37.8 ± 8.4 (IV)	108
Rabbit	2/4 IU	Dimethyl-β-cyclodextrin	640 ± 104	2.8 ± 3 12.9 ± 4.4 [b]	56
Rabbit	28 IU	Sodium polystyrene sulphonate	413 ± 72	6.5 (IM)	63
Rabbit	15.8 IU	Glycofurol	41%[a]	–	65
Rabbit	10 IU	Maltodextrin	3668 ± 82	8.7 ± 2.6 (IV)	107
Sheep	1 IU/kg	Sodium taurodihydro-fusidate	–	16.4 ± 2.4 (IV)	109

[a] Decrease in blood glucose.
[b] Percentage of dose received.

Membrane Effects of Absorption Enhancers

One of the most important considerations in the administration of substances to the nose is that the substance will be safe, both as a one-off dose, as well as after long term repeated application. A nasal formulation that promotes drug absorption will be rendered useless if it causes damage to the nasal mucosa. In general, of most concern in terms of safety is the use of absorption promoters, although it is also important to consider the potential toxicity of any drug carrier, and indeed the drug itself.

Chitosan has been extensively investigated, and has been reported as being generally safe. After histological evaluation and enzyme release experiments one study concluded that chitosan caused "relatively mild and reversible effects" (68) which is supported by other histological studies (40, 68, 69) where recovery of the epithelium was such that it appeared similar to controls. Chitosan has also been found to have no subsequent effect on nasal clearance times in humans (53). Other studies concerning chitosan however have been less favorable (70), and reports on the effects of various other absorption enhancers on the nasal membrane suggest that and suggest that more research is required (71–73).

Bioadhesion

Bioadhesive substances that will adhere to the nasal mucosa are often used in attempts to overcome rapid mucociliary clearance of the formulation from the nasal cavity. Bioadhesion has been defined as the ability of a material to adhere to a biological tissue for an extended period of time (6). The main purpose of a bioadhesive is to bind in some way with the mucous layer that covers the nasal epithelium, thereby decreasing its rate of clearance from the nasal cavity. The resultant prolonged contact time with the mucosa gives the opportunity for increased absorption (6), along with the potential to adapt the formulation for controlled release of the drug if desired (74). Bioadhesion is generally achieved with the use of polymers, and may offer an alternative to the use of absorption promoters in nasal formulations, as increased nasal residence may confer increased opportunity for drug absorption.

The physical process of mucoadhesion is well defined (75), and factors that are thought to be of importance to the overall bioadhesion of a polymer system include hydrophilic functional groups, molecular weight, cross-linking, concentration, viscosity, swelling behavior, and pH (6,72,74,76–83). The degree of swelling appears to affect the extent of bioadhesion by forming a gel structure with increasing flexibility of the polymer chains (74), permitting intimate contact between the polymer and the mucosal surface by allowing the chain sections of the polymer to interact and become entangled with the glycoproteins of the mucus (77,80,84).

Polymers that are bioadhesives generally contain groups such as hydroxyl or carboxyl moieties capable of forming hydrogen bonds (77–79). This promotes the absorption of water by the polymer to produce the swelling thought to be required for mucoadhesion. Hydrophilic groups are also thought to be important for the polymer to hydrogen bond with the glycoproteins present in the mucous layer (82, 84).

In a study by Smart et al. (75) it was found that to obtain greatest bioadhesion with sodium carboxymethylcellulose, a molecular weight of $\geq 78,600$ Da is desirable, and there is general agreement that adhesion increases with molecular weight (6,77).

It is unclear which of these properties is the most important to the performance of a bioadhesive formulation, and it seems probable that most polymers exhibit bioadhesion due to a combination of effects, rather than as a result of any one mechanism in particular.

Conventional Nasal Delivery Systems

Nasal Bougies (Buginaria)

Bougies were medicated pencils intended for insertion into the nostril. They were prepared in the same way as suppositories but differed in shape, resembling a pointed rod, and were usually made from a gelato-glycerin base. Nasal bougies are rarely used today probably due to their size made them uncomfortable for the patient and they were messy after melting. They were typically 1–3 in. long and weighed 10–18 grain (0.66–1 gram) (85–87). They required careful pouring of the molten mass due to the tapered mould geometry.

Nasal Drops

Nasal drops are perhaps the simplest and most convenient means of administration of drugs to the nose. The disadvantage of this dosage form is that an exact amount of the formulation cannot be delivered easily and the formulation may be contaminated by the dropper. A major drawback of all water based dosage forms is microbiological stability, and the required preservatives may also impair mucociliary function (88). Besides micro-biological stability, the chemical stability of the dissolved drug and the short residence time of the formulation in the nasal cavity are other major disadvantages of such liquid formulations (84,89).

Nasal Sprays

Nasal spray devices include the squeezed bottle, the metered-dose spray pump and newer devices known as airless or preservative free spray pumps. Most of the nasal preparations on the market containing solutions, emulsions or suspensions are delivered by metered-dose spray pumps. Compared to squeezed bottles and continuous valve sprays, they allow administration of a defined dose with high dosing accuracy and a typical pattern (90). Dose volumes between 25 and 200 µl are available as standard. Spray characteristics vary according to the mechanical properties of the pump and the physical properties of the product. Viscosity, thixotropic behavior, elasticity and surface tension of the liquid determine the spray pattern, the particle size of the drops, the dose and the dosing accuracy.

Powders

Powders can be considered advantageous compared to solutions as they tend not to require preservatives, and have increased stability. However, the advantage of powder formulations over liquids is highly dependent on the solubility of the drug, its absorption rate, particle size and its irritation potential. Furthermore, the powder properties of size and shape, density and flow characteristics have an influence on distribution in the nose, as only particles of 5–10 µm tend to remain in the nasal cavity when inhaled (4).

Specialized delivery devices are required to deliver accurate metered doses of powders into the nose (90).

Nasal Gels

For most therapeutic indications, the major drawback of the formulations detailed above is that they are susceptible to rapid mucociliary clearance. This allows only a short contact time for topical agents, and a narrow window for absorption of systemically acting compounds. Nasal gels have been investigated as a means of prolonging the contact time of the formulation with the nasal mucosa, however nasal administration of gels can be technically challenging and may require a specialized device. There will also be a limit to the viscosity of gel that can be formulated for convenient nasal administration, and the issue of dosing accuracy is once again a drawback for this type of formulation.

THE NASAL INSERT FORMULATION

Nasal polymer gels are an initial step towards achieving increased bioavailability and improved therapeutic outcome by utilizing bioadhesion to maximize contact time with the nasal mucosa. In order to overcome the drawbacks of the messiness of nasal gels and difficulty of administration, preparation of freeze dried polymer gels may offer a promising solution. Freeze drying, or lyophilization, is a widely used technique for the stabilization and preservation of heat labile substances such as biological products and pharmaceuticals. Its use in the preservation of food stuffs, biological products and pharmaceuticals is well documented (91–95). Lyophilization is the term given to the process whereby ice is sublimed from frozen solutions, generally under reduced pressure, leaving a dry porous mass of approximately the same size and shape as the original frozen mass. As a result of the displacement of ice crystals within the formulation, lyophilized products are porous in nature and have the potential to undergo rapid rehydration when in contact with solvent or a moist surface. This provides an ideal opportunity to lyophilize a bioadhesive polymer gel for nasal administration, which could rehydrate on the mucosal surface to form a more concentrated and viscous gel than could be normally be easily administered to the nasal cavity.

In conventional freeze drying of a pharmaceutical, a glass vial is partially filled with a solution of the substance to be lyophilized and placed in the freeze dryer. The partially stoppered vial remains open throughout the drying process to allow water vapor to escape from the frozen solution and on completion of lyophilization cycle a "plug" of dry material is left in the vial. This technology has therefore been utilized to manufacture a bioadhesive nasal insert formulation that consists of a unit dose, drug containing polymer plug, or nasal insert, with sufficient mechanical strength to be readily handled and easily inserted into the nasal cavity without the requirement for a complex dosing device (96,97). The resultant lyophilized nasal insert (Fig. 2) may also dispense with the need for preservatives that are required for liquid formulation, and greatly reduced concerns over stability on storage. Such a unit dose formulation could also provide more reliable and accurate dosing than many gel formulations, and a method for larger scale industrial manufacture has been proposed (98).

In Vitro Properties

A typical SEM image of the internal structure of a lyophilized nasal insert formulation is shown in Figure 3. It can be seen that the formulation forms a highly porous structure upon lyophilization, which is expected to provide an ideal route for water ingress and subsequent rehydration of the polymer on contact with the moist nasal mucosa. Rapid rehydration of the polymer is an important step in gel formation within the nasal cavity, in order for bioadhesion of the polymer to occur before the formulation is cleared from the mucosal surface by the mucociliary transit system.

Experiments on in vitro water uptake of the inserts from a synthetic mucosal surface have demonstrated hydration of 30–50% within 15 min, well within the timeframe of normal mucociliary clearance. Assuming, therefore that bioadhesion will occur, the resultant hydrated polymer matrix could then be used to deliver drug to the nasal mucosa in a sustained release fashion over an extended period of time. This combination of properties could then potentially be used to both increase bioavailability and reduce administration frequency, providing sustained therapeutic plasma levels. According to the requirements of the compound and disease condition, the polymer carrier may then be adjusted to achieve more rapid, or alternatively, prolonged release profiles. In vitro release data for a nasal insert (Fig. 4) showed that nicotine could be released from the lyophilized formulation over an extended period of time (up to 4 h) in comparison with simple solution and powder formulations (99).

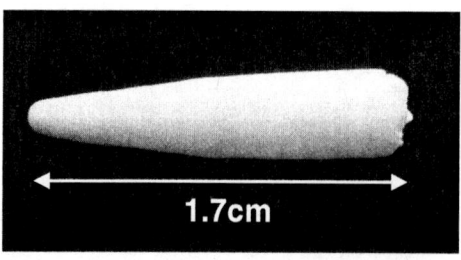

Figure 2 A lyophilized nasal insert.

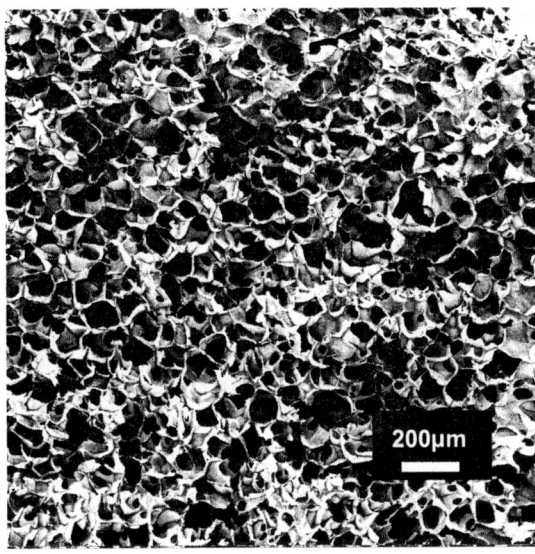

Figure 3 Internal structure of the lyophilized nasal insert. *Source*: From Ref. 107.

In Vivo Assessment

Animals which have been used to obtain in vivo data on nasal drug absorption and pharmacokinetics include rats (100), rabbits (101) and sheep (58,102). The sheep has been used for nasal absorption studies as it has a nasal surface area per kilogram body weight that is similar to that of humans compared to other animals. For example, nasal mucosal surface area per kilogram of bodyweight (cm^2/Kg) varies between 41.6 for rats, 22.0

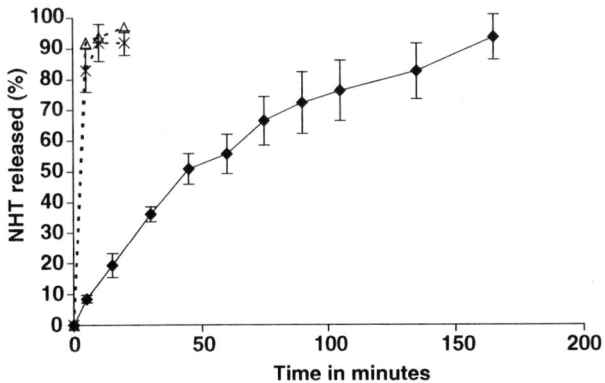

Figure 4 In vitro release of the lyophilized nasal insert (♦) compared to conventional liquid (Δ) and powder (×) formulations (broken lines). *Source*: From Ref. 107.

for dogs, 20.3 for rabbits, 2.5 for humans, and 8.2 for sheep (103,104). It has been reported that nasal bioavailability results obtained in the sheep model closely mirror those obtained in man, and the sheep offers a reliable model for nasal absorption of all types of drug formulation (8). It has also been shown that when assessed using gamma scintigraphy, the clearance rate of a bioadhesive chitosan system in sheep was comparable with that found in humans, making the sheep a good predictive model for humans (105).

Sheep were therefore chosen as a pre-clinical model to assess nasal absorption from the lyophilized nasal insert formulation. Nicotine hydrogen tartrate, known to be readily absorbed from the nasal mucosa in sheep (106), was used as a model compound, and the lyophilized nasal insert formulation was compared with nasally administered conventional nicotine powder and solution formulations (99).

The mean plasma profiles obtained and relevant pharmacokinetic data are shown in Figure 5 and Table 2, respectively. The mean profile for the nasal insert shows a more gradual rise in plasma nicotine in comparison with the conventional nasal formulations, with significant levels sustained over approximately 2 h, followed by a gradual decrease in plasma levels at a slower rate than for the other nasal preparations. The nasal insert formulation gave a relative bioavailability of 83.4%, compared to 51.5% and 27.2% for the nasal spray and powder respectively, a promising increase over the conventional nasal formulations. The extended absorptive phase of the nasal insert suggests prolonged nasal residence of the formulation, and continued absorption of nicotine suggests that the lyophilized insert remains in the nasal cavity releasing nicotine for approximately 2–3 h. In vitro

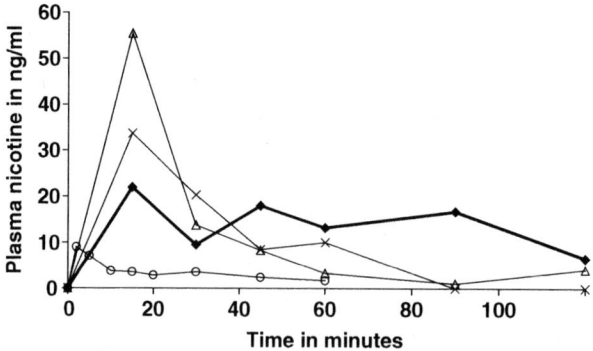

Figure 5 Mean plasma concentration time profiles of different formulations following intranasal and intravenous administration of 4 and 1 mg respectively of nicotine in sheep. (O) IV; (♦) nasal insert; (Δ) nasal spray; (×) nasal powder. *Source*: From Ref. 107.

Table 2 Pharmacokinetic Parameters (mean values ± s.d.) Following Intranasal (4 mg) and Intravenous (1 mg) Administration of Nicotine in Sheep

Formulation	T_{max} (min)	C_{max} (ng/ml)	AUC (ng.h/ml)	$F_{rel}(\%)$
IV	2.80 ± 1.40	9.4 ± 5.8	16.3 ± 32.25	100
Lyophilized insert	39.4 ± 33.0	27.6 ± 23.4	54.4 ± 69.4	83.4
Spray	21.0 ± 13.4	45.6 ± 24.0	33.6 ± 32.7	51.5
Powder	20.0 ± 7.7	36.4 ± 19.4	17.7 ± 6.40	27.2

Abbreviation: AUC, area under the curve.
Source: From Ref. 107.

observations show that a viscous bioadhesive gel is formed on hydration of the dosage form. If this gel is formed on administration of the insert to the sheep nasal cavity, then it would be expected to resist the rapid mucociliary clearance rate by adhering to the mucosa.

The literature reports the preparation of many other formulations designed to achieve bioadhesion for the enhancement of nasal absorption, with examples ranging from microspheres composed from various materials (8,102), resin complexes (106), nanoparticles (7), chitosans (105), and polycarbophil–cysteine conjugates for covalent bonding (68). However the major advantage of the formulation design of the nasal insert is the simple to administer, unit dosing that dispenses with the requirement for complex dosing devices. Lyophilization is a low temperature process and causes minimum damage and loss of activity of delicate heat labile materials, and further more may be used to achieve increased stability in comparison with liquid formulations. Owing to the wide range of pharmaceutical polymers available on the market, it is likely that for the majority of compounds and applications a formulation solution can be reached using the lyophilized nasal insert technology.

CONCLUSION

The nasal route is attractive as a means of delivering compounds for systemic effect, as a result of the highly vascular nature of the mucosa, and avoidance of hepatic first pass metabolism. However, the rapid rate of mucociliary transit in the nasal cavity is a major barrier to drug absorption, leaving only a short window for absorption of drug from conventional liquid and powder formulations. The lyophilized nasal insert has been designed in an attempt to overcome this disadvantage, and pre-clinical data have suggested that prolonged nasal residence and absorption can be achieved using this particular formulation. Furthermore, the formulation strategy allows controlled unit dosing, which is not as messy as conventional bioadhesive gel formulations, and does not require a specialized administration device.

REFERENCES

1. Quadir M, Zia H, Needham TE. Toxicological implications of nasal formulations. Drug Del 1999; 6:227–42.
2. Junginger HE. Bioadhesive polymer systems for peptide delivery. Acta Pharma Technol 1990; 36:110–26.
3. Cauna N. Blood and nerve supply of the nasal lining. In Procter DF, Anderson I, eds. The Nose, Upper airway Physioogy and the Atmospheric Environment. Amsterdam: Elsevier Biomedical Press; 1982:45–66.
4. Behl CR, Pimplaskar HK, Sileno AP, de Meireles VD, Romeo VD. Effects of physicochemical properties and other factors on systemic nasal drug delivery. Adv Drug Del Rev 1998; 29:89–116.
5. Pontiroli AE, Alberetto M, Secchi A, Dossi G, Bosi I, Pozza G. Insulin given intra-nasally induces hypoglycaemia in normal and diabetic subjects. Br Med J 1982; 284:303–6.
6. Dondeti P, Zia H, Needham TE. Bioadhesive and formulation parameters affecting nasal absorption. Int J Pharm 1996; 127:115–33.
7. Dyer AM, Hinchcliffe M, Watts P, et al. Nasal delivery of insulin using novel chitosan based formulations: A comparative study in two animal models between simple chitosan formulations and chitosan nanoparticles. Pharm Res 2002; 19:998–1008.
8. Illum L, Farraj NF, Fisher AN, Gill I, Miglietta M, Benedetti LM. Hyaluronic acid ester microspheres as a nasal delivery system for insulin. J Control Release 1994; 29:133–41.
9. Jones N. The nose and paranasal sinuses physiology and anatomy. Adv Drug Del Rev 2001; 51:5–19.
10. Yang T, Hussain A, Paulson J, Abbruscato TJ, Ahsan F. Cyclodextrins in nasal delivery of low-molecular-weight heparins: In vivo and in vitro studies. Pharm Res 2004; 21:1127–36.
11. Mao S, Chen J, Wei, Liu H, Bi D. Intranasal administration of melatonin starch microspheres. Int J Pharm 2004; 272:37–43.
12. Ugwoke MI, Kaufmann G, Verbeke N, Kinget R. Intranasal bioavailability of apomorphine from carboxymethylcellulose-based drug delivery systems. Int J Pharm 2000; 202:125–31.
13. Leitner VM, Guggi D, Krauland AH, Bernkop-Schnurch A. Nasal delivery of human growth hormone: in vitro and in vivo evaluation of a thiomer/glutathione microparticulate delivery system. J Control Release 2004; 100:87–95.
14. Sloat BR, Cui Z. Strong mucosal and systemic immunities induced by nasal immunisation with anthrax protective antigen protein incorporated in liposome-protamine-DNA particles. Pharm Res 2006; 23:262–9.
15. Jiang GE, Joshi SB, Peek LJ, et al. Anthrax vaccine powder formulations for nasal mucosal drug delivery. J Pharm Sci 2006; 95:80–96.
16. Sing M, Briones M, O'hagan DT. A novel bioadhesive intranasal delivery system for inactivated influenza vaccines. J Control Release 2001; 70:267–76.
17. Chien YW, Su KSE. Intranasal drug delivery for systemic medications. Crit Rev Ther Drug Carr Syst 1987; 4:67–194.

18. Mygind N, Thomsen J. Diurnal variation of nasal protein concentration. Acta Otolaryngol 1976; 82:219–22.
19. Washington N, Washington C, Wilson CG. Physiological Pharmaceutics. London: Taylor and Francis; 2001.
20. Bateman JRM, Pavia D, Clarke SW. The retention of lung secretions during the night in normal subjects. Clin Sci 1978; 55:523.
21. Chou K, Donovan MD. The distribution of local anesthetics into the CSF following intranasal administration. Int J Pharm 1998; 168:137–45.
22. Illum L. Transport of drugs from the nasal cavity to the central nervous system. Eur J Phar Sci 2000; 11(1):1–18.
23. Kao HD, Traboulsi A, Itoh S, Dittert L, Hussain A. Enhancement of the systemic and CNS specific delivery of L-Dopa by the nasal administration of its water soluble prodrugs. Pharm Res 2000; 17:978–84.
24. Westin UE, Bostrom E, Grasjo J, Hammarlund-Udenaes M, Bjork E. Direct nose-to-brain transfer of morphine after nasal administration to rats. Pharm Res 2006; 23:565–72.
25. Kaliner M, Marom Z, Patow C, Shelhamer J. Human respiratory mucus. J Allergy Clin Immunol 1984; 73:318–23.
26. Marttin E, Schipper NGM, Verhoef JC, Merkus FWHM. Nasal mucociliary clearance as a factor in nasal drug delivery. Adv Drug Deliv Rev 1998; 29: 13–38.
27. Mygind N, Dahl R. Anatomy, physiology and function of the nasal cavities in health and disease. Adv Drug Deliv Rev 1998; 29:3–12.
28. Mygind N. Nasal Allergy. Oxford: Blackwell Scientific Publications; 1978.
29. Proctor DF. Mucociliary system. In: Procter DF, Anderson I, eds. The Nose, Upper airway Physioogy and the Atmospheric Environment. Amsterdam: Elsevier Biomedical Press; 1982:245–78.
30. Andersen I, Proctor DF. Measurement of nasal mucocilliary clearance. Eur J Resp Dis 1983; 64:37–40.
31. Hussain AA. Intranasal drug delivery. Aov Drug Del Rev 1998; 29(1,2):39–50.
32. Donovan MD, Huang Y. Large molecule and particulate uptake in the nasal cavity: the effect of size on nasal absorption. Adv Drug Del Rev 1998; 29: 147–55.
33. McMartin C, Hutchinson LEF, Hyde R, Peters GE. Analysis of structural requirements for the absorption of drugs and macromolecules from the nasal cavity. J Pharm Sci 1987; 76:535–40.
34. Chien YW, Su KSE, Chang SF. Anatomy and physiology of the nose. In: Chien YW, Su KSE, Chang SF, eds. Nasal Systemic Drug Delivery. New York: Marcel Dekker Inc.; 1989:1–26.
35. Corbo DC, Huang YC, Chien YW. Nasal delivery of progestational steroids in ovariectomised rabbits. I. Progesterone—Comparison of pharmacokinetics with intraveous and oral administration. Int J Pharm 1998; 46: 133–40.
36. Hinchcliffe M, Illum L. Intranasal insulin delivery and therapy. Adv Drug Del Rev 1999; 35:199–234.
37. Hirai S, Ikenaga T, Matsuzawa T. Nasal absorption of insulin in dogs. Diabetes 1978; 27:296–9.

38. Chien YW. Biopharmaceutics basis for transmucosal delivery. STP Pharma Sci 1995; 5:257–75.
39. Olivier JC, Djilani M, Fahmy S, Couet W. In situ nasal absorption of midazolam in rats. Int J Pharm 2001; 213:187–92.
40. Tengamnuay P, Sahamethapat A, Sailasuta A, Mitra AK. Chitosans as nasal absorption enhancers of peptides: comparison between free amine chitosans and soluble salts. Int J Pharm 2000; 197:53–67.
41. Ohwaki T, Ando H, Watanabe S, Miyake Y. Effects of dose, pH, and osmolarity on nasal absorption of secretin in rats. J Pharm Sci 1985; 74:550–2.
42. Ohwaki T, Ando H, Kakimoto F, Uesugi K, Watanabe S, Miyake Y, Kayano M. Effects of dose, pH, and osmolarity on nasal absorption of secretin in rats. II: Histological aspects of the nasal mucosa in relation to the absorption variation due to the effects of pH and osmolarity. J Pharm Sci 1987; 76:695–8.
43. Pereswetoff-Morath L, Edman P. Influence of osmolarity on nasal absorption of insulin from the thermogelling polymer ethyl(hydroxyethyl) cellulose. Int J Pharm 1995; 125:205–13.
44. Edman P, Bjork E, Ryden L. Microspheres as a nasal delivery systems for peptide drugs. J Control Release 1992; 21:165–72.
45. Harris AS, Svensson E, Wagner ZG, Lethagen S, Nilsson IM. Effect of viscosity on particle size, deposition and clearance of nasal delivery systems containing desmopressin. J Pharm Sci 1988a; 77:405–8.
46. Soane RJ, Frier M, Perkins AC, Jones NS, Davis SS, Illum L. Evaluation of the clearance characteristics of bioadhesive systems in humans. Int J Pharm 1999; 178:55–65.
47. Harris AS, Ohlin M, Lethagen D, Nilsson IM. Effects of concentration and volume on nasal bioavailability and biological response to desmopressin. J Pharm Sci 1988; 77:337–39.
48. Hirai S, Yashiki T, Matsuzawa T, Mima H. Absorption of drugs from the nasal mucosa of rat. Int J Pharm 1981; 7:317–25.
49. Behl CR, Pimplaskar HK, Sileno AP, Xia WJ, Gries WJ, deMeireles JC, Romeo VD. Optimization of systemic nasal drug delivery with pharmaceutical excipients. Adv Drug Del Rev 1998; 29:117–34.
50. Junginger HE, Hoogstraate JA, Verhoef JC. Recent advances in buccal drug delivery and absorption—in vitro and in vivo studies. J Control Release 1999; 62:149–59.
51. Shao Z, Mitra AK. Nasal membrane and intracellular protein and enzyme release by bile salts and bile salt-fatty acid mixed micelles: correlation with facilitated drug transport. Pharm Res 1992; 9:1184–9.
52. Bagger MA, Nielsen HW, Bechgaard E. Nasal bioavailability of peptide T in rabbits: absorption enhancement by sodium glycocholate and glycofurol. Eur J Pharm Sci 2001; 14:69–74.
53. Aspden TJ, Illum L, Skaugrud O. Chitosan as a nasal delivery system: evaluation of insulin absorption enhancement and effect on nasal membrane integrity using rat models. Eur J Pharm Sci 1996; 4:23–31.
54. Fernandez-Urrusuno R, Calvo P, Remunan-Lopez C, Vila-Jato JL, Alonso MJ. Enhancement of nasal absorption of insulin using chitosan nanoparticles. Pharm Res 1999; 16:1576–81.

55. Oechslein CR, Fricker G, Kissel T. Nasal delivery of octreotide: Absorption enhancement by particulate carrier systems. Int J Pharm 1996; 139:25–32.
56. Schipper NGM, Romeijin SG, Verhoef JC, Merkus FWHM. Nasal insulin delivery with dimethyl-beta-cyclodextrin as an absorption enhancer in rabbits: powder more effective than liquid formulations. Pharm Res 1993; 10:682–6.
57. Natsume H, Iwata S, Ohtake K, et al. Screening of cationic compounds as an absorption enhancer for nasal drug delivery. Int J Pharm 1999; 185:1–12.
58. Illum L, Farraj NF, Davis SS, Johansen BR, O'Hagan DT. Investigation of the nasal absorption of biosynthetic human growth hormone in sheep–use of a bioadhesive microsphere delivery system. Int J Pharm 1990; 63:207–11.
59. Shojaei AH, Khan M, Lim G, Khosravan R. Transbuccal permeation of a nucleoside analog, dideoxycytidine: effects of menthol as a permeation enhancer. Int J Pharm 1999; 192:139–46.
60. Dondeti P, Zia H, Needham T. In-vivo evaluation of spray formulations of human insulin for nasal delivery. Int J Pharm 1995; 122:91–105.
61. Wang J, Tabata Y, Morimoto K. Aminated gelatin microspheres as a nasal delivery system for peptide drugs: Evaluation of in vitro release and in vivo absorption in rats. J Control Release 2006; 113:31–37.
62. Sarkar MA. Drug metabolism in the nasal mucosa. Pharm Res 1992; 9:1–9.
63. Takenaga M, Serizawa Y, Azechi Y, Ochiai A, Kosaka Y, Igarashi R, Mizushima Y. Microparticle resins as a potential nasal drug delivery system for insulin. J Control Release 1998; 52:81–87.
64. Pillion DJ, Ahsan F, Arnold JJ, Balusubramanian BM, Piraner O, Meezan E. Synthetic long-chain alkyl maltosides and alkyl sucrose esters as enhancers of nasal insulin absorption. J Pharm Sci 2002; 91:1456–62.
65. Bechgaard E, Gizurarson S, Hjortjaer RK, Sorenson AR. Intranasal administration of insulin to rabbits using glycofurol as an absorption promoter. Int J Pharm 1996; 128:287–9.
66. Morimoto M, Morisaka K, Kamada A. Enhancement of nasal absorption of insulin and calcitonin using poly-acrylic acid gel. J Pharm Pharmacol 1985; 37: 134–6.
67. Dodane V, Khan MA, Merwin JR. Effect of chitosan on epithelial permeability and structure. Int J Pharm 1999; 182:21–32.
68. Bernkop-Schnurch A, Gilge B. Anionic mucoadhesive polymers as auxiliary agents for the peroral administration of poly peptide drugs: influence of the gastric juice. Drug Dev Ind Pharm 2000; 26:107–13.
69. Bernkop-Schnurch A, Schwarz V, Steininger S. Polymers with thiol groups: a new generation of mucoadhesive polymers? Pharm Res 1999; 16:876–81.
70. Illum L, Farraj NF, Davis SS. Chitosan as a novel nasal delivery system for peptide drugs. Pharm Res 1994; 11:1186–9.
71. Carreno-Gomez B, Duncan R. Evaluation of the biological properties of soluble chitosan and chitosan microspheres. Int J Pharm 1997; 148:231–40.
72. Witschi C, Mrsny RJ. In-vitro evaluation of microparticles and polymer gels for use as nasal platforms for protein delivery. Pharm Res 1999; 16:382–90.
73. Dyvik K, Graffner C. Investigation of the applicability of a tensile testing machine for measuring mucoadhesive strength. Acta Pharm Nord 1992; 4: 79–84.

74. Wang J, Sakai S, Deguchi Y, Bi D, Tabata Y, Morimoto K. Aminated gelatin as a nasal absorption enhacer for peptide drugs: Evaluation of absorption enhacing effect and nasal mucosa perturbation in rats. J Pharm Pharmacol 2002; 54:181–8.

75. Alur HH, Pather SI, Mitra AK, Johnston TP. Transmucosal sustained-delivery of chlorpheniramine maleate in rabbits using a novel, natural mucoadhesive gum as an excipient in buccal tablets. Int J Pharm 1999; 188:1–10.

76. Smart JD, Kellaway IW, Worthington HEC. An in-vitro investigation of mucosa-adhesive materials for use in controlled drug delivery. J Pharm Pharmacol 1984; 36:295–299.

77. Lee JW, Park JH, Robinson JR. Bioadhesive-based dosage forms: The next generation. J Pharm Sci 2000; 89:850–66.

78. Mortazavi SA. An in vitro assessment of mucus/mucoadhesive interactions. Int J Pharm 1995; 124:173–82.

79. He P, Davis SS, Illum L. In vitro evaluation of the mucoadhesive properties of chitosan microspheres. Int J Pharm 1998; 166:75–68.

80. Nakamura K, Maitani Y, Lowman AM, Takyama K, Peppas NA, Nagai T. Uptake and release of budesonide from mucoadhesive, pH-sensitive copolymers and their application to nasal delivery. J Control Release 1999; 61: 329–35.

81. Bernkop-Schnurch A, Steininger S. Synthesis and characterisation of mucoadhesive thiolated polymers. Int J Pharm 2000; 194:239–247.

82. Nakamura F, Ohta R, Machida Y, Nagai T. In vitro and in vivo nasal mucoadhesion of some water soluble polymers. Int J Pharm 1996; 134:173–81.

83. Ugwoke MI, Verbeke N, Kinget R. The biopharmaceutical aspects of nasal mucoadhesive drug delivery. J Pharm Pharmacol 2001; 53:3–22.

84. Illum L, Jorgensen H, Bisgaard H, Krogsgaard O, Rossing N. Bioadhesive microspheres as a potential nasal drug delivery system. Int J Pharm 1987; 39: 189–99.

85. Gunn C, Carter SJ. Cooper and Gunn's Dispensing for Pharmaceutical Students, 11 edn. London: Pitman Medical Publishing Co. Ltd.; 1965.

86. Martindale WH, Westcott WW. The Extra Pharmacopoeia, 17 edn. London: H.K. Lewis & Co. Ltd.; 1920.

87. Martindale WH. The Extra Pharmacopoeia, 22 edn. London: The Pharmaceutical Press; 1941.

88. Batts AH, Marriott C, Martin GP, Bond SW. The effect of some preservatives used in nasal preparations on mucociliary clearance. J Pharm Pharmacol 1989; 41:156–9.

89. Hardy JG, Lee SW, Wilson CG. Intranasal drug delivery by spray and drops. J Pharm Pharmacol 1985; 37:294–7.

90. Kublik H, Vidgren MT. Nasal delivery systems and their effect on deposition and absorption. Adv Drug Del Rev 1998; 29:157–77.

91. Carpenter JF, Pikal MJ, Chang, BS, Randolph TW. Rational design of stable lyophilized protein formulations: some practical advice. Pharm Res 1997; 14: 969–75.

92. Couriel B. Freeze drying: past, present, and future. J Parent Drug Assoc 1980; 34:352–7.

93. Goldblith SA, Rey L, Rothmayr WW. Freeze Drying and Advanced Food Technology. London: Academic Press; 1975.

94. Pikal MJ. Freeze drying of proteins. Part II: Formulation selection. Bio Pharm 1990; October:26–30.

95. Pikal MJ. Freeze-drying of proteins. Part I: process design. Bio Pharm 1990; September:18–27.

96. Thapa P. Studies of a lyophilized nasal delivery system. Ph.D Thesis, University of Strathclyde, UK: 2000.

97. McInnes F, Girkin J, McConnell G, Stevens HNE, Baillie AJ, Structure and rate of water ingress of a lyophilized dosage form. AAPS Pharm Sci 5 2003; S1:W4147.

98. Thapa P, Baillie AJ, Stevens HNE. Lyophilisation of unit dose pharmaceutical dosage forms. Drug Dev Ind Pharm 2003; 29:595–602.

99. McInnes FJ, Thapa P, Baillie AJ, et al. In-vivo evaluation of nicotine lyophilized nasal insert in sheep. Int J Pharm 2005; 304:72–82.

100. Jung BH, Chung BC, Chung SJ, Lee MH, Shim CK. Prolonged delivery of nicotine in rats via nasal administration of proliposomes. J Control Release 2000; 66:73–79.

101. Ugwoke MI, Kaufmann G, Verbeke N, Kinget R. Intranasal bioavailability of apomorphine from carboxymethylcellulose-based drug delivery systems. Int J Pharm 2000; 202:125–31.

102. Illum L, Fisher AN, Jabbal-Gill I, Davis SS. Bioadhesive starch microspheres and absorption enhancing agents act synergistically to enhance the nasal absorption of polypeptides. Int J Pharm 2001; 222:109–19.

103. Illum L. Nasal Delivery. The use of animal models to predict performance in man. J Drug Target 1996; 3:427–42.

104. Gizurarson S. Animal models for intranasal drug delivery studies. Acta Pharm Nord 1990; 2:105–22.

105. Soane RJ, Hinchcliffe M, Davis SS, Illum L. Clearance characteristics of chitosan based formulations in the sheep nasal cavity. Int J Pharm 2001; 217: 183–91.

106. Cheng YH, Watts P, Hinchcliffe M, et al. Development of a novel nasal formulation comprising an optimal pulsatile and sustained plasma profile for smoking cessation. J Control Release 2002; 79:243–54.

107. McInnes FJ. In-vitro and In-vivo properties of a lyophilized nasal dosage system. Ph.D Thesis, University of Strathclyde, UK; 2003.

10

Advanced Formulation Strategies for Central Nervous System Drug Delivery

Xudong Yuan and David Taft

Division of Pharmaceutical Sciences, Long Island University, Brooklyn, New York, U.S.A.

INTRODUCTION

Brain and central nervous system (CNS) disorders are major causes of disability. Despite extensive research, clinical treatment options are still limited for the majority of CNS diseases, and therapeutic outcomes are sometimes disappointing. The blood-brain barrier (BBB) presents a formidable obstacle to effective drug delivery to the CNS. Although biologically intended to protect the brain and spinal cord and provide a very stable fluid environment, the presence of a BBB makes treatment of many CNS diseases difficult to achieve, as required therapies cannot be delivered across the barrier in sufficient quantities or at all (1). Moreover, CNS-active agents must pass through the BBB, and overcome drug efflux mechanisms in order to reach the brain and exert therapeutic effects. Therefore, an understanding of the physiology and cell biology of the BBB, rational drug design, and optimized drug structure and physiochemical properties are necessary in order to achieve efficient drug delivery to CNS (2). For some existing CNS medications, advanced formulation and drug delivery strategies can be utilized to improve delivery efficiency and drug therapy.

This chapter provides an overview of CNS drug delivery. It begins with a summary of the physiologic and physicochemical barriers that must be breached to deliver medications to the brain. Next, approaches that have been devised to overcome these barriers are presented together with relevant examples. These approaches include both noninvasive and invasive methods. While much of the research in this area remains at preclinical stages, formulation scientists have successfully applied various technologies to develop delivery systems that have optimized therapeutic outcomes across several CNS diseases. These examples are provided later in the chapter, together with proposed strategies to treat those brain conditions that are presently refractive to drug therapy (e.g., HIV, tumors). The chapter concludes with a look to the future of CNS drug therapy—the challenges that remain and the opportunities that exist for expanded development of clinically effective formulations for treating patients with CNS disorders.

BARRIERS AND OBSTACLES TO CNS DRUG DELIVERY

Physiological Barriers

There are several physiological or pathological factors preventing systemic drug delivery to CNS, including the BBB and the blood–cerebrospinal fluid barrier (BCB). Of these, the BBB is the most important barrier in the drug transport process.

The BBB can be defined as a physiological mechanism that alters the permeability of brain capillaries so that the majority of substances are prevented from entering brain tissue, although some molecules are allowed to enter freely. In the late 19th century, the German bacteriologist Paul Ehrlich observed that the certain dyes administered intravenously to small animals stained all the organs except the brain. His interpretation for the experiment was that the brain had a lower affinity for the dye than the other tissues. However, Edwin E. Goldmann, a student of Ehrlich, injected the dye trypan blue directly into the cerebrospinal fluid of rabbits and dogs. He found that the dye readily stained the entire brain but did not enter the bloodstream to stain the other internal organs. This experiment demonstrated that CNS is separated from the blood by a barrier. It is now well established that the BBB is a unique membranous barrier that tightly segregates the brain from the circulating blood (3,4). Through advances in experimental technology, brain capillaries, which comprise the BBB, can be visually observed using electron microscopy.

Physiologically, the BBB is found in all vertebrate brains, and in humans it is formed within the first trimester of life. The cellular locus of the BBB is the endothelial cell of the brain capillary. The structure of brain capillaries is unique, and this creates a permeability barrier between the blood and the extracellular fluid in brain tissue. Brain capillaries and the spinal cord do not have the small pores that allow solutes to circulate to other organs. These capillaries are lined with a layer of special endothelial cells that lack fenestrations and are sealed with tight junctions (5). There are three types of ependymal cells lining the monolayer of endothelial cells in the BBB: astrocyte foot processes, pericytes, and neurons. Astrocytes form the structural framework for the neurons, and control their biochemical environment.

Astrocyte foot processes or limbs spread out and interconnect to encapsulate the capillaries. Pericytes are mesenchymal-like cells that are associated with the walls of small blood vessels. Neurons (also known as neurones, nerve cells and nerve fibers) are electrically excitable cells in the nervous system that function to process information and transmit signals. The cell biology of the BBB phenomenon is based on interactions among these different cell types, as demonstrated in Figure 1. The tight junctions between endothelial cells produce a very high trans-endothelial electrical resistance which limits aqueous based paracellular diffusion that is observed in other organs (6–8).

In addition to blocking paracellular pathways for solute transport across the BBB, pinocytotic mechanisms and cell fenestrations are virtually nonexistent across the brain capillary endothelium. As a result, transcellular bulk flow of solute does not circulate through the BBB. Under these conditions, solute can access brain interstitium via two pathways: lipid mediation and catalyzed transport. In lipid mediation, lipid-soluble solutes can diffuse through the capillary endothelial membrane and passively cross the BBB. Catalyzed transport is classified into three general categories: carrier-mediated transport, active efflux transport, and receptor-mediated transport.

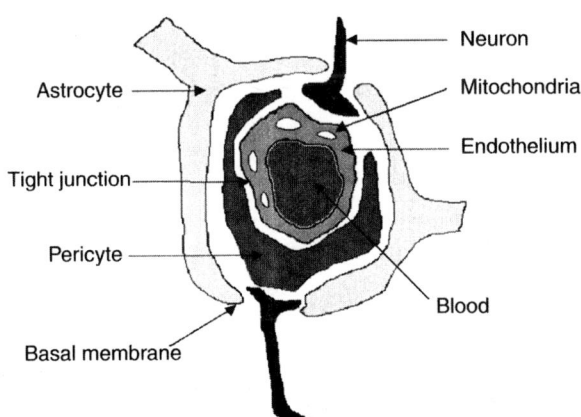

Figure 1 Schematic representation of the brain capillary endothelial cell. These cells form the BBB, and possess tight junctions and low permeability. The cells lack fenestration, have limited pinocytosis, and contain a large volume of mitochondria. Astrocytic endfeet cover the endothelial cell surface. Due to the tightness of the endothelial barrier, paracellular transport of substances is negligible under physiological conditions. Drugs enter the brain only by passive transcellular diffusion, receptor transcytosis or through carrier-mediated transport. *Source*: Reproduced from Ref. 17 with permission.

As a general rule, only the lipophilic molecules less than ~600–700 Da can be transported passively across the BBB. The barrier effectively blocks the majority of hydrophilic molecules and small ions. Unfortunately, as a result, many drugs are excluded from the CNS, making it difficult to treat CNS diseases. Current drug therapies for brain disorders are primarily lipophilic and can readily cross the BBB. However, not all lipophilic drugs can penetrate the BBB. For example, anticancer drugs such as doxorubicin, vincristine and vinblastine are lipophilic molecules, but can hardly pass through the BBB. The presence of efflux transporters such as P-glycoprotein (P-gp) limits access of these medications to the CNS.

There are a number of transport systems on both luminal and abluminal membranes of the endothelial cell that mediate solute transcytosis from blood to brain. The transport systems are involved in the uptake of nutrients such as glucose, amino acids, choline, purine bases, or nucleosides. However, there is also enzymatic activity inside the BBB. Solutes are subject to degradation by enzymes present inside the endothelial cells. These cells contain large densities of mitochondria, metabolically highly active organelles. BBB enzymes also recognize and rapidly degrade most peptides, including naturally occurring neuropeptides (9,10).

The BBB determines whether or not a compound can reach the CNS, either by passive diffusion or through carrier-mediated or receptor-based transport systems. Considerable research has focused on the structural and physicochemical requirements favoring transport across the BBB as related to anatomical and physiological features. Such studies have had a significant effect on the design of CNS drugs with improved permeability across the BBB. The BBB should be regarded as a dynamic rather than a rigid barrier; it can be influenced by astrocytes and probably also by neuronal and hormonal stimuli, and its properties are also affected by diseases of the CNS. This may offer new strategies for targeting drugs to the brain (11).

Besides the BBB, another barrier that limits CNS accumulation of systemically administered is the BCB. The BCB can regulate the movement of molecules into the cerebrospinal fluid (CSF), due to fact that CSF can exchange molecules with the interstitial fluid of the brain parenchyma. Physiologically, the BCB is located in the epithelium of the choroids plexus, which works together with the arachnoid membrane to limit penetration of molecules into the CSF. The passage of substances from the blood through the arachnoid membrane is prevented by tight junctions (12). In addition, there are a number of transport systems in the choroids plexus that extrude compounds from the CSF into the plasma. For example, a variety of therapeutic organic acids (e.g., penicillin, methotrexate, zidovudine) are actively removed from the CSF by an organic anion transporter.

Furthermore, substantial inconsistencies exist between the composition of the CSF and the interstitial fluid of the brain parenchyma, suggesting the presence of what is sometimes called the CSF–brain barrier (13).

The long pathways that comprise this potential barrier may prevent drugs from migrating from the CSF to brain interstitial fluid, thereby preventing these compounds from achieving therapeutically effective concentrations in the brain.

In the case of a CNS tumor, the BBB in the microvasculature is an even more difficult barrier to overcome. While the BBB can be significantly compromised during the progression of disease, additional physiological barriers limit drug delivery into solid tumors. Due to heterogeneous microvasculature distribution, drug delivery to neoplastic cells in a solid tumor is usually not uniform. During the progression of a CNS tumor, decreased vascular surface area and increased diffusion pathway further reduce exchange of molecules. Additionally, high pressure in and adjacent to the tumor make the cerebral microvasculature in adjacent regions of normal brain even less permeable to drugs than normal brain endothelium, leading to exceptionally low extra-tumoral interstitial drug concentrations (14). Therefore, it is critical to optimize drug properties and formulations to overcome BBB and achieve therapeutical concentration inside the tumor.

Efflux Transport Barriers in the CNS

Besides the physiologic barriers discussed above, there are several passive and active efflux mechanisms that will influence drug concentrations in the brain (Fig. 2). Active efflux of drugs from the CNS by these transporters may result in reduced drug concentration, even though drugs possess preferable physicochemical properties for BBB penetration (e.g., lipid solubility). Efflux mechanisms in the CNS include multi-drug resistance protein (MRP), P-gp and breast cancer resistant protein (BCRP), all of which belong to the ATP-binding cassette (ABC) family of transport proteins (15,16). Many lipophilic drugs are substrates for P-gp (Table 1), MRPs or organic anion transporting polypeptides (OATPs) that are expressed in brain capillary endothelial cells and/or astrocytes and are key elements of the molecular machinery that confers special permeability properties to the BBB. The combined action of these carrier systems results in rapid efflux of xenobiotics from the CNS (17).

The MRP family consists of least five isoforms with different levels of expression in different tissues in humans. The MDR gene produces P-gp, which exists in the luminal membranes of the cerebral capillary endothelium and actively effluxes lipid soluble substrates from cells. ABC efflux transporters in the BBB can minimize neurotoxic adverse effects of drugs that otherwise would penetrate into the brain. Modulation of ABC efflux transporters at the BBB forms a novel strategy to enhance the penetration of drugs into the brain and may yield new therapeutic options for drug-resistant CNS disease (18). Therefore, drug delivery to CNS can be increased not only by enhancing influx, but also by inhibiting efflux through the BBB.

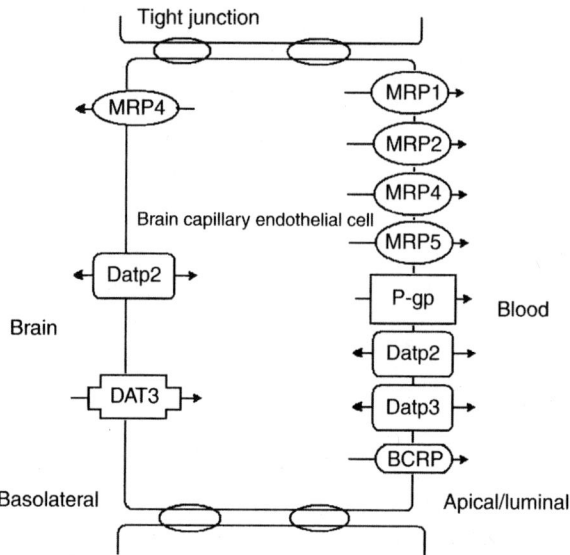

Figure 2 Illustration of drug efflux proteins expressed in the BBB. The proposed direction of the transport is indicated by arrows. Efflux transporters that are localized on the luminal side of the brain capillary endothelium function to restrict brain uptake of xenobiotics. Transporters that mediate intracellular uptake at the basolateral membrane may act in concert with efflux transporters at the apical membrane, thereby enhancing extrusion of drugs from the brain. Included in the figure are members from the ABC family of transporters including p-glycoprotein (P-gp), breast cancer resistant protein (BCRP), multidrug resistance proteins (MRPs). The organic anion transporting polypeptides (OATPs) are also depicted. *Source*: Reprinted from Ref. 17 with permission.

Inhibition can be achieved by co-administering an inhibitor of the efflux transport protein. Transport inhibition can be utilized as a novel therapeutic approach to modulate or by-pass drug efflux transporters at the BBB to treat drug-resistant brain diseases.

Physicochemical Properties of Medications

In terms of physicochemical requirements for good CNS delivery, low molecular weight, lack of ionization at physiological pH, and lipophilicity are the most important (13). As a general rule, maximum CNS delivery potential occurs for compounds with a log octanol-water partition coefficient (log Po/w) near 2 (19). The log Po/w parameter represents an informative physicochemical parameter used in medicinal chemistry, and it been shown to be a fairly useful descriptor of CNS penetration (20). Lipophilicity influences processes such as oral absorption, brain uptake, and

Table 1 Substrates and Inhibitors of P-glycoprotein

Substrates			Inhibitors
Therapeutic class	Examples		
Analgesics	Morphine	1st generation	Verapamil
Antibacterials	Ceftriaxone		Cyclosporin A
	Crythromycin		Quinidine
	Quinolones		Quinine
Anticancer drugs	Doxorubicin		Amiodarone
	Vinblastine		Surfactants
	Vincristine,		(cremophor EL)
	Etoposide		
	Paclitaxel		
	Methotrexate		
Antidepressants	Amitryptyline	2nd generation	PSC-833 (valspodar)
	Venlafaxine		GF120918
	Paroxetine		(elacridar)
Antiepileptic drugs	Phenytoin		VX-710 (biricodar)
	Carbamazepine		dexverapamil
	Lamotrigine		
	Phenobarbital		
	Felbamate		
	Gabapentin		
	Topiramate		
Antifungals	Itraconazole	3rd generation	OC 144-093 (ONT-093)
	Ketoconazole		
Cardiac glycosides	Digoxin		LY335979
Calcium channel blockers	Diltiazem		(zosuquidar)
	Nifedipine		XR9576 (tariquidar)
	Verapamil		R101933 (laniquidar)
HIV protease inhibitors	Amprenavir		GF120918
	Indinavir		
	Saquinavir		
Immunosuppressive agents	Cyclosporin		

Source: From Ref. 18.

various pharmacokinetic parameters. Increasing log Po/w usually increases oral absorption and volume of distribution of a medication. However, hepatic extraction may also through increased metabolism efficiency by cytochrome P450 and other liver enzymes (21–23).

Molecular size (or molecular weight) and hydrogen bonding capacity are two other properties often considered as important for membrane permeation and pharmacokinetics. As stated previously, the molecular weight of compounds that are candidates for CNS delivery should be less than 600–700 Da. If significant BBB transport is desired, molecular weight should be less than 400 Da, regardless of drug lipophilicity (24). Increasing molecular size often results in higher potency, but inevitably also leads to either higher lipophilicity or greater hydrogen bonding capacity, resulting in poor dissolution/solubility and limited oral absorption (25). Brain uptake has been shown to be negatively correlated with hydrogen bonding (26). If a drug forms hydrogen bonds, its ability to permeate the endothelial cell membrane will be compromised, and the drug uptake in the brain will be limited.

TOOLS TO STUDY CNS DRUG DELIVERY

Penetration of the BBB is essential for effective pharmacotherapy of CNS disorders. Various techniques have been used to study the pharmacokinetics and pharmacodynamics of CNS active agents by determining unbound drug concentrations in the extracellular fluid of the brain. In vivo techniques include the brain uptake index (27), the brain efflux index (BEI) (28), brain perfusion (29), the unit impulse response method (30) and micro-dialysis (31).

BEI, an intracerebral microinjection technique, can be used to examine BBB efflux transport mechanisms under in vivo conditions. The BEI method may be employed to determine the apparent in vivo drug efflux rate constant across the BBB and to monitor drug concentration dependency and efflux inhibition (32).

CSF concentrations are commonly used as a surrogate marker of CNS availability of drugs. The equilibrated CSF-to-unbound plasma concentration ratio may be a good indicator of the balance between drug permeability across the blood-CNS barriers and the sink action of CSF turnover. As lipophilicity and membrane permeability increase, the CSF-to-plasma unbound concentration ratio increases toward unity. Deviations are noted for lipophilic drugs that are highly bound to CSF proteins (ratios >1), and lipophilic drugs that are efflux transporter substrates (ratios <1). Despite the complexity of CSF pharmacokinetics, a rapid kinetic equilibrium exists between the CSF and biophase for certain medications. In these cases, drug concentration in the CSF can serve as a proximate reference for detailed investigations of factors affecting the intrinsic pharmacodynamics of a centrally acting drug (33). For the direct measurement of the brain interstitial fluid drug concentration, a determinant of the in vivo effect of a drug in the CNS, brain microdialysis can be a useful tool (34–36). However, intracerebral microdialysis is an invasive technique, which may cause tissue trauma and therefore affect BBB function.

In order to study CNS drug transport more efficiently, some in vitro models that closely mimic the in vivo system have been developed to predict the BBB permeability of drugs. Three types of brain capillary endothelial cell culture, including primary cultures, cell lines and co-culture systems, are currently used as in vitro BBB models (37). Generally, the in vitro BBB model consists of a co-culture of brain capillary endothelial cells on one side of a filter and astrocytes on the other, and a good in vivo and in vitro correlation has been demonstrated for the investigation of the role of the BBB in the delivery of nutrients and drugs to the CNS (38,39). Other BBB models have been developed from cerebral capillary endothelium (porcine brain capillary endothelial cells) or choroid plexus epithelial cells (porcine choroid plexus) (40,41).

The permeability of the BBB is just one of the factors determining the drug bioavailability in the brain. The BBB generally only allows passage of select lipophilic drugs by passive diffusion. As mentioned previously, the transmembrane protein P-gp is one of the carrier systems that transport drugs out of the brain through efflux. P-gp affects the pharmacokinetics of many drugs, and can be inhibited by administration of modulators or competitive substrates. Therefore, identification and classification of CNS drugs as P-gp substrates or inhibitors are of crucial importance in drug development.

Similar to CNS drugs, delivery of diagnostic agents also presents several challenges as a result of the special features of CNS blood vessels and tissue fluids. The anatomy of large vessels can be imaged using bolus injection of X-ray contrast agents to identify sites of malformation or occlusion, and blood flow can be measured using magnetic resonance imaging (MRI) and computed tomography (CT), while new techniques permit analysis of capillary perfusion and blood volume. Absolute quantities can be derived, although relative measures in different CNS regions may be as useful in diagnosis.

Local blood flow, blood volume, and their ratio (mean transit time) can be measured with high speed tomographic imaging using MRI and CT. Intravascular contrast agents for MRI are based on high magnetic susceptibility and include gadolinium, dysprosium and iron. Recent advances in MRI technology permit non-invasive "labeling" of endogenous water protons in flowing blood, with subsequent detection as a measure of blood flow. Imaging the BBB most commonly involves detecting disruptions of the barrier, allowing contrast agents to leak out of the vascular system.

Techniques for imaging the dynamic activity of the brain parenchyma mainly involve Positron Emission Tomography (PET), using a variety of radiopharmaceuticals to image glucose transport and metabolism, neurotransmitter binding and uptake, protein synthesis and DNA dynamics. PET studies can play an important role in the screening process as a follow-up of high-throughput in vitro assays. Several rodent studies have shown the potential value of PET to measure the effect of P-gp on the pharmacokinetics and brain uptake of radiolabeled compounds. By quantitative PET measurement of P-gp function, the dose of modulators required to increase the concentration of CNS drugs may be determined, and this may result in improved drug therapy (42). In addition, PET can be used for assessment of mechanisms underlying drug resistance in epilepsy, examination of the role of the BBB in the pathophysiology of neurodegenerative and affective disorders, and exploration of the relationship between polymorphisms of transporter genes and the pharmacokinetics of test compounds within the CNS (43). PET methods also permit detailed analysis of regional function by comparing resting and task-related images, which is important in improving understanding of both normal and pathological brain function (44).

STRATEGIES AND TOOLS TO ENHANCE CNS DRUG DELIVERY

From the preceding discussion, it is clear that delivering drugs to the CNS at therapeutically effective concentrations is a difficult challenge. There are two general approaches available to overcome the BBB, and to enhance drug delivery to the brain: non-invasive and invasive drug delivery (Table 2). Of these, non-invasive methods are generally preferred and have been extensively studied. Strategies for both non-invasive and invasive drug delivery are presented below.

NonInvasive CNS Drug Delivery

Drug Analogs and Prodrugs

Compound lipophilicity is a fundamental physicochemical property that plays a pivotal role in the absorption, distribution, metabolism, and excretion (ADME) of therapeutic agents. A parabolic relationship often exists between measured lipophilicity and in vivo brain penetration of drugs, where those moderate in lipophilicity often exhibit highest uptake (45). One

Table 2   General Approaches to Enhance CNS Drug Delivery

Noninvasive approaches	Drug analogs and prodrugs
	Transdermal drug delivery systems (patch, cream, gel, and microemulsion)
	Liposomes
	Nanoparticles
	CDS
	Carrier-mediated CNS drug delivery
	Receptor-mediated CNS drug delivery
	Enzyme inhibition and absorption enhancer
	Gene delivery to CNS
	Intranasal delivery
Invasive approaches	BBB disruption
	Direct infusion methods
	Implantable devices

Abbreviations: CDS, chemical delivery systems; CNS, central nervous system.

way to promote drug permeation through the BBB is to improve the lipophilicity of the parent compound. For example, hydrophobic analogues of small hydrophilic drugs usually can be expected to more readily penetrate the BBB. However, increasing lipophilicity is not always successful. If the lipophilicity is too high, brain extraction of lipophilic compounds will decrease. Moreover, lipophilic compounds are generally more susceptible to hepatic metabolism, leading to increased drug clearance. On the other hand, very polar compounds normally exhibit high water solubility and greater excretion by the kidneys. These compounds also contain ionizable functional groups that limit the BBB penetration at physiologic pH. In other words, a delicate balance between cerebrovascular permeability and plasma solubility is required. Specifically, the optimal log Po/w is ~1.5–2.5. It should be noted that log Po/w alone seems to have limited ability in predicting brain/blood concentration ratios. However, in combination with other parameters, log Po/w can still reasonably predict brain–blood partitioning (46,47).

A prodrug strategy has been tested on various compounds, with the goal of improving CNS uptake. Colinidine analogs have been designed as lipophilic and highly selective alpha 2-adrenergic stimulants. The pharmacologic features of these agents are comparable to clonidine and alpha-methylnorepinephrine, the principal metabolite of methyldopa (48). Liphophilicity modification is not only limited to small molecules, but can be applied to peptides as well (49). One study demonstrated that point modifications of a BBB-impermeable polypeptide, horseradish peroxidase (HRP), with lipophilic (stearoyl) or amphiphilic (pluronic block copolymer) moieties considerably enhanced transport across the BBB, resulting in accumulation of the polypeptide in the brain in vitro and in vivo. The enzymatic activity of HRP was preserved following transport. Modifications of the HRP with amphiphilic block copolymer moieties through degradable disulfide links resulted in the most effective transport of the HRP across in vitro brain microvessel endothelial cell monolayers and efficient delivery of HRP to the brain (50).

Endomorphin II (ENDII), an endogenous ligand for the mu-opioid receptor, was investigated as a possible analgesic agent with fewer side effects than morphine. To improve CNS entry of ENDII, structural modification was also examined to determine whether Pro(4) substitution and cationization affected its physicochemical characteristics, BBB transport, and analgesic profile. The study showed that the structural modifications enhanced analgesic activity after intravenous administration, information that should aid in the future development of successful opioid drugs (51).

Parent drug molecules can also be modified to form a prodrug, which is usually pharmacologically inactive. The chemical change is generally intended to improve physicochemical properties of a drug, such as low membrane permeability or water solubility, and therefore enhance its delivery. After administration, the prodrug is able to readily access the receptor site, where it will be retained there. The prodrug is then converted to its bioactive form, usually via a single activation step. There are several approaches involved in prodrug formation, such as esterification or amidation of hydroxy-, amino-, or carboxylic acid- functional groups. Because these modifications will endow the prodrugs with greater lipid solubility, they will enter into the brain more easily. Once inside the CNS, hydrolysis of the modifying group will generate the compound's active form.

In order to dramatically improve lipophilicity, drug can also be conjugated to a lipid moiety, such as fatty acid, glyceride or phospholipid. For example, several acid-containing drugs, including levodopa, GABA, niflumic acid, valproate or vigabatrin have been coupled to diglycerides or modified diglycerides (52).

As mentioned earlier, while the increased lipophilicity may improve movement across the BBB, it also tends to increase uptake into other tissues, causing increased tissue burden and non-target toxicity. Furthermore, in addition to facilitating drug uptake into the CNS, lipophilicity also enhances efflux processes in the brain. This can result in poor tissue retention and a short biological duration of action. In addition, metabolites of some prodrugs may exert side effects or toxicity. As a result, these effects may decrease the therapeutic index of drugs masked as prodrugs. On the other hand, prodrug approaches that target specific membrane transporters have recently been explored. This involves designing a prodrug as a substrate for a specific membrane transporter, such as the amino acid, peptide or glucose transporters (53).

Non-steroidal anti-inflammatory drugs (NSAIDs) have been proposed to prevent or to cure Alzheimer's disease. Potential prodrugs of several NSAIDs have been synthesized in order to increase their access to the brain. Using a chemical delivery approach (described later), the carboxylic group of an NSAID molecule is attached to 1,4-dihydro-1-methylpyridine-3-carboxylate, which acts as a carrier, via an amino alcohol bridge. Experimental measurements show that these prodrugs were more lipophilic compared to their corresponding parent compounds and consequently a better BBB penetration is hypothesized (54). An example is 1,3-diacetyl-2-ketoprofen glyceride, a prodrug of ketoprofen and a model compound that was developed as CNS drug delivery system to treat Alzheimer's disease.(55).

Based on the physicochemical and pharmacological properties of drugs possessing an adamantane skeleton, an adamantane-based moiety was evaluated as a drug carrier for poorly absorbed compounds, including peptides that are active towards the CNS. In one investigation, various [D-Ala2]Leu-enkephalin derivatives were prepared through conjugation with an adamantane-based moiety at the C-terminus or N-terminus, and their biological activities were examined. The results suggest that introduction of the lipophilic adamantane moiety into [D-Ala2]Leu-enkephalin would improve the permeation of the poorly absorbed parent peptide through the BBB without loss of antinociceptive effect (56).

Several azidothymidine (AZT) prodrugs (conjugated with the 1-adamantane moiety via an ester bond) were synthesized to improve the transport of AZT into the CNS. Inclusion of a 1-adamantane moiety to AZT resulted in enhancement of the BBB penetration. This pharmaceutical approach may be beneficial for the efficient treatment of the CNS infection by human immunodeficiency virus (HIV) (57).

Tenofovir disoproxil fumarate, a prodrug of tenofovir (9-[9(R)-2-(phosphonomethoxy)propyl]adenine, PMPA), has been approved for use in the combination therapy of HIV-1 infection. PMPA can cross the choroid plexus and access the CSF, and thus may be effective towards HIV-infected perivascular and meningeal macrophages. However, PMPA is unlikely to reach the infected microglia of deep brain sites (58).

Fosphenytoin sodium (Cerebyx®), a phosphate ester prodrug of phenytoin, was developed as an alternative to parenteral formulations that utilize phenytoin sodium. Fosphenytoin has fewer local adverse effects (e.g., pain, burning, and itching at the injection site) after intramuscular and intravenous administration than parenteral phenytoin. Therapeutic effects are similar for both preparations, although transient paresthesias are more common with fosphenytoin (59,60).

Nelarabine is a water-soluble prodrug of the cytotoxic deoxyguanosine analog ara-G, which is rapidly converted to ara-G in vivo by adenosine deaminase. Nelarabine has shown activity in the treatment of T-cell malignancies, especially T-cell acute lymphoblastic leukemia. The excellent CSF penetration of nelarabine and ara-G supports further investigation into the contribution of nelarabine to the prevention and treatment of CNS leukemia (61).

A metabolically stable and centrally acting analog of pGlu-Glu-Pro-NH2 thyrotropin-releasing hormone ([Glu2]TRH), a tripeptide structurally related to thyrotropin-releasing hormone (TRH), was designed by replacing the amino-terminal pyroglutamyl residue with a pyridinium moiety. The novel analog maintained its antidepressant potency, but showed reduced analeptic action compared to [Glu2]TRH. Thus, an increase in the selectivity of CNS-action was obtained by the incorporation of the pyridinium moiety (62).

A promising prodrug has been developed to treat Parkinson's disease. SPD148903 represents a new type of prodrug. The compound has an enone structure that undergoes oxidative bioactivation to the catecholamine S-5,6-diOH-DPAT, which is delivered enantioselectively into the CNS. S-5,6-diOH-DPAT displays mixed dopamine D(1)/D(2) receptor agonist properties similar to apomorphine. This concept has the potential to improve therapeutic outcomes in Parkinson's disease by complimenting L-dopa therapy, the current treatment of choice (63).

A prodrug approach was used to enhance CNS delivery of the antiviral medications 2′,3′-dideoxyinosine (ddI) and nelfinavir. In addition, inhibition of P-gp efflux was initiated to further increase the CNS delivery of the medications. Overall, preclinical studies in rats showed that this therapeutic strategy increased the brain/plasma ratios of both ddI and nelfinavir (64). Enhanced CNS delivery of certain poorly penetrating 2′,3′-dideoxynucleosides has been achieved by designing prodrugs that are substrates for enzymes, such as adenosine deaminase (ADA), that are present at high activities in brain tissue (65).

It should be noted that not all prodrugs work better than their parent compounds. For example, the esters of chlorambucil did not exhibit superior anticancer activity than equimolar chlorambucil in a rat model of brain-sequestered carcinosarcoma (71). An aza-analogue of furamidine, 6-[5-(4-amidinophenyl)-furan-2-yl] nicotinamidine (DB820), has potent in vitro antitrypanosomal activity. However, the prodrug suffers from poor oral activity because of its positively charged amidine groups (72). In addition, increased lipophilicity alone does not ensure that a given prodrug will deliver higher levels of a parent compound to the CNS. Both the selectivity and absolute rate of bioconversion in the brain are important factors (73).

Transdermal Drug Delivery Systems (Patch, Cream, Gel, and Microemulsion)

Transdermal drug delivery systems (TDDS) or transdermal therapeutic systems (TTS) are convenient dosage forms in terms of application, patient compliance and readily withdrawal of drug (if desired). Several CNS medications have been formulated into various transdermal systems for local and prolonged delivery. Examples of these approaches are detailed below.

Hyoscine (scopolamine), a competitive inhibitor of the muscarinic receptors of acetylcholine, has been shown to be one of the most effective agents for preventing motion sickness. However, a relatively high incidence of side effects and a short duration of action restrict the usefulness of this agent when administered orally or parenterally. To address these drawbacks, a novel transdermal preparation of hyoscine was developed. Pharmacokinetic studies indicate that transdermal administration delivers the drug into the systemic circulation at a controlled rate over an extended period (72 h), providing a means of delivery that is similar to a slow intravenous infusion. Therapeutic trials demonstrated that a single transdermal hyoscine patch is significantly superior to placebo and oral meclizine in preventing motion sickness. Thus, evidence suggests transdermal hyoscine may offer an effective and conveniently administered alternative for the prevention of motion-induced nausea and vomiting in certain situations (70).

Another transdermal therapeutic system for scopolamine (TTS-S) was developed to overcome the adverse effects and short duration of action seen when scopolamine when is administered orally or parenterally. The system contains a drug reservoir (1.5 mg) that is programmed to deliver 0.5 mg over a 3-day period. The TTS provides the approximate functional equivalent of a 72-h slow intravenous infusion of drug (71,72).

Rotigotine (Neupro) is formulated as a transdermal delivery system designed to provide a selective, non-ergot D3/D2/D1 agonist to the systemic circulation over a 24-h period in patients with Parkinson's disease. This formulation is described in detail later in this chapter.

A TDDS containing fentanyl (Duragesic®) is available to treat moderate to severe pain. Transdermally-administered fentanyl exhibits a favorable pharmacokinetic profile, an is a standard therapy for chronic pain associated with cancer and other diseases (73).

Nicotinic receptor dysfunction and impaired semantic memory occur early in patients suffering from Alzheimer's disease. Previous research indicated that the ability of nicotine to enhance alertness, arousal, and cognition in a number of nonclinical populations was a function of its ability to stimulate CNS nicotinic cholinergic receptors. Studies showed that transdermal administration of nicotine increased both regional cerebral glucose metabolism (rCMRglc) and semantic memory (as assessed by verbal fluency) (74). The results point to a possible alternative indication for transdermal nicotine therapy.

A submicron emulsion of the benzodiazepine diazepam, a lipophilic molecule with CNS activity, was investigated for efficacy as a novel transdermal formulation. Diazepam was formulated in various topical regular creams and submicron emulsion creams of different compositions. The different formulations were applied topically and protection against pentamethylenetetrazole-induced convulsive effects in mice was monitored. The efficacy of diazepam applied topically in emulsion creams strongly depended on the oil droplet size and to a lesser degree on formulation composition. Preparing the emulsion with a high-pressure homogenizer caused a drastic reduction in the droplet size, significantly increasing the activity of diazepam following transdermal application. Combined with high-pressure homogenization, the presence of lecithin (an efficient dispersant) into the formulation resulted in effective droplet size reduction to below 1 µm (100–300 nm). Overall, submicron emulsions were found to be effective vehicles for transdermal delivery of diazepam, generating significant systemic activity of the drug compared with regular creams or ointments (75).

Two 5-nitroimidazoles compounds, MK-436 and fexinidazole, were formulated as gels by the addition of hydroxypropylcellulose. When used in combination with melarsoprol, the formulation was able to cure experimental murine CNS-trypanosomiasis with one-day therapy. Likewise, combined melarsoprol/MK-436 gel successfully cured experimental CNS-trypanosomiasis with a single treatment. Topical application of melarsoprol/MK-436 gel also eliminated hind leg paralysis that associated with post-treatment reactive encephalopathy caused by non-curative treatment of CNS-trypanosomiasis (76,77).

Basic fibroblast growth factor (bFGF) is a morphogenic, chemotactic, mitogenic, and angiogenic peptide found within the CNS with potent neurotrophic effects. A potential role in ischemia-induced vascular growth has been suggested for bFGF. Research showed that single, topical administration of human recombinant basic fibroblast growth factor (rbFGF) in the rat cerebral cortex promoted capillary overgrowth, an effect that might mimic the angiogenic response observed after brain ischemia. This finding has potential implications in angiogenic therapy (78).

Liposomes

Another approach used to increase lipophilicity of a CNS-active drug without modifying its molecular structure is liposomes. Liposomes are spherical vesicles with membranes composed of a phospholipid bilayer. For drug delivery, liposomes are able to encapsulate a hydrophilic drug molecule inside its double layer-structure. The transport capacity of liposomes could be enhanced by subsequent conjugation of the liposome to a BBB drug delivery vector. It is well established that liposomes, even small unilamellar vesicles, generally do not undergo significant transport through the BBB in the absence of vector-mediated drug delivery (79). Another disadvantage of liposomes as a drug delivery system is that these structures are rapidly removed from the bloodstream following intravenous administration, owing to uptake by cells lining the reticulo-endothelial system or the mononuclear phagocytic system. Incorporation of PEGylation technology and chimeric peptide technology into the design of these systems helps to mediate BBB transport and inhibit peripheral clearance of liposomes (Fig. 3).

A bi-functional PEG 2000 derivative was synthesized that contained thiolated murine monoclonal antibody (MAb) and a distearoylphosphatidy-lethanolamine moiety, to incorporate into the liposome surface. This combined technology resulted in the construction of PEGylated immuno-liposomes that are capable of receptor-mediated transport through the BBB in vivo (79). After administration, MAb binds to the BBB transferrin receptor, which has been successfully used as a vector in delivery of other large molecules across the BBB. The immuno-liposome delivery system has the ability to dramatically increase brain drug delivery by up to four orders of magnitude. This delivery system may be of significance to brain drug delivery because it permits brain targeting of the liposomally encapsulated drug, and may consequently offer a significant reduction in side effects. Compounds with excellent neuro-pharmacologic potential in vitro that may have been rejected for clinical use because of low brain delivery (or some minor side-effects) may now be reevaluated for potential use in conjunction with this delivery system.

Liposomes can also overcome some disadvantages of prodrugs. Since the liposome capsule undergoes degradation to release its contents, drug is delivered without the use of disulfide or ester linkages, which may significantly

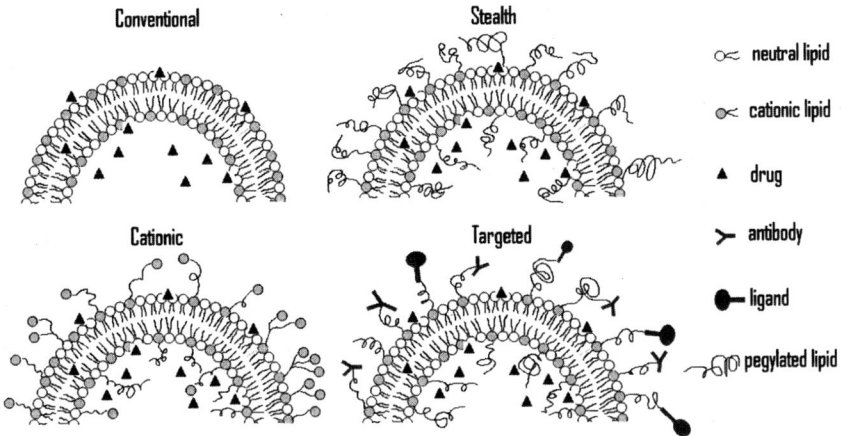

Figure 3 Schematic representation of a liposome. Liposomes are spherical vesicles with membranes composed of a phospholipid bilayer. Liposomes can be used as drug carriers and loaded with a great variety of molecules, such as small drug molecules, proteins, and nucleotides. As shown in the figure, liposomes are extremely versatile and can be used in number of applications, including CNS drug delivery.

affect pharmacological actions (80). The surgical delivery of therapeutic agents into the parenchyma of the brain is problematic because it has been difficult to establish the biological fate of the material after infusion, or to determine a suitable dose. Gadoteridol (GDL) loaded liposomes have been used as tracer agents that allow one to track material in real-time using MRI. MRI allows precise tracking and measurement of liposomes loaded with markers and therapeutics in the CNS (81).

Convection-enhanced delivery (CED) is a recently developed technique for local delivery of agents to a large area of tissue in the CNS. CED was combined with a highly stable nanoparticle/liposome containing CPT-11 (nanoliposomal CPT-11) to provide a dual drug delivery strategy for brain tumor treatment. CED of nanoliposomal CPT-11 greatly prolonged tissue residence while also substantially reducing toxicity, resulting in a highly effective treatment strategy in preclinical brain tumor models (82,83).

Liposomes loaded with GDL, in combination with CED, offer an excellent option to monitor CNS delivery of therapeutic compounds using MRI (84,85). Saposin C is one of four small lipid-binding proteins that derive from a single precursor protein, named prosaposin (PSAP). PSAP has several neuronal effects, including neurite outgrowth stimulation, neuron preservation, and nerve regeneration enhancement. Delivery of saposin C ex vivo into cultured neurons and in vivo into brain neuronal cells in mice across the BBB was accomplished with intravenously administered dioleoylphosphatidylserine liposomes. These studies may yield a new therapeutic approach for neuron protection, preservation, and regeneration (86).

High-dose glucocorticosteroid hormones are a mainstay in the treatment of relapses in multiple sclerosis. Ultrahigh doses of glucocorticosteroids were delivered to the CNS of rats with experimental autoimmune encephalomyelitis (EAE) using a novel formulation of polyethylene glycol (PEG)-coated long-circulating liposomes encapsulating prednisolone (prednisolone liposomes, PL). PL is a highly effective drug therapy for EAE, and it was found to be superior to a 5-fold higher dose of free methylprednisolone, possibly due to drug targeting by PL. These findings may have implications for future therapy of autoimmune disorders such as multiple sclerosis. (87).

Mirfentanil hydrochloride, a novel CNS analgesic with a short duration of action, was successfully encapsulated in liposomes of varying composition. The lipid composition of the formulation was varied to optimize the stabilization of liposomes and the encapsulation of solute. Only 35% of encapsulated drug was released when liposome formulations containing monosialoganglioside (GM1) were incubated with human plasma over a 24-h period, suggesting that these systems could be used for controlled drug delivery in vivo (88).

Nanoparticles

In addition to liposomes, the use of other nanocarriers, such as solid polymeric and lipid nanoparticles, may be advantageous over existing CNS delivery strategies. Not only can these nanocarriers mask the BBB uptake-limiting characteristics of the therapeutic drug molecule, but they also protect the drug from chemical/enzymatic degradation, and provide opportunity for sustained release. Reduction of toxicity to peripheral organs can also be achieved with these nanocarriers (89). Polymeric nanoparticles have been used to enhance therapeutic efficacy and reduce toxicity of a variety of medications (98), as well as to enhance delivery of imaging agents (90). Drugs that have successfully been transported into the brain using this carrier include dalargin (a hexapeptide), kytorphin (a dipeptide), loperamide, tubocurarine, MRZ 2/576 (an NMDA receptor antagonist), and doxorubicin (91). The nanoparticles may be especially helpful for the treatment of the disseminated and very aggressive brain tumors. Intravenously injected doxorubicin-loaded polysorbate 80-coated nanoparticles were associated with a 40% cure rate in rats with intracranially transplanted glioblastomas (92).

The mechanism of the nanoparticle-mediated transport of the drugs across the BBB has not been completely elucidated. The most likely mechanism is endocytosis by the endothelial cells lining the brain blood capillaries. Nanoparticle-mediated drug transport to the brain depends on the overcoating of the particles with polysorbate molecules such as polysorbate 80. Overcoating with these materials presumably results in adsorption of apolipoprotein E from blood plasma onto the nanoparticle surface. As a result, the particles seem to mimic low density lipoprotein (LDL) particles that can interact with the LDL receptor, leading to their uptake by the endothelial cells. Subsequently, the drug may be released in these cells and diffuse into the brain interior, or the particles may be transcytosed. Other processes such as tight junction modulation or P-gp inhibition also may occur. Moreover, these mechanisms may run in parallel or may be cooperative, thus enabling drug delivery to the brain (93–97). Research shows that PEGylated n-hexadecylcyanoacrylate nanoparticles, made by PEGylated amphiphilic copolymer, penetrate into the brain to a larger extent without inducing any modification of BBB permeability (98). Cationic bovine serum albumin (CBSA) conjugated with poly(ethyleneglycol)-poly(lactide) (PEG-PLA) nanoparticle (CBSA-NP) has been studied for brain drug delivery. CBSA-NP was shown to be preferentially transported across the BBB with minimal toxicity, suggesting the possibility to deliver therapeutic agents to the CNS using this carrier (99).

Nanoparticle drug delivery systems have also been studied for delivery of HIV/AIDS medications, to facilitate complete eradication of viral load from reservoir sites in the body (100). Moreover, iron oxide based contrast agents have been investigated as more specific MRI agents for diagnosing CNS inflammation. Ferumoxtran-10, a virus-size nanoparticle, is taken up by reactive cells and is used to visualize phagocytic components of CNS lesions, a useful tool for monitoring therapy in patients (101). Iron oxide nanoparticles can therefore be used as a marker for the long-term noninvasive MR tracking of implanted stem cells (102).

Metal ions accumulate in the brain with aging and in several neurodegenerative diseases. Aside from Wilson's disease (the copper storage disease), recent attention has focused on the accumulation of zinc, copper, and iron in the brain in Alzheimer's disease, and the accumulation of iron in Parkinson's disease. Nanoparticles have been demonstrated to deliver D-penicillamine to the brain, as well as to reduce metal ion accumulation in Alzheimer's disease and other CNS disorders (103).

Nanoparticles have also been studied as transport vectors for peptides. Among these is the enkephalin analog dilargin. Nanoparticles were formulated with colloidal polymer particles of poly-butylcyanoacrylate with dilargin absorbed onto the surface. The particles were then coated with polysorbate 80. Intravenous injections of the vector-dalargin produce analgesia, while dalargin alone does not (104).

Overall, nanoparticle formulations appear to have no effect on primary BBB parameters in established in vitro and in vivo BBB models (105). Additionally, there is little in vivo or in vitro evidence to suggest that a generalized toxic effect on the BBB is the primary mechanism for drug delivery to the brain (106). Using these systems, however, the possibility of a general toxicity is still a serious concern for nanoparticle systems, and this requires further study (107).

Chemical Delivery Systems

Chemical delivery systems (CDS) represent a method to target drug to specific sites or organs in the body based on predictable enzymatic activation. After certain chemical modifications, compounds are transformed into inactive chemical derivatives. The newly attached moieties usually form monomolecular units, and provide site-specific or site-enhanced delivery of the drug through multi-step enzymatic and/or chemical transformations. CDS is similar to a prodrug approach, differing by the fact that multi-step activation and targetor moieties are involved. Using the general CDS concept, successful deliveries have been achieved not only to the brain, but also to the eye, and to the lung (108,109).

By converting a lipophilic drug into a lipid-insoluble molecule in the brain, permeation of the molecule out of the brain is diminished. If the same conversion also takes place in the rest of the body, this accelerates peripheral elimination and improves targeting. It should be emphasized that CDS not only achieves delivery to the brain, but it also provides preferential delivery, which means brain targeting. Ultimately, this should allow administration of smaller doses, reduction of peripheral side effects, and prolonged therapeutic response. Therefore, CDSs can be used not only to deliver compounds that otherwise have no access to the brain (e.g., steroid hormones and peptides), but also to retain lipophilic compounds within the brain (110).

Anionic chemical delivery systems have been developed to deliver testosterone and zidovudine to the brain via phosphonate derivatives (111). When incorporated into a bulky molecule, a peptide unit can be delivered by CDS to provide direct BBB penetration, and avoid recognition by peptidases in the plasma (112–115). To achieve delivery and sustained activity with such complex systems, it is important that the targeted enzymatic reactions take place in a specific sequence. Upon delivery, the first step must be the conversion of the targetor to allow for "lock-in." This must be followed by second step (removal of the L function) to form a direct precursor of the peptide that is still attached to the charged targetor. Subsequent cleavage of the targetor-spacer moiety finally leads to the active peptide. Several CDS have been synthesized for the cholinesterase inhibitor 9-amino-1,2,3,4-tetrahydroacridine (THA). In vivo distribution studies indicate that elevated and sustained levels of the pyridinium quaternary ion derivative were present in the CNS. In addition, THA was generated in the CNS from the quaternary salt precursor at low concentrations, indicating a slow but sustained release. The CDS for THA were found to be less acutely toxic compared to THA alone (116).

A CDS system based on the redox conversion of a lipophilic dihydropyridine to an ionic, lipid-insoluble pyridinium salt has been developed to improve the access of therapeutic agents to the CNS. A dihydropyridinium-type CDS or a redox analog of a drug is sufficiently lipophilic to enter the brain by passive transport, where it then undergoes enzymatic oxidation to an ionic pyridinium compound, which promotes retention in the CNS. At the same time, peripheral elimination of the entity is accelerated due to facile conversion of the CDS in the body (117). This methodology has been extended to deliver neuroactive peptides such as enkephalin to the brain, and has demonstrated promise in laboratory models (118).

There are several other CDS-based approaches of note. A zidovudine-CDS complex is capable of delivering higher zidovudine doses to lymphocytes and neural cells, resulting in improved antiretroviral activity. This represents a potential therapy for AIDS dementia (119). CDSs based on a dihydropyridine-quaternary pyridinium ion redox system, analogous to the naturally occurring NADH-NAD+ system, were synthesized for a group of staphylococcal penicillinase resistant penicillins (e.g., methicillin, oxacillin, cloxacillin, dicloxacillin) in order to improve their penetration of the CNS. In vivo distribution studied in rats and rabbits demonstrated BBB penetration of the compounds by CDS, whereas no drug was detected in the brain following direct drug administration (120).

Carrier Mediated CNS Drug Delivery

Given the nature of the BBB and BCB, drug transport to the brain is a transcellular process, with lipophilicity being a primary determinant of the process. For a number of compounds, however, lipophilicity does not correlate with CNS penetration; that is, drug uptake across the BBB is lower than expected (121). This discrepancy can be explained by the presence of transport systems that may function in CNS uptake and efflux of xenobiotics (Fig. 2). Understanding the key features of these pathways may allow for improved treatment of diseases of the CNS though enhanced uptake of neuropharmaceuticals. Furthermore, CNS-related side effects of medications could be avoided by blocking these mechanisms.

Various transport systems can be targeted to promote drug uptake into the brain. Among these are the moncarboxylic acid transporters (MCT), of which at least eight isoforms have been identified (122). MCTs are involved in bidirectional membrane transport across the BBB. MCT1 has been identified on the basolateral membrane of brain capillary endothelial cells, where it transports lactic acid and other short chain acids. The pH dependence of this acidic system suggests a proton cotransport mechanism is involved. Medications bearing a carboxylic acid moiety appear to be substrates for MCT1. One such class of medications is HMG-CoA reductase inhibitors (e.g., lovastatin and simvastatin). CNS distribution of the less lipophilic drug pravastatin suggests a lower affinity for the transporter, resulting in a reduced incidence of CNS side effects. Overall, an understanding of the role of MCTs in the BBB is still evolving.

Another family of transporters implicated in CNS uptake is the OATP family. Two isozymes that have been identified are OATP1 and OATP2. OATP1 is expressed in the luminal membrane of the BCB, while OATP2 is expressed in the basolateral membrane of the BCB and the brush border membrane of the BBB. Transport across OATP2 is bi-directional (123).

In addition to passive transport (a function of lipophilicity and molecular weight), at least two distinct carrier mediated transport mechanisms have been identified for CNS transport of organic cations. Analogous to MCT1, structural differences among a class of compounds such as the H1-antagonists alters the affinity for these systems and, consequently, the side effect profile of individual medications (122).

Receptor Mediated CNS Drug Delivery

Receptor mediated transport (RMT) is a potential pathway for delivery of large molecule peptides or proteins to the brain. During RMT, the molecule is shuttled across the BBB into brain interstitial fluid by specific receptors. Among the receptors that have been identified is the endothelial transferrin receptor (TfR). TfR is a bidirectional transporter, mediating the transport of halo-transferrin (blood/brain) and apo-transferrin (brain/blood). Additionally, endocytotic mechanisms exist on the BBB that mediate uptake into the CNS. Among these, the type I scavenger receptor (SR-BI) is involved with uptake of low density lipoprotein into the brain (124).

Both absorptive-mediated and receptor medicated endocytosis represent promising routes for peptide delivery. For example, complexation of neuroactive peptide with cationized albumin could facilitate translocation into the CNS through absorptive endocytosis (125). Alternatively, monoclonal antibodies to a specific receptor could be used to enhance CNS uptake via chimerization (126,127). Covalent attachment of a poorly permeable compound to a suitable vector could enhance brain uptake.

Chimeric peptide technology, wherein a non-transportable drug is conjugated to a BBB transport vector, has been proposed. A BBB transport vector is a modified protein or receptor-specific monoclonal antibody that undergoes receptor-mediated transcytosis through the BBB in vivo. Different approaches for linking drugs to transport vectors may be broadly classified as belonging to one of three classes: chemical, avidin–biotin, or genetic engineering (128). Multiple classes of therapeutics have been delivered to the brain with the chimeric peptide technology, including peptide-based pharmaceuticals, such as a vasoactive intestinal peptide analog or neurotrophins such as brain-derived neurotrophic factor, antisense therapeutics including peptide nucleic acids, and small molecules encapsulated within liposomes.

This strategy has been called Molecular Trojan Horse Technology, and it offers potential for ferrying molecules across the BBB (129). Specific substances can be delivered to the brain by attaching them to a protein that is normally able to cross the barrier. Conjugation of a peptide or antisense therapeutic to a BBB molecular Trojan horse has been shown to generate CNS effects in vivo after intravenous dosing compared with control (administration of therapeutic agent alone) (124).

The successful delivery of a drug through the BBB in vivo requires special molecular formulation of the drug. Therefore, it is important to merge CNS drug discovery and delivery as early as possible in drug development process (130). A disulfide linker can be used to deliver the drug after disulfide cleavage in brain (131). A non-cleavable linkage such as an amide bond can also be used to attach the drug to the transport vector. Cleavage is achieved by reduction of the disulfide bond, and all the bonds including amide bonds are ultimately hydrolyzed in the lysosomal compartment. PEGylation technology is used with a longer spacer arm comprised of a PEG moiety. The placement of this long spacer arm between the transport vector and the drug removes any steric hindrance caused by attachment of the drug to the transport vector, and drug binding to the cognate receptor is not impaired (128).

Inhibition of CNS Efflux Mechanisms

As noted above, lipophilicity remains an important determinant of drug permeation through the BBB. However, there a number of compounds that display poor CNS distribution despite high lipophilicity. While plasma protein binding and molecular size of these compounds are possible explanations for these findings, accumulating evidence (e.g., transgenic animal studies) points to the presence of efflux mechanisms in the BBB and BCB that restrict access of these compounds to the CNS. Most notable among these transport mechanisms is P-gp (17).

The identification of P-gp on the apical surface of brain capillary endothelial cells suggests a role of this transport system in limiting access to the CNS. P-gp is also expressed in the apical side of the BCB. Indeed, a number of in vitro and in vivo studies support the general concept that P-gp pumps xenobiotics from the brain interstitial fluid into the blood. Much of the information regarding P-gp-mediated CNS efflux has been generated using transgenic mice lacking P-gp. The various P-gp "knockout" mouse models are mutant strains of mice that are genetically deficient in MDR1a, MDR1b (or both genes). Knockout mice do not display physiologic abnormalities and have life span comparable to normal mice. However, CNS distribution of a number of compounds is increased dramatically in knockout mice. This is perhaps best illustrated by the pesticide ivermectin. While ivermectin has a good safety profile in normal mice, it causes lethal neurotoxicity in knockout strains (122). Studies with knockout mice have demonstrated a role of P-gp in limiting the CNS penetration of medications used to treat HIV (e.g., indinavir, ritonavir), cancer (e.g., doxorubicin, vincristine), cardiovascular disorders (e.g., digoxin, quinidine) and pain (e.g., morphine).

In addition to P-gp, other efflux transporter systems exist in the CNS. Like P-gp, BCRP is drug efflux transporter belonging to the ABC family. Substrate specificity of BCRP shows considerable overlap with P-gp, and this indicates a similar role of this transporter on drug pharmacokinetics (17).

Inhibition of P-gp and/or BCRP is being investigated as a potential strategy to improve CNS penetration and delivery of drugs to the brain. This approach may be particularly useful in the treatment of brain tumors and CNS metastases. Preclinical studies have demonstrated that CNS penetration of anticancer drugs (e.g., paclitaxel, docetaxil) can be improved by co-administrating with a P-gp inhibitor (e.g., valspodar, elacridar). Similar results may be expected for BCRP (132).

Enzyme Inhibition and Absorption Enhancement

In terms of oral drug delivery, new insights regarding the role of the intestine as a selective barrier to drug absorption have emerged. Numerous membrane transport systems are present in the intestine to facilitate the absorption of essential nutrients, systems that may also be responsible for oral absorption of certain classes of medications. Conversely, transporters in the enterocyte also serve as detoxification mechanisms in the body, which limit bioavailability through intestinal exsorption. By understanding the membrane transport mechanisms involved in oral drug absorption, strategies can be developed to enhance drug delivery of poorly bioavailable compounds.

The efflux transporters P-gp and BCRP are known to be expressed in the intestine, where they actively extrude a variety of compounds. Inhibition of these transporters, therefore, is a logical strategy to improve oral bioavailability. Preclinical studies in mice found that the bioavailability of topotecan increased from 40% to 97% upon co-administration with a P-gp inhibitor (132).

Intestinal enzyme inhibition may also be an effective tool to increase the oral bioavailability of compounds that undergo first-pass intestinal metabolism. However, simultaneous systemic enzyme inhibition may be undesirable, and thus should be minimized. For example, 2-Beta-fluoro-2′,

3'-dideoxyadenosine (F-ddA) is an ADA activated prodrug of 2-beta-fluoro-2',3'-dideoxyinosine (F-ddI) that provides enhanced delivery to the CNS. F-ddA has been tested clinically for the treatment of AIDS. Unfortunately, intestinally localized ADA constitutes a formidable enzymatic barrier to the oral absorption of F-ddA. Through careful selection of an enzyme inhibitor, dosage regimen design, and by considering the inhibition vs. drug absorption profiles, local enzyme inhibition can be optimized to achieve local ADA inhibition, with minimal systemic inhibition (133,134).

DeltaG, the 12 kDa active fragment of zonula occludens toxin, has been used as absorption enhancer to increase the brain distribution of MTX and paclitaxel, two commonly used anticancer agents with poor distribution into the brain. DeltaG significantly enhances the brain distribution of MTX (hydrophilic) and paclitaxel (lipophilic) and has the potential to be further developed as adjunct therapy to increase delivery of poorly permeable chemotherapeutic and other CNS targeted compounds (135).

Gene Delivery to CNS

Gene transfer offers the potential to explore basic physiological processes and to intervene in human disease. The extension of gene therapy to the CNS, however, faces the delivery obstacles of a target region that is postmitotic and isolated behind the BBB. Approaches to this problem have included grafting genetically modified cells to deliver novel proteins, or introducing genes by viral or synthetic vectors geared toward the CNS cell population. Invasive approaches such as direct inoculation and bulk flow, as well as osmotic and pharmacological disruption, have also been used to circumvent the BBB's exclusionary role. Once the gene is delivered, a myriad of strategies have been tested to affect a therapeutic result. Gene-activating prodrugs are the most common antitumor approach. Other approaches focus on activating immune responses, targeting angiogenesis, and influencing apoptosis and tumor suppression. At this time, therapy directed at neurodegenerative diseases has centered on ex vivo gene therapy for supply of trophic factors to promote neuronal survival, axonal outgrowth, and target tissue function. Despite early promise, gene therapy for CNS disorders will require advancements in methods for delivery and long-term expression before becoming feasible for human disease (136).

Gene therapy has already shown promise as a tool for brain protection and repair from neuronal insults and degeneration in several animal models, and is currently being tested in clinical trials. The choice of an appropriate vector system for transferring the desired gene into the affected area of the brain is an important issue for developing a safe and efficient gene therapy approach for the CNS. Both viral and non-viral vectors have been used in gene delivery to treat brain disorders (137,138).

In non-viral vector gene delivery, lipid-based vectors such as liposomes are commonly used vectors in mediating the delivery of therapeutic genes to the CNS (139). Promising results have been obtained in terms of the level and duration of gene expression, particularly in cortical neurons that were transfected with the Tf-associated lipoplexes. These findings highlight the potential utility of these lipid-based carriers for delivering genes within the CNS (140,141). A simple and highly efficient lipofection method was used for primary embryonic rat hippocampal neurons (up to 25% transfection) that exploits the M9 sequence of the non-classical nuclear localization signal of heterogeneous nuclear ribonucleoprotein A1 for targeting beta(2)-karyopherin (transportin-1). This technique can facilitate the implementation of promoter construct experiments in post-mitotic cells, stable transformant generation, and dominant-negative mutant expression techniques in CNS cells (142). Multi-lipofection provides a mild and efficient means of delivering foreign genes into astrocytes in a primary culture, making astrocytes good candidate vehicle cells for gene/cell therapy in the CNS (143). DC-Chol (dimethylaminoethane-carbamoyl-cholesterol) liposome-mediated NGF gene transfection may have therapeutic potential for treatment of brain injury (144).

Sufficient gene transfer into CNS-derived cells is the critical step for developing strategies for gene therapy. Research shows that liposome-mediated gene transfer is an efficient method for gene transfer into CNS cells in vitro, but the transfection efficiency into the rat brain in vivo is low (145). Alternative gene transfer techniques, such as using cationic liposomes to achieve therapeutically useful levels of expression of neurotrophins in the CNS, could provide new strategies for treating a traumatically injured CNS (146). Likewise, antisense oligonucleotides (AS-ODNs) offers a precise and specific means of knocking down expression of a target gene, and is a major focus of research in neuroscience and other areas. It has become increasingly obvious, however, that there are a number of hurdles to overcome before antisense can be used effectively in the CNS, most notably finding suitable nucleic acid chemistries and an effective delivery vehicle to transport AS-ODNs across the BBB to their site of action. Despite these problems, a number of potential applications of AS-ODNs in CNS therapeutics have been validated in vitro and, in some cases, in vivo (147).

The efficient and targeted transfer of genes is the goal of gene therapy. In the CNS, this is challenging due in part to the exquisite anatomy of the brain. Viral vectors have better transfection rates but a higher incidence of deleterious effects than non-viral vectors. Herpes simplex virus (HSV) vectors are particularly amenable to CNS therapies as they are capable of transducing a variety of cells, have a large transgene capacity and can exist as either oncolytic or non-immunogenic vectors. The versatility and therapeutic use of this vector platform has been used in two CNS disorders, Alzheimer's disease and malignant brain tumors (148). Herpes simplex virus type 1-thymidine kinase (HSV1-TK) in combination with the prodrug ganciclovir (GCV) represents an efficient suicide gene approach in brain tumor gene therapy. The effectiveness of HSV1-DeltaTK in preventing brain tumor growth in vivo, combined with its reduced cytotoxicity, both in vivo and in primary cultures of CNS cells, could represent an advantage for treatment of brain tumors using gene therapy (149,150). A hemagglutinating virus of Japan (HVJ)-liposome vector has been used to deliver oligodeoxynucleotides in the CNS in vivo and in vitro. Thus it is an efficient method for ODN transfer and holds promise as a gene delivery method in the CNS (151). (HVJ)-AVE liposome, an anionic type liposome with a lipid composition similar to that of HIV envelopes and coated by the fusogenic envelope proteins of inactivated HVJ, has been used for gene delivery into the CSF space (152).

Gene delivery could be optimized by a combination of drug-loaded liposomes, polymeric nanoparticles, non-viral DNA complexes, viruses and therapeutic CED to overcome particle binding and clearance by cells within the CNS (153). CNS gene transfer could provide new approaches for modeling neurodegenerative diseases and devising potential therapies for these diseases. One such disorder is Parkinson's disease, in which dysfunction of several different metabolic processes has been implicated (154).

Intranasal Delivery

Drug delivery through the nose, also known as the olfactory pathway, has also been explored as an alternative strategy to deliver CNS drugs. Drugs delivered intranasally are transported along olfactory sensory neurons, and can yield significant concentrations in the CSF and olfactory bulb. Hydroxyzine and triprolidine have both been reported to reach the CNS following nasal administration. In vitro experiments have been conducted to evaluate the effect of directionality, donor concentration and pH on the permeation of hydroxyzine and triprolidine across excised bovine olfactory mucosa. The studies demonstrated that bidirectional flux (mucosal→ submucosal and submucosal→mucosal flux) of hydroxyzine and triprolidine across the olfactory mucosa was linearly dependent upon the donor concentration, without any evidence of saturable transport. The lipophilicity of these compounds, coupled with their ability to inhibit P-gp, enable them to freely permeate across the olfactory mucosa. Despite the presence of a number of protective barriers such as efflux transporters and metabolizing enzymes in the olfactory system, lipophilic compounds such as hydroxyzine and triprolidine can access the CNS primarily by passive diffusion when administered via the nasal cavity (155).

In a previous study, the plasma pharmacokinetics of nipecotic acid and its n-butyl ester were compared after intranasal and intravenous administration. Intranasal administration of an ester formulation was crucial for delivery of nipecotic acid to the brain, as the evidence suggested that ester hydrolysis is rate limiting to nipecotic acid brain delivery. The formed nipeoctic acid displayed tissue trapping in brain. The study also demonstrated that parenteral dosing of nipecotic acid esters is unnecessary for systemic or brain delivery of nipecotic acid, the finding may apply to other CNS active zwitterion esters (156).

The nose has also been studied as a possible route for the systemic delivery of water soluble prodrugs of L-dopa. L-dopa prodrugs had improved solubility and lipophilicity, with relatively rapid in vitro conversion in rat plasma. Following intranasal administration, prodrug absorption was rapid and complete, with a bioavailability of approximately 90%. Thus, utilization of water soluble prodrugs of L-dopa via the nasal route in the treatment of Parkinson's disease may have therapeutic advantages such as improved bioavailability, decreased side effects, and potentially enhanced CNS delivery (157).

In general, there are several obstacles that must be overcome nasal delivery to be successful. These include enzyme activity in nasal epithelium, low pH, mucosal irritation, and regional variability in absorption caused by nasal pathology. However, this method is convenient and relatively noninvasive compared to some other strategies. Nevertheless, more research is needed in order to intranasal delivery systems and to achieve therapeutic drug concentrations in CNS using this pathway.

Invasive Drug Delivery

BBB Disruption

In recent years, various strategies have been designed to circumvent the BBB. Transient BBB disruption and osmotic BBB disruption are such strategies, and have been studied extensively in preclinical and clinical studies. After the BBB is being weakened, drugs can undergo enhanced extravasation rates in the cerebral endothelium, resulting in increased parenchymal drug concentrations. A variety of techniques that transiently disrupt the BBB have been investigated, such as administration of dimethyl sulfoxide or ethanol and metals such as aluminium; X-irradiation; and the induction of pathological conditions including hypertension, hypercapnia, hypoxia or ischemia.

Encouraging results have been obtained by disturbing the BBB in the treatment of brain tumors (158). Intracarotid injection of an inert hypertonic solution such as mannitol or arabinose has been employed to initiate endothelial cell shrinkage and opening of BBB tight junctions for a period of a few hours, permitting delivery of antineoplastic agents to the brain (159). Intra-arterial chemotherapy in combination with BBB disruption increases drug delivery to tumors 2- to 5-fold, and to surrounding brain tissue by 10- to 100-fold compared with intravenous administration of chemotherapy. Primary CNS lymphoma (PCNSL) is an excellent model for studying dose intensity because PCNSL is a highly infiltrative, chemosensitive, primary CNS malignancy in which the integrity of the BBB is highly variable. In patients with PCNSL, a chemotherapy-responsive tumor type, survival time is highly associated with total drug dose delivered, even in analyses designed to control for potential survival biases (160).

Delivery of chemotherapy to treat malignant brain tumors is markedly enhanced when given in conjunction with osmotic disruption of the BBB. Osmotic opening or disruption of the BBB is achieved while the patient is under general anesthesia, by the infusion of mannitol into the internal carotid or vertebral artery circulation. Mannitol infusion is followed by administration of intraarterial chemotherapy (161,162). In one study, radiographic tumor response and survival were evaluated in the pediatric and young adult population with germ cell tumor, primary CNS lymphoma, or primitive neuroectodermal tumor receiving intra-arterial carboplatin- or methotrexate-based chemotherapy with osmotic BBBD (163).

The concept of a blood-brain or blood–tumour barrier blocking the passage of lipid insoluble cytoreductive drugs from blood into brain-tumor, led to a protocol of intraarterial injection of high dose methotrexate (MTX) during reversible osmotic BBBD (164). Osmotic disruption of the BBB has also been used to deliver recombinant adenoviral vectors, and magnetic resonance imaging agents. (165,166).

Some relatively safer biochemical techniques have also been developed to disrupt the BBB. Selective opening of brain tumor capillaries by the intracarotid infusion of leukotriene C4 was achieved without concomitant alteration of the adjacent BBB (167). In contrast to osmotic disruption methods, biochemical opening is based on the novel observation that normal brain capillaries appear to be unaffected when vasoactive leukotriene treatments are used to increase their permeability. However, brain tumor capillaries or injured brain capillaries appear to be sensitive to treatment with vasoactive leukotrienes, and the permeation is dependent on molecular size (168).

Direct Infusion Methods

Intralumbar injection or intreventricular infusion methods deliver drugs directly into the CSF. Drugs can be infused intraventricularly using an Ommaya reservoir, a plastic, dome-shaped device, with a catheter used to deliver chemotherapy to the CNS. Intra-CSF administration bypasses the BCB and immediately produces high drug concentrations in the CSF. As a result, a smaller dose is required. Additionally, drug protein binding and enzymatic activities are also lower in the CSF compared to plasma. Therefore, drugs have longer half-life in the CSF. However, this method provides a slow rate of drug distribution within the CSF, and it may increase intracranial pressure, and cause hemorrhage, CSF leaks, neurotoxicity, and CNS infections. The success of this approach is further limited by the CSF–brain barrier, composed of barriers to diffusion into the brain parenchyma. The greatest utility of this delivery methodology has been in cases where high drug concentrations in the CSF and/or the immediately adjacent parenchyma are desired, such as in the treatment of carcinomatous meningitis or for spinal anesthesia/analgesia (169).

Intrathecal and intracerebral drug administration differs fundamentally from systemic drug administration in terms of pharmacokinetic characteristics determining brain tissue concentration, where the available dose reaching the target organ is 100%. But the concentration distribution is not even among different brain tissue and result in steep concentration gradient (170,171). Intrathecal drug delivery has been used in the intracerebroventricular administration of glycopeptide and aminoglycoside antibiotics in meningitis, intraventricular treatment of meningeal metastasis, intrathecal injection of baclofen for treatment of spasticity and the infusion of opioids for severe chronic pain.

Implantable Devices

Polymer based drug delivery to CNS and biodegradable polymer implants are evolving approaches for treating brain diseases (172,173). Depots, wafers and microspheres are the mostly commonly used implants, aimed at bypassing the BBB and delivering CNS-active drugs inside the brain.

A depot is an implanted formulation that serves as drug reservoir to release the drugs in a sustained and controlled manner. A depot formulation of liposomal cytarabine (DepoCyte®) has proven to be useful as intrathecal treatment of neoplastic and lymphomatous meningitis. DepoCyte has been used to treat patients diagnosed with acute leukemia with CNS involvement, and side effects were mild and manageable in all patients (174). As discussed later in this chapter, a mutivesicular liposome formulation provides sustained epidural delivery of morphine to give prolonged analgesia (175).

In addition to depots, wafers can be implanted that deliver the drugs to the brain interstitium, and provide therapeutic levels of medication to its intracranial target site in a controlled manner. Gliadel® is a white, dime-sized wafer made up of a biocompatible polymer that contains the cancer chemotherapeutic drug, carmustine (BCNU). It provides localized delivery of chemotherapy directly to the site of the tumor and is an FDA approved brain cancer treatment. After a neurosurgeon removes a high-grade malignant glioma, up to eight wafers can be implanted in the cavity where the tumor resided. Once implanted, Gliadel slowly dissolves, releasing high concentrations of BCNU at the tumor site, targeting microscopic tumor cells that sometimes remain after surgery. The specificity of Gliadel minimizes drug exposure to other areas of the body (176,177).

CNS drugs can also be loaded into plolymeric biodegradable microspheres to achieve local and sustained drug delivery to the brain. Microparticles can also be easily implanted by stereotaxy in discrete, precise and functional areas of the brain without damaging the surrounding tissue. This type of implantation avoids the inconvenient insertion of large implants by open surgery and can be repeated if necessary (178). Although the implant devices provide local drug delivery, the drugs still can possibly reach the nearby sites in brain (179). Mathematical models have been used to describe the drug release from the implant formulations using the basic principles of diffusion, convection, binding and elimination (180).

Besides of depot and polymeric implants mentioned above, several other techniques have been developed for local interstitial CNS drug delivery to achieve high intracranial drug concentrations. Implantable pumps have been used to provide sustained drug infusion through intraventricular or intrathecal route. The pump can be refilled by sub-cutaneous injection after implantation. The rate of delivery can be controlled by a computer or microchip. Different pumps have been developed that rely on various technologies to provide drug delivery. For example, the Infusaid™ pump relies on the vapor pressure of compressed Freon to deliver drug solution at a constant rate. The MiniMed PIMS™ system uses a solenoid pumping mechanism, and the Medtronic SynchroMed™ system delivers drugs via a peristaltic delivery.

Biological tissue that secretes a desired therapeutic agent can also be implanted into the brain to achieve interstitial drug delivery. This approach has been most extensively studied in the treatment of Parkinson disease (46). Transplanted tissue often does not survive well in vivo, due to a lack of neovascular innervation. An alternative extension of this method is to use gene therapy to develop optimized biological tissue for interstitial drug delivery. This approach has been studied in the treatment of brain tumors (181). Besides survival upon implantation, repulsion of foreign tissues by the brain is a major concern of this approach. However, the survival of tissue grafts may be improved by advancements in techniques for culturing distinct cell types.

UTILIZATION OF ADVANCED FORMULATION DESIGN TO OPTIMIZE THERAPEUTIC OUTCOMES IN CNS DISEASES

Disorders of the brain and CNS are a leading cause of disability, accounting for more hospitalizations and long-term care than all other diseases combined (182). As discussed previously in this chapter, transport across the BBB is the major obstacle to effective drug delivery to the brain. Accordingly, the BBB is regarded as the bottleneck in development of neurotherapeutics (124). Extensive research over the years has identified various strategies to enhance brain uptake of medications. These approaches were detailed in the previous section, and a number of examples were provided.

Formulation scientists have successfully devised new drug formulations and delivery systems to improve therapeutic outcomes for patients suffering with brain and CNS disorders (Table 3). As our understanding of the physiology of the brain continues to increase, new strategies are being investigated to enhance drug delivery to the brain. A summary of recent developments in drug therapy for specific CNS disorders is provided below. For certain diseases (e.g., ADHD, depression), the impact of novel formulations on therapeutic outcomes are discussed. For those conditions that are largely refractory to drug therapy (HIV, brain tumors), strategies that are proposed to address therapeutic shortcomings associated with these diseases.

Attention-Deficit/Hyperactivity Disorder

Attention-deficit/hyperactivity disorder (ADHD) is a common neurobehavioral disorder that is often recognized in early childhood, and can persist into adulthood. The principal characteristics of ADHD are inattention, hyperactivity, and impulsivity. In the United States, between 3% and 6% of school-aged children meet accepted diagnostic criteria for ADHD (183). Over 60% of children with a diagnosis of ADHD continue to exhibit ADHD symptoms through their adult life (184).

Methylphenidate (MPH), a psychostimulant medication, has been used for the last 50 years to treat children with ADHD, and is the most widely prescribed medication in child psychiatry (185). Immediate release formulations of MPH are short-acting and patients are typically dosed 3–4 times per day to achieve and maintain adequate ADHD symptom control. Consequently, children with ADHD require one or more doses during school hours. Beyond issues of storage and handling, adherence to a multiple dose regimen is problematic in this patient population.

To overcome the challenges of short-acting MPH with multiple-daily-dose regimens, long-acting, stimulant formulations were introduced decades ago for use in treating ADHD. However, these first-generation formulations sustained release MPH formulations were not well accepted in clinical practice due to slower onset of action, variable response, and overall reduced efficacy compared to IR formulations (186). It was hypothesized that the therapeutic failure of initial SR MPH formulations was caused by the inability of a zero-order drug release design to maintain efficacy throughout the day.

Recently, a number of second generation formulations have been developed to overcome the shortcomings of earlier SR products and optimize therapeutic outcomes for ADHD patients. The technology of these

Table 3 Advanced Formulation Strategies: Examples of Drug Products Used to Treat CNS Diseases and General Approaches to Enhance CNS Drug Delivery

CNS Disease	Product(s)	Description
Attention deficit hyperactivity disorder (ADHD)	Concerta® Metadate CD® Adderall XR®	Extended-release formulations of methylphenidate (MPH) providing once-a-day therapy
	Daytrana®	Transdermal delivery system of MPH approved for use in children
Parkinson's disease	Neupro®	Transdermal delivery system of rotigitine for patients in early stages of disease
Pain	Actiq® Bema™ Fentanyl	Oral formulations designed to provide buccal delivery of fentanyl for patients with breakthough pain
	DepoDur®	Extended release liposome formulation of morphine sulfate. Epidermal administration after surgery provides pain relief for up to 48 h
Epilepsy	Cerebyx®	Injectable formulation of fosphenytoin (a phosphate ester prodrug of phenytoin) produces fewer adverse reactions that other phenytoin salts
	Diastat®	Diazepam rectal gel used provides at home treatment of breakthrough seizures
CNS Tumors	Depocyte®	Depot liposomal formulation of cytarabine used as an intrathecal treatment of neoplastic and lymphomatous meningitis
	Gliadel®	Wafer formulation containing carmustine that provides localized delivery of chemotherapy directly to tumor site. Wafers are implanted are removal of malignant glioma, releasing high concentrations of drug targeting microscopic tumor cells that sometimes remain after surgery

Abbreviations: MPH, methylphenidate.

new "once daily" ER formulations is based on the strategy of delivering an initial burst dose of MPH in addition to providing an extended release of drug. The burst dose provides an initial bolus delivery of MPH that is expected to achieve peak plasma concentrations and have rapid onset of efficacy within 2 h of dosing. This is followed by an extended, controlled delivery of MPH that is designed to achieve higher plasma concentrations during later times and maintain efficacy for an extended period of time (187). While these new MPH products share the same therapeutic goal, they utilize various formulation technologies, as describe below.

Concerta® as is a once-daily MPH medication that provides effective treatment for ADHD patients throughout the day (188). The formulation is based on the ALZA's OROS® Tri-Layer Technology (Fig. 4). Drug is delivered via a dual process of aqueous dissolution of the drug overcoat and osmotic delivery of the core drug. When the tablet is ingested, 22% of the dose is released by dissolution of the drug overcoat layer. Upon dissolution of the overcoat, an osmotic gradient is establishing across the rate controlling membrane, providing controlled drug delivery for approximately 10 h (189,190). Clinical studies have demonstrated comparable plasma exposure for 12 h following single dose administration of Concerta compared with IR MPH administered three times daily (Fig. 5) (185).

Like Concerta, Metadate CD® is a once daily formulation of MPH. The product formulation is based on Diffucaps® technology (Fig. 6), which provides a modified release profile of MPH for optimal therapeutic results (191). The biphasic drug release profile is created by layering MPH onto sugar sphere, followed by a rate-controlling, polymer membrane, yielding beads less than 1 mm in diameter. Metadate CD is a hard gelatin capsule containing MPH in both rapid release and continuous release beads such

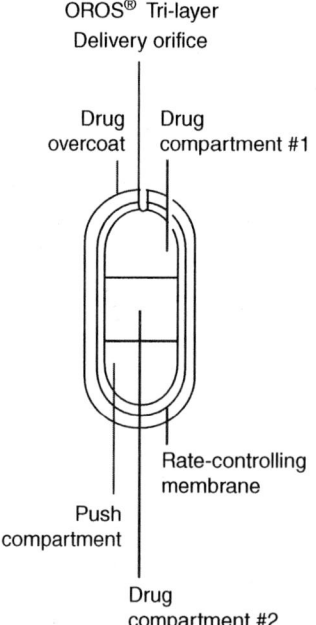

Figure 4 OROS®-Trilayer Technology. Drug is delivered via a dual process of aqueous dissolution of the drug overcoat and osmotic delivery of the core drug. Upon dissolution of the overcoat, an osmotic gradient is establishing across the rate controlling membrane, providing controlled drug delivery for approximately 10 h. *Source*: From Ref. 189.

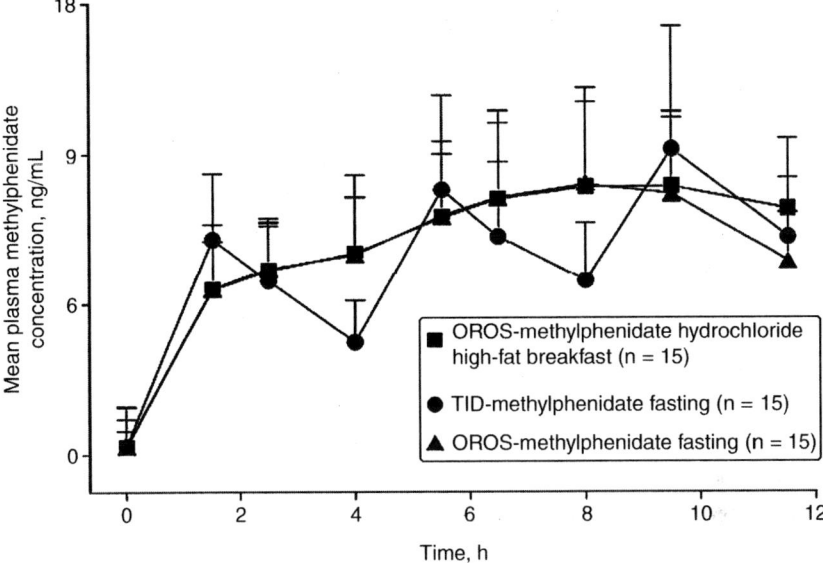

Figure 5 The Diffucaps™ system for drug delivery. Drug release profiles are created by layering active drug onto a neutral core such as sugar spheres, crystals or granules followed by a rate-controlling, functional membrane. By incorporating beads of differing drug release profiles into hard gelatin capsules, combination release profiles can be achieved. *Source*: Reprinted from Ref. 185 with permission.

that 30% of the dose is rapidly released and 70% of the dose is continuously released. With this biphasic delivery, Metadate CD capsules provide school-day-long control of symptoms associated with ADHD with a single morning dose (192).

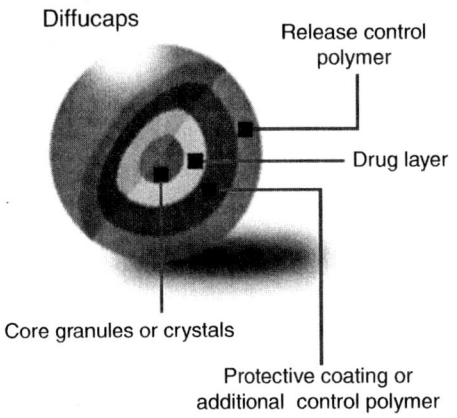

Figure 6 Pharmacokinetic profiles comparing an OROS–methylphenidate formulation administered with a high-fat breakfast and without (fasting) food, and TID (3 times daily)—immediate release methylphenidate administered in the fasting state. *Source*: Reprinted with permission from Eurand.

Adderall XR® is yet another extended release formulation for MPH. Unlike most MPH products that contain the hydrochloride salt of MPH (including those described above), Adderall XR contain equal amounts of four amphetamine salts: dextroamphetamine sulfate, dextroamphetamine saccharate, racemic amphetamine aspartate monohydrate, and racemic amphetamine sulfate. These salts provide a 3:1 ratio of d-amphetamine to l-amphetamine. The once-daily capsule formulation two types of drug-containing beads designed to give a double-pulsed delivery of amphetamine. Immediate-release beads release the first half of the dose upon ingestion, and the delayed-release beads begin to release the second half of the dose 4 hours later (193,194).

A methylphenidate transdermal patch (Daytrana®) has been approved by the US FDA for use in children aged 6–12 years. This delivery system permits sustained absorption of MPH through the skin and into the bloodstream. After morning application to the hip, the formulation is designed to MPH continuously throughout the day; the patch is removed after 9 h and its effects continue for 3 h thereafter (195). In well controlled trials in children with ADHD, patients administered MTS patches showed significantly greater improvements in their ADHD symptoms than placebo recipients (196).

ADHD was once considered to be a condition that children outgrew before reaching their teens. Current data indicate that up to 80% of children with ADHD continue to experience symptoms into their teen years, and up to 65% of children with ADHD have symptoms well into adulthood. A single morning dose of second generation MPH SR products provide symptomatic coverage in children throughout the day. However, adolescents and adults with ADHD may require symptom control that extends beyond the 12 h provided by existing formulations.

To better meet adult-patient needs for prolonged ADHD symptom control with once-daily dosing, an enhanced extended-release amphetamine formulation has been designed to provide coverage for up to 16 h (194). The formulation is composed of a single morning capsule containing three types of beads. The first two type of beads are similar to those found is Adderall XR, and provide a double-pulsed delivery of mixed amphetamine salts. The third bead provides an additional extended-release dose of amphetamine during the later part of the day. Clinical studies have demonstrated similar a dose-augmentation strategy consisting of Adderall XR supplemented by administration of IR MPH 8 h later.

Parkinson's Disease

Parkinson's disease (PD) is a chronic, progressive disorder of the CNS that belongs to a group of conditions called motor system disorders. Approximately 4 million patients worldwide suffer from this condition, which results from a lack of dopamine, a chemical messenger responsible for transmitting signals within the brain. Loss of dopamine causes neurons to fire out of control, leaving patients unable to direct or control their movement in a normal manner (197).

As described previously, a number of approaches have been studied in an effort to improve PD therapy. While there is no cure for PD, medications are available to provide symptomatic treatment. Levodopa, the most commonly administered therapy, is an example of a BBB targeting strategy (189). Following oral administration, levodopa absorbed into the bloodstream, and ultimately is taken up by the brain, where it is converted into dopamine. However, despite it widespread use in patients suffering from PD, long-term therapy with levodopa is associated with the development of dyskinesias and variable patient response, also known as being "on" or "off" (198).

Delayed onset of drug activity following oral dosing of levodopa is thought to results from slowed gastric emptying and poor solubility of levodopa in the GI tract. Researchers have adopted a prodrug strategy to increase the rate of levodopa systemic absorption. Etilevodopa (ethyl ester) and melevodopa (ethylester) are highly soluble prodrugs of levodopa (199,200). Compared with standard levodopa, the prodrugs are more readily dissolved in the stomach. The prodrugs pass unchanged into the small intestine, where they are rapidly hydrolyzed and absorbed into the circulation as levodopa. Clinical studies demonstrated that levodopa peak plasma levels were significantly increased following treatment with etilevodopa tablets compared with levodopa tablets, resulting in improved patient response (199). Improved clinical efficacy (defined as shorter latency to "on" and "on" duration) has been demonstrated for melevodopa compared to standard therapy (200).

Issues related to variable efficacy ("on" and "off") and side effects of PD medications are due in part to the fluctuating plasma profile of medication following oral administration. Based on studies suggesting that continuous dopaminergic stimulation may prevent or delay the onset of dyskineasia (abnormal movements) in PD patients, a transdermal formulation of rotigotine (Neupro®), a D3/D2/D1 dopamine agonist, has been developed (201). Pharmacokinetic data in humans have shown that the system provides steady-state plasma levels of rotigotine throughout the 24-h patch application. Results of clinical trials indicate transdermal ritigotine system is an effective monotherapy for patients in the early stages of PD (202,203). Rotigotine transdermal application may also be useful in patients with advanced disease as a supplement to levodopa therapy. By allowing patients to be effectively treated with a lower dose of levodopa, toxic effects of the drug can be avoided (204).

Pain

Pain is a major healthcare problem worldwide. According to the International Association for the Study of Pain, chronic pain afflicts one in five adults (205). Chronic pain is defined as pain that persists or recurs for more than 3 months. The financial cost of pain is comparable to cancer and cardiovascular disorders. Accordingly, drug therapy for pain relief is an ever expanding market, generating annual sales of over \$24 billion worldwide.

Chronic pain is a major challenge for patients with terminal disease such as cancer patients. These patients typically experience two types of pain: persistent pain and breakthrough pain. Persistent cancer pain is characterized as continuous pain present for long periods of time. Breakthrough pain is caused either by the cancer itself or the cancer treatment and is connected to activities such as walking. Severe chronic pain is typically treated with potent opioid analgesics such as morphine and fentanyl. Drug formulations are on the market that provide continuous around-the-clock opioid administration for an extended period of time. Among these are a number of sustained release formulations for morphine (oral administration) and transdermal fentanyl patches (described previously). For years, these products have been providing palliative pain relief for terminal cancer patients.

To complement existing products for chronic pain sufferers, novel formulations are being developed that provide rapid delivery of opioid analgesics. These formulations are designed to deliver a drug across mucous membranes for time-critical conditions such as breakthrough cancer pain, or in emergency situations where intravenous drug delivery is unavailable.

Oral transmucosal fentanyl citrate (ACTIQ®) is a solid formulation (lozenge) of fentanyl citrate specifically developed and approved for control of breakthrough pain in cancer patients. The formulation is a solid drug matrix that is designed to be dissolved slowly in the mouth, facilitating transmucosal (buccal) absorption (206,207). Buccal drug delivery avoids presystemic metabolism of fentanyl, resulting in enhanced bioavailability compared to oral dosing. Actiq® has a rapid onset of action and a short duration of effect, and clinical studies have demonstrated its an effective alternative over intravenous opioids to rapidly titrate analgesia in cancer patients experiencing severe pain (208,209).

In addition to Actiq, an alterative fentanyl delivery system is being developed and is currently in clinical trials. The device is based on BEMA™ drug delivery technology, and consists of a dissolvable, dime-sized polymer disc for application to mucosal (inner lining of cheek) membranes. The disc is designed to deliver a rapid, reliable dose of fentanyl across the buccal membrane. Upon completion of drug delivery, the disc is designed to disintegrate in the mouth leaving no drug residue (210,211).

Besides chronic pain, acute pain is a major problem after surgery and trauma. Liposome technology has generated interest for the development of sustained release delivery systems of anesthetics and analgesics (212). For example, lidocaine loaded liposome formulations are the recommended treatment of cutaneous analgesia in the pediatric patients undergoing venipuncture (213,214). Liposomes also represent a novel approach for providing post-operative analgesia in orthopedic surgery. Encapsulation of morphine into multivesicular liposomes offers a novel approach to sustained-release drug delivery upon epidural administration.

DepoDur® is an extended-release liposome injection of morphine sulfate based on DepoFoam™ technology. The delivery system consists of microscopic, spherical particles (Fig. 7). The particles are tens of microns in diameter and have large trapped volume, thereby affording delivery of large quantities of drugs in the encapsulated form in a small volume of injection. Since the formulation consists of synthetic analogs of naturally occurring lipids, DepoFoam is biocompatible and biodegradable (215,216). Clinical studies have shown that epidermal administration of the formulation provides up to 48 h of pain relief following hip surgery (217).

Epilepsy

Epilepsy is a neurological condition that makes people susceptible to seizures. It is a common CNS disorder that affects 2.7 million people in the United States, with 181,000 new cases diagnosed each year (218). Epilepsy

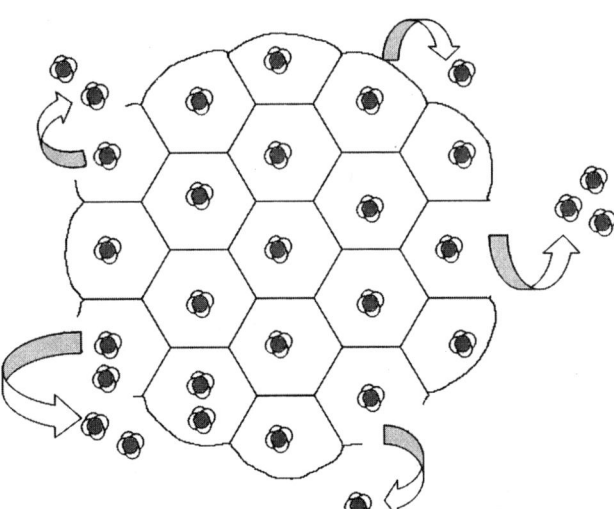

Figure 7 The DepoFoam™ drug delivery system. Drug is encapsulated in microscopic, spherical particles composed of hundreds of non-concentric aqueous chambers that are separated by bilayer lipid membranes. Drug release from a DepoFoam formulation can be controlled by modification of the lipid components, aqueous excipients and manufacturing process. The DepoFoam system provides either local site or systemic delivery. *Source*: From Ref. 216.

accounts for annual healthcare costs of $12.5 billion. Epilepsy is typically treated with medications to prevent and/or control seizures. Accordingly, there are a number of products on the market that can be administered orally (including controlled release formulations) and parenterally (including fosphenytoin, described previously).

While medications are generally effective in the majority of patients, breakthrough seizures can occur. The term refers to seizure activity experienced by a patient with epilepsy, on a stable regimen of anti-epileptic drugs. Breakthrough seizures are intermittent and periodic episodes of markedly increased seizure activity. Treatment guidelines indicate acute seizure treatment within 5–10 min to avoid an increased risk of life-threatening complications (218). For patients, this typically means emergency treatment in a medical center. However, a rapid acting anti-convulsant formulation has been developed that offers at-home treatment for select patients who experience bouts of increased seizure activity.

Diazepam rectal gel (DIASTAT® AcuDial™) is a novel formulation approved by the US FDA to control bouts of increased seizure activity for refractory patients (both adults and children) with epilepsy on stable regimens of antiepilepsy medications. The formulation is packaged in pre-filled syringes that are designed to deliver diazepam gel rectally in patients. The formulation is stable at room temperature, making it a convenient delivery system (219,220).

Human Immunodeficiency Virus

HIV is the virus that causes acquired immune deficiency syndrome (AIDS). HIV kills or damages cells in the body's immune system, gradually destroying the body's ability to fight infection and certain cancers. An estimated 1,039,000 to 1,185,000 persons in the United States were living with HIV/AIDS at the end of 2003 (221).

Antiretroviral drugs are medications for the treatment of infection by retroviruses, primarily HIV and the most common class of agents in the treatment of HIV infection. Since the introduction of zidovudine (AZT) in 1987, a relatively large number of compounds have been developed for the treatment of HIV-induced AIDS. Antiretroviral drugs act at different stages of the HIV life cycle and are classified based on their structure and site of action. The four classifications are as follows: (1) fusion or entry inhibitors, (2) reverse transcriptase inhibitors, (3) integrase inhibitors, and (4) protease inhibitors (PIs). Reverse transcriptase inhibitors are sub-classified as nucleoside reverse transcriptase inhibitors, non-nucleoside reverse transcriptase inhibitors, and nucleotide reverse transcriptase inhibitors. Standard treatment of HIV infection involves a combination of several antiretroviral drugs and is known as highly active anti-retroviral therapy (HAART). Combination HIV therapy, where multiple medications with anti-HIV activity are combined in one fixed dose, are widely used for treatment of HIV and AIDS in order to avoid increasing expression of resistance to the drugs when they are administered alone. Because of HAART therapy, the medial survival time of HIV/AIDS patients now exceeds 10 years.

Despite the progress made in the treatment of HIV infection, the virus can become sequestered certain regions of the body such as the CNS. While HAART therapy can eradicate the systemic viral load in HIV patients undergoing treatment, discontinuation of therapy is associated with a high incidence of relapse of re-infection from these reservoir sites and the potential for resistance development. Furthermore, HIV infection causes devastating structural damage to the brain, and as the disease progresses viral load in the brain ultimately leads to dementia (222). Current HAART medications are unable to achieve sufficient drug levels in the CNS due to poor permeability across the BBB. These drugs are substrates for BBB efflux transporters, thereby limiting access of these medications to the brain. Thus, delivering therapeutically relevant levels of antiviral medications to the CNS is of paramount importance for treating AIDS patients.

To address this deficiency and improve therapeutic outcomes in HIV patients, investigators are looking towards nanotechnology-based drug delivery systems. The potential impact of nanoparticulate drug carriers on HIV therapy are described in recent reviews (223,224). These formulations are designed to facilitate complete eradication of viral load from the CNS and other reservoir sites. Multilayer nanosystems (formulations containing particles 500 nm) are capable of encapsulating a variety of compounds, improving oral bioavailability, and promoting efficient delivery and residence in tissues to allow for drug internalization. These lipophilic systems can be modified in terms of size, surface charge, and surface composition to control drug release and target specific regions of the body. The application of nanosystems to deliver HIV medications to the CNS was discussed previously in this chapter.

Nanocarrier-based drug delivery systems have been shown to improved antiviral drug delivery to intracellular reservoirs (lymphocytes, macrophages) as well as anatomical areas such as the lymph nodes (223). For CNS delivery, nanoparticulate formulations could be designed to combine HAART medications with efflux transporter inhibitors to enhance BBB uptake. This is an area of interest for formulation scientists, where the ultimate goal is to achieve complete eradication of HIV in infected patients.

CNS Tumors

CNS tumors are a worldwide health problem. Data indicate that 176,000 new cases of primary CNS tumors (tumors originating in the CNS) were diagnosed globally in 2000, with an estimated mortality of 128,000 (225). This data does not include the number of patients who die each year from brain metastases associated with non-CNS cancer (secondary brain tumors).

The World Health Organization classifies primary brain tumors into nine categories, are based on the types of cells in which the tumors originate. Gliomas are primary brain tumors that are made up of glial cells, the cells that provide important structural support for the nerve cells in the brain. Infiltrative astrocytoma and glioblastoma multiforme account for nearly 85% of all brain tumors, with the remainder spread among the other seven types (226).

Treatment of CNS tumors typically involves surgery to excise the tumor, followed by radiation therapy. Despite these aggressive approaches, the prognosis for brain tumor patients is generally poor. This is due, in part, to the limited effectiveness of chemotherapy in treating CNS tumors. Anticancer agents must overcome the various barriers to drug delivery discussed earlier in this chapter. As a result, few compounds are able to achieve cytotoxic concentrations in the CNS using conventional therapeutic approaches.

One strategy to enhance CNS penetration is BBB disruption (an invasive method), but this approach has not improved therapeutic outcomes. Regional administration of chemotherapeutic agents has shown promise in treating CNS tumors. Likewise, novel drug delivery systems such Depocyte® and Gliadel wafers (described above) can be implanted release drug at the tumor site. Since many anticancer agents are substrates for BCRP and P-gp, combination therapy with an efflux transport modulator may enhance CNS permeation of these compounds (227).

Overall, the use of anticancer agents in primary and secondary CNS is generally ineffective in prolonging overall patient survival. Effective BBB drug delivery strategies must be developed in order to improve therapeutic outcomes in this patient population. Among approaches presently being studied, bypassing drug efflux transporters at the BBB is a promising novel chemotherapeutic strategy (132).

CONCLUSIONS: PERSPECTIVES AND FUTURE DIRECTIONS

Pharmacotherapy of brain disorders is a formidable challenge. Only a handful of CNS diseases (ADHD, depression, pain, epilepsy) are effectively treated with drug therapy. The majority of CNS disorders including cancer, stroke, autism, neuron-HIV, and Alzheimer's disease are refractory to drug therapy. The blame for the bottleneck in CNS drug development does not fall on the brain discovery sciences, as research in molecular neuroscience has led to the identification of new therapeutic molecules to treat the brain (124). Unfortunately, research aimed at providing answers to the BBB problem is lagging behind these advances. In other words, we know what the potential CNS therapies are, but we are presently unable to deliver them.

The BBB poses physiological, metabolic, and biochemical obstacles to drug delivery. However, the advanced formulation and delivery strategies outlined in this chapter can be utilized to overcome these barriers. Both the pharmaceutical industry and academia must make an accelerated effort to develop CNS drug delivery programs based on a thorough molecular understanding of the BBB. In doing so, efficient drug delivery to target sites in the CNS can be achieved. With these BBB delivery solutions in place, drug therapies could be developed that would advance therapeutic outcomes for most CNS diseases. Otherwise, many CNS disorders will remain untreatable.

REFERENCES

1. Begley DJ. Understanding and circumventing the blood-brain barrier. Acta Paediatr Suppl 2003; 92(443):83–91.
2. Misra A, Ganesh S, Shahiwala A, Shah SP. Drug delivery to the central nervous system: a review. J Pharm Pharm Sci 2003; 6(2):252–73.
3. Begley DJ. The blood-brain barrier: principles for targeting peptides and drugs to the central nervous system. J Pharm Pharmacol 1996; 48:136–46.
4. Schlossauer B, Steuer H. Comparative anatomy, physiology and in vitro models of the blood-brain and blood-retina barrier. Curr Med Chem 2002; 2:175–86.
5. Crone C, The blood-brain barrier: a modified tight epithelium. In: Suckling AJ, Rumsby MG, Bradbury MWB (eds), The Blood–Brain Barrier in Health and Disease. Chichester: Ellis Harwood; 1986:17–40.
6. Brightman M. Ultrastructure of brain endothelium, in Bradbury MWB (ed) Physiology and pharmacology of the blood-brain barrier. Handbook of experimental pharmacology 103, Berlin: Springer-Verlag; 1992:1–22.
7. Lo EH, Singhal AB, Torchilin VP, Abbott NJ. Drug delivery to damaged brain. Brain Res Rev 2001; 38:140–8.
8. Davson H, Segal MB. Physiology of the CSF and blood-brain barriers. Florida: CRC Press; 1995.
9. Brownless J, Williams CH. Peptidases, peptides and the mammalian blood-brain barrier. J Neurochem 1993; 60:1089–96.
10. Witt KA, Gillespie TJ, Huber JD, Egleton RD, Davis TP. Peptide drug modifications to enhance bioavailability and blood-brain barrier permeability, Peptides 2001; 22:2329–43.
11. De Boer AG, Breimer DDJ. The blood-brain barrier: clinical implications for drug delivery to the brain. R Coll Physicians Lond 1994; 28(6):502–6.
12. Nabeshima S, Reese TS, Landis DM, Brightman MW. Junctions in the meninges and marginal glia. J Comp Neurol 1975; 164(2):127–69.
13. Pardridge WM. Recent advances in blood brain–barrier transport. Ann Rev Pharmacol Toxicol 1988; 28:25–39.
14. Cornford EM, Braun LD, Oldendorf WH, Hill MA. Comparison of lipid-mediated blood-brain barrier penetrability in neonates and adults. Am J Physiol 1982; 243:C161–8.

15. Cole SPC, Bhardwaj G, Gerlach JH, et al. Over expression of a transporter gene in a multidrug-resistant human lung cancer cell line. Science 1992; 258: 1650–4.

16. Taylor EM, The impact of efflux transportersin the brain on the development of drugs for CNS disorders. Clin Pharmacokinet 2002; 41:81–92.

17. Loscher W, Potschka H. Role of drug efflux transporters in the brain for drug disposition and treatment of brain diseases. Prog Neurobiol 2005; 76(1): 22–76.

18. Loscher W, Potschka H. Blood-brain barrier active efflux transporters: ATP-binding cassette gene family. NeuroRx 2005; 2(1):86–98.

19. Gupta SP, QSAR studies on drugs acting at the central nervous system. Chem Rev 1989; 89:1765–800.

20. Hansch C, Leo A, Hoekman D. Exploring QSAR. Hydrophobic, Electronic, and Steric Constants. Washington: American Chemical Society; 1995.

21. van de Waterbeemd H, Smith DA, Beaumont K, Walker DK. Property-based design: optimization of drug absorption and pharmacokinetics. J Med Chem 2001; 44:1313–33.

22. Lin JH, Lu AY. Role of pharmacokinetics and metabolism in drug discovery and development. Pharmacol Rev 1997; 49:403–49.

23. Lewis DFV, Dickins M. Substrate SARs in human P450s. Drug Discov Today 2002; 7:918–25.

24. Levin VA. Relationship of octanol/water partition coefficient and molecular weight to rat brain capillary permeability. J Med Chem 1980; 23:682–4.

25. van de Waterbeemd H, Smith DA, Jones BC. Lipophilicity in PK design: methyl, ethyl, futile. J Comput Aided Mol Des 2001; 15(3):273–86.

26. Cornford EM, Oldendorf WH. Epilepsy and the blood-brain barrier. Adv Neurol 1986; 44:787–812.

27. Oldendorf WH. Measurement of brain uptake of radiolabeled substances using a tritiated water internal standard. Brain Res 1970; 24:1629–39.

28. Kakee A, Terasaki T, Sugiyama Y. Brain efflux index as a novel method of solute analyzing efflux transport at the blood-brain barrier. J Pharmacol Exp Ther 1996; 277:1550–9.

29. Begley DJ, Squires LK, Zlokovic BV, et al. Permeability of the blood-brain barrier to the immunosuppressive cyclic peptide cyclosporine A. J Neurochem 1998; 55:1222–30.

30. Jaehde U, Langemeijer MWE, de Boer AG, Breimer DD. Cerebrospinal fluid transport and deposition of the quinolones ciprofloxacin and pefloxacin. J Pharmacol Exp Ther 1992; 263(3):1140–6.

31. de lange ECM, de Boer AG, Breimer DD. Monitoring in-vivo BBB drug transport: CSF sampling, the unit impulse response method and, with special reference, intracerebral microdialysis. STP Pharm Sci 1997; 7(1):17–28.

32. Kakee A, Terasaki T, Sugiyama Y. Brain efflux index as a novel method of solute analyzing efflux transport at the blood-brain barrier. J Pharmacol Exp Ther 1996; 277:1550–9.

33. Shen DD, Artru AA, Adkison KK. Principles and applicability of CSF sampling for the assessment of CNS drug delivery and pharmacodynamics. Adv Drug Deliv Rev 2004; 56(12):1825–57.

34. Terasaki T, Deguchi Y, Kasama Y, Pardridge WM, Tsuji A. Determination of in vivo steady-state unbound drug concentration in the brain interstitial fluid by microdialysis. Int J Pharm 1992; 81:143–52.

35. Deguchi Y, Morimoto K. Application of an in vivo brain microdialysis technique to studies of drug transport across the blood-brain barrier. Curr Drug Metab 2001; 2(4):411–23.

36. Menacherry S, Hubert W, Justice JB. In vivo calibration of microdialysis probes for exogenous compounds. Anal Chem 1992; 64:577–83.

37. Hurst RD, Fritz IB. Properties of an immortalized vascular endothelial/glioma cell coculture model of the blood-brain barrier. J Cell Physiol 1996; 167:81–8.

38. Dehauck MP, Dehouck B, Schluep C, Lemaire M, Cecchelli R. Drug transport to the brain: comparison between in-vitro and in-vivo models of the blood-brain barrier. Eur J Pharm Sci 1995; 3:357–65.

39. Lohmann C, Huwel S, Galla HJ. Predicting blood-brain barrier permeability of drugs: evaluation of different in vitro assays. J Drug Target 2002; 10(4):263–76.

40. Tewes B, Franke H, Hellwig S, Hoheisel D, Decker S, Griesche D, Tilling T, Wegener J, Galla HJ. Preparation of endothelial cells in primary cultures obtained from the brains of 6-month old pigs. In: de Boer AG, Sutanto W (eds), Tramsport across the blood-brain barrier: In-vitro and in-vivo techniques. Amsterdam: Academic publishers, 1997:91–7.

41. Gath U, Hakvoort A, Wegener J, Decker S, Galla HJ. Porcine choroids plexus cells in culture: Expression of polarized phenotype, maintenance of barrier properties and apical secretion of CSF-components. Eur J Cell Biol 1997; 74:68–78.

42. Elsinga PH, Hendrikse NH, Bart J, van Waarde A, Vaalburg W. Positron emission tomography studies on binding of central nervous system drugs and p-glycoprotein function in the rodent brain. Mol Imaging Biol 2005; 7(1): 37–44.

43. Elsinga PH, Hendrikse NH, Bart J, Vaalburg W, van Waarde A. PET Studies on P-glycoprotein function in the blood-brain barrier: how it affects uptake and binding of drugs within the CNS. Curr Pharm Des 2004; 10(13):1493–503.

44. Abbott NJ, Chugani DC, Zaharchuk G, Rosen BR, Lo EH. Delivery of imaging agents into brain Adv Drug Deliv Rev 1999; 5;37(1–3):253–77.

45. Waterhouse RN. Determination of lipophilicity and its use as a predictor of blood-brain barrier penetration of molecular imaging agents. Mol Imaging Biol 2003; 5(6):376–89.

46. Madrid Y, Langer LF, Brem H, Langer R. New directions in the delivery of drugs and other substances to the central nervous system. Adv Pharmacol 1991; 22:299–324.

47. Crivori P, Cruciani G, Carrupt P. Predicting blood-brain barrier permeation from three-dimensional molecular structure. J Med Chem 2000; 39:4750–5.

48. Sweet CS. New centrally acting antihypertensive drugs related to methyldopa and clonidine. Hypertension. 1984; 6(5 Pt 2):II51–6.

49. Prokai-Tatrai K, Prokai L. Modifying peptide properties by prodrug design for enhanced transport into the CNS. Prog Drug Res 2003; 61:155–88.

50. Batrakova EV, Vinogradov SV, Robinson SM, Niehoff ML, Banks WA, Kabanov AV. Polypeptide point modifications with fatty acid and amphiphilic

block copolymers for enhanced brain delivery. Bioconjug Chem 2005; 16(4): 793–802.

51. Hau VS, Huber JD, Campos CR, Lipkowski AW, Misicka A, Davis TP. Effect of guanidino modification and proline substitution on the in vitro stability and blood-brain barrier permeability of endomorphin II. J Pharm Sci 2002; 91(10):2140–9.

52. Lambert DM. Rationale and applications of lipids as prodrug carriers. Eur J Pharm Sci 2000; 11:S15–27.

53. Han HK, Amidon GL. Targeted prodrug design to optimize drug delivery. AAPS PharmSci 2000; 2:E6.

54. Perioli L, Ambrogi V, Bernardini C, Grandolini G, Ricci M, Giovagnoli S, Rossi C. Potential prodrugs of non-steroidal anti-inflammatory agents for targeted drug delivery to the CNS. Eur J Med Chem 2004; 39(8):715–27.

55. Deguchi Y, Hayashi H, Fujii S, Naito T, Yokoyama Y, Yamada S, Kimura RJ. Improved brain delivery of a nonsteroidal anti-inflammatory drug with a synthetic glyceride ester: a preliminary attempt at a CNS drug delivery system for the therapy of Alzheimer's disease. Drug Target 2000; 8(6):371–81.

56. Kitagawa K, Mizobuchi N, Hama T, Hibi T, Konishi R, Futaki S. Synthesis and antinociceptive activity of [D-Ala2] Leu-enkephalin derivatives conjugated with the adamantane moiety. Chem Pharm Bull (Tokyo) 1997; 45(11): 1782–87.

57. Tsuzuki N, Hama T, Kawada M, et al. Adamantane as a brain-directed drug carrier for poorly absorbed drug 2. AZT derivatives conjugated with the 1-adamantane moiety. J Pharm Sci 1994; 83(4):481–4.

58. Anthonypillai C, Gibbs JE, Thomas SA. The distribution of the anti-HIV drug, tenofovir (PMPA), into the brain, CSF and choroid plexuses. Cerebrospinal Fluid Res 2006; 3;3:1.

59. Browne TR, Kugler AR, Eldon MA. Pharmacology and pharmacokinetics of fosphenytoin. Neurology 1996; 46(6 Suppl 1):S3–7.

60. Knapp LE, Kugler AR. Clinical experience with fosphenytoin in adults: pharmacokinetics, safety, and efficacy. J Child Neurol 1998; 13(Suppl 1):S15–8.

61. Berg SL, Brueckner C, Nuchtern JG, Dauser R, McGuffey L, Blaney SM. Plasma and cerebrospinal fluid pharmacokinetics of nelarabine in nonhuman primates. Cancer Chemother Pharmacol 2006; Sep 5(Epub ahead of print).

62. Prokai-Tatrai K, Teixido M, Nguyen V, Zharikova AD, Prokai L. A pyridinium-substituted analog of the TRH-like tripeptide pGlu-Glu-Pro-NH2 and its prodrugs as central nervous system agents. J Med Chem 2005; 1(2): 141–52.

63. Venhuis BJ, Rodenhuis N, Wikstrom HV, et al. A new type of prodrug of catecholamines: an opportunity to improve the treatment of Parkinson's disease. J Med Chem 2002; 45(12):2349–51.

64. Savolainen J, Edwards JE, Morgan ME, McNamara PJ, Anderson BD. Effects of a P-glycoprotein inhibitor on brain and plasma concentrations of anti-human immunodeficiency virus drugs administered in combination in rats. Drug Metab Dispos 2002; 30(5):479–82.

65. Johnson MD, Chen J, Anderson BD. Investigation of the mechanism of enhancement of central nervous system delivery of 2′-beta-fluoro-2′,

3′-dideoxyinosine via a blood-brain barrier adenosine deaminase-activated prodrug. Drug Metab Dispos 2002; 30(2):191–8.

66. Handelmann GE, Nevins ME, Mueller LL, Arnolde SM, Cordi AA. Milacemide, a glycine prodrug, enhances performance of learning tasks in normal and amnestic rodents. Pharmacol Biochem Behav 1989; 34(4):823–8.

67. Greig NH, Genka S, Daly EM, Sweeney DJ, Rapoport SI. Physicochemical and pharmacokinetic parameters of seven lipophilic chlorambucil esters designed for brain penetration. Cancer Chemother Pharmacol 1990; 25(5): 311–9.

68. Ansede JH, Voyksner RD, Ismail MA, Boykin DW, Tidwell RR, Hall JE. In vitro metabolism of an orally active O-methyl amidoxime prodrug for the treatment of CNS trypanosomiasis. Xenobiotica 2005; 35(3):211–26.

69. Morgan ME, Chi SC, Murakami K, Mitsuya H, Anderson BD. Central nervous system targeting of 2′,3′-dideoxyinosine via adenosine deaminase-activated 6-halo-dideoxypurine prodrugs. Antimicrob Agents Chemother 1992; 36(10):2156–65.

70. Clissold SP, Heel RC. Transdermal hyoscine (Scopolamine). A preliminary review of its pharmacodynamic properties and therapeutic efficacy. Drugs 1985; 29(3):189-207.

71. Nachum Z, Shupak A, Gordon CR. Transdermal scopolamine for prevention of motion sickness: clinical pharmacokinetics and therapeutic applications. Clin Pharmacokinet 2006; 45(6):543–66.

72. Kim EJ, Jeong DU. Transdermal scopolamine alters phasic REM activity in normal young adults. Sleep 1999; 15;22(4):515–20.

73. Gourlay GK, Kowalski SR, Plummer JL, et al. The efficacy of transdermal fentanyl in the treatment of postoperative pain: a double-blind comparison of fentanyl and placebo systems. Pain 1990; 40(1):21–8.

74. Parks RW, Becker RE, Rippey RF, et al. Increased regional cerebral glucose metabolism and semantic memory performance in Alzheimer's disease: a pilot double blind transdermal nicotine positron emission tomography study. Neuropsychol Rev 1996; 6(2):61–79.

75. Schwarz JS, Weisspapir MR, Friedman DI. Enhanced transdermal delivery of diazepam by submicron emulsion (SME) creams. Pharm Res 1995; 12(5):687–92.

76. Jennings FW, Atouguia JM, and Murray M. Topical chemotherapy for experimental murine African CNS-trypanosomiasis: the successful use of the arsenical, melarsoprol, combined with the 5-nitroimidazoles, fexinidazole or MK-436. Trop Med Int Health 1996; 1(3):363–6.

77. Jennings FW, Chauviere G, Viode C, Murray M. Topical chemotherapy for experimental African trypanosomiasis with cerebral involvement: the use of melarsoprol combined with the 5-nitroimidazole, megazol. Trop Med Int Health 1996; 1(5):590–8.

78. Cuevas P, Gimenez-Gallego G, Carceller F, Cuevas B, Crespo A. Single topical application of human recombinant basic fibroblast growth factor (rbFGF) promotes neovascularization in rat cerebral cortex. Surg Neurol 1993; 39(5):380–4.

79. Huwyler J, Wu D, Pardridge WM. Brain drug delivery of small molecules using immunoliposomes. Proc Natl Acad Sci 1996; 93:14164–9.

80. Iyer M, Mishra R, Han Y. Predicting blood-brain barrier partitioning of organic molecules using membrane–interaction QSAR analysis. Pharm Res 2002; 19:1611–21.
81. Krauze MT, Forsayeth J, Park JW, Bankiewicz KS. Real-time imaging and quantification of brain delivery of liposomes. Pharm Res 2006; 23(11):2493–504.
82. Noble CO, Krauze MT, Drummond DC, Yamashita Y, Saito R, Berger MS, Kirpotin DB, Bankiewicz KS, Park JW. Novel nanoliposomal CPT-11 infused by convection-enhanced delivery in intracranial tumors: pharmacology and efficacy. Cancer Res 2006; 66(5):2801–6.
83. Saito R, Krauze MT, Noble CO, et al. Tissue affinity of the infusate affects the distribution volume during convection-enhanced delivery into rodent brains: implications for local drug delivery. J Neurosci Methods 2006; 30;154(1–2):225–32.
84. Krauze MT, Mcknight TR, Yamashita Y, et al. Real-time visualization and characterization of liposomal delivery into the monkey brain by magnetic resonance imaging. Brain Res Brain Res Protoc 2005; 16(1–3):20–6.
85. Mamot C, Nguyen JB, Pourdehnad M, et al. Extensive distribution of liposomes in rodent brains and brain tumors following convection-enhanced delivery. J Neurooncol 2004; 68(1):1–9.
86. Chu Z, Sun Y, Kuan CY, Grabowski GA, Qi X. Saposin C: neuronal effect and CNS delivery by liposomes. Ann NY Acad Sci 2005; 1053:237–46.
87. Schmidt J, Metselaar JM, Wauben MH, Toyka KV, Storm G, Gold R. Drug targeting by long-circulating liposomal glucocorticosteroids increases therapeutic efficacy in a model of multiple sclerosis. Brain 2003; 126(Pt 8):1895–904.
88. Betageri GV, Segers J, Spaulding TC, Pejaver SK, Brittain HG. Liposomal formulation of mirfentanil hydrochloride. PDA J Pharm Sci Technol 1996; 50 (4):219–22.
89. Ali A, Kolappa Pillai K, Jalees Ahmad F, Dua Y, Iqbal Khan Z, Vohora D. Comparative efficacy of liposome-entrapped amiloride and free amiloride in animal models of seizures and serum potassium in mice. Eur Neuropsychopharmacol 2007; 17(3):227–9.
90. Tiwari SB, Amiji MM. A review of nanocarrier-based CNS delivery systems. Curr Drug Deliv 2006; 3(2):219–32.
91. Couvreur P, Dubernet C, Puisieux F. Controlled drug delivery with nanoparticles: Current possidilities and future trends. Eur J Pharm Biopharm 1995; 41:2–13.
92. Muldoon LL, Tratnyek PG, Jacobs PM, Doolittle ND, Christoforidis GA, Frank JA, Lindau M, Lockman PR, Manninger SP, Qiang Y, Spence AM, Stupp SI, Zhang M, Neuwelt EA. Imaging and nanomedicine for diagnosis and therapy in the central nervous system: report of the eleventh annual Blood-Brain Barrier Disruption Consortium meeting.
93. Golden PL, Maccagnan TJ, Pardridge WM. Human blood-brain barrier leptin receptor: Binding and endocytosis in isolated human brain microvessels. J Clin Invest 1997; 99:14–8.
94. Tamai I, Sai Y, Kobayashi H, Kamata M, Wakamiya T, Tsuji A. Structure-internalization relationship for adsorptive-mediated endocytosis of basic peptides at the blood-brain barrier. J Pharmacol Exp Ther 1997; 280:410–5.
95. Kreuter J. Nanoparticulate systems for brain delivery of drugs. Adv Drug Deliv Rev 2001; 23;47(1):65–81.

96. Tamai I, Sai Y, Kobayashi H, Kamata M, Wakamiya T, Tsuji A. Structure-internalization relationship for adsorptive-mediated endocytosis of basic peptides at the blood-brain barrier. J Pharmacol Exp Ther 1997; 280: 410–5.

97. Kreuter J, Alyautdin RN, Kharkevich DA, Ivanov AA. Passage of peptides through the blood-brain barrier with colloidal polymer particles (nano-particles). Brain Res 1995; 674:171–4.

98. Pilar C, Bruno G, Helene C, Didier D, Jean A, Jene-Pierre N, Dominique G, Elias F, Jean AP, Patrick C. Long-circulating PEGylated polycyanoacrylate nanoparticles as new drug carrier for brain delivery. Pharm Res 2001; 18(8): 1157–66.

99. Lu W, Tan YZ, Hu KL, Jiang XG. Cationic albumin conjugated pegylated nanoparticle with its transcytosis ability and little toxicity against blood-brain barrier. Int J Pharm 2005; 13;295(1–2):247–60.

100. Vyas TK, Shah L, Amiji MM. Nanoparticulate drug carriers for delivery of HIV/AIDS therapy to viral reservoir sites. Expert Opin Drug Deliv 2006; 3(5): 613–28.

101. Manninger SP, Muldoon LL, Nesbit G, Murillo T, Jacobs PM, Neuwelt EA. An exploratory study of ferumoxtran-10 nanoparticles as a blood-brain barrier imaging agent targeting phagocytic cells in CNS inflammatory lesions. Am J Neuroradiol 2005; 26(9):2290–300.

102. Jendelova P, Herynek V, Urdzikova L, et al. Magnetic resonance tracking of transplanted bone marrow and embryonic stem cells labeled by iron oxide nanoparticles in rat brain and spinal cord. J Neurosci Res 2004; 76(2):232–43.

103. Cui Z, Lockman PR, Atwood CS, Hsu CH, Gupte A, Allen DD, Mumper RJ. Novel D-penicillamine carrying nanoparticles for metal chelation therapy in Alzheimer's and other CNS diseases. Eur J Pharm Biopharm 2005; 59(2):263–72.

104. Pardridge WM. Receptor-mediated peptide transport through the blood-brain barrier. Endocrine Rev 1986; 7:314–30.

105. Lockman PR, Koziara J, Roder KE, Paulson J, Abbruscato TJ, Mumper RJ, Allen DD. In vivo and in vitro assessment of baseline blood-brain barrier parameters in the presence of novel nanoparticles. Pharm Res 2003; 20(3):409–16.

106. Kreuter J, Ramge P, Petrov V, et al. Direct evidence that polysorbate-80-coated poly(butylcyanoacrylate) nanoparticles deliver drugs to the CNS via specific mechanisms requiring prior binding of drug to the nanoparticles. Pharm Res 2003; 20(3):409–16.

107. Kreuter J. Transport of drugs across the blood-brain barrier by nanoparticles, Curr Med Chem 2002; 2:241–9.

108. Bodor N, Buchwald P. Drug targeting via retrometabolic approaches. Pharmacol Ther 1997; 76:1–27.

109. Palomino E, Kessel D, Horwitz JP. A dihydropyridine carrier system for sustained delivery of 1′, 3′-dideoxynucleosides to the brain. J Med Chem, 1989; 32:622–5.

110. Porkai L. Peptide drug delivery into the central nervous system. Prog Drug Res 1998; 51:95–131.

111. Somogyi G, Nishitani S, Nomi D, Buchwald P, Prokai L, Bodor N. Targeted drug delivery to the brain via phosphonate derivatives. I: Design, synthesis,

and evaluation of an anionic chemical delivery system for testosterone. Int J Pharm 1998; 166:15–26.

112. Boder N, Prokai L, Wu W-M. A strategy for delivering peptides into the central nervous system by sequential metabolism. Science 1992; 257:1698–700.

113. Bodor N, Prokai L. Molecular packaging: peptide delivery to the central nervous system by sequential metabolism. In Taylor M, Amidon G, eds. Peptide-Based Drug Design: Controlling Transport and Metabolism. Washington: American Chemical Society; 1995:317–37.

114. Chen P, Bodor N, Wu W-M, Prokai L. Strategies to target kyotorphin analogues to the brain. J Med Chem 1998; 41:3773–81.

115.. Wu J, Yoon S-H, Wu W-M, Bodor N. Synthesis and biological evaluation of a brain targeted chemical delivery system of [Nva2]-TRH, J pharm Pharmacol 2002; 54:945–50.

116. Pop E, Prokai-Tatrai K, Scott JD, Brewster ME, Bodor N. Application of a brain-targeting chemical delivery system to 9-amino-1,2,3,4-tetrahydroacridine. Pharm Res 1990; 7(6):658–64.

117. Prokai L, Prokai-Tatrai K, Bodor N. Targeting drugs to the brain by redox chemical delivery systems. Med Res Rev 2000; 20(5):367–416.

118. Bodor N, Farag HH, Brewster ME. Site-specific sustained release of drugs to the brain. Science 1981; 214(18):1370–2.

119. Mizrachi Y, Rubinstein A, Harish Z, Biegon A, Anderson WR, Brewster ME. Improved brain delivery and in vitro activity of zidovudine through the use of a redox chemical delivery system. AIDS 1995; 9(2):153–8.

120. Pop E, Wu WM, Bodor N. Chemical delivery systems for some penicillinase-resistant semisynthetic penicillins. J Med Chem 1989; 32(8):1789–95.

121. Sun H, Dai H, Shaik H, Elmquist WF. Drug efflux transporters in the CNS. Adv Drug Del Rev 2003; 55:83–105.

122. Taylor EM. The impact of efflux transporters in the brain on the development of drugs for CNS disorders. Clin Pharmacokinet 2002; 41:81–92.

123. Tamai I, Tsuji A. Drug delivery through the blood brain barrier. Adv Drug Del Rev 1996; 19:401–24.

124. Pardridge WM. The blood-brain barrier: Bottleneck in brain drug development. NeuroRx 2005; 2:301–14.

125. Pardridge WM, Triguero D, Buciak J, Yang J. Evaluation of cationized rat albumin as a potential blood-brain barrier drug transport vector. J Pharmacol Exp Ther 1990; 255:893–9.

126. Pardridge WM, Kumagai AK, Eisenberg JB. Chimeric peptides as a vehicle for peptide pharmaceutical delivery through the blood-brain barrier. Biochem Biophys Res Commun 1987; 146:307–15.

127. Saito Y, Buciak J, Yang J, Pardridge WM. Vector-mediated delivery of 125I–labeled-β-amyloid peptide A-β 1-40 through the blood-brain barrier and binding to alzheimer disease amyloid of the A-1-40/vector complex. Proc Natl Acad Sci USA 1995; 92:10227–31.

128. Pardridge WM. Vector-mediated drug delivery to the brain. Adv Drug Deliv Rev 1999; 5;36(2–3):299–321.

129. Pardridge WM. Drug and gene targeting to the brain with molecular Trojan horses. Nat Rev Drug Discov 2002; 1:131–9.

130. Pardridge WM. Peptide Drug Delivery to the Brain. New York: Raven Press 1991; 1–357.
131. Bickel U, Yoshikawa T, Landaw EM, Faull KF, Pardridge WM. Pharmacologic effects in-vivo in brain by vector- mediated peptide drug delivery. Proc Natl Acad Sci USA 1993; 90:2618–22.
132. Breedveld P, Beijen JH, Schellens JHM. Use of p-glycoprotein and BCRP inhibitors to improve oral bioavailability and CNS penetration of anticancer drugs. Trends Pharm Sci 2006; 27:17–24.
133. DeGraw RT, Anderson BD. Enhanced oral bioavailability of 2'- beta-fluoro-2',3'-dideoxyadenosine (F-ddA) through local inhibition of intestinal adenosine deaminase. Pharm Res 2001; 18(9):1270–6.
134. Singhal D, Morgan ME, Anderson BD. Role of brain tissue localized purine metabolizing enzymes in the central nervous system delivery of anti-HIV agents 2'-beta-fluoro-2',3'-dideoxyinosine and 2'-beta-fluoro-2',3'-dideoxyadenosine in rats. Pharm Res 1997; 14(6):786–92.
135. Menon D, Karyekar CS, Fasano A, Lu R, Eddington ND. Enhancement of brain distribution of anticancer agents using DeltaG, the 12 kDa active fragment of ZOT. Int J Pharm 2005; 306(1–2):122–31.
136. Tang G, Chiocca A. Gene transfer and delivery in central nervous system disease. Neurosurg Focus 1997; 3(3):e2.
137. de Lima MC, da Cruz MT, Cardoso AL, Simoes S, de Almeida LP. Liposomal and viral vectors for gene therapy of the central nervous system. Curr Drug Targets CNS Neurol Disord 2005; (4):453–65.
138. Berry M, Barrett L, Seymour L, Baird A, Logan A. Gene therapy for central nervous system repair. Curr Opin Mol Ther 2001; 3(4):338–49.
139. da Cruz MT, Cardoso AL, de Almeida LP, Simoes S, de Lima MC. f-lipoplex-mediated NGF gene transfer to the CNS: neuronal protection and recovery in an excitotoxic model of brain injury. Gene Ther 2005; 12(16):1242–52.
140. da Cruz MT, Simoes S, de Lima MC. Improving lipoplex-mediated gene transfer into C6 glioma cells and primary neurons. Exp Neurol 2004; 187(1):65–75.
141. Murray KD, Etheridge CJ, Shah SI, Matthews DA, Russell W, Gurling HM, Miller AD. Enhanced cationic liposome-mediated transfection using the DNA-binding peptide mu (mu) from the adenovirus core. Gene Ther 2001; 8(6):453–60.
142. Ma H, Zhu J, Maronski M, Kotzbauer PT, Lee VM, Dichter MA, Diamond SL. Non-classical nuclear localization signal peptides for high efficiency lipofection of primary neurons and neuronal cell lines. Neuroscience 2002; 112(1):1–5.
143. Wu BY, Liu RY, So KL, Yu AC. Multi-lipofection efficiently transfected genes into astrocytes in primary culture. J Neurosci Methods 2000; 30;102(2):133–41.
144. Zou LL, Huang L, Hayes RL, et al. Liposome-mediated NGF gene transfection following neuronal injury: potential therapeutic applications. Gene Ther 1999; 6(6):994–1005.
145. Kofler P, Wiesenhofer B, Rehrl C, Baier G, Stockhammer G, Humpel C. Liposome-mediated gene transfer into established CNS cell lines, primary glial cells, and in vivo. Cell Transplant 1998; 7(2):175–85.

146. Yang K, Clifton GL, Hayes RL. Gene therapy for central nervous system injury: the use of cationic liposomes: an invited review. J Neurotrauma 1997; 14(5):281–97.

147. Godfray J, Estibeiro P. The potential of antisense as a CNS therapeutic. Expert Opin Ther Targets 2003; 7(3):363–76.

148. Maguire-Zeiss KA, Bowers WJ, Federoff HJ. HSV vector-mediated gene delivery to the central nervous system. Curr Opin Mol Ther 2001; 3(5):482–90.

149. Cowsill C, Southgate TD, Morrissey G, et al. Central nervous system toxicity of two adenoviral vectors encoding variants of the herpes simplex virus type 1 thymidine kinase: reduced cytotoxicity of a truncated HSV1-TK. Gene Ther 2000; 7(8):679–85.

150. Rosolen A, Frascella E, di Francesco C, et al. In vitro and in vivo antitumor effects of retrovirus-mediated herpes simplex thymidine kinase gene-transfer in human medulloblastoma. Gene Ther 1998; 5(1):113–20.

151. Amada K, Moriguchi A, Morishita R, et al. Efficient oligonucleotide delivery using the HVJ-liposome method in the central nervous system. Am J Physiol 1996; 271(5 Pt 2):R1212–20.

152. Sagihara Y, Saitoh Y, Kaneda Y, Kohmura E, Yoshimine T. Widespread gene transfection into the central nervous system of primates. Gene Ther 2000; 7(9): 759–63.

153. MacKay JA, Deen DF, Szoka FC Jr. Distribution in brain of liposomes after convection enhanced delivery; modulation by particle charge, particle diameter, and presence of steric coating. Brain Res 2005; 1035(2):139–53.

154. Corso TD, Torres G, Goulah C, et al. Assessment of viral and non-viral gene transfer into adult rat brains using HSV-1, calcium phosphate and PEI-based methods. Folia Morphol (Warsz) 2005, 64(3):130–44.

155. Kandimalla KK, Donovan MD. Transport of hydroxyzine and triprolidine across bovine olfactory mucosa: role of passive diffusion in the direct nose-to-brain uptake of small molecules. Int J Pharm 2005; 302(1–2):133–44.

156. Wang H, Hussain AA, Wedlund PJ. Nipecotic acid: systemic availability and brain delivery after nasal administration of nipecotic acid and n-butyl nipecotate to rats. Pharm Res 2005; 22(4):556–62.

157. Kao HD, Traboulsi A, Itoh S, Dittert L, Hussain A. Enhancement of the systemic and CNS specific delivery of L-dopa by the nasal administration of its water soluble prodrugs. Pharm Res 2000; 17(8):978–84.

158. Fortin D, Desjardins A, Benko A, Niyonsega T, Boudrias M. Enhanced chemotherapy delivery by intraarterial infusion and blood-brain barrier disruption in malignant brain tumors: the Sherbrooke experience. Cancer 2005; 103(12):2606–15.

159. Neuwelt EA, Dahlborg SA. Blood–brain barrier disruption in the treatment of brain tumors: clinical implications. In Neuwelt EA, ed. Implications of the Blood Brain Barrier and its Manipulation: Clinical Aspects, Vol. 2. New York: Plenum Press, 1989:195–262.

160. Kraemer DF, Fortin D, Doolittle ND, Neuwelt EA. Association of total dose intensity of chemotherapy in primary central nervous system lymphoma (human non-acquired immunodeficiency syndrome) and survival. Neurosurgery 2001; 48(5):1033–40.

161. Doolittle ND, Petrillo A, Bell S, Cummings P, Eriksen S. Blood–brain barrier disruption for the treatment of malignant brain tumors: The National Program. J Neurosci Nurs 1998; (2):81–90.

162. Neuwelt EA, Goldman DL, Dahlborg SA, et al. Primary CNS lymphoma treated with osmotic blood-brain barrier disruption: prolonged survival and preservation of cognitive function. J Clin Oncol 1991; (9):1580–90.

163. Dahlborg SA, Petrillo A, Crossen JR, et al. The potential for complete and durable response in nonglial primary brain tumors in children and young adults with enhanced chemotherapy delivery. Cancer J Sci Am 1998; 4(2): 110–24.

164. Heimberger K, Samec P, Podreka I, et al. Reversible opening of the blood-brain barrier in the chemotherapy of malignant gliomas. Wien Klin Wochenschr 1987; 99(11):385–8.

165. Neuwelt EA, Wiliams PC, Mickey BE, Frenkel EP, Henner WD. Therapeutic dilemma of disseminated CNS minoma and the potential of increased platinum-based chemotherapy delivery with osmotic blood-brain barrier disruption. Pediatr Neurosurg 1994; 21:16–22.

166. Doran SE, Ren XD, Betz AL, Pagel MA, Neuwelt EA, Roessler BJ, Davidson BL. Gene expression from recombinant viral vectors in the central nervous system after blood-brain barrier disruption. Neurosurgery 1995; 36:965–70.

167. Miller G. Breaking down barriers. Science 2002; 297:1116–8.

168. Chio CC, Baba T, Black KL. Selective blood–tumor pro-barrier disruption by leukotrienes. J Neurosurg 1992; 77:407–10.

169. Harbaugh RE, Saunders RL, Reeder RF. Use of implantable pumps for central nervous system drug infusions to treat neurological disease. Neurosurgery 1988; 23(6):693–8.

170. Blasberg RG, Patlak C, Fenstermacher JD. Intrathecal chemotherapy: Brain tissue profiles after ventriculocisternal perfusion. J Pharmacol Exp Ther 1975; 195:73–83.

171. Huang TY, Arita N, Hayakawa T, Ushio Y. ACNU, MTX and 5-FU penetration of rat brain tissue and tumors, J Neurooncol 1999; 45:9–17.

172. Brem H, Langer R. Polymer based drug delivery to the brain. Sci Med 1996; 3(4):1–11.

173. Brem H, Gabikian P. Biodegradable polymer implants to treat brain tumors. J Control Release 2001; 74:63–7.

174. Sancho JM, Ribera JM, Romero MJ, Martin-Reina V, Giraldo P, Ruiz E. Compassionate use of intrathecal depot liposomal cytarabine as treatment of central nervous system involvement in acute leukemia: report of 6 cases. Haematologica 2006; 91(3):ECR02.

175. Yaksh TL, Provencher JC, Rathbun ML, et al. Safety assessment of encapsulated morphine delivered epidurally in a sustained-release multi-vesicular liposome preparation in dogs. Drug Deliv 2000; 7(1):7–36.

176. Fung LK, Ewend MG, Sills A, et al. Pharmacokinetics of interstitial delivery of carmustine, 4-hydroperoxycyclophosphamide and paclitaxel from a biodegradable polymer implant in the monkey brain. Cancer Res 1998; 58(4):672–84.

177. http://www.mgipharma.com/wt/page/gliadel(accessed February 2007).

178. Dang W, Colvin OM, Brem H, Saltzman WM. Covalent coupling of methotrexate dextran enhances the penetration of cytotoxicity into a tissue like matrix. Cancer Res 1994, 54:1729–35.

179. Krewson CE, Klarman ML, Saltzman WM. Distribution of nerve growth factor following direct delivery to brain interstitium. Brain Res 1995; 680:196–206.

180. Hanes J, Batycky RP, Langer R, Edwards DA. A theoretical model of erosion and macromolecular drug release from biodegrading microspheres. J Pharm Sci 1997; 86(12):1464–77.

181. Lal B, Indurti RR, Couraud PO, Goldstein GW, Laterra J. Endothelial cell implantation and survival within experimental gliomas. Proc Natl Acad Sci USA 1994; 91(21):9695–9.

182. Ambikanandan M, Ganesh S, Aliasgar S, Shrenik P, Shah J. Drug delivery to the central nervous system: a review. J Pharm Pharmaceut Sci 2003; 6(2): 252–73.

183. Goldman LS, Genel M, Bezman RJ, Slanetz PJ. Diagnosis and treatment of attention-deficit/hyperactivity disorder in children and adolescents. Council on Scientific Affairs, American Medical Association. J Am Med Assoc 1998; 279:1100–7.

184. Rowland AS, Lesesne CA, Abramowitz AJ. The epidemiology of attention-deficit/hyperactivity disorder (ADHD): A public health view. Merit Retard Dev Disabil Res Rev 2002; 8:162–70.

185. Swanson J, Gupta S, Lam A, et al. Development of a new once-a day formaultion of methylphenidate for the treatment of attention–deficit/hyper-activity disorder. Arch Gen Psychiatry 2003; 60:204–11.

186. Pelham WE Jr, Sturges J, Hoza J, et al. Sustained release and standard methylphenidate effects on cognitive and social behavior in children with ADHD. Pediatrics 1987; 80:491–501.

187. Swanson JM, Wigal SB, Wigal T, et al. A comparison of once-daily extended-release methylphenidate formulations in children with attention-deficit/hyperactivity disorder in the laboratory school (the Comacs Study). Pediatrics 2004; 113:206–16.

188. Concerta® [package insert], McNeil Pediatrics, a Division of McNeil-PPC, Inc. Ft. Washington PAE; 2000–2007. http://www.concerta.net/html/concerta/hcp/prescribing_info.jsp? (accessed February 2007).

189. http://www.alza.com/alza/oros(accessed February 2007).

190. Concerta® Clinical Pharmacology and Biopharmaceutics Review, June 2000. http://www.fda.gov/cder/foi/nda/2000/21–121_Concerta_biopharmr.pdf.

191. http://www.eurand.com/page.php?id = 22 (accessed February 2007).

192. Anderson VR, Keating GM. Spotlight on methylphenidate controlled-delivery capsules (equasymtrade markxl, metadate cdtrade mark) in the treatment of children and adolescents with attention-deficit hyperactivity disorder. CNS Drugs 2007; 21(2):173–5.

193. Adderal® [package insert], Shire US, Wayne PA; 2006. http://www.adderallxr.com/assets/pdf/prescribing_information.pdf.

194. Ermer JC, Shojaei A, Pennick M, Anderson CS, Silverberg A, Youcha SH. Triple-bead mixed amphetamine salts (spd465), a novel, enhanced extended-release amphetamine formulation and a dose-augmentation strategy of mixed

amphetamine salts extended release plus mixed amphetamine salts immediate release: a comparison of bioavailability. Curr Med Res Opin 2007.

195. Anderson VR, Scott LJ. Methylphenidate transdermal system: In attention-deficit hyperactivity disorder in children. Drugs 2006; 66(8):1117–26.

196. Daytrana® website, http://www.daytrana.com/Consumers/Default.aspx (accessed February 2007).

197. Michael J. Fox Foundation website, http://www.michaeljfox.org/website (accessed February 2007).

198. Poewe W, Wenning G. Levodopa in Parkinson's disease: mechanisms of action and pathophysiology of late failure. In: Jankovic J, Tolosa E, eds. Parkinson's Disease and Movement Disorders. 4th edn. Philadelphia, PA: Lippincott Williams & Wilkins, 2002:104–15.

199. Djaldetti R, Giladi N, Hassin-Baer S, Shabtai H, Melamed E. Pharmacokinetics of etilevodopa compared to levodopa in patients with Parkinson's disease: an open-label, randomized, crossover study. Clin Neuropharmacol 2003; 26(6):322–6.

200. Stocchi F, Fabbri L, Vecsei L, Krygowska-Wajs A, Monici Preti PA, Ruggieri SA. Clinical efficacy of a single afternoon dose of effervescent levodopa-carbidopa preparation (CHF 1512) in fluctuating Parkinson disease. Clin Neuropharmacol 2007, 30(1):18–24.

201. http://www.aderis.com/products/rotigotine.htm (accessed February 2007).

202. The Parkinson Study Group. A controlled trial of rotigotine monotherapy in early Parkinson's disease. Arch Neurol 2003; 60(12):1721–8.

203. Watts RL, Jankovic J, Waters C, Rajput A, Boroojerdi B, Rao J. Randomized, blind, controlled trial of transdermal rotigotine in early Parkinson disease. Neurology 2007; 68:272–6.

204. Zareba G. Rotigotine: A novel dopamine agonist for the transdermal treatment of Parkinson's disease. Drugs Today (Barc). 2006; 42(1):21–8.

205. http://www.painreliefhumanright.com/pdf/04a_global_day_fact_sheet.pdf (accessed February 2007).

206. Actiq® [package insert], Cephalon, Inc. Salt Lake City, UT 84116, 2006. http://www.actiq.com/pdf/package_insert.pdf.

207. Mystakidou K, Katsouda E, Parpa E, Vlahos L, Tsiatas ML. Oral transmucosal fentanyl citrate: overview of pharmacological and clinical characteristics. Drug Deliv 2006; 13(4):269–76.

208. Burton AW, Driver LC, Mendoza TR, Syed G. Oral transmucosal fentanyl citrate in the outpatient management of severe cancer pain crises: a retrospective case series. Clin J Pain 2004, 20(3):195–7.

209. Aronoff GM, Brennan MJ, Pritchard DD, Ginsberg B. Evidence-based oral transmucosal fentanyl citrate (OTFC) dosing guidelines. Pain Med 2005; 6(4): 305–14.

210. Biodelivery Sciences International website, http://www.bdsinternational.com/pipeline/index.html (accessed February 2007).

211. http://www.in-pharmatechnologist.com/news/ng.asp?n = 67543-biodelivery-sciences-bema-actiq (accessed February 2007).

212. Couvreur P, Vauthier C. Nanotechnology: intelligent design to treat complex disease. Pharm Res 2006; 23:1417–50.

213. Eidelman A, Weiss JM, Lau J, Carr DB. Topical anesthetics for dermal instrumentation: a systematic review of randomized, controlled trials. Ann Emerg Med 2005; 46(4):343–51.

214. Migdal M, Chudzynska-Pomianowska E, Vause E, Henry E, Lazar J. Rapid, needle-free delivery of lidocaine for reducing the pain of venipuncture among pediatric subjects. Pediatrics 2005; 115:393–8.

215. Mantripragada S. A lipid based depot (DepoFoam technology) for sustained release drug delivery. Prog Lipid Res 2002; 41(5):392–406.

216. DepoDur® website, http://www.depodur.com/depodurmain.html (accessed February 2007).

217. Viscusi ER, Kopacz D, Hartrick C, Martin G, Manvelian G. Single-dose extended-release epidural morphine for pain following hip arthroplasty. Am J Ther 2006; 13(5):423–31.

218. Epilepsy Foundation website, http://www.epilepsyfoundation.org/

219. Pellock JM. Safety of Diastat, a rectal gel formulation of diazepam for acute seizure treatment. Drug Saf 2004; 27(6):383–92.

220. Diastat® website, http://www.diastat.com/HTML-INF/index.htm (accessed February 2007).

221. Glynn M, Rhodes P, Estimated HIV prevalence in the United States at the end of 2003. National HIV Prevention Conference. Atlanta: Abstract 595 2005.

222. Nath A, Sacktor N. Influence of highly active antiretroviral therapy on persistence of HIV in the central nervous system. Curr Opin Neurol 2006; 19(4):358–61.

223. Amiji MM, Vyas TK, Shah LK. Role of nanotechnology in HIV/AIDS treatment: potential to overcome the viral reservoir challenge. Discov Med 2006; 6(34):157–62.

224. Vyas TK, Shah L, Amiji MM. Nanoparticulate drug carriers for delivery of HIV/AIDS therapy to viral reservoir sites. Expert Opin Drug Deliv 2006; 3(5): 613–28.

225. Parkin DM, Bray F, Ferlay J, Pisani P. Estimating the world cancer burden: Globocan 2000. Int J Cancer 2001; 15; 94(2):153–6.

226. http://www.oncologychannel.com/braincancer/types.shtml (accessed February 2007).

227. Motl S, Zhuang Y, Waters CM, Stewart CF. Pharmacokinetic considerations in the treatment of brain tumors. Clin Pharmacokinet 2006; 45:871–903.

11

Cardiovascular Pharmacology

Tariq Javed
Research, Enterprise, and Regional Affairs, University of Greenwich, London, U.K.
Ghassan F. Shattat
Health and Life Sciences Department, Coventry University, Coventry, U.K.

INTRODUCTION

The prevalence of cardiovascular diseases will undoubtedly increase as the mean age of the populations throughout the world increases. Heart disease is generally complex and multifactorial. Despite considerable advances in therapy, the understanding of the underlying mechanisms are still rather poor. Efforts are being directed towards the development of new drugs through the modification of the physicochemical properties of drug molecules, and the design and synthesis of new drugs with particular emphasis on drug delivery systems, pharmaceutical materials research, controlled drug release and targeting of drugs to the site of action to enhance the therapeutic effects.

The perspectives of the user enables the pharmacist and clinicians in understanding the factors leading to the suboptimal use of drugs in society. Identifying, solving and preventing drug-related problems in order to improve health and quality of life through pharmaceutical care in the community are therefore also important.

Equally, the optimal utilization of pharmacist and clinician know-how will bring tremendous benefits to the patients from the viewpoint of social pharmacy in the promotion of safe and efficient use of drugs.

In this chapter, key research developments in cardiovascular medicine covering the area of congestive heart failure, cardiac arrhythmias, hypertension, ischemic heart disease and a variety of related disorders of the cardiovascular system are critically reviewed. The physical, chemical, and biological aspects of structure–activity relationships and pharmaceutics is also briefly presented to highlight the importance of drug delivery systems and the importance of clinical pharmacy from a pharmacological and pharmaceutical perspective.

THE ACTION OF THE HEART AND BLOOD CIRCULATION

The heart acts as a muscular pump and beats about 70 times every minute. The heart pumps blood around the body at a rate of 5 L/min, that is, about 180 million gallons during a lifetime's pumping. The blood pressure is dependent upon the heart's constant pumping of both blood and the size of all its blood vessels through which the blood passes.

The pumping of the heart or the heartbeat is caused by alternating contractions and relaxations of the myocardium. These contractions are stimulated by electrical impulses from the sinoatrial or SA node located in the muscle of the right auricle. An impulse from the SA node causes the two auricles to contract, forcing blood into the ventricles. Contraction of the ventricles is controlled by impulses from the atrioventricular or AV node located at the junction of the two auricles (Fig. 1).

Following contraction, the ventricles relax and allow the pressure to fall. Blood again flows into the auricles and an impulse from the SA node repeats the cycle again; this process is called the cardiac cycle. The period of relaxation is called a diastole and the period of contraction is called a systole. Diastole is the longer of the two phases so that the heart can rest between the contractions. Hence, systolic pressure and diastolic pressure are observed. The rate of SA impulse is subconsciously regulated by the autonomic nervous system. In this way the heartbeat is either accelerated or slowed down in response to physical activity and other factors.

The force of blood pumped through the arteries exerts its pressure on the arterial walls. The body monitors blood pressure by means of receptors present in the main arteries of the heart and then controls it through changes in heartbeat and the flow of blood. Vasomotor effectors increase or decrease the diameter of the blood vessels and also regulate blood pressure. Physical exertion and digestion of food places a heavy demand on the heart's blood output, which can increase as much as ten-fold to meet the special needs.

There are two main systems which operate to maintain arterial blood pressure. These are the sympathetic nervous system and the renin–angiontensin–aldosterone system. One of the processes that tends to worsen hypertension is renal glomerular sclerosis. The control system fails because narrowing of the renal vessels upsets the normal relationship between renal blood flow and arterial pressure.

Blood pressure is governed by the following *five* elements:

1. cardiac output
2. volume of blood
3. viscosity of blood

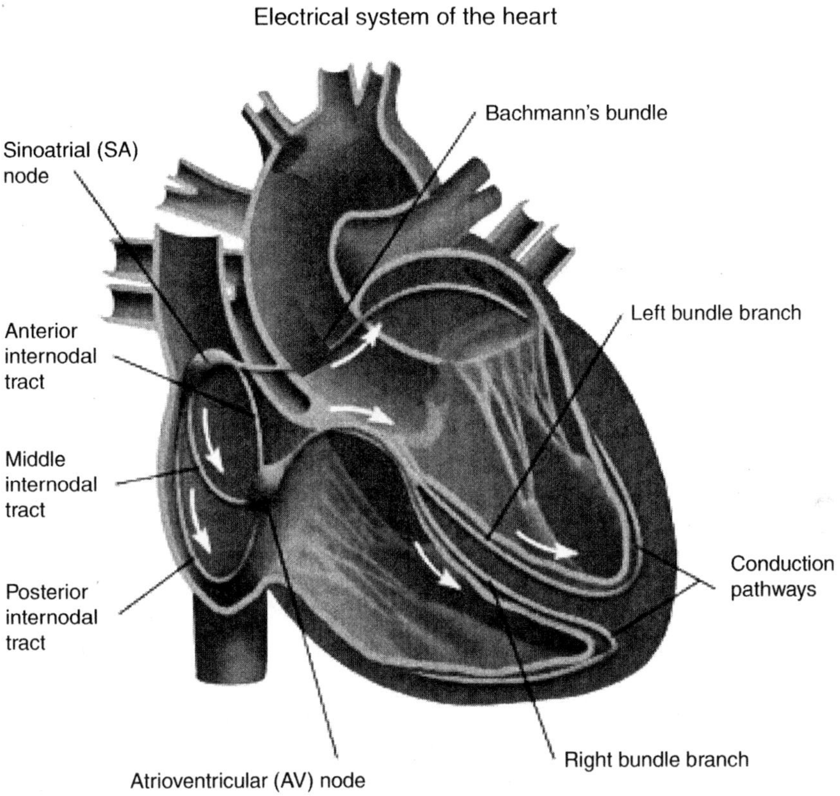

Electrical system of the heart

Figure 1   The conducting system of the heart.

4.   resistance of the arterioles
5.   elasticity of the arterial walls.

Abnormality of any of these elements can cause blood pressure to be too high or too low.

CARDIAC ARRHYTHMIAS AND CLINICAL THERAPY

Arrhythmia

Despite considerable progress in management over the recent years, coronary artery disease (CAD) remains the leading cause of death in the industrialized world. It is estimated that CAD is responsible for causing 152,000 deaths per year in the United Kingdom. 400,000 deaths per year in the United State of America, and one in eight deaths worldwide. Many of these deaths are attributed to the development of cardiac arrhythmias, especially ventricular fibrillation during periods of myocardial ischemia or infarction. Myocardial ischemia is characterized by ionic and biochemical alterations, creating an unstable electrical substrate capable of initiating and sustaining arrhythmias, and infarction creates areas of electrical inactivity (1–3) and blocks conduction, which also promotes arrhythmogenesis.

Arrhythmias are dysfunctions which cause abnormalities (4–6) in impulse formation (automaticity), conduction (conductivity), or both in the myocardium. Most disturbance of cardiac rhythm arises when a small group of cardiac cells competes with or replaces the normal pacemaker. Any circumscribed region that acts in such a way is called an ectopic focus, meaning that it is not located in the usual pacemaker region, the sinoatrial node. The result can be abnormal ectopic beats and rhythms. The three basic mechanisms or causes responsible for ectopic beats (6–9) and rhythms (arrhythmias) are enhanced automaticity, reentry, and triggered activity.

Abnormal Automaticity

The sinoatrial node shows the fastest rate of phase 4 depolarization and therefore, exhibits a higher rate of discharge than that occurring in other pacemaker cells exhibiting automaticity. The SA node normally sets the pace of contraction for the myocardium, and latent pacemakers (5–7,9,10) are depolarized by impulses coming from the SA node. So, if cardiac sites other than the SA node show enhanced automaticity, they may generate competing stimuli, and arrhythmias may arise.

Common causes of enhanced automaticity are an increase in chatecholamines, digitalis toxicity, hypoxia, potassium imbalance (5,10), and myocardial ischemia or infarction.

Abnormalities in Impulse Conduction

A common mechanism by which abnormal conduction causes arrhythmias is called reentry excitation, which occurs if a unidirectional block is caused by myocardial injury or prolonged refractory period. With normal conduction, the electrical impulse moves freely down the conduction system until it reaches recently excited tissue that is refractory to stimulation. This causes the impulse to be extinguished. The SA node then recovers and fires spontaneously, and the conduction process starts all over again. Reentry excitation means that an impulse continues to reenter an area of the heart rather than coming to an end. For this to occur, the impulse must encounter an obstacle in the normal conducting pathway. The obstacle is usually an area which allows conduction (5,7,9,10) in only one direction and causes a circular movement of the impulse. Re-entry is the most common cause of arrhythmias (5) and this can occur in any level of the cardiac conduction system.

Triggered Activity

Triggered activity is an abnormal conduction of latent pacemaker and myocardial cells in which the cells may depolarize more than once following stimulation by a single electrical impulse. The level of membrane action potential spontaneously and rhythmically increases after depolarization until it reaches threshold potential, causing the cells to depolarize.

This phenomenon, called after depolarization, can occur immediately following repolarization early, in phase 3 (early after depolarization), or late in phase 4 (delayed after depolarization).

Triggered activity can result in atrial or ventricular ectopic beats occurring singly, in groups of two (paired or coupled beats), or in bursts (7,10) of three or more beats (paroxysms of beats or tachycardia).

Common cause of triggered activity (10) include an increase of catecholamines, digitalis toxicity, hypoxia, and myocardial ischemia.

Types of Arrhythmias

Sinus Tachycardia

Sinus tachycardia is an arrhythmia originating in the SA node (10,11) characterized by rate of over 100 beats per minute. Sinus tachycardia may caused by numerous conditions such as fever, hypotension, heart failure, thyrotoxicosis, stimulation of the sympathetic nervous system, and life style drugs such as alcohol, caffeine, and nicotine. Treatment of sinus tachycardia (6,10) should be directed towards identifying the underlying cause of the arrhythmia. The P, QRS, and T deflection are all normal, but cardiac cycle (4,10) duration (the PP interval) is altered.

Sinus Bradycardia

Sinus bradycardia is an arrhythmia originating in the SA node (10,11), characterized by a rate of less than 60 beats per minute.

Sinus bradycardia may occur with excessive vagal stimulation, deficient sympathetic tone, and sinus node dysfunction. It often occurs in healthy young adults, especially athletes and during sleep. Other conditions associated with sinus bradycardia include hypothyroidism, hypothermia, and drugs such as beta-blocker agents, amiodarone, diltiazem, and verapamil. A symptomatic sinus bradycardia does not require treatment. Acute, symptomatic sinus bradycardia can be treated with atropine or a temporary transvenous pacemaker. Chronic symptomatic sinus bradycardia (6,10) requires insertion of permanent pacemaker. The P, QRS, and T deflection (4,10) are all normal, but cardiac cycle duration (the PP interval) is altered.

Sinus Arrest and Sinoatrial Exit Block

Sinus arrest is an arrhythmia caused by episodes of failure in the automaticity of the SA node, resulting in bradycardia, asystole, or both. Sinoatrial exit block is an arrhythmia caused by a block in the conduction of the electrical impulse from the SA node to the atria, resulting in bradycardia, a systole, or both. Sinus arrest and SA exit block may be precipitated by an increase in vagal (parasympathetic) tone on the SA node, hypoxia, hyperkalemia, or excessive dose of digitalis or propranolol. SA exit block may also result from quinidine toxicity.

Premature Atrial Contractions

Is an extra atrial contraction (4,10,11) consisting of an abnormal (sometimes normal) P wave followed by a normal or abnormal QRS complex, occurring earlier than the next expected beat of the underlying rhythm.

Common cause of premature atrial contractions (PACs) include an increase in catecholamines and sympathetic tone, infections, emotion, stimulants (alcohol, caffeine, and tobacco), lack of sleep, digitalis toxicity. Hypoxia (10,11). The electrophysiological mechanism responsible for PACs is either enhanced automaticity (10) or reentry. The P wave superimpose QRS complex on PACs.

Atrial Tachycardia

Atrial tachycardia is an arrhythmia originating in an ectopic pacemaker in the atria or the site of a rapid reentry circuit in the AV node with a rate between 160 and 240 beats per minute. It includes non-paroxymal atrial tachycardia and paroxymal atrial tachycardia (PAT). PAT is often called paroxymal supraventricualr tachycardia (4,10) when the site of origin (either atria or AV junction) cannot be determined with certainty.

Atrial tachycardia commonly starts and ends abruptly, occurring in paroxysms (paroxymal atrial tachycardia). When atrial tachycardia does not start and end abruptly, it is called (nonparoxymal atrial tachycardia). By definition three or more consecutive premature atrial contractions are considered to be atrial tachycardia.

The causes of atrial tachycardia are essentially the same as those of premature atrial contractions. Non-paroxymal atrial tachycardia is most commonly caused by digitalis toxicity. The electrophysiological mechanism which is most likely to be responsible for non-paroxymal atrial tachycardia (10) is enhanced automaticity, for paroxymal atrial tachycardia, it is a reentry mechanism.

Atrial Fibrillation

Atrial fibrillation is an arrhythmia arising in multiple ectopic pacemakers or sites of rapid reentry circuits in the atria characterized by an irregular (10), often rapid ventricular response. So it is disorganized, tremor-like movements of the atria. This lack of effective atrial contraction impairs ventricular filling, decreases cardiac output (6,11) and may lead to the formation (6,10,12) of atrial thrombi. Another characteristic is a very rapid atrial rate (400–600 beats per minute). Atrial fibrillation is a major cause of morbidity in the aging population. Initially preponderant in men over 60, it ultimately becomes a disease of women (13,14) in part perhaps because of their greater longevity.

Atrial fibrillation is commonly associated with advanced rheumatic heart disease, hypertension, or coronary heart disease. The electrophysiological mechanism responsible for atrial fibrillation is either enhanced automaticity or reentry.

P waves do not appear in the ECG; they are replaced by continuous irregular fluctuations (4,11) of potential, called F waves, and QRS complexes is normal.

Atrial Flutter

Atrial flutter is an arrhythmia arising in ectopic pacemaker or sites of a rapid reentry circuits in the atria, characterized by rapid abnormal atrial flutter F waves (6,10) and a slower, regular ventricular response. Atrial flutter (6,10–12) is characterized by a rapid (270–230 atrial beats per minute).

Chronic atrial flutter is most commonly seen in middle-age or elderly people with advanced rheumatic heart disease, particularly if mitral or tricuspid valvular disease is present, and in those with coronary or hypertensive heart disease. The electrophysiological mechanism responsible for atrial flutter is either enhanced automaticity (10) or reentry. Atrial flutter is characterized (11) by two or three P waves (atrial contraction) for every QRS complex (ventricular contraction).

Premature Junctional Contractions

A premature junctional contraction (PJC) is an extra ventricular contraction that originates in an ectopic pacemaker in the AV junction, occurring before the next expected beat of the underlying rhythm. Premature junctional contractions are also called premature junctional beats or complexes.

PJCs are a result of digitalis toxicity, increase in parasympathetic tone on the SA node, an excessive dose of certain drugs (quinidine and procainamide) or sympathomimetic drugs, hypoxia and congestive heart failure. The electrophysiological mechanism responsible for premature junctional contractions is either enhanced automaticity or reentry. PJC consists of a normal or abnormal QRS complex with or without an abnormal P wave.

Junctional Escape Rhythm

Junctional escape rhythm is an arrhythmia originating in an escape pacemaker in the AV junction with a rate of 40–60 beats per minute.

Junctional escape rhythm is a normal response of the AV junction when the rate of impulse formation in the SA node become less than that of the escape pacemaker in the AV junction, or when the electrical impulse from the SA node or atria fail to reach the AV junction because of a sinus arrest, sinoatrial exit block, or third-degree AV block. Generally, when an electrical impulse fails to arrive at the AV junction within 1.0–1.5 s, the escape pacemaker in the AV junction begins to generate electrical impulses at its inherent firing rate of 40–60 beats per minute.

Non-Paroxymal Junctional Tachycardia

Nonparoxymal junctional tachycardia (accelerated junctional rhythm) is an arrhythmia originating in an ectopic pacemaker in the AV junction with a regular rhythm and a rate of 60–150 beats per minute.

Non-paroxymal junctional tachycardia is most commonly a result of digitalis toxicity. Other common causes are excessive administration of catecholamines and damage to the AV junction from myocardial infarction or rheumatic fever. The electrophysiological mechanism responsible for non-paroxymal junctional tachycardia (10) is enhanced automaticity.

Paroxymal Junctional Tachycardia

Paroxymal junctional tachycardia (PJT) is an arrhythmia originating in an ectopic pacemaker or the site of a rapid reentry circuit in the AV junction with a rate between 160 and 240 beats per minute. PJT is often called paroxymal supraventricular tachycardia (4) when the site of origin cannot be determined with certainty.

PJT may occur without apparent cause in healthy persons of any age with no apparent underlying heart disease. It may precipitate by an increase in catecholamines and sympathetic tone, stimulants or emotional stress. The electrophysiological mechanism responsible for PJT is a reentry (10).

First-Degree AV Block

First-degree AV block is an arrhythmia in which there is a constant delay in the conduction of the electrical impulse (6,10) usually through the AV node. It is characterized by abnormally prolonged PR intervals (10,11,15) that are greater than 0.20 s usually about 0.28 s and constant.

First-degree AV block occurs commonly in acute inferior myocardial infarction because of the effect of an increase in parasympathetic tone, ischemia on the AV node and inflammation of the AV bundle.

Second-Degree AV Block

Second-degree AV block is an arrhythmia in which a block of conduction of the electrical impulse occurs. The sites of block may be located above or below the His bundle.

A block below the bundle is usually more serious than one above the bundle, because the former is more likely to evolve to third-degree AV block.

It is characterized by abnormally prolonged PR interval 0.25–0.45 s; some P waves trigger QRS complexes and others do not; 2:1, 3:2 P wave/QRS complex ratios may occur. Second-degree AV block occurs commonly due to excessive vagal stimulation.

Third-Degree AV Block (Complete AV Block)

Third-degree AV block is the complete absence of conduction of the electrical impulses through the AV node, bundle of His, or bundle branches (4,6,10) characterized by independent beating of the atria and ventricles. Atrial rhythm is usually approximately 100 beats per minute and ventricular rhythm is less than 40 beats per minute.

Third-degree AV block occurs commonly due to ischemia of AV nodal fibres associated with myocardial infarction, increased vagal tone, electrolyte imbalance (10,11) and digitalis toxicity.

Premature Ventricular Contractions

Premature ventricular contraction (PVCs) is an extra ventricular contraction consisting of an ectopic pacemaker in the ventricles. It occurs earlier than the expected beat of the underlying rhythm (4,10) and is usually followed by a compensatory pause. PVCs is also defined as discrete and identifiable premature QRS (10,16) complexes.

PVCs may occur in healthy persons with apparently healthy hearts and without apparent cause. PVCs frequently occur in acute myocardial infarction; they may also be caused by an increase in sympathetic or parasympathetic tone (1,4,10) lack of sleep and with coronary thrombosis. The electrophysiological mechanism (10) responsible for PVCs is either enhanced automaticity or reentry. PVCs usually characterized by prolonged QRS complex (10,11) and inverted T wave.

Ventricular Tachycardia

Ventricular tachycardia is an arrhythmia originating in an ectopic pacemaker (6,10) on the ventricles with a rate between 110 and 250 beats per minute. It is also defined as a run of four or more consecutive ventricular (9,16) premature beats.

Ventricular tachycardia may occur in paroxysms of three or more PVCs separated by the underlying rhythm (non-sustained ventricular tachycardia (1,4,10) or paroxymal ventricular tachycardia) or persist for a long period of time (sustained ventricular tachycardia).

Ventricular tachycardia usually occurs in the presence of significant cardiac disease. Most commonly it occurs in coronary artery disease like in acute myocardial infarction. Digitalis toxicity is also a common cause (1,10) of ventricular tachycardia. The electrophysiological mechanism responsible for ventricular tachycardia (10) is usually characterized by abnormally wide and bizarre QRS complexes.

Trosade de Pointes

Trosade de pointes, a French term meaning twisting of points, is a syndrome characterized by clinically serious ventricular arrhythmia similar to fibrillation, with rapid asynchronous complexes, which is always spontaneously reversible (unlike ventricular fibrillation which is usually not reversible).

It is usually associated with a prolonged QT interval in the ECG, which may be congenital (in congenital long QT syndrome) or drug induced (by classes IA and III antiarrhythmics).

Ventricular Fibrillation

Ventricular fibrillation is an arrhythmia arising in numerous ectopic pacemakers in the ventricles, produces ineffective myocardial contraction (1,6,10) so that there is no cardiac output. It is defined as a signal for which individual QRS deflections can no longer be distinguished from one another, and for which a rate can no longer be measured.

Ventricular fibrillation is a lethal arrhythmia if it lasts for more than few minutes and it is the main mechanism (1,6,9) of sudden cardiac death.

Ventricular fibrillation usually occurs in the presence of significant cardiac disease, most commonly in coronary artery disease, myocardial ischemia, acute myocardial infarction, and in third-degree AV block. The electrophysiological mechanism responsible for ventricular fibrillation is either enhanced automaticity or reentry. The PR intervals and the QRS complexes (10,11) are absent in ventricular fibrillation.

Accelerated Idioventricular Rhythm

Accelerated idioventricular rhythm is an arrhythmia originating in an ectopic pacemaker in the ventricles with heart rate less than 120 beats per minute.

Accelerated idioventricular rhythm1 (7,10) is relatively common in acute myocardial infarction although it may also result from digitalis toxicity. The electrophysiological mechanism responsible for accelerated idioventricular rhythm is probably enhanced automaticity.

Ventricular Escape Rhythm

Ventricular escape rhythm is an arrhythmia originating in an escape pacemaker in the ventricles with a rate of less than 40 beats per minute.

Ventricular escape rhythm usually occurs as a result of sinus arrest, sinoatrial block, or third-degree AV block. The PR intervals are absent in ventricular escape rhythm, and the QRS complexes exceed 0.12 s and are irregular.

Ventricular a Systole

Ventricular a systole is the absence of all electrical activity within the ventricles, and it is one of the common causes of cardiac arrest.

Ventricular a systole usually occurs in third-degree AV block. In the dying heart, ventricular a systole is usually the final arrhythmia following ventricular tachycardia, ventricular fibrillation, and ventricular escape rhythm. The P waves may be present or absent in ventricular a systole, the PR intervals and the QRS complexes (10) are absent.

TREATMENT OF ARRHYTHMIAS

The objective of treating an arrhythmia is to provide an effective drug at an appropriate concentration that can be tolerated by the patient and is free of adverse effects. The concentration needed will depend on the patient, in particular, on their particular arrhythmia profile.

Class IA: Sodium Channel Blockers

Quinidine

Quinidine is the prototype class IA drug. At high doses, it can precipitate arrhythmias, which can lead to fatal ventricular fibrillation. Because of quinidine's toxic potential, calcium antagonists, such as verapamil, are increasingly replacing this drug in clinical use (5). In addition to class IA, quinidine depresses the contractility of the myocardium (negative inotropic effect), and reduces vagus nerve activity (9) on the heart (antimuscarinic effect). Quinidine is most commonly used (15) in the United States.

Mechanism of action: Quinidine binds to open and inactivated sodium channels and prevents sodium influx, thus slowing the rapid upstroke during phase 0. It also decreases the slope of phase 4 spontaneous depolarization.

Therapeutic uses: Quinidine is useful in the treatment of a wide variety of arrhythmias including atrial, AV junctional and ventricular tachyarrhythmias.

Pharamacokinetics: Quinidine sulfate is rapidly and almost completely absorbed (5) after oral administration. The remainder of the drug is excreted (17,18) unchanged by the kidneys.

Adverse effects: Quinidine must never be used alone to treat atrial fibrillation or flutter as its antimuscarinic action enhances AV conduction and the heart rate may accelerate. The negative inotropic action of quinidine may result in hypotension and cardiac failure. Cardiotoxic effects are exacerbated by hyperkalemia (5). Plasma digoxin concentration is raised by quinidine (displacement from tissue binding and impairment of renal excretion) (9). A potential adverse effect of quinidine (or any antiarrhythmic drug) is exacerbation of the arrhythmia.

Procainamide

Procainamide is a derivative of the local anesthetic procaine. The mechanism of action and the therapeutic uses of procainamide are similar to those of quinidine (5).

Pharmacokinetics: Procainamide is more than 75% bioavailable after oral administration. The intravenous preparation is relatively frequently used but can cause hypotension if rapidly administered. Procainamide (17,18) has a relatively short half-life of 2–3 h.

Available portion of the drug (17) is acetylated in the liver to N-acetyl procainamide (NAPA). NAPA unlike the parent drug has properties of a class III drug. NAPA is eliminated via the kidneys (5).

Adverse effects: The most common problem with procainamide is lupus-like syndrome. Sixty-five percent of patients will develop antibodies within 12 months; only 12% will show symptoms that are reversible when the drug is stopped (19). Rare central nervous system side effects include depression, hallucinations, and psychosis, but gastrointestinal intolerance is less frequent than with quinidine (17).

Disopyramide

Disopyramide is most commonly used in Europe (15). The mechanism therapeutic uses of disopyramide (5,6,17) are similar to those of quinidine.

Pharmacokinetics: About one half of the orally ingested drug is excreted unchanged by the kidneys. Approximately 30% of the drug is converted by the liver (5,17,18) to the less active mono-N dealkylated metabolite. Drug half-life is 5–8 h (6).

Adverse effects: Most of the side effects seen with disopyramide relate to its anticholinergic activity (dry mouth, blurred vision, urinary retention, and constipation) (5,15,17).

Class IB: Sodium Channel Blockers

Lidocaine

Lidocaine, a local anesthetic, is the prototype of class IB (5,6).

Mechanism of action: Lidocaine, shorten phase 3 repolarization and decrease the duration of action potential. Unlike quinidine, which suppresses arrhythmias caused by increased normal automaticity, lidocaine suppresses arrhythmia caused by abnormal automaticity.

Therapeutic uses: It is the drug of choice for treating serious ventricular arrhythmias associated with acute myocardial infarction, cardiac surgery and electrical cardioversion (5,17). Lidocaine is also the drug of choice for emergency treatment of cardiac arrhythmias (5).

Pharmacokinetics: Lidocaine is given intravenously because of extensive first-pass transformation by the liver, which precludes oral administration. The drug is dealkylated and eliminated almost entirely by the liver, so dosage adjustments are necessary in the presence of hepatic dysfunction (5,6,17,18).

Adverse effects: Central nervous system side effects predominate; drowsiness, slurred speech, agitation, convulsions, and confusion (5,17).

Mexiletine and Tocainide

Mexiletine and tocainide are oral analogues of lidocaine with similar pharmacological actions. They are used to suppress ventricular fibrillation or ventricular tachycardia. They are well absorbed from the gastrointestinal tract. Adverse effects are similar to those of lidocaine (5,6,17,19). Tocainide has pulmonary toxicity (5,6) which may lead to pulmonary fibrosis.

Phenytoin

Phenytoin, an anticonvulsant, may be used to treat arrhythmias produced by digoxin intoxication. The antiarrhythmic properties of phenytoin generally resemble those of lidocaine. Phenytoin decreases automaticity and improves conduction through the AV node. Decreased automaticity helps control arrhythmias, whereas enhanced conduction may improve cardiac function, further, because heart block may result from digoxin, quinidine, or procainamide. Phenytoin may relieve arrhythmias (6,17,20) without intensifying heart block. Phenytoin is given orally.

Class IC: Sodium Channel Blockers

These drugs are approved only for refractory ventricular arrhythmias. However, recent data have cast serious doubts on the safety of class 1C drugs.

Flecainide

Mechanism of action: Flecainide suppresses phase 0 upstroke in Purkinje and myocardial fibres. This causes marked slowing of conduction in all cardiac tissues, with a minor effect on the duration of the action potential and refractoriness.

Automaticity is reduced by an increase in the threshold potential rather than a decrease in the slope of phase 4 depolarization.

Therapeutic uses: Flecainide is useful in treating refractory ventricular arrhythmias. It is particularly useful in suppressing premature ventricular contractions. Flecainide (5,17) has a negative inotropic effect and can cause worsening of congestive heart failure. The cardiac arrhythmia suppression trial investigators concluded that antiarrhythmic therapy with flecainide (21), in post myocardial infarction patients increased the risk of sudden death although the agent was able to suppress premature ventricular complexes.

Pharmacokinetics: Flecainide is absorbed orally, undergoes minimal biotransformation, and has a half-life of 16–20 h (5,17).

Adverse effects: Flecainide can cause dizziness, blurred vision, headache, and nausea. Like other class 1C drugs, flecainide can aggravate preexisting arrhythmias or induce life-threatening ventricular tachycardia that is resistant to treatment (5,17,22).

Propafenone

Propafenone has similar actions and therapeutic uses to flecainide (5,17).

Pharmacokinetics: Propafenone undergoes extensive first-pass transformation to hydroxylated metabolite with reduced electrophysiological uses.

Adverse effects: Propafenone can cause nausea, weakness, and metallic taste. It may also have severe proarrhythmic effects (17), similar to those of the other agents with class 1C properties.

Class II: Beta-Adrenegic Blockers

These agents exert antiarrhythmic effects by blocking sympathetic nervous system stimulation of beta receptors in the heart and decrease risks of ventricular fibrillation (5,6,9,15). These drugs diminish phase 4 depolarization, thus depressing automaticity, prolonging AV conduction, and decreasing heart rate and contractility (4,7). Blockage of receptors in the SA node and ectopic pacemaker decrease automaticity, and blockage of receptors in the AV node increase the refractory period.

These drugs are effective for the treatment of supraventricular arrhythmias and those resulting from excessive sympathetic activity. Thus they are most often used to slow the ventricular rate of contraction in supraventricular tachyarrhythmias. They are also used for atrial flutter and fibrillation (5,6,12).

As a class, beta-blockers are being used more extensively because of their effectiveness and their ability to reduce mortality in a variety of clinical settings, including post-myocardial infarction and heart failure (6). With the exception of beta-blockers, none of the available antiarrhythmic drugs (23) have significantly improved survival after myocardial infarction.

These drugs include propranolol which is a β-adrenegic antagonist that is most widely used in the treatment of cardiac arrhythmias (5,24).

Propranolol reduces the incidence of sudden arrhythmic death after myocardial infarction. Esmolol is a very short-acting beta blocker used for intravenous administration in acute arrhythmias occurring during surgery or in emergency situations. Other useful drugs are pindolol, metoprolol, acebutolol, sotalol, atenolol, and timolol and nadolol (5,6,12,17,25).

Pharmacokinetics of Class II: For long-term use, any of the oral preparations of beta-blockers is suitable. In emergencies propranolol or esmolol may be given intravenously.

Adverse effects: Adverse cardiac effects from over dose include heart block or even cardiac arrest. Heart failure may be precipitated (12) when a patient is dependent on sympathetic drive to maintain output.

Class III: Potassium Channel Blockers

Potassium channel blockers (sotalol, amiodarone, bretylium, ibutilide, and dofetilide) block the outward-flowing potassium channel, and decrease potassium conductance (15,19,26) which prevent or delay repolarization (prolongation of action potential duration and end resting potential).

Sotalol

Sotalol has both beta-adrenergic blocking and potassium channel blocking activity (5,6,17). Beta-blocking effects predominate at lower doses and class III effects predominate (6) at higher doses.

Mechanism of action: Sotalol blocks rapid outward potassium current; this blockade prolongs both repolarization and the duration of the action potential, thus lengthening the effective refractory period.

Therapeutic uses: Sotalol is approved for treatment of ventricular tachycardia and fibrillation. It has also been used in smaller doses, to prevent atrial fibrillation (5,17). Sotalol is more effective in preventing arrhythmia recurrence and in decreasing mortality than mexiletine, procainamide, propafenone, and quinidine (5) in patients with sustained ventricular- tachycardia.

Pharmacokinetics: Sotalol has a half-life of about 12 h after oral administration. It is eliminated by the kidneys (5,17).

Adverse effects: Sotalol had the lowest rate of acute or long-term adverse effects. The symptoms of trosade de pointes presents serious potential adverse effects (5) typically seen in 3–4 % of patients. Fatigue, dizziness, and insomnia can occur.

Amiodarone

Amiodarone contains iodine and is related structurally to thyroxine. It has complex effects of classes I–IV actions. Amiodarone has antianginal as well as antiarrhythmic activity (5,6,17,27).

Mechanism of action: Its dominant effect is prolongation of the action potential duration and refractory period (5,27).

Therapeutic uses: Amiodarone is effective in the treatment of severe refractory supraventricular and ventricular tachyarrhythmias. Its clinical usefulness (5,6,12,17) is limited by its toxicity.

Pharmacokinetics: Amiodarone is incompletely absorbed after oral administration (5,6,17).The drug is extensively deethylated in the liver to N-desethylamiodarone, full clinical effects may not be achieved for up to 6 weeks after initiation of treatment (17).

Adverse effects: Amiodarone shows a variety of toxic effects. Some of the more common effects include gastrointestinal tract intolerance, tremor, dizziness, hyper, or hypo thyroidism, liver toxicity, muscle weakness, and blue skin discoloration caused by iodine accumulation in the skin. More than one half of the patients (5,6,17) show side effects sufficiently severe to prompt its discontinuation.

Bretylium

Mechanism of action: Bretylium has a number of direct and indirect electrophysiological actions, the most prominent of which are prolongation of the refractory period and raising of the intensity of the electrical current necessary to induce ventricular fibrillation (5,6) in the His–Purkinje system.

Therapeutic uses: Bretylium is reserved for life-threatening ventricular arrhythmias, especially recurrent ventricular fibrillation or tachycardia.

Pharmacokinetics: Bretylium is poorly absorbed from the gastrointestinal tract and is therefore generally administrated parenterally. The drug is excreted (5,17) unchanged in the urine.

Adverse effects: Bretylium can cause severe postural hypotension. Nausea and vomiting (5,6,17) may occur after rapid intravenous administration.

Class IV: Calcium Channel Blockers

Calcium channel blockers block the movement of calcium into conductile and contractile myocardial cells. As antiarrhythmic agents, they act primarily against tachycardias at SA and AV nodes because the cardiac cells and slow channels that depend on calcium influx are found mainly at these sites. Thus, reduce automaticity of the SA and AV nodes, slow conduction, and prolong the refractory period in the AV node (5,6,12,17,28,29).

Verapamil and Diltiazem

Verapamil shows greater action on the heart than on vascular smooth muscle, whereas nifidipine, a calcium channel blocker is used to treat hypertension exerts a stronger effect on vascular smooth muscle than on the heart. Diltiazem is intermediate in its action (5).

Mechanism of action: Calcium channel blockers, such as verapamil and diltiazem, are more effective against the voltage-sensitive channels, causing a decrease in the slow inward current that triggers cardiac contraction. Verapamil and diltiazem bind only to open, depolarized channels, thus preventing repolarization until the drug dissociates from the channel. By decreasing the inward current carried by calcium, verapamil and diltiazem slow conduction and prolong the effective refractory period in tissues dependent on calcium currents, such as the AV node.

Therapeutic uses: Verapamil and diltiazem are more effective against atrial than ventricular arrhythmias. They are useful in treating reentrant supraventricular tachycardia and reducing ventricular rate in atrial flutter and fibrillation (5,6,17).

Pharmacokinetics: Verapamil and diltiazem are absorbed after oral administration. They are metabolized by the liver, and the metabolites (5,6) are primarily excreted by the kidneys.

Adverse effects: Both drugs may cause bradycardia and therefore may be contraindicated in patients with preexisting depressed cardiac function. These drugs can also cause a decrease in blood pressure (5,6,17) caused by peripheral vasodialtion.

Class V: Other Antiarrhythmic Drugs

Digoxin

Digoxin shortens the refractory period in atrial and ventricular myocardial cells while prolonging the effective refractory period and diminishing conduction velocity in Purkinje fibres. Digoxin is used to control the ventricular response rate in atrial fibrillation and flutter. At toxic concentration, digoxin causes ectopic ventricular beats that may result in ventricular tachycardia and fibrillation (5,17,29).

Adenosine

Adenosine is a naturally occurring nucleoside, but at high doses the drug decreases conduction velocity, prolongs the refractory period, and decreases automaticity (5,6,17) in the AV node. Intravenous adenosine is the drug of choice for abolishing acute supraventricular tachycardia. It has low toxicity (5,29,30) but causes flushing, chest pain and hypotension.

Magnesium Sulfate

Magnesium sulfate is given intravenously in the treatment of several arrhythmias, including prevention of recurrent episodes of trosades de pointes and treatment of digitalis induced arrhythmias. Its antiarrhythmic effects (6,10) may derive from imbalances of magnesium, potassium, and calcium.

Non-Pharmacological Treatment of Arrhythmias

Non-pharmacologic treatment is preferred, at least initially for several arrhythmias. For example, sinus tachycardia usually results from such disorders as infection, dehydration, or hypotension and treatment should be aimed toward relieving the underlying causes. For ventricular fibrillation, immediate defibrillation (6,24,31–34) by electrical counter shock (electro-version or cardioversion) is the initial treatment of choice. The technique consists of applying the shock, by means of a defibrillator to the chest wall of the patient, through two electrodes, one overlapping the lower border of the left scapula and the other over the third right intercostals space. The energy of the brief (2 ms) shocks used ranges from 100 to 400 joules (35).

The impetus for non-pharmacologic treatments developed mainly from studies demonstrating that antiarrhythmic drugs could worsen existing arrhythmias, cause new arrhythmias, and cause higher mortality rates in clients receiving the drugs than clients not receiving the drugs. Current technology allows clinicians to insert pacemakers and defibrillators, for example, the implantable cardioverter-defibrillators (ICD) have been useful to control brady arrhythmias or tachy arrhythmias. Similarly, radio waves have been used in radio frequency catheter ablation or surgery to deactivate ectopic foci (6).

The ICD are designed to deliver an electric shock when ventricular fibrillation or tachycardia occurs. They are increasingly being recognized as most efficacious and cost-effective in initial therapy (36) for the prevention of sudden cardiac death. The major goals of defibrillator therapy are detection and termination of ventricular tachyarrhythmias, prevention of sudden death, reduction in patient mortality (37) and in improving the quality of life.

STRUCTURE–ACTIVITY RELATIONSHIPS OF SOME CLASS I ANTIARRHYTHMIC DRUGS

The chemical structure of compounds found to possess antiarrhythmic activity covers a wide range, and because of this great structural heterogenicity, efforts to determine a definitive structure have not been too successful. Even some of the more recently discovered and currently used antiarrhythmic drugs were synthesized for quite different purposes, and routine pharmacological screening and clinical trials revealed their antiarrhythmic activity.

The first extensive reviews on the SAR of antiarrhythmic drugs were by Szekeres and Papp (38) and Conn (39). These investigations indicated that most antiarrhythmic drugs possess a tertiary amine group that appears to be a key component for activity. The presence of aromatic ring system is important, and this moiety is usually connected to the tertiary amine via a hydroxy substituted alkyl chain ester or amide group. The methoxy group on the aromatic ring may enhance activity. Since these early studies the number of compounds synthesized and tested for antiarrhythmic activity has increased dramatically. The structure activity relationships and approaches to new antiarrythmic agents of some class I antiarrhythmic agents are highlighted below.

Class IA: Antiarrhythmic Agents

Procainamide Hydrochloride

Procainamide hydrochloride; *p*-Amino-N-[2-(diethylamino)ethyl]benzamide monohydrochloride. It was developed in the course of research for compounds structurally similar to procaine, which had limited effect as an antiarrhythmic agent because of its central nervous system side effects and short-lived action resulting from the rapid hydrolysis of its ester linkage by plasma esterases. Procainamide hydrochloride is also more stable in water than procaine because of its amide structure.

Metabolism of procainamide hydrochloride occurs through the action of N-acetyltransferase. The product of enzymatic metabolism of procainamide hydrochloride is NAPA, which possesses only 25% of the activity of the parent compound (40). A study of the disposition of procainamide hydrochloride showed 50% of the drug was excreted unchanged in the urine, with 7% to 24% recovered as NAPA (41,42).

Disopyramide Phosphate

Disopyramide phosphate. α-[2-(Diisopropylamino)ethyl]-α-phenyl-2-pyridineacetamide phosphate (Fig. 2) is an oral and intravenous class lA antiarrhythmic agent. Oral administration of the drug produces peak plasma levels within 2 h. the drug is approximately 50% bound to plasma protein and has a half-life of 6.7 h in humans. More than 50% is excreted unchanged in the urine. Disopyramide phosphate commonly exhibits side effects of dry mouth, constipation, urinary retention, and other cholinergic-blocking actions because of its structural similarity to anticholinergic drugs.

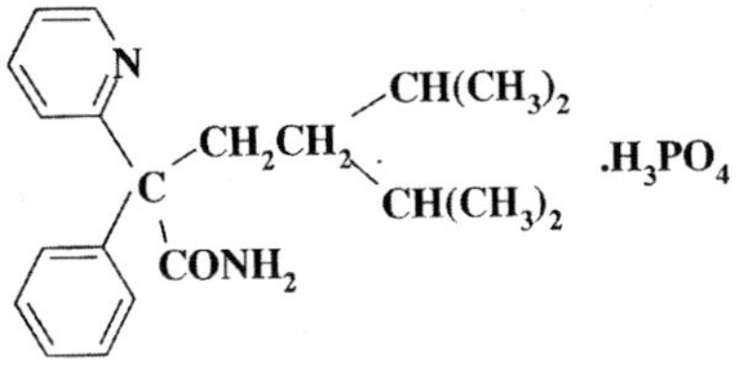

Figure 2 Disopyramide phosphate.

Class IB: Antiarrhythmic Agents

Lidocaine Hydrochloride

Lidocaine hydrochloride. 2-(Diethylamino)-2′,6′-acetoxylidide monohydrochloride. This drug was conceived as a derivative of gramine (3-dimethylaminomethylindole) and introduced as a local anesthetic, is now being used intravenously as a standard parenteral agent for suppression of arrhythmias associated with acute myocardial infarction and cardiac surgery.

Lidocaine is the drug of choice for the parenteral treatment of premature ventricular contractions. Lidocaine hydrochloride administration is limited to the parenteral route and is usually given intravenously, though adequate plasma levels are achieved after intramuscular injections. Lidocaine hydrochloride is not bound to any extent to plasma proteins and is concentrated in the tissues. It is metabolized rapidly by the liver. The first step is deethylation, with the formation of monoethylglycinexylidide, followed by hydrolysis of the amide (43). Metabolism is rapid, the half-life of a single injection ranging from 15 to 30 min. Lidocaine hydrochloride is a popular drug because of its rapid action and its relative freedom from toxic effects on the heart, especially in the absence of hepatic disease. Monoethylglycinexylidide, the initial metabolite of lidocaine, is an effective antiarrhythmic agent; however, its rapid hydrolysis by microsomal amidases prevents its use in humans.

Tocainide Hydrochloride

2-Amino-2′,6′-propionoxylidide hydrochloride. Tocainide hydrochloride (pKa 7.7) is an analogue of lidocaine. It is orally active and has electrophysiological properties similar to lidocaine (44). Total body clearance of tocainide hydrochloride is only 166 mL/min, suggesting that hepatic clearance is not large (Fig. 3).

Because of low hepatic clearance, the hepatic extraction ratio must be small; therefore, tocainide hydrochloride is unlikely to be subject to a substantial first-pass effect. The drug differs from lidocaine in that it lacks two ethyl groups, which provides tocainide hydrochloride some protection from first-pass hepatic elimination after oral ingestion. Tocainide hydrochloride is hydrolyzed in a manner similar to lidocaine. None of its metabolites are active.

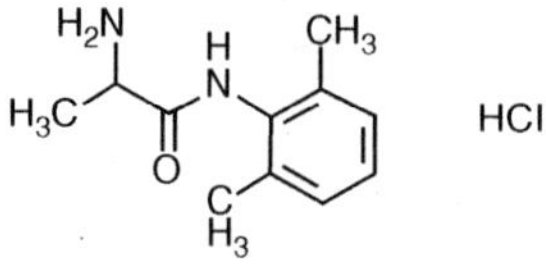

Figure 3   Tocainide hydrochloride.

Class IC: Antiarrhythmic Agents

Encainide Hydrochloride

Encainide hydrochloride. 4-Methoxy-N-[2-(1-methyl-2-piperidinyl) phenyl] benzamide monohydrochloride. Encainide hydrochloride is a benzanilide derivative that has local anesthetic properties in addition to its class le antiarrhythmic action.

The drug is metabolized extensively, producing products that also have antiarrhythmic properties (45). The metabolite, 3-methoxy-O-demethylencainide, is about equipotent to encainide hydrochloride. O-Demethylencainide is considerably more potent than the parent drug.

The half-life of encainide hydrochloride is 2–4 h. The active metabolites have longer half-lives, estimated as being up to 12 h, and may play an important part in the use of this drug in long-term therapy. Encainide hydrochloride also undergoes N-demethylation to form N-demethylencainide.

Flecainide Acetate

Flecainide acetate. *N*-(2-Piperidinylmethyl)-2,5-bis(2,2,2-trifluoroethoxy) benzamide monoacetate. It is a chemical derivative of benzamide. The drug undergoes biotransformation, forming a meta-O-dealkylated compound, the antiarrhythmic properties of which are one-half as potent as those of the parent drug, and meta-O-dealkylated lactam of flecainide (46) with little pharmacological activity. Flecainide acetate has some limitations because of central nervous system side effects (Fig. 4).

Lorcainide Hydrochloride

Lorcainide hydrochloride. *N*-(4-Chlorophenyl)-N-[1-(1-methylethyl)-4-piperidinyl]benzeneacetamide monohydrochloride. Lorcainide hydrochloride undergoes metabolic N-dealkylation to produce norlorcainide. Metabolism is the product of the first-pass clearance after oral administration. The basis for this observation is that norlorcainide is not produced in significant amounts in the body following intravenous administration. Norlorcainide is an important metabolite of the parent drug, as it is cleared slowly from the body and has a half-life that is approximately three times longer. Accumulation of norlorcainide is of considerable clinical importance because the metabolite is equipotent to the original drug (Figs. 5 and 6).

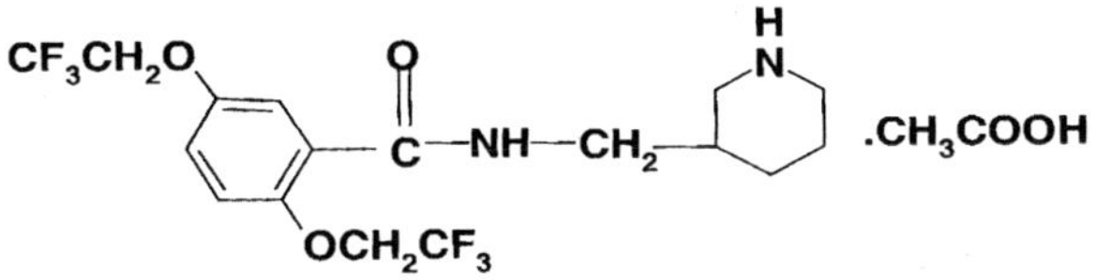

Figure 4 Flecainide acetate.

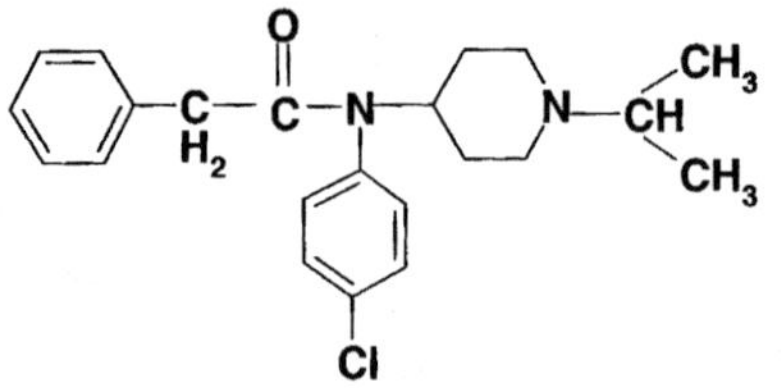

Figure 5   Lorcainide hydrochloride.

Moricizine

Moricizine, ethyl 10-(3-morpholinopropionyl)phenothiazine-2-carbamate, is a phenothiazine derivative used for the treatment of malignant ventricular arrhythmias.

New Chemical Entities

SUN-1165

SUN-1165 contains the highly basic pyrolizidinyl moiety of the less basic diethylamine group of lidocaine (47). Electrophysiological studies in vitro revealed that SUN-1165 is a class 1B drug (48,49). Compared to lidocaine and disopyramide, the compound had less central nervous system (CNS) toxicity and anticholinergic activity respectively (Fig. 7) (50).

Recainam (WY-423262)

Recainam (WY-423262) is class IB lidocaine-like drug. Therapeutic intravenous doses failed to produce CNS or negative inotropic (51) side effects in dogs, and the drug safely and effectively reduced premature ventricular contractions (52) in man (Fig. 8).

Droxicainide (ALS-1249)

Droxicainide (ALS-1249) was reported to have antiarrhythmic and local anesthetic properties (53,54) that are qualitatively similar to lidocaine, but

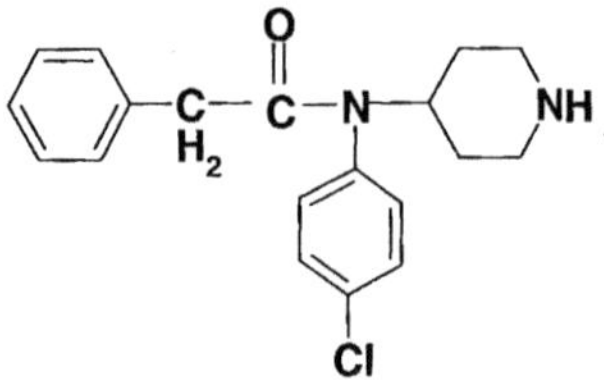

Figure 6   Norlorcainide.

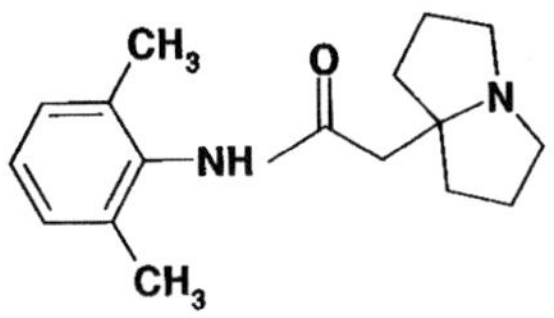

Figure 7 SUN.

with a better therapeutic index over 24 hour period in infarct dogs (55) than lidocaine. The piperidine and azepine analogues of droxicainide (56) which lack the hydroxyethyl moiety are also active (Fig. 9).

ACC-9358

Electrophysiological studies showed that ACC-9358 is a class I agent (Fig. 10).

Propisomide (CM-7857)

Disopyramide continues to serve as a prototype for new antiarrhythmics. A series in which the phenyl group of disopyramide was replaced with alkyl moieties (57) was synthesized. One of the compounds in this series, propisomide (CM-7857), was reported to possess a longer duration of action, fewer neurologic and gastrointestinal (58) and less anticholinergic activity than disopyramide. As with disopyramide, the major metabolite (59) in both plasma and urine was the mono-N-dealkylated analogue. A single dose of propisomide was compared to a variety of other antiarrhythmics in 10 patients. The compound (60) was effective in 6; it increased PQ intervals, and had no effects on QRS or QT intervals (Fig. 11).

AHR-10718

AHR-10718 is the optimal compound in a series of aminoethylureas that suppressed arrhythmias in the ouabain-intoxicated and Harris dog models. The drug caused a use-dependent decrease in action potential upstroke velocity (V_{max}), Purkinje fibre (61) conduction on velocity and APD (Fig. 12).

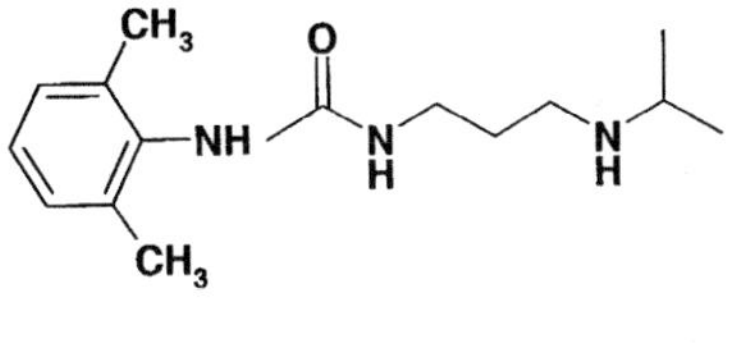

Figure 8 Recainam (WY-423262).

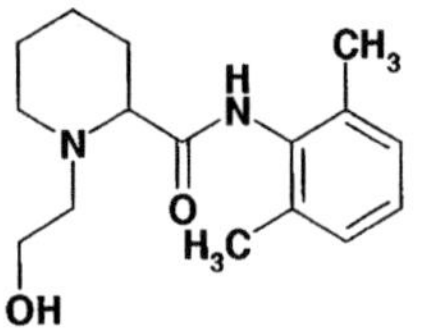

Figure 9   Droxicainide (ALS-1249).

Carocainide (MD77020)

Carocainide, a benzofuran antiarrhythrnic, decreased V_{max} in isolated papillary muscle and Purkinje fibres: it decreased the plateau amplitude and APD. The antagonized digitalis and infarction induced arrhythrnias (62–64) in dogs (Fig. 13).

E-0747

E-0747 was more potent than quinidine, disopyramide, lidocaine, or phenytoin. Moreover, at therapeutic doses, the compound appeared to have low potential (65) for cardiodepression (Fig. 14).

Indecainide (L Y135837)

Indecainide, is a potent class IC antiarrhythmic agent (66,67) with an exceptionally long half-life time (52 seconds) for recovery (66) from sodium channel block. This compound was more potent than either aprindine or disopyramide against ouabain or Harris arrhythmias in dogs (Fig. 15).

ROLE OF PROTEINS IN CARDIOLOGY AND INTRACELLULAR SIGNALLING

Ischemia injury can result in cell death and irreversible loss of function in a variety of biological systems. The sequential events that produce this cardiac dysfunction include a decreased endothelial release of nitric oxide (NO), up-regulation of adhesion molecules on the endothelial surface leading to enhanced leukocytes–endothelium interaction and release of superoxide radicals. These radicals are largely responsible for producing cardiac dysfunction and enhanced necrosis. The time course of these events starts at 2.5–5 min post-reperfusion.

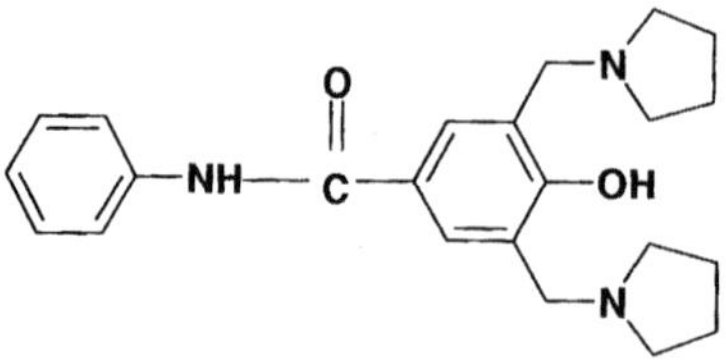

Figure 10   ACC-9358.

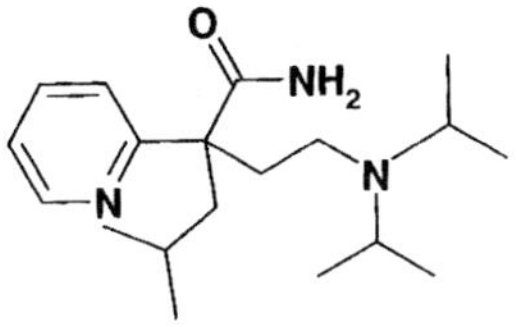

Figure 11 Proplsomide (CM-7857).

Protein Signalling Pathways

An understanding of the intracellular signalling mechanisms by which cells protect themselves against ischemia-induced damage bears great clinical significance with respect to the treatment and prevention of tissue injury. Cells start to generate agonists especially those that signal via $G\alpha_l$, such as adenosine, acetylcholine, opioids and bradykinin, which bind to G protein coupled receptors (GPCR) and initiate a signalling cascade that involves activation of phosphoinositide-3-kinase (PI3K), endothelial NO synthase, tyrosine kinase, protein kinase c, glycogen synthase, mitogen-activated protein kinases, and other signalling pathways.

Activation of these signalling pathways along with generation of reactive species leads to alterations in the activity of key mitochondrial proteins such as mitochondrial ATP-sensitive K^+ channels and the mitochondrial permeability transition pore. Alterations on these mitochondrial proteins results in altered metabolism and inhibition of cell death, thus resulting in cardioprotection (Fig. 4).

Activation of GPCR leads to signalling via $G\alpha_l$ and $G\beta\gamma$ and initiates a signalling cascade. Acetylcholine-induced protection leads to activation of PI3K pathways via GPCR transactivation of the epidermal growth factor, which also signals through $G\beta\gamma$ and lead to activation of PI3K. At the same time activation of GPCR lead to activation of mitogen-activated protein kinase (MAPK) pathway via $G\alpha$-dependant signalling, by transactivation of the epidermal growth factor; MAPK pathway operate through sequential phosphorylation events to phosphorylate transcription factors and regulate gene expression (Fig. 4).

PI3K generates phosphoinositides that localize kinases, such as phosphoinositide-dependant kinase (PDK), with substrates, leading to

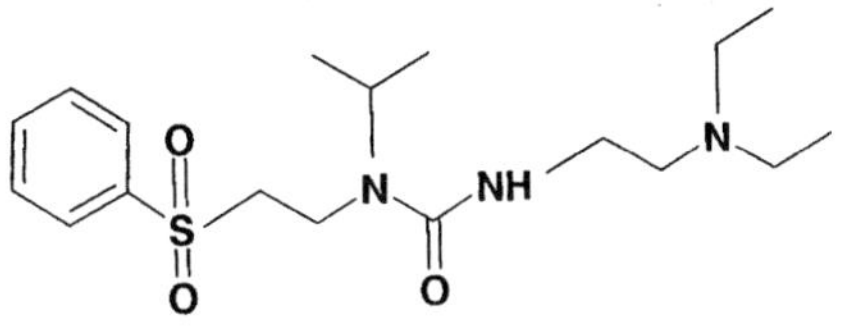

Figure 12 AHR-10718.

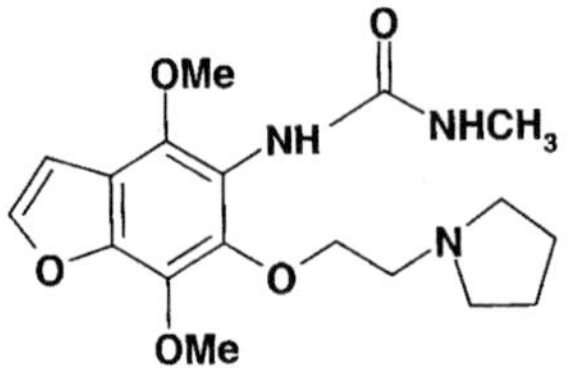

Figure 13 Carocainide (MD770207).

activation of downstream kinases such as protein kinase B (PKB, also known as Akt), mammalian target of rapamycin (mTOR) and p70S6-kinase. Both PDK1 and PKB play an important role in the activation of p70S6 kinase. PDK1 phosphorylation of thr 229 is required for activation of p70S6K. Prior phosphorylation of thr 389 by mammalian target of rapamycin (mTor) is necessary before PDK1 can phosphorylate thr 229.

At the same time PKB phosphorylates and activates endothelial NO synthase (eNOS) and phosphorylates and inactivates GSK3β and the proapoptic BAD. PI3K also play a role in the activation of protein kinase C (PKC), but the precise role of PI3K has not been determined; some studies suggested that PI3K activation of PKC_ε occurs via an eNOS-mediated mechanism.

However, another study (68) suggested that the reactive oxygen species (ROS), is generated by preconditioning is involved in activation of PKC. With regards to PKC there is increasing evidence for functional coupling of PKC to tyrosine kinase in the heart. Both PKC and tyrosine kinase show a cardioprotective effect via activation of transcription regulatory protein nuclear factor (NF-$_\kappa$B) which occur through both tyrosine and serine phosphorylation of $I_\kappa B\alpha$ (Fig. 16).

Furthermore PKC activation has been shown to be important in activation of the mitoKATP channel, at the same time NO has been shown to activate mitoKATP channel. PKC_ε also forms a complex with the components of the mitochondrial permeability transition (MPT) pore. However, the association of PKC_ε with components of the MPT pore does not demonstrate that this association is important in modulating MPT or cardioprotection. This study also reported that ERK and PKC_ε are contained in a multimeric mitochondrial signalling complex and that PKC_ε may lead to activation of ERK in this complex. One study reported that KATP channel

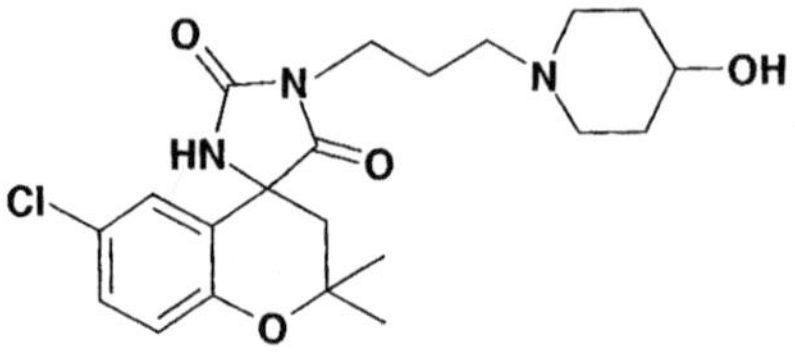

Figure 14 E-0747.

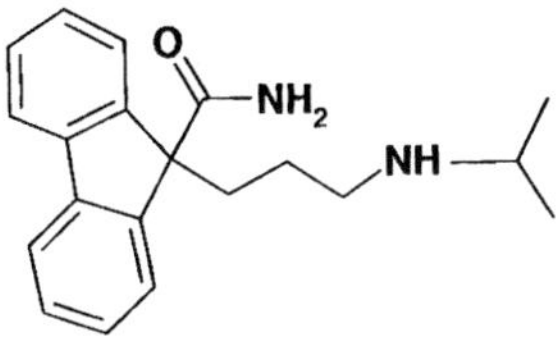

Figure 15 Indecainide (LY135837).

opening can lead to activation of ERK, which is likely to be secondary to ROS generated opening of the mitoKATP channel (Fig. 16).

CARDIOVASCULAR DISEASES AND CLINICAL THERAPY

Ischemic Heart Disease

This disease is usually caused by atherosclerosis which narrows the coronary arteries. Common risk factors include hypertension, genetic disposition, smoking, diabetes, and hyperlipidemia amongst a variety of other conditions (69). The manifestations of ischemic heart disease frequently results in sudden death, myocardial infarction or angina pectoris.

Nitrates

The nitrates in the body dilate blood vessels in three areas: viz. (i) venous circulation, which decreases venous return and preload on the heart, (ii) the arterioles, which reduce peripheral resistance and afterload. Both reduce stress on the myocardial wall and lower oxygen demand, (iii) the coronary arteries, especially in the presence of coronary spasm to improve the oxygen supply.

A large number of patients tend to be tolerant to the anti-anginal effects of the nitrates as a result of depletion of the essential thio (-SH) groups especially if they are taken for prolonged periods (usually for more than 24 h). Whenever it is necessary to prescribe, a nitrate-free period must be built-in into the treatment regimen on an individual basis for each patient.

The nitrates are converted in the body to NO. This in combination with sulphydryl (-SH) groups forms nitrosothiols which thus activate the enzyme guanylyl cyclase to produce the secondary messenger, cyclic GMP. This causes the smooth muscle to relax with vasodilatation.

Typical examples of nitrates: Glyceryl trinitrate is well absorbed sublingually but, if taken orally, it is broken down by the liver by first pass metabolism. Isosorbide dinitrate (can be taken sublingually, orally or intravenously although, the absorption rate is slow, its duration of action is much longer). Isosorbide mononitrate may be taken orally but in general, its use is limited for practical reasons.

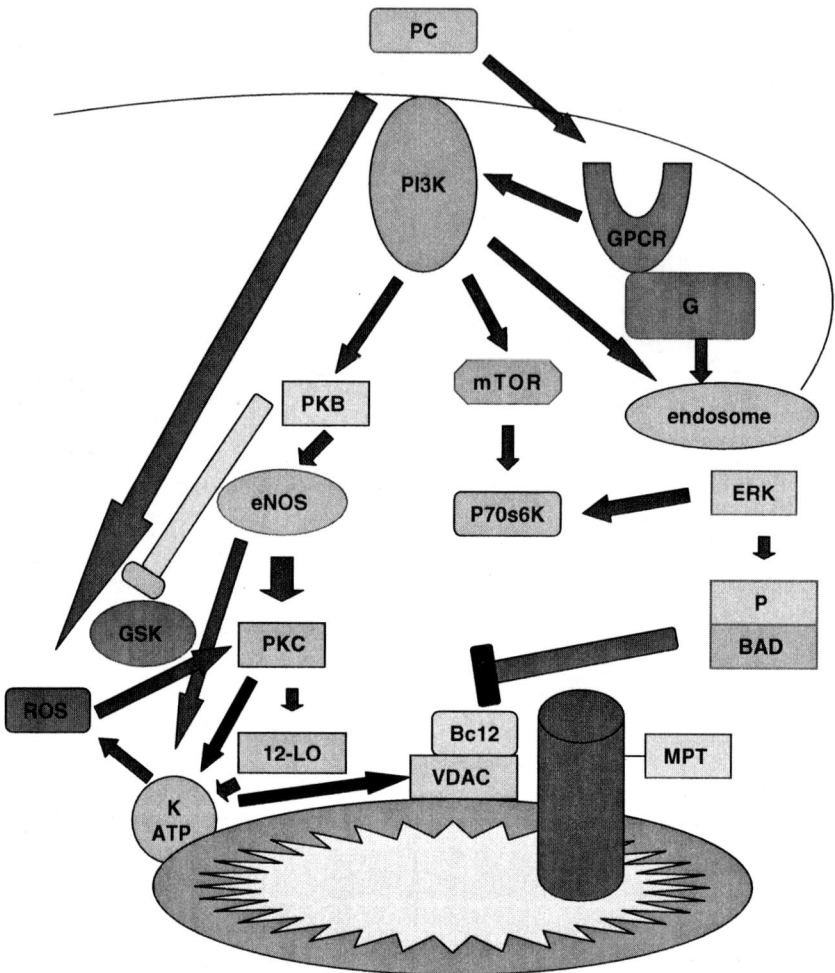

Figure 16 Primary protein signalling pathways.

Beta-Blockers

These types of drugs block the beta-adrenoceptors, that is, both beta-1 and beta-2, which are present in the heart to reduce the production of cyclic AMP. The beta-blockers may be either non-selective and block both types of beta-receptors, or sufficiently selective to bind to beta-1 receptors. The beta-blockers slow the heart and thereby reduce its force of contraction and decrease oxygen demand. On the other hand, they may also increase oxygen supply in response to a diastole when most of the blood flow in the coronary arteries is extended. Typical examples of beta-blockers: Atenolol: is a hydrophobic beta-1-selective blocker. Propranolol, is lipophilic and a non-selective beta-blocker and Metaprolol, is a lipophilic beta-1-selective blocker.

Clinical uses include:

1. hypertension,
2. angina pectoris,
3. following myocardial infarction, beta-1-selective blockers cause a reduction in risk of recurrence,
4. useful in anxiety and manifestations of anxiety such as tremor and depression,
5. arrhythmia.

Calcium-Channel Blockers

The calcium-channel blockers (antagonists) act by antagonising the movement of calcium through channels in the cell membranes. In cardiac muscle or vascular smooth muscle, any change in membrane potential causes calcium to enter the cell through the calcium channels that are voltage-dependent. The calcium channel blockers reduce the force of contraction of the heart and thus slow the heart rate in vitro. They also reduce smooth muscle contraction, and act as vasodilators of coronary and systemic arteries and alter cardiac rhythm.

Activity of calcium antagonists: Calcium antagonists inhibit the slow inward current caused by entry of extra cellular calcium through the cell membrane of excited cells, particularly arteriolar smooth muscles, and cardiac arterial cells.

Calcium ions play an essential role in the regulation of skeletal and smooth muscle contractility in the performance of the normal and diseased heart. The treatment of hypertension with the introduction of calcium channel blockers (CCBs), was a major breakthrough was infact approached when it was discovered that vascular smooth muscle is linked to the movement of calcium ions with the muscle cells from extra-corporeal circulation.

The channels in the cell membrane are normally occupied by calcium ions bound to receptor storage sites and function as voltage-dependant ionic gates. Their opening is controlled by the electronic gradient across the cell membrane. The gates open and allow the calcium ions to enter the vascular muscle cells upon stimulation by noradrenaline, which is released from the nerve endings. This causes the muscle cell system to contract in order to maintain blood pressure. Excessive stimulation, increases the heart rate and myocardial oxygen demand resulting in chronic hypertension and coronary spasm, causing a severe chest pain (angina).

Regulation of calcium channel blockers: There are three major types of calcium channels blockers, viz. (1) voltage-dependent, (2) receptor-operated, and (3) stretch operated.

1.   Voltage (potential)-dependent Ca^{2+}-channels (homologous to Na^+ and K^+ channels, consist of at least three types in the body, L, T, and N). L-type channels:
     i.   have large sustained conductance, become slowly inactive and are widespread in the cardiovascular system;
     ii.   are responsible for the plateau phase (slow inward current) of action potential;
     iii. may trigger release of internal Ca^{2+};
     iv.   are sensitive to Ca^{2+} channel blockers.

Cardiac L-channels are regulated by cAMP-dependent protein kinase.

     T-type channels:
     i.   are structurally similar to L-type channels;
     ii.   become inactive rapidly;
     iii. are involved in cardiac pacemaker activity, growth regulation and triggering contraction of vascular smooth muscles.

T-type channels are not very sensitive to most of the L-type Ca^{2+}-channel blockers.

     N-type channel:
They are found only in neuronal cells and are not very sensitive to Ca^{2+}-channel blockers used for treating cardiovascular disorders.

2.   Receptor-operated Ca^{2+}-channels (e.g. alpha-adrenergic receptors): not very sensitive to Ca^{2+}-channel blockers.
3.   "Stretch" operated Ca^{2+} channels are not particularly important in maintaining vascular smooth muscle tone as they do not appear to be sensitive to Ca^{2+}- channel blockers.

Two types of calcium channel blockers are used in clinical situations: those which are selective for L-type (long-lasting, large-currents, or slow) and those that are non-selective. In clinical practice, selective agents are primarily used.

The selective calcium channel blockers share a similar antihypertensive mechanism of action: they inhibit the influx of extracellular calcium through L-type channels, resulting in relaxation of vascular smooth muscle and reduction in vascular resistance.

The reason why calcium antagonists are useful as drugs is probably because they are a heterogeneous group of compounds which have marked differences in chemical structure, binding sites, tissue selectivity, clinical activity, and therapeutic effect.

Calcium antagonists are chemically and pharmacologically a diverse groups of drugs. These include verapamil and its analogues, the benzo-1,5-thiazepine related to diltiazem and the 1,4-dihydropyridines, for example, nifedipine. These were briefly discussed earlier.

Typical examples of calcium-channel blockers: There are generally three types of calcium-channel blockers. They bind to different although related receptor sites on the calcium channel. Their effects vary with different tissues.

1. Verapamil is less effective on peripheral circulation but more active on the heart. It slows the heart and reduces the force of contraction and can be taken orally or intravenously.
2. Diltiazem is more effective on the heart compared with nefedipine and is more active on peripheral circulation than verapamil. It is metabolised by the liver.
3. Dihydropyridines: Examples are nifedipine, nitrendipine, nimodipine, nislodipine and felodipine. They mainly cause arterial vasodilatation and reduce blood pressure and afterload on the heart i.e. the load against which the heart ejects blood. It increases heart rate and the force of contraction. Nefedipine is orally well absorbed and is metabolised by the liver. It can be used in combination with beta-blockers in angina or hypertension. Although rare, combination therapy may precipitate heart failure. Verapamil in combination with beta-blockers has been reported to be dangerous especially when administered intravenously. It can cause severe hypotension and bradycardia.

The pharmacological effects of the newer types of dihydropyridine drugs show that they can improve efficacy and vascular selectivity with a longer duration of action.

Clinical uses include:

1. hypertension;
2. angina, especially due to coronary artery spasm;
3. Raynaud's phenomenon (nefedipine);
4. supraventricular tachycardia (verapamil).

The binding sites for all three chemical types of calcium channel blockers are present in many tissues, including myocardium, smooth muscle, skeletal muscle, and glandular tissue. Each of the three selective calcium channel blockers interact with a specific receptor domain found on a large membrane-spanning protein that constitute a substantial portion of the L-type, voltage dependent calcium channel.

These receptor sites are all located on the surface on the alpha subunit of the channel. The 1,4-dihydropyridine receptor is most accessible and is located on the surface of the channel. This receptor has been the most widely studied of the three groups, and therefore a relatively greater number of dihydropyridines have been designed to bind and inhibit at that site.

Congestive Heart Failure

Heart failure usually results from damage to the myocardium typically from ischemic heart disease, cardiomyopathy or myocarditis. Hypertension also contributes to cardiac failure especially if it is poorly controlled or results from damage to the heart valves and fluid overload. Apart from drugs, metabolic diseases such as, for example, thyrotoxicosis also makes a contribution. The overall result is that cardiac failure results in poor perfusion of vital organs and tissues and causes congestion of the lungs.

Treatment of Congestive Heart Failure is mainly carried out by:

1.   diuretics,
2.   cardiac stimulants,
3.   angiotension converting enzyme inhibitors (ACE) inhibitors,
4.   angiotensin II receptor antagonists,
5.   cardiac glycosides, digoxin and digitalis.

Diuretics

These are still the mainstay of treatment in cardiac failure, the most potent diuretics are loop diuretics such as frusemide and bumetanide. They act by promoting sodium and chloride excretion. That is, they promote water loss to relieve systemic and pulmonary congestion. They can act at different sites in the nephron in different ways. Other than diuresis, diuretics may work through different mechanisms. For example, thiazide diuretics which are less potent and more appropriate for elderly patients for maintenance therapy only partially effect lowering of blood pressure via diuresis. This effect is to enlarge the result of vasodilatation. In acute cardiac failure diuretics are used intravenously. Some examples of these types of drugs include: spironolactone, amiloride, trimterene, thiazides, frusemide, butetanide, ethacrynic acid, mercurials, mannitol, and xanthines. In general there are four major groups:

1.   thiazides,
2.   loop diuretics,
3.   potassium-sparing diuretics,
4.   osmotic diuretics.

Thiazides and related drugs: Thiazides inhibit the absorption of sodium ions and chloride ions at the distal convoluted tubule. Water is lost with the sodium ions. More sodium ions then reach the distal tubule, where it is exchanged with potassium ions. Thiazides also cause potassium loss. Magnesium is also lost while Ca^{2+} is retained. The principal sites of action are the distal nephron, distal tubule, Loop of Henle, and proximal tubule.

The maximal hypotensive response to thiazides can be reached at relatively low doses. They are ineffective in the presence of renal impairment. At one time, thiazides were used in extremely high doses with a high incidence of adverse effects. Nowadays, thiazides are used in low doses as they are equally effective in hypertension and result in fewer problems.

Typical examples are: Bendrofluazide which is well absorbed after oral administration. It causes diuresis within approximately 1–2 h, with lasting effect for up to 12 h and is metabolized by the liver.

Clinical uses:

1. congestive heart failure,
2. hypertension,
3. congestive cardiac failure.

A number of adverse reactions can occur such as dehydration; this is a common risk in the use of all diuretics especially in the elderly patients. Other effects include: hypokalemia (usually mild), hyperuricemia/gout, impaired glucose tolerance/diabetes mellitus, insulin resistance (significance is uncertain), Hypercholesterolemia (usually in short-term therapy rather than in long-term therapy), impotence, and hypercalcemia. Contraindications are gout, diabetes mellitus, and renal disease.

Loop diuretics: Loop Diuretics inhibit the resorption of chloride ions in the ascending limb of the loop of Henle with loss of chloride ions and sodium ions. Water loss with electrolytes then follows. The responsive action of loop diuretics is dose related, and is mainly effective even in renal impairment, unlike thiazides. They have a much greater diuretic effect compared with the thiazides. Their use in acute pulmonary oedema causes venodilatation when administered intravenously. They also reduce preload before diuresis starts. Typically, like the thiazides, the loop diuretics cause potassium loss although the unlike effect is that they cause Ca^{2+} loss.

Typical examples are: frusemide or bumetamide which act within 10 minutes of an IV dose. After oral administration, their effects tend to peak at about 1 hour with lasting effects of up to 4–6 hours.

Clinical uses:

1. edema, and
2. congestive heart failure.

Typical adverse effects include: hypokalemia, renal impairment (which results from dehydration and pre-renal failure and direct toxic effects on the kidneys), and ototoxicity (usually deafness with frusemide).

Typically, drugs such as probenecid and the non-steroidal anti-inflammatory drugs interfere with the action of diuretics. Loop diuretics can usually be combined with potassium sparing diuretics in order to reduce potassium loss. Similarly, they can be combined with ACE inhibitors in hypertension or heart failure. However, diuretics can reduce excretion of lithium ions which can potentially lead to serious toxicity.

Potassium-sparing diuretics: These types of diuretics are chemically a diverse group that, unlike other diuretics, retain potassium ions. They tend to act in the distal convoluted tubule where potassium and sodium ions are exchanged, and are thus considered to be weak diuretics.

Their main use is to treat or prevent hypokalemia which is caused by thiazides or loop diuretics. In the past, potassium supplements were widely recommended, although very high doses are needed and this apparently reduces patient compliance. It is not normally necessary to treat or prevent a diuretic-induced hypokalemia, but if needed, potassium sparing diuretics have been indicated for clinical use.

Spironolactone is an aldosterone antagonist. Aldosterone itself causes sodium ion retention with potassium loss. In contrast, spironolactone displays the opposite effect. Amiloride and Triamterene both interfere with sodium and/ or potassium ion exchange. All these diuretics are administered orally.

Clinical uses: Use of spironolactone diuretic is limited due to its carcinogenicity, which can cause:

1.   ascites (from liver disease);
2.   resistant congestive heart failure (with loop diuretics);
3.   primary hyperaldosteronism (Conn's syndrome).

Amiloride and Triamterene are useful in:

1.   diuretic-induced hypokalemia,
2.   oedema.

Typical adverse effects are: hyperkalemia and gynecomastia (with spironolactone). Again, combinations with potassium supplements or ACE inhibitors may lead to serious hyperkalemia.

Osmotic diuretics: These drugs are only employed under restricted circumstances. The compounds are filtered in the glomerulus and excreted. Due to their osmotic effects water is retained within the tubule. This causes diuresis of water with some loss of electrolytes.

Clinical uses:

Mannitol is given by IV for the treatment of oliguria in incipient renal failure in order to lower intracranial pressure. It is also used in glaucoma.

Generally, hypernatremia, hyperkalemia, or dehydration are common side-effects. Plasma volume is increased which may precipitate cardiac failure.

Angiotension Converting Enzyme Inhibitors

The renin–angiotensin–aldosterone is involved in the regulation of vascular tone.

Although, their precise mode of action is still not entirely clear it is thought that they exert their clinical effect at least in part by vasodilation. ACE inhibitors are widely distributed in the body; although the systemic effects of the inhibitors and their effects on local angiotensin production in the tissues is unclear. They appear to prevent the formation of angiotensin II and thus block its vasoconstrictor effects and decrease the production of aldosterone and fluid retention. They breakdown inflammatory peptides (typically bradykinin), this can contribute to its anti-hypertensive effect.

Typical examples are: captopril which was the first ACE-inhibitor is given orally. It is cleared by the kidneys. Enalapril is a prodrug which is also given orally and is activated to the active, enalaprilat by the liver. Clearance of the drug occurs via the kidneys. Enalapril is usually given once daily.

Clinical uses:

1. *Cardiac failure*: Angiotension converting enzyme inhibitors reduce mortality.
2. *Hypertension (widely employed)*: ACE inhibitors may be especially useful in the treatment of hypertension in diabetes patients, due to a delay in the onset of microalbuminuria. Both captopril and enalapril appear to be well tolerated.
3. *Myocardial infarctian*: Particularly apparent in patients who experience heart failure as a complication. The mechanism is somewhat unclear but appears to involve a reduction in left ventricular enlargement.

A number of diverse side-effects have been observed. Some examples include:

1. *Severe hypotension*: Especially in cardiac failure, on first dose, or patients on high-dose diuretics. The dose is usually reduced as much as possible prior to taking ACE inhibitors.
2. *Hyperkalemia*: Potassium losing effects of aldosterone are blocked.
3. Impairment of renal function through decrease in renal perfusion and intra renal hemodynamics.

4. *Cough*: Although very common but not serious, probably as a result of potentiation of bradykinin.
5. *Angioedema*: This is more serious and may be caused by potentiation of bradykinin.
6. Other adverse effects, which usually resolve when treatment is stopped, includes rashes, loss of taste and gastro-intestinal disturbances.

Contraindications include:

1. Renal artery stenosis is often asymptomatic but should be suspected in patients with hypertension and peripheral vascular disease.
2. ACE inhibitors should be used with caution in other forms of renal disease, as they have been reported to precipitate a reduction in renal function.

Coadministering with potassium sparing diuretics or potassium supplements presents a risk of serious hyperkalemia. ACE inhibitors are most effective in hypertension when used with a diuretic, although they increase dependence on the renin-angiotensin system.

Angiotensin II Receptor Antagonists

These are a group of drugs which block the receptor at which angiotensin II acts. Their effects are generally the same as ACE inhibitors. Other benefits, include maintenance of renal blood flow, although this has not been proven. Losartan was the first drug of this type and is currently used for hypertension.

Cardiac Glycosides, Digoxin and Digitalis

At the cellular level, digitalis compounds generally block the exchange of intracellular sodium ions for extracellular potassium by inhibiting Na^+/K^+-ATPase in the cell membrane; This increases intracellular sodium ions and encourages exchange for Ca^{2+} in order to raise its intracellular Ca^{2+} levels to increase the force of cardiac contraction (i.e. a positive inotropic action). These effects cause characteristic changes in the ECG.

In the body, digitalis and related compounds stimulate the vagus and increase vagal tone. These oral inotropes are most used in clinical practice. Digoxin is one of the most commonly prescribed. The principle effect of digoxin is to increase myocardial contractility. Some evidence suggests that in patients with a large dilated heart, digoxin reduces wall tension, improves oxygen usage, educes the end-diastolic volume and improves cardiac function. In chronic failure, digoxin is of modest value and is useful to control heart rate in the patient with atrial fibrillation.

Digoxin is well absorbed and can be taken orally or parenterally. It has a long pharmacological half-life of 36–38 h and is excreted unchanged by the kidneys with a very narrow therapeutic ratio (the relationship between clinical and toxic effects) and therefore, it is useful to monitor the plasma levels of digoxin. For rapid effects the dose can be loaded initially and followed by a maintenance dose. Digoxin is metabolised by the liver.

Clinical uses:

1. Atrial fibrillation (this is to slow the ventricular response)
2. Congestive heart failure. This is probably useful in sinus rhythm, but there is some uncertainty).

About 15% of patients on long term treatment will experience toxicity at some time. Severe overdoses can lead to hyponatremia and hyperkalemia due to cellular effects. Hypokalemia or hypercalcemia both increase the risk of digitalis toxicity. Old age or Hypothyroidism also increases sensitivity to standard doses.

The first manifestation of digoxin toxicity is usually gastro-intestinal disturbance, that is, anorexia, nausea, vomiting, anorexia, and occasionally diarrhoea. Thereafter any type of cardiac arrhythmia or conduction problem may occur.

Neurological effects include: fatigue, depression, malaise, insomnia altered colour vision (green/yellow are predominant), and confusion.

Cardiovascular effects include any form of cardiac arrhythmia especially ventricular bigeminy and bradycardias and complete heart block. Electrocardiogram may show first-degree heart block at an early stage, although serious arrhythmias may occur without prior warning.

Gynecomastia is also another unpleasant side effect.

Contraindications associated with digitalis compounds include:

1. myocardial ischemia, which may increase oxygen demand;
2. renal impairment, which may occur especially with digoxin;
3. precipitation of heart block;
4. disease of the AV node which may cause heart block.

Interaction with diuretics may worsen digoxin toxicity and cause hypokalemia. Amiodarone, quinidine and verapamil all increase the levels of plasma digoxin levels by causing a reduction in its clearance. Additionally, interaction with cholestyramine may block absorption of digoxin.

Clinical treatment of digoxin toxicity mainly depends on severity but generally, involves the following measures:

1. withdrawal of drug;
2. intravenously administer correct low serum K^+;

3. use lignocaine or phenytoin to treat ventricular arrhythmias;
4. in life-threatening situations, digoxin-specific can be used since anti-
    bodies will bind to and inactivate digoxin.

Cardiac Arrhythmias

The cardiac arrhythmias have already been well covered, some of which, do
not require treatment; for example, atrial or ventricular ectopics rarely need
treatment. Some of the others are more serious and require treatment for
prophylaxis or termination including correction of the underlying cause.
Antiarrhythmic drugs are usually classified according to the Williams
system. This is based on the changes in the action potential that is produced
by drugs in isolated cardiac cells. A number of drugs may have mixed
properties and do not fit neatly into any particular class. The classification is
not especially useful and is of limited clinical value.

Hypertension

Hypertension (high blood pressure) is considered to be one of the major
medical problems of the world. In the United States nearly 25 million cases
are diagnosed each year. Research shows that effective control of high blood
pressure can reduce associated morbidity (notably stroke, kidney, and
congestive heart failure) and prolong life. It may be argued whether
hypertension is a disease or merely a risk factor towards disease.
Nevertheless, hypertension increases the chance of developing organ
damage and is a major cause of death and disability. Diagnosis of
hypertension must be made carefully and with consideration since once
made, it carries life-long implications in terms of medical care and
management. Blood pressure may be initially high due to patient
apprehension and unfamiliarity with the surroundings. Repeat readings
usually show a fall although, subsequently steady values are obtained. This
is the basal blood pressure upon which the diagnosis of hypertension is
based and decisions are made as to what treatment should be adopted.

In hemodynamic terms:

Blood pressure = cardiac output × peripheral resistance

Hypertension can therefore be a consequence of either a raised cardiac
output and /or increased systemic vascular resistance. Whilst no single
causative mechanism can be identified, the following factors are all thought
to play a part in the development of hypertension.

It has been suggested that this mechanism is upset in hypertension due to
increased resistance of blood vessels in the kidneys, which in turn upsets the
filtration of water and increases the absorption of sodium (Na^+). Thus, blood

volume expands and blood pressure increases to set up a new steady-state in terms of sodium. The blood pressure at this point is significantly high.

In order to influence the regulatory mechanisms a variety of nervous and chemical stimuli are required. However, it should be remembered that the stimulus from nerve endings is chemical in nature and nervous control of blood pressure occurs through the autonomic nervous system. Neural impulses in parasympathetic or sympathetic fibres achieve their effects by discharge of specific neurotransmitter substances, acetylcholine and noradrenaline, respectively. The beta-receptors play a key role in this mechanistic pathway.

Baroreceptors and chemoreceptors are also involved in nervous control of the heart and have an inverse relationship between arterial blood pressure and heart rate.

The World Health Organization has suggested that blood pressure levels which are consistently above 160/95 mmHg should be regarded as "hypertension". However, many clinicians now feel that these levels are too high and that the threshold for diagnosis should be lower. In general, levels persistently above 140/90 mmHg may be regarded as being consistent with established hypertension. In this respect, the use of the word persistent is important as some patients have occasional high readings which are interspersed with normal values commonly known as "labile" hypertension. Around 25% of patients with this type of hypertension are likely to develop "established" hypertension. The prevalence of hypertension (70) is often asymptomatic and carries both morbidity and mortality. It is associated with increased risks from ischemic heart disease and cerebrovascular disturbances, and in severe cases, with heart and renal failure.

The therapeutic advances in the treatment of hypertension and related conditions are now described.

Blood Pressure

Blood pressure is determined by a combination of genetic and environmental factors. The chances of being hypertensive are greater within a family with a history of hypertension. The most constant environmental influence on hypertension is obesity. The mechanisms by which obesity appears to cause high blood pressure are not fully understood. However, it is not certain whether it is due to the high intake of nutrients although blood pressure is generally found to be lower in vegetarians. A decrease in the daily intake of salt may also reduce high blood pressure, but there is still much conjecture on this subject. Other factors such as alcohol consumption or stress also increase blood pressure.

Hypertension presents a unique problem in therapeutics. It is usually a life long disease that causes few symptoms until it reaches an advance stage. Pharmacological treatment varies depending on the severity of hypertension. In mild cases a single drug is usually sufficient such as a beta-blocker or a diuretic.

The definition of hypertension remains controversial. Unfortunately, there is no natural dividing line between "hypertension" and "normotension." Blood pressure varies widely over 24 h, both in those who have high and normal blood pressure.

Effects of Hypertension

In the United Kingdom, high blood pressure is one of the most common medical condition that requires long-term treatment. About one sixth of all men and women have high blood pressure. With increasing age, the blood pressure levels rise considerably. Annual deaths in the United Kingdom from hypertensive disease have been reported to be approximately 500 per million of the population.

Since high blood pressure has no warning signs and no clear symptoms, it is often called a silent killer. People may not be aware of it until they have trouble with their heart, brain, or kidneys. If high blood pressure is untreated, it may cause:

1. The heart to enlarge and lead to heart failure.
2. Aortic aneurysm (consistently high levels of blood pressure accompanied by degenerative changes in the vessel wall leading to a dilation of the aorta. The dilation may form an aneurysm, that is, like a bump or swelling protruding from the vessel resulting in a thin endothelial wall. Persistent hypertension if left untreated.
3. May precipitate rupture of the aorta and cause death.
4. The blood vessels in the kidneys to narrow and cause kidney failure.
5. The arteries in the body to "harden" faster, especially those associated with the heart, brain and the kidneys and lead to a heart attack or stroke.
6. Peripheral arterial disease (hypertension increases the incidence of atheroma of the legs. Intermittent claudication and ischemia of the feet are common clinical manifestations.

Etiology of Hypertension

Hypertension is classified according to the causes responsible for rise in blood pressure. Hypertension is often called essential or primary hypertension if there is no obvious underlying cause (i.e. idiopathic). More than 90% of all high blood pressure cases are considered to be "essential or primary hypertension." Conversely where there is an apparent causal link, for example to an endocrine disorder or kidney disease, it is then classed as secondary hypertension.

There are several factors which contribute to high blood pressure such as:

1. Genetic factors which are often associated with a family history of high blood pressure.
2. *Environmental factors*: Numerous factors have been related to the development of hypertension, but only the following appear to be important:
   i.   *Obesity*: Blood pressure rises with increasing obesity.
   ii. *Alcohol intake*: Consumption of alcohol acutely raises blood pressure.
   iii. *Salt intake*: Recent studies have shown that high-sodium intake is directly associated with higher blood pressure.
   iv. *Smoking*: Smoking increases blood pressure leading to damage of the blood vessels and the heart.
   v.   Stress and lack of regular physical exercise.

Several factors are often involved in elevating blood pressure. Apart from family history, changes in the above factors can help to treat blood pressure.

Management of Hypertension

Treatment with drugs is one of the effective methods for attaining larger reduction in blood pressure. Using drugs to control hypertension without producing unacceptable side effects is important.

In many countries management of hypertension is now based on the "stepped care" approach i.e. a drug regimen which aims to start the patient off with as low a dose as possible with, for example, a diuretic or beta-blocking agent. If this single drug therapy fails to control blood pressure (i.e. mild hypertension: blood pressure 140–160/90–105 mmHg) adequately then two types of drugs are given together to reduce blood pressure (i.e. moderate hypertension: blood pressure 160–200/105–115 mmHg). Should this combination fail to give acceptable clinical results then a third drug is added to the regimen to reduce blood pressure [i.e. severe, accelerated hypertension: blood pressure greater than 115 mmHg (diastolic)]. With the introduction of other drugs such as angiotension converting enzyme inhibitors and calcium antagonists, this traditional approach is now being challenged. However, as a general framework of the understanding of the overall approach to the treatment of hypertension, "stepped care" represents a good starting point for treatment of hypertension.

Action of Antihypertensive Drugs

The history of high blood pressure drugs goes back to 1950s when the first ganglionic blocking drugs were introduced. These drugs interfered with the constriction of blood vessels by nervous control and were the first successful treatment for malignant hypertension. A new generation of drugs have been discovered recently. Many patients can be satisfactorily treated with a single antihypertensive drug; the choice of the drug therapy is determined by safety convenience and freedom form side effects. Another group of patients will require a combination of two or three antihypertensive agents to give a good control with fewer side effects as previously described.

The principal drugs used in single drug treatments of hypertension are diuretics, beta-adrenoceptors, ACE inhibitors and calcium antagonists. Some of the most commonly used drugs to treat high blood pressure are now briefly reviewed:

1. *Diuretics*: Sometimes called "water pills", these drugs flush excess water and sodium from the body by increasing urination. This reduces the amount of fluid in the blood and flushes sodium from blood vessels so that they can open wider, with increased blood flow and reduction of pressure against the vessels. Diuretics are often used in combination with other blood pressure drugs.

2. *Beta-blockers*: These drugs slow the heartbeat by blocking the effects of nerve impulses to the heart and blood vessels, thereby lessening the burden on the heart.

3. *ACE (angiotensin-converting enzyme) inhibitors*: They inhibit formation of the hormone angiotensin II and cause the blood vessels to narrow and thereby increase the blood pressure.

4. *Calcium channel blockers*: They prevent calcium from entering into the muscle cells of the heart and blood vessels, in order to relax the blood vessels and decrease the blood pressure.

5. *Alpha-beta blockers*: These combine the actions of alpha-blockers (which relax the blood vessels) and the beta-blockers (which slow the heartbeat). Their dual effect reduces the amount and pressure of the blood which passes through the blood vessels.

6. *Vasodilators*: These agents directly dilate the blood vessels by relaxing the smooth muscle.

Treatment in Special Circumstances: Hypertensive Emergencies

Lowering of blood pressure urgently is only indicated in patients with hypertensive encephalopathy, or hypertension-induced renal or heart failure although, these conditions are rare.

High blood pressure in the absence of the above conditions do not warrant urgent blood pressure reduction, as it may be dangerous to reduce cerebral or coronary perfusion. In cases where urgent treatment is required, the patient should be admitted to hospital for rest and very carefully monitored. Usually oral therapy with beta-blockers, calcium antagonists or ACE inhibitors may be adequate.

In very rare cases where parenteral therapy is necessary, sodium nitroprusside is used by intravenous infusion. Sodium nitroprusside is a powerful arterial and venous vasodilator. It acts very rapidly with a very short duration of action. Blood pressure rises within a minute of the infusion being stopped. It allows control of blood pressure while other slower-acting drugs are being introduced. Prolonged use for periods of more than 24 h. causes the production of cyanide (a toxic metabolite). Labetalol, a combined alpha- and beta-blocker may also be administered intravenously.

Resistant Hypertension can occur with a failure to respond to triple therapy with standard drugs. This must be fully investigated to exclude the possibility of secondary hypertension. Minoxidil may be used in resistant hypertension. It is a vasodilator which causes fluid retention and tachycardia. It should only be used with diuretics and a beta-blocker. A unusual side effect of the drug is the stimulation of male pattern hair growth. Therefore, it is generally best to avoid the use of this drug in women.

Angina Pectoris

Angina can be managed by treating the acute attack by:

1. Resting the patient.
2. Administering an oral dose of glyceryl trinitrate tablets (GTN) or spray of medication. Patients who do not respond to attacks within minutes of taking the first or second tablet must seek medical attention without delay as this could indicate the early stages of a myocardial infarction (heart attack).

Prophylactic therapy must be taken regularly. There are generally three types of drugs employed: beta-blockers, calcium-channel blockers, and oral nitrates.

Each type of drug differs in its mode of action, although a combination of drugs can be used for the same patient. Aspirin is frequently used as long as there are no contraindications. Aspirin reduces the risk of threat of any serious thrombotic complications that may arise. For patients who previously had myocardial infarction, there is a good chance that aspirin may reduce the risks of further infarctions.

Myocardial Infarction (Heart Attack)

This usually occurs when a blood clot blocks a coronary artery and causes coronary thrombosis in a ruptured atherosclerotic plaque, occluding a coronary artery, so that part of the myocardium ceases to function. The patient will feel a severe chest pain. In addition to loss of myocardium, the patient is likely to be at risk from potentially fatal arrhythmias. Therefore, prompt action should be taken to reduce further loss of myocardium and to improve survival of the patient. About one in ten patients who have unstable angina may develop acute myocardial infarction.

Management of myocardial infarction often requires alternative drugs in order to deal more effectively with further complications. For example, the following measures have been indicated:

1. Use of sublingual glyceryl trinitrate (if a severe attack of angina occurs, as opposed to myocardial infarction);
2. Use of aspirin (oral): providing there is no contraindication. This can improve survival of the patient.

Other treatments involve:

1. Pain relief by IV administration of opiates usually with an anti-emetic drug.
2. Use of thrombolytic drugs if no contraindication exists.
3. Other types of treatments may also be employed to decrease mortality although they are generally less widely adopted, including the use of atenolol by IV or the use of nitrates.

CONCLUSION

The cardiovascular system is a complex network of blood vessels with the heart at its center. The function of the system is to circulate oxygenated blood and nutrients to the various tissues in the body and return deoxygenated blood to the lungs for reoxygenation. The pharmacology of the sympathetic nervous system is important in cardiovascular medicine in view of the stimulating action of the sympathetic receptors. The beta-blockers are among some of the most commonly used drugs. They cause the heart rate to slow down and decrease cardiac contractility.

Systemic hypertension is a major cause of death and disability because it is prevalent, often asymptomatic, and carries morbidity and mortality. The problem of identifying and treating hypertension particularly in the mild to moderate category arises from the fact that uncomplicated hypertension is largely considered to be a symptomless condition. However, hypertension is usually a consequence of either raised cardiac output and/or increased systemic vascular resistance. Complications of raised blood pressure, arrhythmia, and atrial fibrillation can lead to further impairment and cause congestive heart failure.

Although clinical manifestations of coronary heart disease do not differ between hypertensive and normotensive patients, the degree of severity of atherosclerosis is generally greater. The classical clinical feature of coronary heart disease is angina pectoris, the course of which can result in myocardial infarction. There is now an increasing interest in the clinical pharmacology of drugs in the treatment cardiovascular diseases.

Recognition of the central role and the pharmacological action of some of the most commonly used drugs in cardiovascular therapy, such as the antihypertensives, calcium channel blockers, beta-blockers, alpha-blockers, angiotension converting enzyme (ACE) inhibitors, ganglion and adrenergic neurone blockers, vasodilators, diuretics, cardiac glycosides, and drugs for combating arrhythmias, have all been well covered herein to demonstrate their pharmacological importance and widespread use in clinical practice as drug therapies in cardiac and advanced congestive failure.

REFERENCES

1. Ghuran AV, Camm AJ. Ischaemic heart disease presenting as arrhythmias. Br Med Bull 2001; 59:193–210.
2. Rees SA, Curtis MJ. Selective Ik blockade as an antiarrhythmic mechanism: effects of UK 66,914 on ischaemia and reperfusion arrhythmias in rat and rabbit hearts. Br J Pharm 1993; 108(1):139–45.
3. Johnson PJ, Newton JC, Rollins DL, Knisley SB, Ideker RI, Smith WM. Intelligent multichannel stimulator for the study of cardiac arrhythmias. Ann Biomed Eng 2002; 30(2):180–91.
4. Berne RM, Levy MN, Koeppen BM, Stanton BA. Physiology, 4th edn. St. Louis, MO, London: Mosby; 1998:342–57.
5. Mycek MJ, Havey RA, Champe PC. Pharmacology, 2nd edn. Philadelphia: Lippincott Williams & Wilkins; 2000.
6. Abrams AC. Clinical Drug Therapy: Rationales for Nursing Practice, 6th edn. Philadelphia: Lippincott, 2000: 766–76.
7. Milnor RW. Cardiovascular Physiology. New York, Oxford: Oxford University press; 1990:69.
8. Jacob LS. Pharmacology, 4th edn. Philadelphia, PA: Williams & Wilkins: London; 1996.
9. Page C, Curtis M, Sutter M, Walker M, Hoffman B. Integrated Pharmacology. London: Mosby; 1997:159–63.
10. Huszar RJ. Basic Dysrhythmias Interpretation & Management, 2nd edn. St. Louis: London: Mosby Lifeline; 1994:126–202.
11. http://content.answers.com/main/content/wp/en/thumb/e/e8/300px-Sinus RhythmLabels.png, http://www.jm.com/assets/popups/electsys.gif; www.unmc. edu/Physiology/Mann/pix_3/f3-17.gif
12. Laurence DR, Bennett PN and Brown MJ. Clinical Pharmacology, 8th edn. New York: Churchill Livingstone; 1997:459–69.
13. Rosen MR. Mechanism of cardiac arrhythmias: focus on atrial fibrillation. J Gender Specific Med 2001; 4(3):37–47.
14. Wyse DG, Love JC, Yao Q, et al. Atrial Fibrillation: A risk factor for increased mortality – an AVID registry analysis. J Intervent Cardiac Electrophysiol 2001; 5(3):267–73.

15. Palladi RT, Perry MA, Campbell TJ. Proarrhythmic effects of an oxygen-derived free radical generating system on action potentials recorded from guinea pig ventricular myocardium: a possible cause of reperfusion induced arrhythmias. Circulat Res 1987; 61:50–54.

16. Walker MJA, Curtis MJ, Hearse DJ, et al. The Lambeth Conventions: guidelines for the study of arrhythmias in ischaemia, infarction, and reperfusion. Cardiovas Res 1988; 22:447-55.

17. W.Kalant H, Roschalau WHE. Principles of Medical Pharmacology, 6th edn. New York: Oxford University; 1988:458–75.

18. Katzung BG. Basic and Clinical Pharmacology, 8th edn. New York: Lange Medical Books, McGraw-Hill; 2001:225–6.

19. Hollinger MA. Introduction to Pharmacology, 2nd edn. London: Taylor & Francis; 2003:256–261.

20. Burgen ASV, Mitchell JF. Gaddum's Pharmacology, 9th edn. Oxford: Oxford University Press; 1985:127.

21. Fagbemi SO, Chi L, Lucchesi BR. Antifibrillatory and profibrillatory actions of selected class I antiarrythmic agents. J Cardiovas Pharm 1993; 21: 709–19.

22. Scottish Intercollegiate. Guidelines Network. Secondary prevention of coronary heart disease following myocardial infarction. 1st edn. 2000:17–18.

23. Beatch GN, Barrett TD, Plouvier B, et al. Ventricular fibrillation, uncontrolled arrhythmia seeking new targets. Drug Dev Res 2002; 55(1):45–52.

24. Andres and Goth. Medical Pharmacology: Principles and Concepts, 11th edn. St. Louis, MO: Mosby; 1984:447.

25. Smith C and Reynard A. Essential Pharmacology. Philadelphia, PA: Saunders; 1995: 270–72.

26. Craig CR, Stitzel RE. Modern Pharmacology with Clinical Applications, 5th edn. Boston, MA: Little, Brown and Co; 1997: 177–85.

27. Brody M, Larner J, Minneman KP. Human Pharmacology Molecular to Clinical, 3rd edn. St. Louis: Mosby; 1998:209–10.

28. Hamer J. Drugs For Heart Disease, 2nd edn. London: Chapman and Hall; 1987:7–20.

29. British Medical Association. Royal Pharmaceutical Society of Great Britain. British National Formulary 2001; 62.

30. Rosenfeld GC, Loose-Mitchell DS. Pharmacology, 3rd edn. Baltimore, MD: Williams & Wilkins; 1998:88–93.

31. Katzung BG, Trevor AJ, Examination & Board Review Pharmacology, 5th edn. Stamford, CO: Appleton & Lange; 1998; 121.

32. Ciccone CD (1990) Pharmacology in Rehabilitation. Philadelphia, Pa.: F.A. Davis, pp. 258 - 260.

33. Janse MJ, De Bakker JMT. Arrhythmia substrate and management in hypertrophic cardiomyopathy: from molecules to implantable; cardioverter-defibrillators, European Heart Journal Supplements , 2001; 3(L):L15–20.

34. Morena H, Janes MJ, Fiolet JWT, Krieger WJT, Crijns H, Durrer D. Comparison of the effects of regional ischaemia, hypoxia, hyperkalemia, and acidosis on intracellular and extracellular potentials and metabolism in the isolated procine heart. Circulation Res 1980; 46:634–46.

35. Bowman WC, Rand MJ. Textbook of Pharmacology, 2nd edn. London: Blackwell Scientific Publication; 1980.
36. Morganroth J. Antiarrhythmic effects of beta-adrenergic drugs. J Cardiol 1987; 60:10–14D.
37. Kendall MJ, Horton RC. Preventing Coronary Artery Disease Cardio-protective Therapeutics in Practice, 2nd edn. London: Dunitz.
38. Papp GJ, Seekers L. Expt cardiac arrhyth and arrhyth drugs. Budapest Hungary: Akademiai Kiado; 1971.
39. Conn HL. The myocardial cell's structure, function and modification by cardiovascular drugs. Philadelphia, PA: Pennsylvania University Press; 1966.
40. Sic Hondeghem LM, Katzung BG. Antiarrhythmic agents: the modulated receptor mechanism of action of sodium and calcium channel-blocking drugs. Annu Rev Pharmacol Toxicol 1984; 24:387–423.
41. Koch-Wester J, Ann NY. Pharmacokinetic of procainamide in man. Acad Sci 1971; 179:370.
42. Giardinia EV, Dreyfuss J, Bigger JT, Shaw JM, Schreiber EC. Metabolism of procainamide in normal and cardiac patients Clin Pharmacol Therap 1976; 19: 339.
43. Hollunger G. On the metabolism of lidocaine. II. The biotransformation of lidocaine. Acta Pharmacol 1960; 17:365–71.
44. Anderson JL, Mason JW, Roger MD. Clinical electrophysiology effects of tocainide. Circulation 1978; 57:685–90.
45. Schwarz JB, Keefe D, Kates RE, Kirsten E, Harrison DC. Acute and chronic pharmacodynamic interaction of verapamil and digoxin in atrial fibrillation. Circulation 1982; 65:1163–70.
46. Guehler J, Gornick CC, Tobler HG, et al. Electrophysiologic effects of flecainide acetate and its major metabolites in the canine heart. Am J Cardiol 1985; 55(6):807–12.
47. Miyano S, Sumoto K, Satoh F, et al. New antiarrhythmic agents, N-aryl-8-pyrrolizidine alkanamides. J Med Chem 1985; 28:714–7.
48. Yatani A, Akaike N. Effect of a new antiarrhythmic compound SUN 1165 [N-(2,6-dimethylphenyl)-8-pyrrolizidineacetamide hydrochloride] on the sodium currents in isolated single rat ventricular cells. Arch Pharmacol 1984; 326(2): 163–8.
49. Hidaka T, Hamasaki S, Aisaka K, et al. N-(2,6-dimethylphenyl)-8-pyrrolizi-deneacetamide hydrochloride hemihydrate (SUN 1165) a new antiarrhythmic agent: effects on cardiac conduction. Arzneimittel Forschung J 1985; 35(9): 1381–6.
50. Aisaka K, Hidaka T, Inomata N, Hamasaki S, Ishihara T, Morita M. N-(2,6-dimethylphenyl)-8-pyrrolizidineacetamide hydrochloride hemihydrate (SUN 1165): a new potent and long-acting antiarrhythmic agent. Arzneimittel Forschung J 1985; 35(8):1239–45.
51. Bergey JL, Sulkowsky T, Much DR and Wendt RL. Antiarrhythmic, hemo-dynamic and cardiac electrophysiological evaluation of N-(2,6-dimethylphenyl)-N-[3-(l-methylethylamino)propyl]urea, Arzneimittel Forschung J 1983; 33(9): 1258–68.

52. Luceri RM, Castellanos A, Zaman L, Selby T, Myerburg R. Electrophysiology of recainam. Clin Res 1984; 32:244.

53. Helgesen KG, Kristiansen O, Refsum H. Comparison of electrophysiological and mechanical effects of droxicanide and lidocaine on heart muscle isolated from rats. Acta Pharmacol Toxicol 1984; 55(4):303–7.

54. Aberg G, Ekenstam B, Smith E. Pharmacological studies on droxicainide, a new antiarrhythmic agent. Arzneimittel Forschung J 1983; 33(5):706–10.

55. Aberg G, Ronfeld R, Smith E. The acute antiarrhythmic effects of droxicainide and lidocaine in unanesthetized dogs. J Cardiovas Pharm 1984; 6(2):355–60.

56. Tejerina T, Barrigon S, Tamargo J. Comparison of three beta-aminoanilides: IQB-M-81, lidocaine and tocainide, on isolated rat atria, Eur J Pharm 1983; 95: 93–9.

57. Berhart CA, Condamine C, Demarne H, et al. Synthesis and antiarrhythmic activity of new [(dialkylamino) alkyl] pyridylacetamides, Journal of Medicinal Chemistry, 1983; 26:451–5.

58. Gautier P, Escande D, Bertrand JP, Seguin J, Guiraudou P. Electrophysiological effects of penticainide (CM 7857) in isolated human atrial and ventricular fibers. J Cardiovasc Pharmacol. 1989; 13(2):328–35.

59. Houin G, Jeanniot JP, Ledudal P, Barre J, Tillement JP. Liquid-chromatographic determination of propisomide and its mono-N-dealkylated metabolite in plasma urine, J Pharm Sci 1985; 31(7):1222–4.

60. Priori S, Facchini M, Bonazzi O, Varisco T, Zuanetti G, Schwartz PJ, Cuspidic. Evaluation of flecainide in the therapy of chronic ventricular arrhythmia using the acute oral method. G Ital Cardiol 1985; 15(3):273–82.

61. Damiano BP, Le-Marec H, Rosen MR (1985) Electrophysiological effects of AHR 10718 on isolated cardiac tissues. Fed Proc Eur J Pharm 1985; 108(3): 243–55.

62. Pourrias B, Huerta F, Santamaria R, Versailles JT. Effects of MD 770207 (carocainide) on experimental ventricular arrhythmias, Arch Int Pharmacodyn Ther 1983; 363(1):85–102.

63. Huerta F, Pourrias. Effects of carocainide (770207) on canine cardiac automaticity, Arch Int Pharmacodyn Ther 1984; 268(2):216–24.

64. Huerta F, Santamaria R, Pourrias B. Electrophysiological effects of (770207) on the dog heart. Arch Int Pharmacodyn Ther 1984; 267(2):289–98.

65. Mitsuhashi H, Akiyama K, Hashimoto K, Sawa Y, Hatori Y. Antiarrhythmic effects of a new drug, E-0747, on canine ventricular arrhythmia models. Jpn J Pharmacol 1987; 44(2):155–62.

66. Steinberg MI, Wiest SA. Electrophysiological studies of indecainide hydrochloride, a new antiarrhythmic agent, in canine cardiac tissues. J Cardiovas Pharm 1984; 6(4):614–21.

67. Dennis PD, Williams EMV. Effects on rabbit cardiac potentials of aprindine and indecainide, a new antiarrhythmic agent, in normoxia and hypoxia, Br J Pharm 1985; 85(1):11–9.

68. Shattat FG. Synthesis and pharmacological evaluation of novel ethyl-1H indole-2-carboxamides as potential cardiovascular agents, Ph.D Thesis, Coventry University; 2004.

69. Peter Winstanley P, Walley T. Pharmacology, Churchill's Mastery of Medicine, A Clinical Core Text for Integrated Curricula with Self-Assessment. Edinburgh: Churchill Livingstone; 1996:26–34.

70. Vaughan-Thomas H. Health Care Education Services Ltd, Edinburgh. Nottingham, England: The Boots Company PLC; 1985:1–60.

12

Immunosuppressant Drugs

Troy Purvis

College of Pharmacy, University of Texas at Austin, Austin, Texas, U.S.A.

Kirk A. Overhoff

Schering-Plough Research Institute, Kenilworth, New Jersey, U.S.A.

Prapasri Sinswat

Department of Pharmacy, Chulalongkorn University, Bangkok, Thailand

Robert O. Williams III

College of Pharmacy, University of Texas at Austin, Austin, Texas, U.S.A.

INTRODUCTION

Within the past few decades, the number of patients on organ transplant waiting lists has increased dramatically while the number of available organs has remained constant (1). This shortage has led to a reduction in the acceptable criteria for organ transplants ("marginal" donor) and a higher risk for organ rejection. Coupled with the rise in incidence of autoimmune diseases, safer and more effective immunosuppressant medication is needed. Immunosuppressants are used to modulate and in some cases inhibit the cascade of reactions leading to an immune response. Classification of immunosuppressants is based on their mechanism of action, and each class has its own set of formulation challenges. Researchers have used several strategies to improve formulation design of these drugs to overcome some of these formulation challenges. Nanoparticle engineering [Nanocrystals®, solution based dispersion by supercritical fluids (SEDS), and emulsification techniques] is one example of formulation design that can overcome formulation challenges such as poor aqueous solubility. Choice of excipients and different routes of administration of the drug can also improve bioavailability of immunosuppressants hindered by p-glycoprotein (PGP) efflux and cytochrome P450 (CYP 3A) metabolism. Changing the route of administration of these drugs can improve targeting and absorption of the drugs (e.g. inhalation administration for lung transplantation), leading to improved therapeutic outcomes. Novel approaches to formulation design of immunosuppressant drugs have lead to enhanced therapeutic outcomes of these drugs for various disease states. While immunosuppressants are primarily used to treat solid organ (liver, kidney, heart, and lung) and tissue (bone marrow) transplant rejection, other disorders such as multiple sclerosis, psoriasis, ulcerative colitis, and asthma are all being aggressively treated with immunosuppressant drugs. Advanced design of drug delivery systems for delivering immunosuppressants shows promise in treating these disease states along with other autoimmune diseases. Immunosuppressant drugs are classified into one of four categories based on their mechanism of action in the body and the type of immunosuppressive effect that is observed in vivo. The categories of immunosuppressants are: glucocorticoids, immunophilin binders, cytostatics, and other immunosuppressant drugs (including monoclonal antibodies, interferons, and other proteins) (2). For the purposes of this review, novel formulations of "small" molecule drugs, including glucocorticoids, immunophilin binders, and cytostatics, will be discussed.

Over a period of time, doses of the drug can be reduced in transplant patients as the chance of transplant rejection decreases, but most transplant patients must maintain administration of immunosuppressants for life. Therapeutic drug monitoring of blood drug concentrations typically accompanies long-term immunosuppressant therapy due to intrapatient variability of drug absorption. Normalizing the bioavailability of these medications and acquiring more stable blood levels has been the goal of formulation scientists. Frequently, patients are administered combinations of different types of immunosuppressants in order to modulate all components of the immune response cascade.

Using novel drug delivery systems, immunosuppressant drugs can show increased efficacy in vivo. Various strategies have been used to develop drug delivery systems of immunosuppressants with enhanced therapeutic outcomes. Reducing primary particle size of the drug has shown increased absorption with various different types of immunosuppressants. Nanocrystal wet-milling technology has shown promise in delivery of the poorly water soluble crystalline drugs (3), including immunosuppressants (4). Various types of immunosuppressant-loaded microparticles and nanoparticles have been produced, showing promising therapeutic results in vivo. Solid lipid nanoparticles (5, 6), biodegradable encapsulating nanoparticles (7), and diffusion controlled release nanoparticles (8) have shown enhanced modes of delivery for various routes of administration. Overall, this chapter summarizes advanced formulation designs of immunosuppressants resulting in enhanced therapeutic outcomes of these drugs with improved targeting and delivery.

GLUCOCORTICOIDS

The corticosteroids or glucocorticoids were the first type of drug found to have potent immunosuppressive effects (9) and current formulations containing glucocorticoids include those for pulmonary, oral, topical, and parenteral administration. These drugs include prednisone, prednisolone, hydrocortisone, betamethasone, budesonide, and other similar compounds produced by the adrenal glands. They act to reduce the inflammatory response along with modulation of the immune response. Glucocorticoids were first used to treat acute transplant rejection and graft-vs-host disease in transplant patients, but now are frequently used in combination with other medications for transplant maintenance. Other indications for glucocorticoids include treatment of allergic reactions of varying types, and autoimmune disorders. The mechanism of action of the glucocorticoids is owed to their broad spectrum ability to suppress cell-mediated immunity (9). These drugs inhibit the production of cytokines and the interleukins IL-1, IL-2, IL-3, IL-4, IL-5, IL-6, IL-8, and tumor necrosis factor gamma by inactivating the genes that code for them (10). Reduction in the amount of free cytokines, specifically IL-2, reduces the rate of T-cell proliferation, leading to an overall reduction in the immune response. Humoral immunity results from the smaller amounts of IL-2 and IL-2 receptors expressed after glucocorticoid administration. Anti-inflammatory effects of glucocorticoids result from the synthesis and release of lipocortin-1 which binds to cell membranes and prevents phospholipase A2 from producing eicosanoids from arachadonic acid. Lipocortin-1 also acts to inhibit emigration, chemotaxis, and the release of immune system mediators from neutrophils, macrophages, and mast cells. Cyclooxygenases (COX-1 and COX-2) are also suppressed by glucocorticoids, adding to their anti-inflammatory action (10). This broad spectrum of immunomodulatory activity has led to the use of glucocorticoids in treating a wide variety of chronic disorders including but not limited to asthma, rheumatoid and juvenile arthritis, lupus erythematosus, and inflammatory bowel disease (IBD). However, because of their broad immunomodulatory activity, glucocorticoids can reduce the patient's ability to fight infection while inhibiting the reparative processes in the body.

The physicochemical properties of the glucocorticoids have been studied extensively. In general, the glucocorticoid molecules contain a central steroidal backbone with differing functional groups which impart the anti-inflammatory action to the compound. The steroidal backbone includes three fused six-membered ring structures with a five-membered ring fused to it (11). As a result, the glucocorticoids are generally lipophilic in nature with low aqueous solubility, and chemical alteration of these drugs can be performed to improve their solubility and potency. For example, prednisone is a derivative of the naturally occurring compound cortisone, having an extra degree of unsaturation in the steroidal structure. Despite the low aqueous solubility of these drugs in their native forms, steroids may be considered to have good intestinal absorption, having high apparent permeability (log P_{app}) values and showing good passive transport permeability. Formulation of steroids into salt forms generally increases their aqueous solubility and allows for increased bioavailability. Oral bioavailability of the glucocorticoids, however, can be negatively influenced by their interaction with p-glycoprotein (P-GP) and the cytochrome P450 3A4 isoenzymes (CYP 3A4) in the intestines (11). The physicochemical properties of some drugs of interest have been summarized in Table 1. These physicochemical properties must be considered when devising novel drug delivery systems for these types of drugs.

Beclomethasone and Betamethasone

Beclomethasone and betamethasone are two very closely related glucocorticoid immunosuppressants with identical structures, except in betamethasone, the heteroatom substituent on the steroidal backbone is fluorine (12), while in beclomethasone, the heteroatom substituent is chlorine (13). The chemical structures of beclomethasone and betamethasone dipropionate are depicted in Figure 1. Frequently, these two drugs are administered as salt forms (dipropionate, sodium phosphate) due to their low aqueous solubility in their native forms. Typically, betamethasone is administered parenterally, topically, or via the ophthalmic route, while beclomethasone is administered topically or inhaled.

Beclovent® Aerosol for Inhalation (Glaxo Smithkline, Research Triangle Park, NC) is a commercial formulation of beclomethasone dipropionate indicated for asthma and other pulmonary autoimmune conditions (14). Beclomethasone dipropionate is sparingly water soluble and poorly mobilized from the injection site for subcutaneous or intramuscular administration, so pulmonary formulation of this drug is more efficacious than the parenteral formulation. This formulation is composed of microcrystalline beclomethasone dipropionate with oleic acid excipients added as a surfactant (14). Beclovent Aerosol is formulated in a pressurized metered dose inhaler (pMDI) device with chlorofluorocarbons (CFCs) such as trichloromonofluoromethane and dichlorodifluoromethane used as propellant agents. The propellants offer an additional advantage, forming a clathrate with beclomethasone dipropionate, trapping the drug molecule around a "cage" of propellant molecules through dispersion interactions. After actuation of the pMDI, most of the drug is deposited in the mouth and throat areas, while only a fraction is deposited in the deep lung tissue. A considerable amount of the formulation is swallowed, indicating that the particle size of the drug after actuation is larger than the respirable fraction

Table 1 Selected Immunosuppressant Drugs' Physicochemical Properties

Drug	BCS classification	Aqueous solubility[a]	log P app	PGP interaction	CYP 3A4 interaction
Prednisolone	1 (for salt form)	481.1 mg/L [9]	1.4 (9)	PGP substrate (9)	CYP3A4 substrate (9)
Hydrocortisone	1 (for salt form)	896.6 mg/L (9)	1.7 (9)	PGP substrate (9)	CYP3A4 substrate (9)
Dexamethasone	1 (for salt form)	254.8 mg/L (9)	1.8 (9)	PGP substrate (9)	CYP3A4 substrate (9)
Cyclosporin	2	0.028 mg/mL (36)	3.0 (36)	PGP substrate (35)	CYP3A4 substrate (35)
Tacrolimus	2	0.012 mg/mL (42)	4.6 (42)	PGP substrate (35)	CYP3A4 substrate (35)
Sirolimus	2	0.026 mg/mL (53)	3.6 (53)	PGP substrate (35)	CYP3A4 substrate (35)
Azithioprine	1 (for sodium salt)	272 mg/L (56)	0.1 (56)	none known	none known
Methotrexate	1 (for sodium salt)	2600 mg/L (56)	−1.08 (56)	none known	none known

Note: BCS classification is based on aqueous solubility. PGP interaction and CYP 3A4 interaction is qualitative. Log P values are calculated.
[a] Solubility is given as the value in water at 25°C.

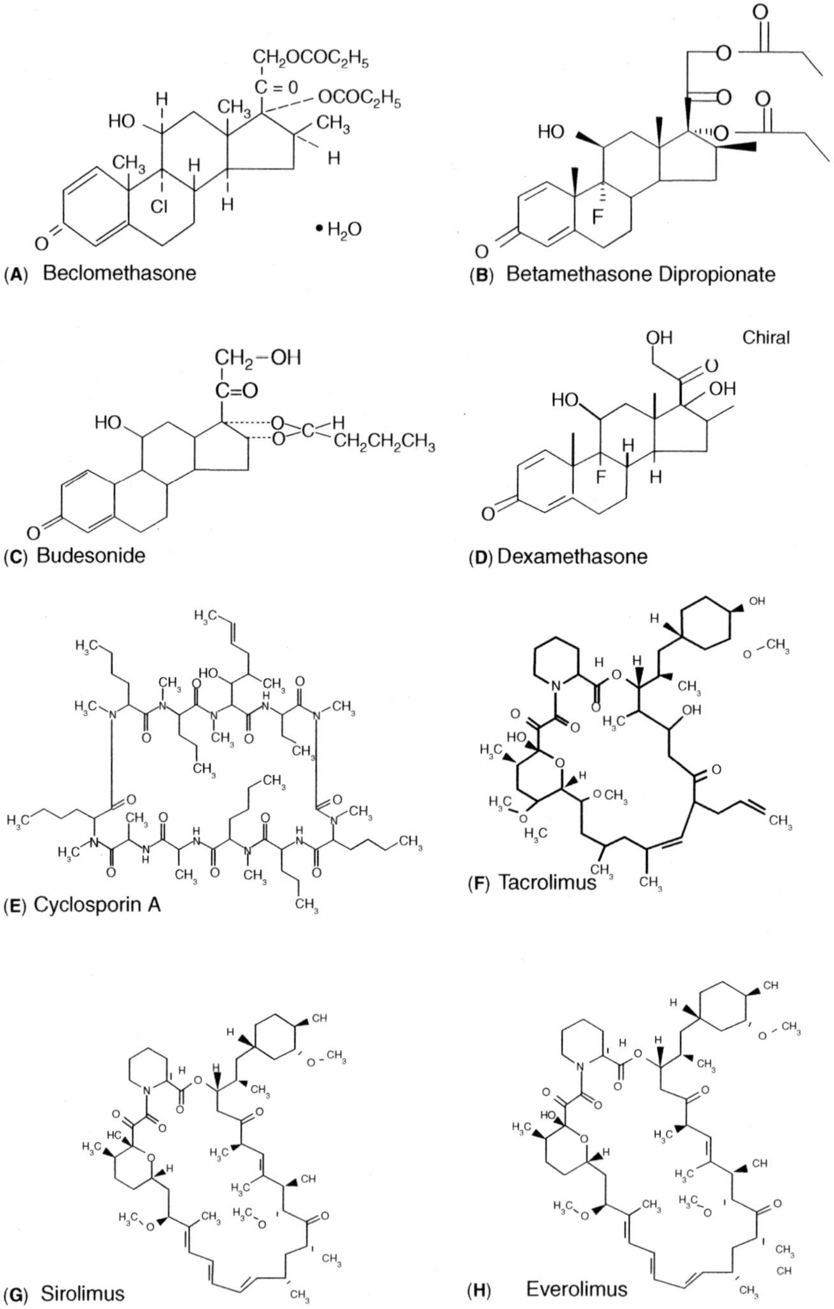

(A) Beclomethasone

(B) Betamethasone Dipropionate

(C) Budesonide

(D) Dexamethasone

(E) Cyclosporin A

(F) Tacrolimus

(G) Sirolimus

(H) Everolimus

Figure 1 *See caption on page 415.*

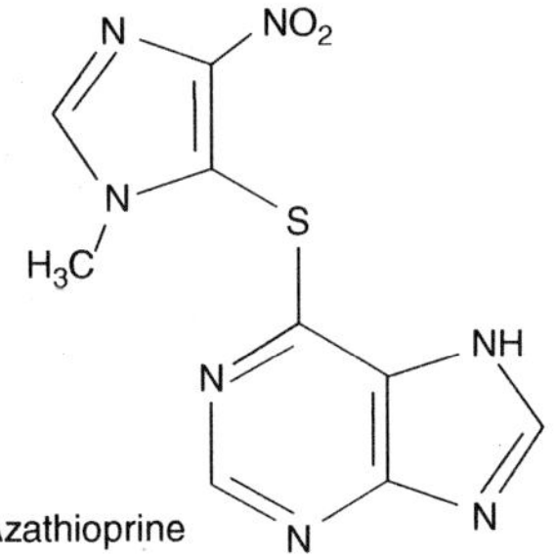

Figure 1 The chemical structures of immunosuppressant drugs detailed in this article. (**A**) Beclomethasone, (**B**) Betamethasone Dipropionate, (**C**) Budesonide, (**D**) Dexamethasone, (**E**) Cyclosporin A, (**F**) Tacrolimus, (**G**) Sirolimus, (**H**) Enverolimus, (**I**) Azathioprine, (**J**) Methotrexate.

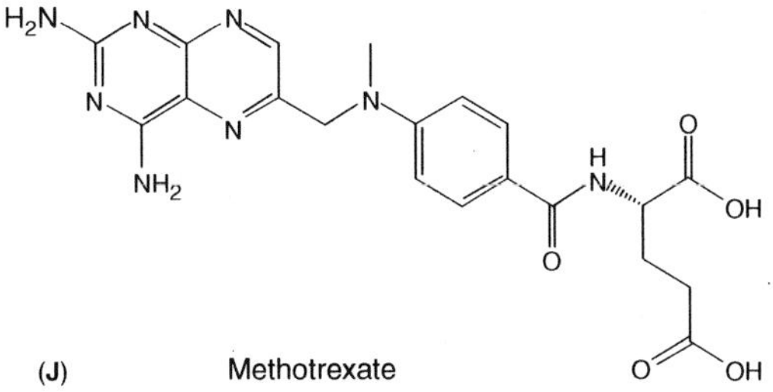

(3–5 µm) needed for lung deposition. Additionally, the Beclovent Aerosol for Inhalation contains ozone-depleting chlorofluorocarbons which are being phased out of production (14).

Other pMDI formulations of beclomethasone have been reported which do not contain ozone-depleting chlorofluorocarbons. Rocca-Serra et al. have studied the efficacy and tolerability of beclomethasone pMDI formulations which contain hydrofluroalkane (HFA-134a) propellants as an alternative to CFCs (15). These propellants do not show a propensity to form clathrates with the drug, but they do effectively propel the drug into the patients' respiratory system in the desired respirable fraction. This study involved monitoring asthmatic patients' in two treatment groups: one group using the commercial Beclojet® with CFC containing propellants and the other using the beclomethasone/HFA-containing propellant. The study did not involve a placebo group because of ethical concerns about harm to human asthmatic subjects not receiving treatment for their conditions. A double-blind study was designed with randomized groups receiving one of the two treatments over a 6-week study period. Blood cortisol concentrations, adverse events, and treatment efficacy were studied. Peak expiratory flow (PEF), the measure of breathing efficiency, was the main measure of efficacy of the formulations. PEF was measured both in the morning and in the evening following treatment with the two different beclomethasone pMDI's. In the protocol study group, the average starting baseline PEF for patients was measured at 392.6 L/min for the HFA group and 407.3 L/min for the CFC group. Final PEF measurements showed 403.9 L/min in the HFA group and 414.7 L/min in the CFC group. The protocol study's baseline to final PEF ratios were similar, showing a ratio of 1.06 for the HFA group and 1.05 for the CFC group. The results of the study also showed that the particle sizes of aerosols produced by the

HFA formulation were smaller than those produced by the CFC formulation, resulting in the need to half the dose when switching from the CFC to the HFA formulation. The present study demonstrated that beclomethasone HFA was not statistically inferior to the beclomethasone CFC formulation for adjusting morning PEF, evening PEF, or PEF variability. The study constituted 449 patients, with a total number of adverse events involving the respiratory system being 57 in the HFA group, compared with 35 in the CFC group, most of the adverse events being mild to moderate. The study concluded that non-extra fine beclomethasone HFA-134a formulation was equivalent to the CFC formulation used in the study with regard to clinical efficacy and tolerability in patients with mild to severe asthma (15).

Another study involving the use of HFA propellant over CFC propellant in pMDI formulations of beclomethasone was conducted, evaluating the quality of life of asthmatic patients using these treatments. Juniper et al. reported data using clinical indexes of patients switched from conventional beclomethasone treatment to approximately half the dose of extra fine beclomethasone aerosol (16). This study involved human asthmatic volunteers with a total number of 152 male patients and 40 female patients. The formulation of the HFA-BDP pMDI was altered in one group to produce extra fine aerosols with an average particle size of 1.1 μm. The baseline (pre-treatment) quality of life scores were measured by a 7-point value score with instance of severe asthma symptoms occurring (1 = severe asthma all the time to 7 = severe asthma none of the time). The average baseline score for male patients was 5.45, while the average baseline score for female patients was 5.36. Patients were randomized in a 3:1 HFA-BDP group (400–1600 μg/day), a 3:1 CFC-BDP group (400–1600 μg/day), and an HFA-BDP group with approximately half the dose of BDP administered (200–800 μg/day). The duration of the study was 12 months, and the mean change from baseline was analyzed. Results showed that the overall change in baseline quality of life was increased to about 0.18 with the HFA (200–800 μg/day) group at 2 months, while the CFC group (400–1600 μg/day) increased only 0.8 over the 2 month time period. At 4 months, the HFA group (200–800 μg/day) showed increased quality of life scores of about 2.4 and the CFC group (400–1600 μg/day) increased to 1.6, while at 8 months, the HFA group (200–800 μg/day) increased to 2.8, with the CFC group (400–1600 μg/day) decreasing to 0.7. At the final time point of 12 months, the HFA (200–800 μg/day) the scores had increased to 0.32 from baseline, while the HFA (400–1600 μg/day) remained constant at about 0.33. The study concluded that patients switched from CFC-BDP to approximately half the dose of HFA-BDP extra fine aerosol experienced significant improvement in asthma-specific quality of life. The reason for the discrepancy between quality of life and clinical outcomes was that HFA-BDP spray was deposited in more in the peripheral airways as well as alveolar sacs. Also, improvement in quality of life seen in the present study developed progressively over the 12 month time period, suggesting that improvement could be increased with even longer term use of HFA-BDP (200–800 μg/day). The difference in quality of life score was significant between the HFA-BDP group (200–800 μg/day) versus the CFC (800–1600 μg/day) group by 0.24 standard deviations. The study concludes that switching from CFC based BDP aerosols to extra fine aerosol mist with HFA-BDP (200–8000 μg/day) may experience clinically important improvements in quality of life. This study also confirmed from shorter studies that asthma control can be maintained with approximately half the dose of HFA-BDP versus CFC based BDP treatment (16).

NanoSystems® particle size reduction technology has been used to produce drug formulations of immunosuppressants with improved physicochemical properties (17). The Nanocrystal process was initially developed by NanoSystems, a division of Elan Pharmaceutical Company. This process involves the use of media milling technology to formulate poorly water soluble drugs into nanocrystalline particles which offer improved drug delivery (17). In this process, large micron sized crystalline particles of poorly water soluble compounds are milled in aqueous solution containing water-soluble stabilizer solutions. This process produces physically stable dispersions consisting of nanometer sized drug crystals that do not spontaneously reaggregate due to the surface adsorption of stabilizing excipients. High energy wet milling of the crystals in the presence of the stabilizing excipients causes the drug crystals to fracture to nano-sized particles. Polymeric excipients then adsorb to the crystal surface of the particles, inhibiting aggregation and particle growth of the crystals and providing stable dispersions of the drug crystals. Following milling for 30–60 min, unimodal distribution profiles and mean particle diameters of < 200 nm have been reported. This type of processing is especially useful in formulating drugs that are poorly water soluble and do not dissolve readily in the stabilizer-containing aqueous media in which they are milled (17). A schematic of the Nanocrystal technology is shown in Fig. 2 (17). A comparison of the size of beclomethasone dipropionate crystals both before and after Nanocrystal processing is shown in Fig. 3 (4).

The use of Nanocrystal technology for enhancing delivery of beclomethasone dipropionate (BCD) is an example of novel formulation design of immunosuppressant drugs. An initial study involved the preparation of pulmonary BCD using polyvinyl alcohol (PVA) as a stabilizer. Unmilled BCD showed a mean particle size of 10.5 μm, and after Nanocrystal processing of BCD in 2.5% PVA, the mean particle size was reduced to 267 ± 84 nm (18). The particle size of the crystals remained constant throughout the study, and following 7 months of storage at room temperature, the mean size was found to increase only slightly to 282 ± 73 nm (18). A second study was conducted, using BCD wet-milled

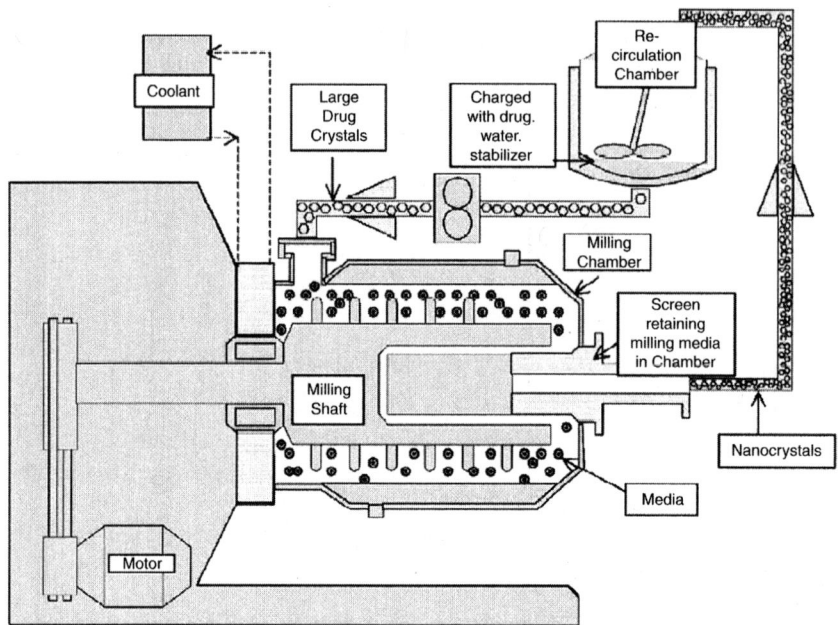

Figure 2 The media milling process is shown in a schematic representation. The milling chamber charged with polymeric media is the active component of the mill. The mill can be operated in a batch or recirculation mode. A crude slurry consisting of drug, water, and stabilizer is fed into the milling chamber and processed into a nanocrystal dispersion. The typical residence time required to generate a nanometer-sized dispersion with a mean diameter of < 200 nm is 30–60 min. *Source*: From Ref. 17.

with Tyloxapol 2% w/w solution in water (4). Following production of the BCD aqueous dispersion, the formulation was nebulized using an Omron Micro-air® NEU-30 nebulizer. The droplet size of the nebulized dispersion was controlled within a range of 1–7 µm, and the nebulized dispersion was analyzed using Andersen cascade impaction (4). When viewed as a percentage of emitted dose through the mouthpiece, the respirable fraction ranged from 56 to 72% for the nanocrystalline formulation versus 36% for the propellant system. In addition, the throat deposition was 9–10% of the emitted dose for the novel dispersion, as compared to 53% for the commercial product. Therefore, the novel dispersion technology provides greater deposition of drug to the conducting airways and deep lung tissue both in quantity and as a percent of emitted dose than the commercial BCD product, lowering the dose needed for therapy and decreasing potential side effects. Also, the use of nebulized aqueous dispersions of nanocrystalline BCD represents an environmentally sound alternative to the use of chlorofluorocarbon based propellant formulations (4).

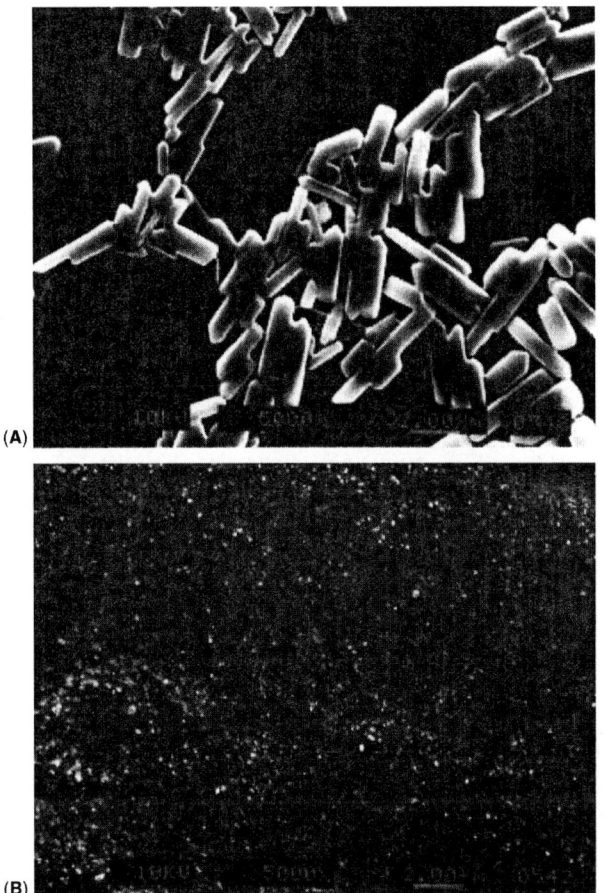

Figure 3 SEM of (**A**) micronized bulk beclomethasone dipropionate; and (**B**) beclomethasone dipropionate media milled using the Nanosystems® process. *Source*: From Ref. 4.

Dry powder pulmonary formulations of BCD within mucoadhesive microspheres were studied in order to increase the residence time in the lungs and decrease dosing frequency and overall dose (19). The microspheres were spray-dried from either an aqueous suspension or ethanolic solution containing BCD and the mucoadhesive polymer hydroxypropyl cellulose (HPC). Formulations produced by aqueous suspension showed crystalline characteristics, while those produced from ethanolic solution showed amorphous character. Amorphous beclomethasone dipropionate and hydroxypropylcellulose (aBCD/HPC) microspheres showed rapid absorption of the drug. Crystalline beclomethasone dipropionate and hydroxypropylcellulose (cBDC/HPC) microspheres showed increased residence time, being retained in the lung longer (with 86% remaining after 180 min), and showed sustained drug release. Bulk crystalline BCD (cBCD) administered without formulation into HPC microspheres and aBCD/HPC showed less than 17% dose remaining and less than 5% dose remaining, respectively, after 180 min. The aBCD/HPC formulation, with a higher apparent solubility and dissolution rate, was able to be absorbed faster than the crystalline forms of the drug, despite being formulated with a mucoadhesive carrier. Both of the HPC microsphere formulations (cBCD/HPC and aBCD/HPC) showed better efficacy against eosinophil accumulation than did the bulk drug substance cBCD. Inhaled cBCD/HPC microspheres showed a 50–60% decrease in the accumulation of lung eosinophils at the 6 and 24 h time points after dosing compared to the aerosolized bulk drug substance (19), even with overall dosing of the BCD reduced from 480 μg of the bulk BCD to 88 μg of the cBCD/HPC formulation. It was shown that five times the dosing of BCD bulk was only effective from 1 to 6 h compared to 24-h efficacy with the cBCD/HPC formulations. Therefore, by controlling the release and retention of BCD in the airways, mucoadhesive BCD microspheres are able to prolong the pharmacokinetic/pharmacodynamic profiles of this drug without increasing drug dose, and, hence unwanted side-effects (19).

Betamethasone sodium phosphate (BSP) encapsulated into poly(lactic/glycolic acid) (PLGA) nanoparticles was evaluated for its effectiveness in the treatment of experimentally induced arthritis. Higaki et al. used the biodegradable nanoparticle/steroid formulation for adjuvant induced arthritis (AA) and antibody induced arthritis (AbIA) (7). The PLGA/BSP nanoparticles were administered intravenous (IV), while the bulk BSP was administered subcutaneously at the site of induced arthritis. The inflammation rate in the test subject animals was evaluated against two control groups treated only with blank PLGA nanoparticles and saline. In the AA mice, at 1 day post treatment, the PLGA/BSP nanoparticles showed 64% inflammation rate, while the bulk BSP administered at 100 μg and 300 μg showed inflammation rates of 68% and 78%, respectively. All three treatment groups showed decreased inflammation rates compared with the control groups which still showed 100% inflammation. In the AA mice, at 7 days post treatment, the PLGA/BSP nanoparticles inflammation rate was 76% vs. 84% and 87% for the bulk BSP groups (100 μg and 300 μg dosing), while the control groups still had 100% inflammation rates. A 36-24% decrease in inflammation was obtained after 1 day and maintained for 1 week with a single injection of 100 μg of PLGA-nanosteroid. The PLGA-nanosteroid particles were also effective against AbIA in mice. The inflammation in the joints of mice induced with AbIA was scored for inflammation on a scale of 0–5 (0 = no inflammation; 5 = maximum inflammation). Twelve days after treatment, the arthritis score for the vehicle alone (PLGA) was 3.9, while the arthritis score for the PLGA/BSP was 1.3 and the BSP 100 μg dose was 2.4. The authors state that the PLGA nanoparticles may protect BSP against conversion and degradation in circulation, preventing the rapid and extensive tissue distribution that occurs with free BSP. The results of this study indicate that a single IV dose of hydrophilic BSP encapsulated in PLGA nanoparticles can lead to rapid, complete, and durable resolution of arthritis inflammation, owing to the enhanced preferential localization of the BSP in the synovial tissue. The observed strong therapeutic benefit obtained with PLGA-nanosteroid is due to the sustained release of the BSP over the longer time period as compared to injection of the bulk BSP. Therefore, targeted drug delivery of BSP using a sustained release PLGA nanosteroid delivery system shows successful treatment against experimental arthritis.

Budesonide

Budesonide (BDS) is a structural analog of the naturally occurring mineralocorticoid, cortisone, produced by the adrenal glands. The chemical structure of budesonide is depicted in Figure 1. This drug was determined not to have the undesirable side effects of water retention, high blood pressure, and muscle weakness associated with other glucocorticoids like cortisone. Budesonide is most often given by oral, topical, and pulmonary administration.

Budesonide has been used to treat a number of autoimmune and autoinflammatory diseases, including those diseases which affect the digestive tract. Orally dosed budesonide has shown effectiveness against Crohn's disease, ulcerative colitis (UC), and IBD, all of which are believed to have an autoimmune component as part of their etiology (20). Drug delivery systems containing budesonide have been used to effectively target the drug to the site of action, avoiding the side effects that could accompany systemic delivery of the drug. One study evaluated the colon specific delivery of budesonide in an enteric formulation (21). This drug delivery system is a microparticulate formulation consisting of hydrophobic budesonide-containing cores of cellulose acetate butyrate (CAB) which are enterically coated with Eudragit® S. Scanning electron micrographs of this formulation are shown in Figure 4. Drug release profiles of the formulation showed no release of budesonide in the acid phase, with sustained release of the drug occurring after the buffer stage. Figure 5 shows the in vitro release profiles of the formulation. These formulations were then tested with rats in vivo to determine the degree of inflammation that the microparticulate budesonide formulation could inhibit. These experiments were designed to mimic the inflammation that occurs with ulcerative colitis using 2,4,6-trinitrobenzene-sulfonic acid (TNBS) to induce inflammation. Histological evaluations of the intestinal tissue of rats showed substantially decreased inflammation,

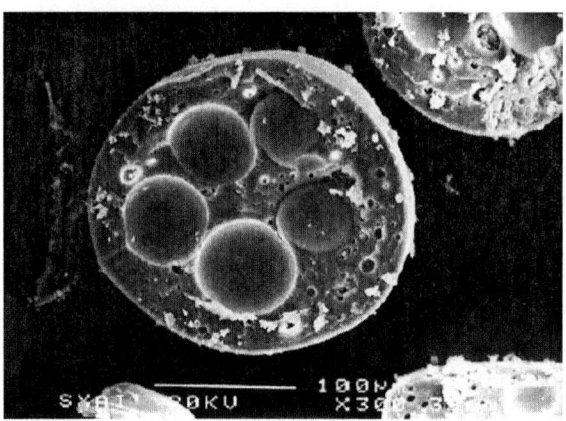

Figure 4 SEM of a cross-section of Eudragit S microparticles containing budesonide-loaded cellulose acetate butyrate cores. *Source*: From Ref. 21.

resulting in lower microscopic histological damage scores. Micrographs of rat intestinal mucosa showed much lower degrees of inflammation when the budesonide CAB microparticles were used (Fig. 6). This colonic delivery system improved the efficacy, at the site of action, of budesonide in the healing of induced colitis in rats, demonstrating that the effects of budesonide were generally improved compared to those obtained with budesonide enteric microparticles for upper intestinal drug release.

Based on previous research, development of the Entocort EC capsule for delivery of budesonide to the site of action for Crohn's disease patients has been reported by AstraZeneca Pharmaceuticals (22). Entocort EC has a pH- and time-dependent release profile for budesonide, developed to

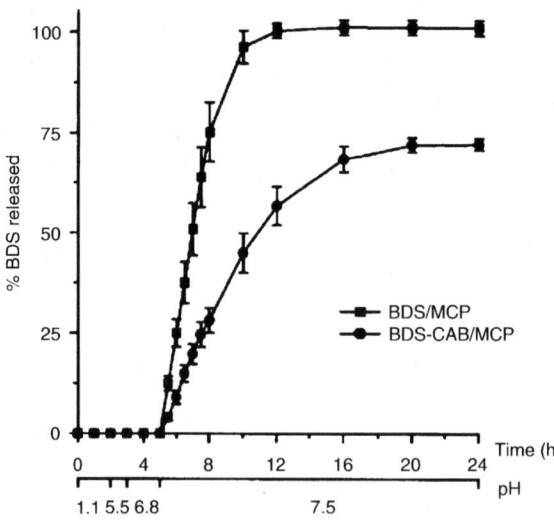

Figure 5 In vitro release profiles obtained from Eudragit S microparticles containing budesonide (BDS) directly encapsulated (BDS/MCP) or included in CAB cores (BDS-CAB/MCP). Data are mean ± standard deviation $n = 4$. *Source*: From Ref. 21.

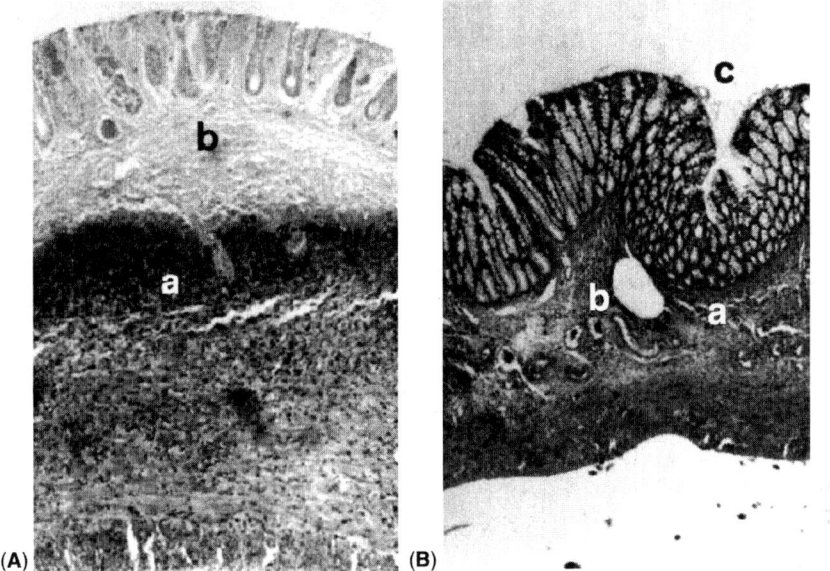

Figure 6 Optical micrographs of the colon of (**A**) a TNBS-treated rat after oral administration of blank microparticles, showing mucosa with severe inflammatory infiltrate (*a*) extensive areas of necrosis (*b*). This colon was given a damage score of 6; and (**B**) optical micrograph of colon of a TNBS-treated rat after oral administration of Eudragit S microparticles containing budesonide loaded CAB cores, showing mucosa with mild inflammatory infiltrate (*a*), vascular congestion (*b*), and well-conserved mucosa (*c*). This colon was given a tissue damage score of 2. *Source*: From Ref. 21.

optimize drug delivery to the ileum, the site of inflamed tissue in Crohn's patients. The dosage form is a capsule with beads consisting of a sugar core, coated first with a layer containing budesonide, ethylcellulose and surfactants, followed by an enteric polymer coating (Eudragit L). This formulation design allows for both the pH- and time-dependent release of the drug. pH-dependent release of the drug was shown to be caused by the Eudragit-L enteric coating, while time-release of the drug was caused by the ethylcellulose/surfactant coating layer. Since ethylcellulose is water-insoluble, the presence of the surfactantants in this layer allowed for partial dissolution and pore-formation in this coating allowing for the time-dependent release of the drug. Scintigraphic methods confirmed that the Entocort® formulation delays absorption and prolongs the rate of elimination, while maintaining complete absorption. This study evaluated the regional deposition and uptake of budesonide in the Entocort capsules and the immediate release capsules (22). Delivery of budesonide from Entocort capsules was evaluated when the formulation was given both before and after a meal to determine absorption with varying intestinal transit times (23). The time-to-peak (t_{max}) plasma concentration was significantly increased with controlled-release budesonide when compared to the immediate release formulation (before breakfast 4.5 vs. 1.8 h; after breakfast 5.2 vs. 2.9 h). When given after breakfast, the controlled release formulation was associated with a mean residence time 1.6 h longer than seen with the immediate release formulation (23). The rate of absorption of budesonide from standard and controlled-release capsules taken before and after breakfast evaluated in this study. The proportions of the budesonide dose absorbed from immediate release and controlled-release capsules in the upper and lower gastrointestinal tract are shown to vary according to the dose delivery system. Controlled-release budesonide, therefore, effectively delivers most of the budesonide dose to the ileum and colon, the regions that are most often affected by IBD. In addition, the time of food intake had little effect on the site of absorption or the bioavailability of the controlled-release formulation (23).

Pulmonary formulations of budesonide have also been reported for use in other chronic inflammatory conditions such as asthma. Waldrep et al. reported inclusion of various types of immunosuppressants in liposomal formulations for aerosol delivery to the lungs (24,25). Liposomal formulations of cyclosporin, beclomethasone, and budesonide were prepared using different types of phospholipids with varying drug to phospholipid ratios. Aerosol particle size analysis demonstrated that the mass median aerodynamic diameter (MMAD) of aerosolized liposomal formulations of these drugs increased minimally with higher liposome concentrations, remaining in the desirable respirable fraction for pulmonary formulations (2–5 μm in diameter) (24). Lobo et al. from Nektar Therapeutics has reported budesonide dry powder formulations for inhalation (26). The powders were manufactured by dissolving the drug in acetone before processing via the SEDS technique (26). These SEDS processed powders were characterized by their low density and a MMAD (2.4 μm average) within the respirable size range. The performance of the budesonide powders were evaluated in a Turbospin® and Eclipse® dry powder inhaler (DPI) device. The capsules of the inhaler devices were filled with powder, and their emitted doses (ED), defined as the relative amount of powder loaded in the capsule that leaves the device, were studied. The SEDS processed powders dispersed well in the DPI devices, exhibiting high ED's (70–80%) and relatively low variability (RSD 8–13%). Regardless of the device, the SEDS processed powders outperformed both the micronized drug and the commercial powder while showing good batch-to-batch reproducibility (RSD < 5%) (26). Both of these formulation techniques offer more effective dosing options for the treatment of asthma. Clinical trials of these formulations are on-going, but preliminary results showed that SEDS powders formulated into DPI devices, using the Turbospin and Eclipse DPI's deposit drug in the deep lung, leading to enhanced treatment of asthmatic conditions.

Dexamethasone

Dexamethasone (DEX) is a synthetic adreno-corticosteroid with similar properties to the naturally occurring hydrocortisone and cortisol, being about 20–30 times more potent than hydrocortisone in its anti-inflammatory effect. This compound, however, has been chemically modified with a fluorine substituent on the steroid ring structure, allowing the drug to have reduced side-effects. The drug is a potent anti-inflammatory medication, lacking the unwanted effect of sodium retention by the body. The chemical structure of dexamethasone is depicted in Figure 1. Typically, dexamethasone is administered orally or topically, but it can also show effectiveness when administered to the lungs of asthmatics.

Formulations of dexamethasone palmitate (DEXP) have been produced using liposomal formulations of the drug (27). Liposomes were produced using DEXP, phosphatidylcholine, and cholesterol in proportions 4:3:0.3, and these formulations showed high drug loading (70%). These formulations were intratracheally administered to the rat model to determine the retention of the liposomes, their release profiles, and the resulting blood levels due to drug release. Lymphocyte and interferon production were shown to be comparable between liposomal and free DEXP, showing no loss of activity when the drug is formulated as a liposome. Liposomes of DEX in proportions of 9:1 DEX:PC were prepared, and pulmonary retention of these DEX in the liposomal formulations in rats was 50%, compared with 26% retention for free DEX after 1.5 hours (28). Liposomal DEX was also shown to have a prolonged action on the proliferation of white blood cells after 72 h, while free DEX showed no further anti-inflammatory effect after 24 h. This study showed that pulmonary dexamethasone was significantly retained in the lungs of the rat model and that the anti-inflammatory action in the lung was prolonged using the liposomal formulations compared to the free DEX, and side effects of the liposomal DEX formulation were reduced. Blood levels of liposomal DEX were reported to be lower in the rat model after 24 h (about 0.8 % recovery), compared with free DEX (about 0.3% recovery), however lung levels were much higher (10% recovery in the lungs vs. less than 1% in the lungs for free DEX), and remained high over a period of time. A lag time in lung to blood partitioning was noted in the liposomal DEX formulation over the free DEX, which is to be expected. In another study, biomarkers of pulmonary inflammation were studied following prophylactic administration of the liposomal DEX formulations and free DEX (29). Liposomal

DEX provided more protection from inflammation than did free DEX, inhibiting the increase in inflammatory mediator activity. Liposomal DEX reduced myleoperoxidase activity by 15%, elastase activity by 68%, and chloramine activity by 50% compared to free DEX. Of the pro-inflammatory mediators studied, liposomal DEX inhibited phospholipase A2 (62% vs. free DEX 45%), leukotriene B4 (76% vs. free DEX 64%), and thromboxane B2 (76% vs. free DEX 64%) (29). Overall, liposomal treatment with DEX showed sustained release of the drug, when administered via the intratracheal route, with lower total dosage needed and a lower side effect profile, with increased prophylaxis against inflammation.

Incorporation of immunosuppressant drugs into the structure of implantable devices has been investigated due to the biofouling and rejection of these devices by the body due to the body's immune response. Dexamethasone has been encapsulated into PLGA microspheres to determine the feasibility of this drug to inhibit inflammation around implantable devices (30). In this study, PLGA microspheres loaded with DEX were prepared with an O/W emulsion technique using polyethylene glycol (PEG) or PVA as aqueous stabilizers. Solvent evaporation and lyophilization yielded microspheres, which were sized with an average diameter of $11 \pm 1\,\mu m$. These microspheres did not show sufficient initial drug release, so microspheres were also pre-degraded to show the effect of weakening the PLGA, and hence fast initial drug release. These microspheres were shown to have a mean diameter of $12 \pm 2\,\mu m$. In vitro studies evaluated the DEX release from the microspheres. The non-degraded and pre-degraded microspheres were mixed to allow for optimum release of drug, and release kinetics of pre-degraded microspheres showed an initial burst release followed by approximately zero order release rate for 1 month (30). Delivery of dexamethasone at the site of the implant has been shown to reduce the inflammatory response and reduce the formation of pericardial adhesions of implants (31). Chorny et al. examined efficacy of DEX loaded polylactide-polyethylene glycol (PLA-PEG) films implanted for inhibition of the formation of pericardial lesions in rabbits. In vitro results showed biphasic drug release rates in serum, with 69% drug released in 72 h. The implants produced sustained drug release at the implantation site after that with low distribution in peripheral tissues. Tendency and density of scarring was evaluated 21 days post-op, and rabbits treated with blank implants were compared to DEX loaded implants. Epicardial adhesions' formation was reduced and the anatomy was preserved in treated animals. It is concluded that local delivery of DEX from biodegradable implants provides a new approach to prevention of scarring, notably pericardial adhesions, while minimizing drug distribution in surrounding tissues and systemic delivery (31).

IMMUNOPHILIN BINDING COMPOUNDS

The immunophilin binding compounds are a class of immunosuppressant drugs consisting of different medications from a broad range of natural and synthetic sources. These drugs include: cyclosporin, tacrolimus, sirolimus, everolimus, and other similar macrolide lactone molecules. Although differences in the individual drugs' mechanisms of action are observed, each of these drugs binds with cytosolic immunophilin proteins to produce complexes which exert their immunosuppressive effects. In the case of cyclosporin and tacrolimus, the first phase of T-lymphocyte activation is halted. Because the immunophilin/drug complex inhibits the production of calcineurin, the dephosphorylation of key transcription factors for the production of interleukins is reduced (32). Phosphorylated factors cannot cross the nuclear membrane, and the production of key factors for lymphocyte activation and proliferation is halted. In this way, cyclosporin and tacrolimus prevent lymphocytes from entering the G0 to G1 cell growth phases. Along with reducing interleukin release, the immunophilin/drug complex reduces the function of effector T-cells by reducing lymphokine production (32). Cyclosporin (Sandimmune® (33), Neoral®) has been used to treat acute graft rejection in liver, kidney, and bone marrow transplant patients. Neoral (cyclosporin microemulsion preconcentrate) (34) has also been used to treat severe psoriasis, rheumatoid arthritis, multiple sclerosis, and myesthenia gravis. Tacrolimus (Prograf® capsules, oral liquid) (35) has shown effectiveness against host-vs-graft disease in kidney and liver transplant patients, and the topical formulation of the drug (Protopic® ointment) (36) is used to treat psoriasis.

In the case of sirolimus and everolimus, the second phase of T-lymphocyte activation is halted by reducing signal transduction and clonal proliferation. Sirolimus and everolimus bind to immunophilin, but the drug/immunophilin complex does not inhibit calcineurin, instead inhibiting another related protein (mammalian target of rapamycin, m-TOR) (37). Sirolimus and everolimus inhibit the binding of m-TOR to interleukin growth factors necessary for the production of kinases and phosphatases in the T-lymphocytes (38). mRNA for key proteins required for the G1 cycle are blocked, and progress of T-lymphocytes from the G1 to S phase in the cell cycle is inhibited. Sirolimus and everolimus act synergistically with the calcineurin inhibitors (cyclosporin and tacrolimus), and can be administered as combination therapies without increasing side effects. For example, sirolimus (Rapamune® tablets, oral liquid) and everolimus have been used in combination with cyclosporin and glucocorticoids in the treatment for long term transplant maintenance. Sirolimus is also effective against psoriasis.

The immunophilin binding compounds vary widely in their molecular structure, but they share common physicochemical properties, and they undergo similar metabolism in the body. Since all of the immunophilin binding compounds have high lipophilic character, shown by their high log P values, they are all substrates for the P-GP efflux pump (39). This class of immunosuppressants has a common characteristic of being P-GP substrates inherent in their hydrophobic nature, containing positively or neutrally charged domains with a large planar molecular structure. Since these compounds are removed from the systemic circulation by P-GP and other multi-drug resistance systems, the immunophilin binding compounds show erratic and extremely variable bioavailability (39). Additionally, the immunophilin binding compounds are metabolized by the CYP 3A4 family of isoenzymes. The main function of these enzymes is to convert non-polar foreign chemicals into more polar (and therefore more excretable) substances. Oral bioavailability of these drugs is adversely affected by the intestinal and first-pass metabolic effects of these enzymes. The P-GP efflux mechanism coupled with the intestinal metabolic activity of CYP 3A4 work synergistically, contributing to the low oral bioavailability of these drugs (39).

Cyclosporin A

Cyclosporin A (CSA) is a peptide made up of 11 amino acids, derived from the *Trichoderma polysporum* fungus, and having highly lipophilic character and low aqueous solubility (40). The chemical structure of CSA is depicted in Figure 1. As discussed previously, this compound is a calcineurin inhibitor type immunosuppressant. Selected physicochemical properties of this drug are shown in Table 1.

The commercial formulations of CSA, including Sandimmune and Neoral (Novartis, East Hanover, NJ), have found widespread therapeutic applications for a variety of immunosuppressive indications. Sandimmune and Neoral are approved for treatment of kidney, liver, and heart transplant patients as well as patients with rheumatoid arthritis and psoriasis. These formulations are available as oral liquids and liquid-containing capsules, and they allow for increased bioavailability of this poorly water soluble drug. Dressman et al. have determined that oral absorption of poorly water soluble and highly permeable drugs is limited by their dissolution rate (41), so formulation of CSA into a more soluble form allows for increased absorption of the drug. Neoral and Sandimmune are oral formulations where the drug is essentially presolubilized in the formulations. The Sandimmune formulation contains CSA dissolved in corn oil and polyoxyethylated glycolyzed glycerides (33) and will not be discussed in depth. However, the Neoral formulations are produced using advanced formulation design. Neoral contains dehydrated alcohol, corn oil mono and di-glycerides, and polyoxyl 40 hydrogenated castor oil (34) formulated as a fine emulsion system. Neoral has shown increased bioavailability over Sandimmune because Neoral contains increased amounts of solubilizers, allowing better absorption than the Sandimmune. For this reason, the two medications cannot easily be substituted for one another without therapeutic drug monitoring. The absolute bioavailability of cyclosporin administered as Sandimmune is population dependent, estimated to be less than 10% in liver transplant patients and as great as 89% in some renal transplant patients. The absolute bioavailability of cyclosporin administered as Neoral has not been determined in adults. In studies of renal transplant, rheumatoid arthritis and psoriasis patients, the mean CSA AUC was approximately 20–50% greater and the peak blood CSA concentration (C_{max}) was approximately 40–106% greater following administration of Neoral compared to following administration of Sandimmune. The dose normalized AUC in de novo liver transplant patients administered Neoral 28 days after transplantation was 50% greater and C_{max} was 90% greater than in those patients administered Sandimmune. AUC and C_{max} are also increased (Neoral relative to Sandimmune) in heart transplant patients, but data are very limited. Although the AUC and C_{max} values are higher on Neoral relative to Sandimmune, the pre-dose trough concentrations are similar for the two formulations. Cyclosporin in the Neoral formulation shows high volume of distribution (3–5 L/kg in liver transplant patients), with excretion of the drug occurring mainly from the biliary route (34).

The main challenges with administration of the commercial CSA formulations are absorption variability and nephrotoxicity. While Neoral is considered to be a substantial improvement over Sandimmune, the emulsion formulation still suffers from considerable absorption variability and nephrotoxicity (34). The commercial formulations exhibit high inter- and intra-patient absorption variability due to various factors. In addition, high oral doses of CSA needed for relief of dermal psoriasis symptoms can lead to nephrotoxicity. Other formulation techniques have been developed to attempt to deal with these problems. The effects of adding micellar solubilizing agents, like the bile acid tauroursodeoxycholate (TUDC), to the oral formulations was investigated for improving oral dosage of CSA (42). Oral doses of cyclosporin in the TUDC-monoolein micellar formula and co-administration of Sandimmune with TUDC was studied in comparison with standard Sandimmune administration for promoting and regulating CSA bioavailability in the rat. Pharmacokinetic parameters were determined in the fasted state with either an IV injection of CSA 5 mg/kg or a single oral dose of CSA with and without TUDC 10 mg/kg. Compared to Sandimmune, the micellar solution of cyclosporin improved the drug's bioavailability by 160% and decreased interindividual variability expressed as a percent of variation from 32% to 15%. The plasma drug concentration vs time curves are shown in Figure 7. Bioavailability of the drug slightly improved in rats receiving Sandimmune plus co-administered TUDC, though not significantly. Data indicate that the carriers used in delivery systems for cyclosporin greatly affect drug bioavailability, and that

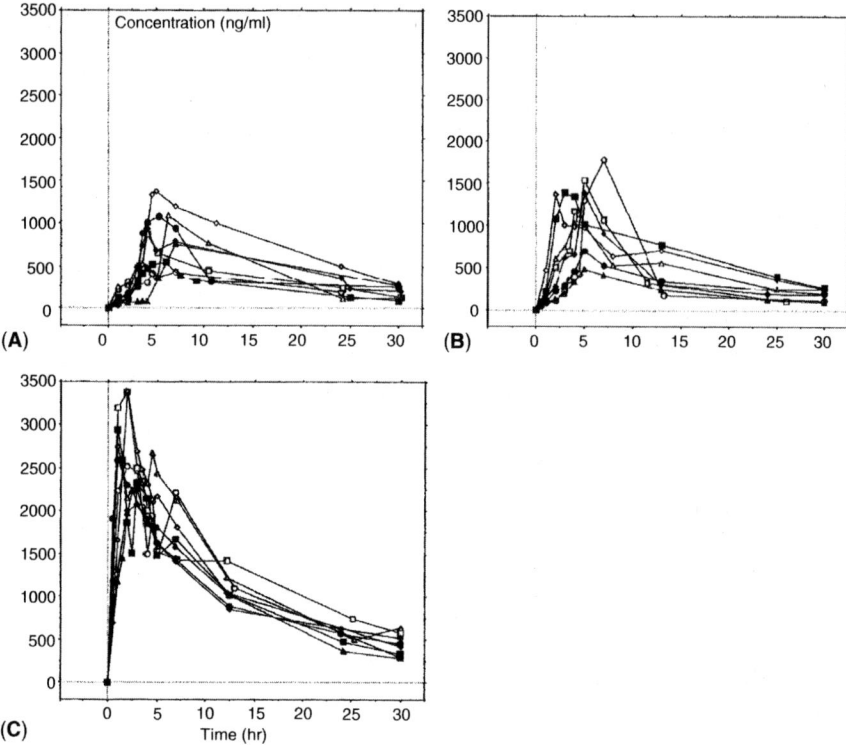

Figure 7 Individual cyclosporin concentrations vs. time after single oral adminis-
tration of 10 mg/kg in rats. (**A**) Group II: Sandimmune oral oily formulation.
(**B**) Group III: Oral coadministration of Sandimmune + TUDC. (**C**) Group IV:
TUDC-monoolein-cyclosporin micellar solution. *Source*: From Ref. 42.

aqueous micellar vehicles provide for high cyclosporin absorption with low
variability (42).

Parenteral delivery of CSA has been studied using microparticulate
and nanoparticulate formulations of the drug. Sanchez et al. produced CSA
loaded polyester nanoparticles for IV administration and studied the
immunosuppressive and nephrotoxic properties of these formulations (43).
Compared with the commercial Sandimmune injectable product, the
microparticulate system investigated in the study offered extended release
of CSA over the study time period. The study evaluated the in vivo release of
CSA from these particles in mice, showing differences in release dependant
on the size of the particles – from 0.2 to 1 to 30 μm in size. These particles
showed increased drug AUC compared to the commercial product. Among
the formulations investigated, the 30-μm microspheres provided constant
levels of CSA to be released over the 3-week test period. Figure 8 shows
the blood-concentration versus time curve for the microparticulate

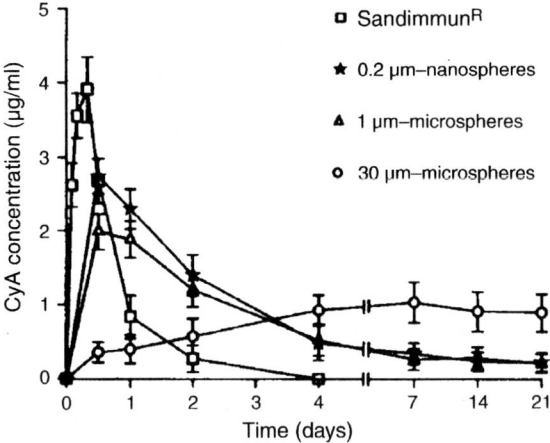

Figure 8 Blood-concentration versus time profiles observed after a single subcutaneous injection of free (Sandimmune) or microencapsulated cyclosporin. Each point represents the average ± SD of four determinations. *Source*: From Ref. 43.

formulations, showing the differences between the formulations particle sizes. The Sandimmune formulation shows rapid increase in blood levels following administration, followed by rapid blood concentration decline, undetectable after 4 days. The nanosphere and microsphere formulations increased blood CSA concentrations to a t_{max} of 0.5 days, followed by a slow decline in drug concentration, depending on the size of the particle. Particles of sized 0.2–1 µm follow the same trend, having t_{max} values of 0.5 days and extending the CSA release from up to 21 days. Particles sized 30 µm show slower drug release, with a t_{max} of 7 days, constantly releasing drug for over 21 days. Further investigations showed CSA preferentially accumulates in the liver, kidney, spleen, and adipose tissues, with lower levels detected in the blood. These data show that CSA therapy by this route can decrease the likelihood of rejection on liver and kidney transplants. Figure 9 shows the length of time immunosuppression can be extended with the microparticulate formulations. Incorporation of CSA into these sustained-release formulations could offer better therapeutic outcomes by reducing the peak drug concentrations and reducing the likelihood of nephrotoxic events associated with high drug concentrations. The authors conclude that release of CSA from microparticles is dependent on the size of the microparticle and its catastrophic degradation rate. A single subcutaneous dose of encapsulated CSA can maintain long-term levels of the drug in tissues which are most frequently transplanted, indicating the modulating ability of the immunosuppression therapy. Also, the sustained-release nature of these formulations can reduce the frequency of IV dosing and improve patient compliance.

In order to improve delivery of CSA to the site of action in lung transplant and asthma patients, pulmonary delivery of the drug has been explored. Mitruka et al. have experimented with pulmonary delivery of CSA

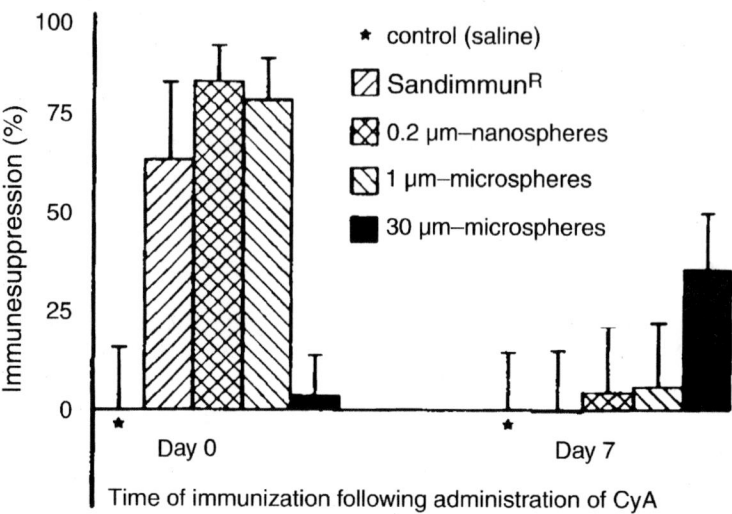

Figure 9 Immunosuppressive effect corresponding to Day 0 and Day 7 postintraperitoneal injection into mice of a single dose (70mg/kg) of free cyclosporin (Sandimmune®) and microencapsulated cyclosporin. Each value represents the average ± SD of four determinations. The mean number of direct PFC per spleen of control animals was taken as 100% of immune response (0% immunosuppression). *Source*: From Ref. 43.

using aerosolized solutions of the drug in ethanol, delivered to rats (44, 45). Aerosol delivery of CSA resulted in higher and more rapid peak drug levels in lung tissue than did systemic delivery. CSA AUC was three times higher in lung tissue with aerosol delivery than with intramuscular delivery (477,965 and 157,706 n h/g, respectively). The lung tissue to blood ratio was higher in the aerosol groups (27.3:1 and 17.4:1) when compared with the intramuscular groups (8.1:1 and 9.4:1). CSA concentrations in blood as a function of time following the first 24 h of drug administration are shown in Figure 10, and CSA concentrations in lung tissue as a function of time is shown in Figure 11. Area under the concentration vs. time curve (AUC)/ dose of CSA (mg/kg) for each of the 4 study groups is shown in Figure 12. Therefore, local aerosol inhalation delivery of CSA provided a regional advantage over systemic intramuscular administration by providing higher peak drug concentrations and lung tissue exposure (45). Additionally, local delivery of CSA by aerosol inhalation dose-dependently prevented acute pulmonary allograft rejection. Effective graft levels and low systemic drug delivery required significantly lower doses than systemic therapy alone (44). Also, expression of inflammatory cytokines involved with graft rejection was suppressed by CSA therapy (44). These data indicate that pulmonary delivery of CSA can be effective at lower systemic levels, reducing the

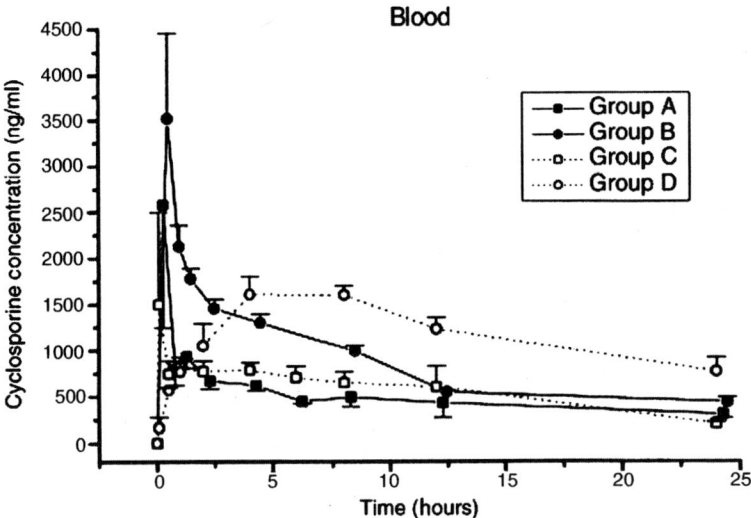

Figure 10 Cyclosporin concentrations in blood as a function of time following the first 24 h after administration of the drug. Group A, aerosol cyclosporin 3 mg/kg; Group B, aerosol cyclosporin 5 mg/kg; Group C, intramuscular cyclosporin 5 mg/kg; Group D, intramuscular cyclosporin 15 mg/kg. *Source*: From Ref. 45.

possibility for nephrotoxic side effects seen with high dose immunosuppressant therapy.

Tacrolimus

Tacrolimus (TAC) or FK506 is a macrolide lactone derived from the bacteria *Streptomyces tsukubaensis*, extracted as one of three tautomeric forms of the molecule from bacterial fermentation. The chemical structure of TAC is depicted in Figure 1. As discussed previously, TAC is a calcineurin inhibitor similar in its effects to cyclosporin. It exhibits higher potency than cyclosporin, but it is similarly lipophilic and poorly water soluble (46). Selected physicochemical properties of this drug are shown in Table 1.

The commercial formulations of TAC include the Prograf capsules for oral administration (35) and the Protopic ointment for dermal application (36). Protopic ointment is a simple mixture of tacrolimus in a base of white petrolatum, mineral oil, paraffin, propylene carbonate, and white wax, and since the design of this formulation is not novel, it will not be discussed. Prograf capsules, however, are produced using a novel solid dispersion processing technique. The production of this solid dispersion formulation (SDF) is described by Yamashita et al. (47). TAC is dissolved in

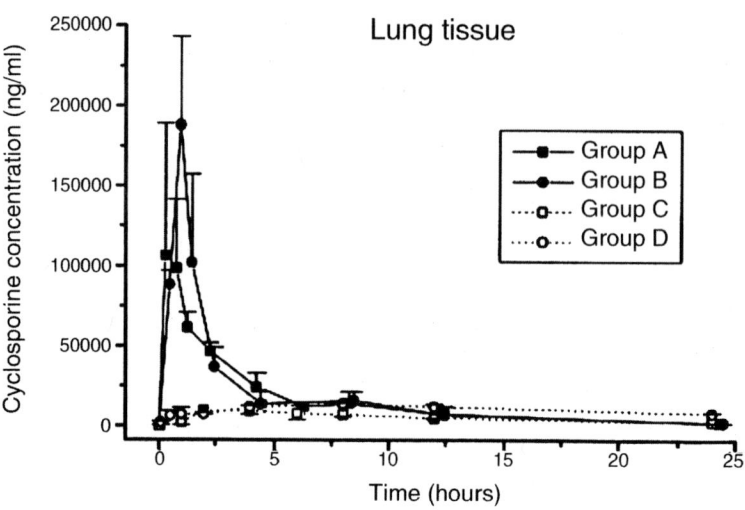

Figure 11 Cyclosporin concentrations in lung tissue with respect to time following the first 24 h after administration of drug. Group A, aerosol cyclosporin 3 mg/kg; Group B, aerosol cyclosporin 5 mg/kg; Group C, intramuscular cyclosporin 5 mg/kg; Group D, intramuscular cyclosporin 15 mg/kg. *Source*: From Ref. 45.

ethanol and a stabilizing hydrophilic polymer is then added. After complete dissolution or swelling of the polymer, the ethanol is removed under reduced pressure and increased heat leading to formation of the SDF. Studies show that the SDF produced using this process contains tacrolimus in an amorphous state, which exhibits higher apparent solubility in dissolution media than does the corresponding crystalline form (47). The resulting SDF yields high supersaturation drug concentrations in vitro that are maintained for an extended period of time. The ability of these formulations to supersaturate the dissolution media highlights the possibility for increased bioavailability from an amorphous drug form. The meta-stable solubility of amorphous drug form may be as high as 100-times greater than its crystalline form (48). If the concentration of drug in solution is significantly increased, the higher chemical potential will lead to an increase in flux across an exposed membrane. This would lead to much higher blood levels for an amorphous drug form, compared to an identical crystalline formulation (49). The blood concentration of tacrolimus after oral administration of the SDF with hydroxypropyl methylcellulose (HPMC) to beagle dogs was measured against oral administration of the crystalline TAC powder (Fig. 13). The pharmacokinetic parameters derived from the analysis of blood concentration versus time curves are shown in Table 2. The AUC (0–8 h) of the SDF composition is almost 10-times higher than that of crystalline TAC, and the C_{max} of the SDF composition shows similar 10-fold

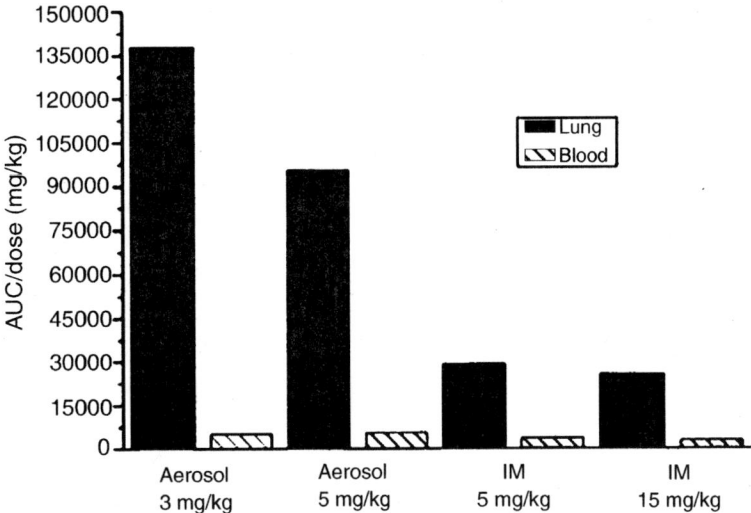

Figure 12 Area under the concentration vs. time curve (AUC)/dose of cyclosporin (mg/kg) for each of the four study groups. *Abbreviation*: IM, intramuscular. *Source*: From Ref. 45.

increases in magnitude compared to the crystalline form (47). Honbo et al. compared tacrolimus SDF's with oily ethanol formulations (OEF) of TAC for efficacy against allograft rejection (50). A rat skin allograft transplantation study was conducted to determine the length of time the graft

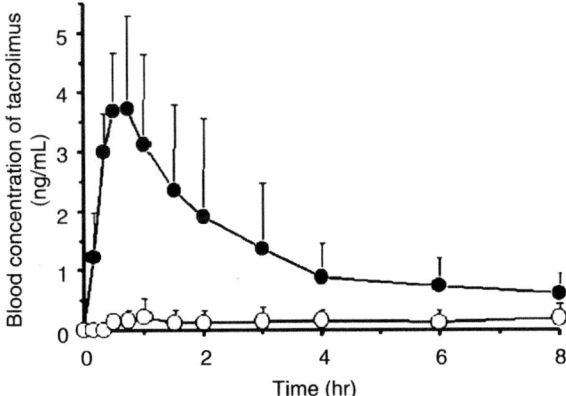

Figure 13 Blood concentration of tacrolimus after oral administration of SDF with HPMC to beagle dogs. (O) tacrolimus crystalline powders; (•) SDF of tacrolimus with HPMC. Values are expressed as the mean with a vertical bar showing SE of six animals. Each dosage form was administered at the dose of 1 mg tacrolimus. *Source*: From Ref. 47.

Table 2 Pharmacokinetic Parameters of Tacrolimus after its Oral Administration to Dogs as Crystalline Powders or SDF of Tacrolimus with HPMC

Sample	$AUC_{0\text{-}8h}$ (ng h/mL)	C_{max} (ng/mL)	T_{max} (h)	MRT (h)
Crystalline powder	1.1 ± 1.4	0.4 ± 0.3	3.1 ± 3.0	3.3 ± 0.9
SDF with HPMC	10.9 ± 6.1^a	4.0 ± 1.2^a	0.6 ± 0.2	2.7 ± 0.1

Note: Each value represents the mean ± SE of six animals. Each dosage form was administered at the dose of 1 mg as tacrolimus.
$^a = P < 0.05$, compared to the corresponding parameter of crystalline powder.
Source: From Ref. 47.

could remain viable when administered doses of oral TAC of 0, 1, 3.2, 10, and 32 mg/kg/day. At 1 and 3.2 mg/kg/day TAC, the animals in both OEF and SDF groups survived for over 30 days. The mean survival time was 14.4 days in the SDF-treated group and 11.2 days in the OEF treated group (50). Survival graphs are shown in Figure 14.

For oral absorption, the major site of absorption for TAC was identified to be the upper small intestine (35). Following administration, TAC has a high volume of distribution and binds extensively into red blood cells and blood proteins. Clearance occurs mainly through the biliary route. Prograf is rapidly absorbed in some patients, with t_{max} occurring within 0.5–3 h, while in other patients, a relatively flat absorption profile was observed, indicating continuous and prolonged absorption. The poor dissolution of TAC in gastric fluids, its low aqueous solubility, and variable

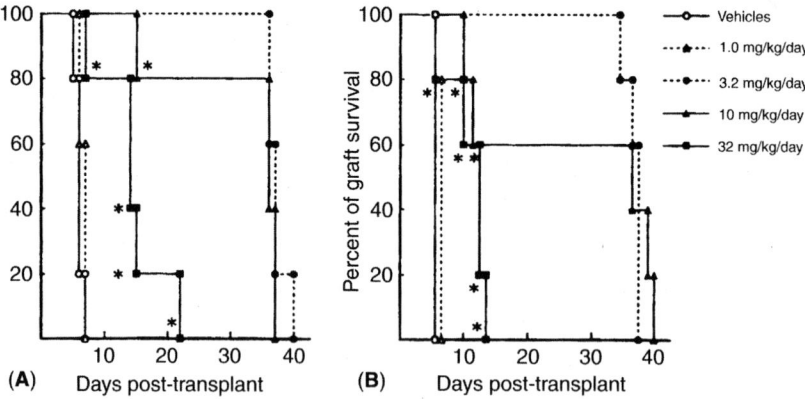

Figure 14 Results of (**A**) SDF-tacrolimus and (**B**) OEF-tacrolimus in the secondary rat skin allograph transplantation study. Dose 0, 1, 3.2, 10, and 32 mg/kg/day; dosage period, once daily for 30 days; number of animals, 5. * - died with living graft. *Source*: From Ref. 50.

gastric motility is thought to be responsible for this observation. For kidney transplant patients, single oral doses of 0.1, 0.15, and 0.2 mg/kg resulted in C_{max} of 19.2, 24.2, and 47.9 ng/mL, respectively. The t_{max} for these measurements varied from 0.7 to 6 h. The mean bioavailability of Prograf capsules was 21.8% for liver transplant patients, 20.1% for kidney transplant patients, and 14.4–17.4% for healthy subjects. Oral bioavailability was reduced when co-administered with a fatty meal. Decreases in AUC (-35% in whole blood) and C_{max} (-57% in whole blood) and increases in t_{max} (173% in whole blood) were observed, showing that both rate and extent of absorption is affected by food (35). Oral absorption of TAC in Prograf, therefore, is incomplete and highly variable, indicating problems with the commercial dosage form.

In addition to its primary use in inhibiting organ transplant rejection, TAC has been used for treatment in a variety of other disorders where reduction of an inflammatory/immune response is critical. Targeted delivery of TAC to the large intestine was investigated by Lamprecht et al. with sustained local release of TAC for treatment of ulcerative colitis (51–53). Microspheres, manufactured using Eudragit P-4135F, show pH-dependent release in the colon using an emulsification/solvent extraction method. In vitro release studies showed the TAC microspheres limited drug release at pH 6.8 and below (less than 10% release over 6 hours) and rapid release (100% within 30 min) at $pH \geq 7.4$ (51). Myeloperoxidase activity in the rat animal model was used to indicate the severity of colitis, with decreasing activity showing less colonic inflammation. TAC microspheres were shown to be as effective in mitigation of colitis induced inflammation when administered subcutaneously, and they are more effective than when administered orally as a solution. The TAC subcutaneous group, however, showed increased levels of adverse effects, including kidney impairment indicated by lower creatinine clearance, than the tacrolimus microsphere formulation. This was mainly due to the higher bioavailability of the TAC, due to the avoidance of the PGP efflux and CYP 3A4 effects seen with the oral administration of TAC (53).

Sirolimus and Everolimus

Sirolimus (SRL) or rapamycin is a macrolide lactone derived from the bacteria *S. hygroscopicus*, found in the soil on the south Pacific island of Rapa Nui (54). As discussed previously, SRL has a mechanism of action different from that of cyclosporin and tacrolimus type calcineurin inhibitors. The chemical structure of SRL is depicted in Figure 1. Everolimus (EVR) is structurally very similar to SRL, with a 2-hydroxyethyl substitution at position 40 (55), and it can be produced by semisynthesis of SRL (56). SRL, along with EVR, is structurally much different than tacrolimus, but it shares common pharmacophores with tacrolimus and other macrolides (57). SRL and EVR are similarly poorly water soluble and highly lipophilic (58). Selected physicochemical properties of sirolimus are shown in Table 1.

The commercial formulations of SRL include the Rapamune tablets for oral administration and Rapamune oral solution (59). There is currently no approved commercial formulation for EVR, but this drug shows promise with its immunosuppressive effects. Rapamune is currently approved for prophylaxis of rejection for kidney allograft patients. The oral solution contains SRL dissolved in ethanol with solubilizers (Phosal® PG and polysorbate 80) present to allow for the drug to remain in solution following exposure to aqueous media and oral administration. Rapamune oral liquid requires refrigeration and must be mixed with water or juice prior to administration, so an alternative tablet form of the drug was developed (59). The SRL drug substance for the Rapamune tablet formulation is manufactured according to the Nanosystems wet-milling process described previously (3,17). For Rapamune tablets, crystalline sirolimus is wet-milled with polymeric stabilizers (polyethylene glycol 8000, povidone, poloxamer 188, and polyethylene glycol 20,000) to decrease the primary particle size of the crystals to below 400 nm. The Nanosystems process stabilizes the surface of the nanosized SRL crystals using these excipients to prevent aggregation of the particles, allowing for increased dissolution rates afforded by the smaller particle size. The resulting aqueous dispersion of SRL behaves like a solution, which is then processed into the finished tablet dosage form. In the case of Rapamune tablets, the drug dispersion is then applied onto placebo tablet cores using spray-drying, followed by an overcoat of pharmaceutical glaze.

Following administration of Rapamune Oral Solution, SRL is rapidly absorbed, with a mean t_{max} of approximately 1 hour after a single dose in healthy subjects and approximately 2 h after multiple oral doses in renal transplant recipients (59). The systemic availability of SRL was estimated to be approximately 14% after the administration of Rapamune Oral Solution. The mean bioavailability of SRL after administration of the tablet is about 27% higher relative to the oral solution. Rapamune oral tablets show improved bioavailability relative to the oral solution, and, therefore, are not bioequivalent to the oral solution (59).

Inclusion of immunosuppressant drugs within the exterior coating of implant devices has also been investigated to improve drug efficacy (drug delivery vehicle) and biocompatibility of the device (prevention of an immune response). Biosensors International developed an implantable stent covered by a resorbable "composite" coating containing a model immunosuppressive drug (the SRL-analog EVR and TAC) within a poly-hydroxyacid biodegradable polymer matrix for the local inhibition of in-stent restenosis (55). Preclinical and clinical trials of the stents were conducted in coronary patients, and safety and efficacy of the new stent design was evaluated against the traditional bare metal S-stent. Stents were implanted into coronary arteries, and patients were monitored for various conditions related to stent implantation. Following preclinical examinations, both biocompatibility and efficacy of the EVR-eluting stents compared to the bare metal stent at 30 days. Clinical evaluation of the stents showed that no major adverse cardiac events or stent thromboses occurred within either group. The everolimus stent group showed an 87% reduction in neointimal volume. The binary restenosis rate at follow-up and 3 months was 0% for the everolimus-eluting stent and 9.1% for the bare metal stent. Release of drug from the stent, therefore, disrupts the restenosis cascade, while allowing sufficient neointimal growth to promote healing and avoid late stent thromboses. Future studies and post-operative follow-up will allow for determination of long-term stent biocompatibility (55).

CYTOSTATICS

Cytostatic compounds are most often used in chemotherapy against carcinoma, but smaller doses can be used as immunosuppressant agents. These drugs inhibit cell division, reducing the proliferation of T- and B-lymphocytes. Antimetabolite drugs are the most common cytostatic compounds used for immunosuppression. These are typically analogs of naturally occurring compounds, like nucleic acid bases, which interfere with DNA and RNA synthesis. Azathioprine and 6-mercaptopurine are analogs of purine, while methotrexate is an analog of folic acid. Alkylating agents such as nitrosoureas, platinum compounds, and cyclophosphamides are cytostatic drugs which work differently from the antimetabolite analog compounds, but also function as immunosuppressant cytostatics. Certain antibiotics are cytostatic and cytotoxic agents which have been used for immunosuppression. These include the antibiotics dactinomycin, mitomycin C, mitramycin, and bleomycin.

Methotrexate

In normal metabolism, folic acid is converted to dihydrofolate, followed by its conversion to tetrahydrofolate, a substance essential for the de novo synthesis of nucleic acids. Methotrexate (MTX), as stated previously, is an analog of folic acid, inhibits the enzyme responsible for folic acid conversion (dihydrofolic acid reductase) by competitively binding to the enzyme with a higher affinity than folic acid, thereby inactivating the enzyme. Therefore, MTX interferes with DNA and RNA synthesis, cell repair, and cellular replication. Active proliferation of cells, like those in the immune system, is subsequently halted, and immunosuppression results. MTX has been shown to be effective against transplant rejection as well as other autoimmune conditions such as rheumatoid arthritis and psoriasis (60). MTX free acid shows some aqueous solubility, but the drug is most commonly formulated as the sodium salt. Unlike other drugs mentioned in this review, MTX oral absorption is actually enhanced by drug transporters in the gastrointestinal tract. Since it is chemically similar to folic acid, the folic acid transporter is utilized to increase MTX bioavailability (39). Trexall® is the trade name for commercially available tablets and injectable formulations of MTX (61). The chemical structure for MTX is shown in Figure 1, and relevant physicochemical properties of the drug are shown in Table 1.

Microspheres of MTX were developed for intra-articular delivery of drug to the synovial fluid for treatment of rheumatoid arthritis. Biodegradable poly-lactic acid (PLLA) microspheres were produced with entrapped MTX which were then injected into the joint cavity of animals with induced arthritis (62). The purpose of this study was to minimize the systemic bioavailability of the drug to reduce toxic side-effects, while maximizing the immunosuppressive properties of the drug to alleviate symptoms of rheumatoid arthritis. MTX plasma concentrations were analyzed and tolerability of the microspheres as well as their degradation and drug release rates were evaluated in vivo. Biocompatibility was evaluated by observing the swelling of the joints of the rabbits, and histological analysis was performed following the injection of the microspheres. Encapsulation efficiency (EE) of the microspheres increased with increasing molecular weight (MW) of PLLA used. Encapsulation efficiencies of 63.6–68% were achieved with 2,000 MW PLLA, while a PLLA polymer MW of 50,000 showed an EE of 89.2%. The mean particle diameter of microspheres increased from 79.5 μm with 2000 MW polymer to 187.6 μm for 100,000 MW polymer. Degradation of the microspheres occurred fastest with the 2000 MW polymer and slowed as polymer molecular weight increased. Microspheres loaded with MTX and 2000 MW PLLA showed a rapid burst phase followed by a slow release rate in vitro. In vivo results showed sustained release of MTX into the systemic circulation in the microsphere formulation, compared to high plasma MTX levels followed by decreasing MTX upon elimination. The t_{max} values of both the microsphere MTX and free MTX showed similar results with a time of 0.5 hours. The C_{max} values, however varied significantly, with free MTX reaching a peak plasma concentration of about 0.6 μg/mL at the t_{max}, while the microsphere MTX formulation only achieved a C_{max} of about 0.05 μg/mL. The free MTX showed an exponential decline in plasma concentration following administration, while the microspherical MTX showed nearly constant release over the 8-h time period. Both formulations showed plasma levels of 0.04 μg/mL at the end of the study time period of 8 h. This shows that nearly zero-order release of MTX is observed using the microsphere formulation. Other in vivo results showed that, after sacrificing the rabbits, no MTX was detected in knee joints of animals dosed with free MTX, while approximately 0.6 μg MTX was detected in knee joints of rabbits treated with MTX microspheres. The MTX concentrations in plasma vs. time were plotted to evaluate the systemic bioavailability of free MTX and MTX microspheres. The plasma concentration vs. time curve is shown in Figure 15. Although the MTX microspheres showed lower systemic bioavailability, histological scoring showed decreased inflammation and joint swelling due to rheumatoid arthritis in animals treated with MTX microspheres over free MTX. Increasing attention is being paid to development of polymeric microspheres for delivery of drugs locally to the joints for sustained drug release. Following intra-articular injection, these microspheres demonstrated good biocompatibility and decreased the clearance rate of MTX from the joint cavity, providing for decreased inflammation in rheumatoid arthritis (62).

Other studies have been conducted to improve delivery of methotrexate to the joints of rheumatoid arthritis patients. Wunder et al. have used albumin-based formulation design for the delivery of MTX (63). They state the main features of appropriate drug carriers are high accumulation in target tissues, low uptake rates by normal tissue, low toxicity, and biochemical potential to be linked to drugs. This research group has found that albumin acts as a good carrier for MTX because it satisfies all of these characteristics when considering tumor tissue as the target. They stated that albumin might also be a suitable drug carrier for drug targeting to inflamed joints of patients with rheumatoid arthritis. Using albumin labeled with aminofluorescein, the study showed that albumin did, indeed, preferentially accumulate in inflamed joints with rheumatoid arthritis. This observation was confirmed using radiolabeled human serum albumin (HSA), which showed significant accumulation in joints of collagen-induced arthritis (CIA)

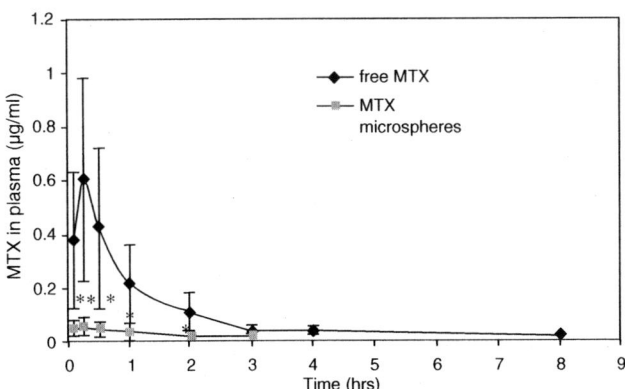

Figure 15 MTX concentrations in rabbit plasma after a single intra-articular injection of either free MTX or 25 mg of MTX loaded microspheres in 200 μL PBS. The dose of MTX injected was 1.5 mg. (*) Indicates statistical difference between MTX plasma concentrations of rabbits injected with free MTX and MTX-loaded microspheres by paired t-test ($p < 0.05$). *Source*: From Ref. 62.

rats. This study evaluated the uptake kinetics of radiolabeled HSA and radiolabeled MTX in the paws of mice both with and without CIA. Significant albumin amounts accumulated in arthritic hind paws, exceeding the uptake in non-inflamed hind paws by 6- to 7-fold. In contrast, uptake of radiolabeled MTX in arthritic hind paws was found to be significantly less, decreasing rapidly over time (63). Different dosages of MTX and MTX-HSA were then administered to rats that had been given CIA, and the incidence of arthritis in these animals was observed. Treatment was started two weeks before onset of disease, and the animals were treated for 4 weeks and received two injections weekly. MTX-HSA was significantly more effective in the suppression of the onset of arthritis than were comparable doses of MTX. At least a 5-fold higher dose of MTX was required to achieve the levels of arthritis suppression seen with MTX-HSA. The effect of different doses of MTX and MTX-HSA on the development of CIA in mice is evaluated in this study. These results indicate that MTX, in contrast with albumin, is rapidly removed from blood circulation and does not accumulate in inflamed paws. When complexed together, however, MTX and HSA target inflammation in CIA mice, allowing for delivery of the drug to the site of action.

Methotrexate formulations for the treatment of psoriasis were developed using solid lipid nanoparticles (SLN) loaded with the drug (64). Topical application of methotrexate was developed to minimize the systemic bioavailability of MTX, while using the immunosuppressive properties of the drug to treat psoriasis. MTX-SLNs were prepared by melting the lipid components and adding the drug to the lipid phase (containing Tween® 80), preparing an aqueous phase using a co-surfactant, followed by emulsification of the two phases with heating and mechanical stirring to form an O/W emulsion. SLNs were then formed by cooling the emulsion and dispersing it in a mixture of cold water and propylene glycol. They were further purified using dialysis to remove any unentrapped drug. Particle size distribution and other in vitro tests were performed to characterize the SLNs. The SLNs were then incorporated into a carbopol gel for application to psoriatic lesions. Deposition of the drug was evaluated on cadaver skin. The tolerability of the formulations as well as their ability to heal psoriatic lesions was evaluated in psoriasis patients. Average amount of MTX deposited on the skin increased from 36.4% for free MTX to 39.1% for the lipid/MTX physical mixture to 74.5% for the MTX-SLN formulation. In human subjects afflicted with psoriasis, the average percent improvement in lesion healing was evaluated, and consistently, the MTX-SLN formulation showed much improved healing over the plain MTX and MTX/lipid physical mixture. Table 3 shows the gel formulations evaluated in this study, with plain MTX, MTX/lipid physical mixtures, and MTX-SLN's detailed. Table 4 shows the average percent improvement in healing of psoriatic lesions using the different MTX formulations. In vivo skin deposition of the free MTX and the physical mixture of MTX/lipids on the skin than the SLN

Table 3 Code of Different MTX Gels Evaluated in the Study

Gel code	Content[a]
MTxG1	Plain MTx (marketed)
MTxG2	MTx and Lipid Physical Mixture
MTxG3	MTx-SLN

[a] Content of MTx in gels were 0.9–1.1% w/w.
Source: From Ref. 64.

formulation, but upon washing of the applied areas, > 99% of the MTX was removed from the skin. The SLN formulations, however, showed higher drug retention on the skin surface. Patient scoring was conducted after treatment with MTX, MTX/lipid mixture, and MTX SLN formulations, from 0 to 4 (0 = no degree of erythema and scaling, and 4 = severe erythema and scaling). Baseline scoring showed high degrees of psoriatic symptoms (ranging from 3.8 to 4). Treatment was administered using the three formulations for 6 weeks, and scoring for the psoriatic symptoms decreased in all treatment groups. Most notably, however, the MTX SLN group showed the fastest recovery time from psoriatic symptoms, showing a score of 0 at the 3 week time period. The free MTX showed a score of 1.9, while the physical mixture showed a scoring of 2.2. The trend continued through the end of the study, (6 weeks), when the final scoring for the MTX was 0.4, the scoring for the MTX/lipid mixture was 0.5, and the scoring for the MTX SLN continued to show 0. While plain MTX is shown to be effective against psoriasis, the physicochemical properties of the drug hinder the drug's ability to penetrate the epidermis. The SLN formulations appear to allow for penetration of the drug deeper into the skin surface, and, therefore lend themselves to more effective treatment against topical skin maladies such as

Table 4 Average Percent Improvement of Healing of Psoriasis Lesions

Time (weeks)	Mean % (± SEM) for formulations tested[a]		
	MTxG1	MTxG2	MTxG3
1	14.1 (0.14)	14.3 (0.26)	33.7 (0.46)
2	29.8 (0.29)	31.1 (0.73)	66.5 (0.38)
3	50.7 (0.95)	52.7 (0.49)	89.1 (0.78)
4	67.6 (1.13)	69.3 (1.09)	99.4 (0.93)
5	79.4 (0.89)	80.4 (0.99)	99.8 (0.24)
6	86.1 (1.16)	86.9 (1.21)	99.8 (0.35)

[a] Average of the eight patients (n = 8).
Source: From Ref. 64.

psoriasis. The findings of this investigation, therefore, demonstrate the promising role of MTX SLN in treatment of psoriasis (64).

Azathioprine

Azathioprine (AZA) is a cytostatic drug, which, as stated previously, is a prodrug analog of the purine molecule, a component of nucleic acids. Until the discovery of cyclosporin in 1978, the combination of azathioprine and corticosteroid medications was the standard for treatment of transplant rejection. In the body, the AZA molecule is cleaved and converted to 6-mercaptopurine, the active form of the drug. This metabolite then goes on to antagonize purine synthesis which halts the synthesis of DNA, RNA, and some proteins (65). This drug therefore inhibits mitosis and cell metabolism, reducing the proliferation of lymphocytes, leukocytes, and mast cells, and subsequently reducing the body's immune response. AZA has also been shown to reduce the amount of free antibodies in circulation, stemming from the reduced number of immune-related cells (66). As a result of its mechanism of action, this drug has been shown to be toxic in large amounts and over long therapeutic dosing schedules. For this reason, therapeutic drug monitoring should accompany the administration of azathioprine. AZA free acid is poorly soluble in water, and it shows varying solubility with pH. The sodium salt of azathioprine is typically used in formulating dosage forms because it shows increased aqueous solubility compared to the free acid (66). AZA, with the commercial trade name Imuran®, is traditionally administered orally as a 50 mg AZA tablet or parenterally as a reconstituted aqueous solution of the drug produced from its lyophilized sodium salt (65). The chemical structure of AZA is depicted in Figure 1, and relevant physicochemical properties are shown in Table 1.

AZA has been studied for its immunosuppressive effect on a number of autoimmune conditions, including Crohn's disease. However, use of this drug has been limited by concerns over its systemic toxicity (66). Because systemic delivery of immunosuppressants can cause opportunistic infection or other harmful side-effects, in many instances, these drugs are desired to be delivered locally, with limited systemic absorption. Colonic delivery has been evaluated as a strategy limit the systemic delivery of AZA thereby reducing much if its toxicity. Zins et al. have developed a delayed-release oral formulation of AZA which shows lower systemic bioavailability than standard AZA tablets, retaining the effectiveness against autoimmune intestinal disorders while decreasing toxicity (67). This study aimed to determine the bioavailability and pharmacokinetic parameters of delayed-release oral AZA capsules at doses of 200, 400, and 600 mg while comparing them to the 100 mg standard capsule. The relative bioavailabilities after ileocolonic administration via delayed-release capsules were 15%, 15%, and 12% for the 200, 400, and 600 mg capsules, compared to 100% for the standard 100 mg dose. Ileocolonic delivery of AZA by this formulation, therefore, reduced the systemic bioavailability, while alleviating the symptoms of Crohn's disease. The therapeutic potential of this ileocolonic delivery formulation, which can limit toxicity by local delivery of AZA, should be investigated in patients with refractory IBD (67).

In addition, AZA was formulated into either a hydrophilic or hydrophobic foam for rectal administration for the treatment of IBD (68). The foams were evaluated for their pharmacokinetic parameters and compared to the delayed-release and standard capsule formulations mentioned above. The bioavailabilities of the drug after colonic AZA administration via an oral delayed release form, a hydrophobic rectal foam, and a hydrophilic rectal foam (7%, 5%, and 1%, respectively) were significantly lower than the bioavailability of AZA after oral administration with the immediate release capsule. The study concluded that AZA delivered directly to the colon by either delayed release oral or rectal foam formulations considerably reduced the systemic AZA bioavailability. The therapeutic potential of these colonic delivery methods, which can potentially limit toxicity by local delivery of high doses of AZA could be beneficial in treatment of IBD (68).

CONCLUSION

Immunosuppressant drugs have been used to treat a wide variety of autoimmune conditions and have shown promise in suppressing the immune response for prevention of organ rejection post-transplantation. Overcoming the drug's physicochemical properties of low water solubility, PGP or CYP 3A4 interaction, or cellular toxicity are needed to produce safe and effective drug products. Some drugs, however, benefit from lower systemic bioavailability and higher drug concentration at the site of action. As shown herein, advanced formulation design has allowed for more efficient delivery of these drugs to their site of action, leading to higher efficacy, reduced toxicity, and lower dosing required to reach a therapeutic response. Alternative routes of administration, such as pulmonary, topical, and oral enteric routes have all shown significant improvement in alleviation of various disease states over conventional treatments. Several currently marketed commercial products utilize novel formulation design in delivery of immunosuppressant drugs, but more research is needed to develop new drug products with yet more enhanced therapeutic outcomes. Future studies of these drugs on solid organ transplants can improve the survival of these transplants, allowing patients to live more normal lives. Improved formulation design could allow for less frequent dosing of these medications to transplant patients, while lessening the occurrence of opportunistic infections which could occur in patients on immunosuppressant therapy. Continuing research in the area of delivery of immunosuppressants is required so that transplantation techniques can improve long-term patient survival. Along with transplant patients, other patients with autoimmune disorders can benefit from immunosuppressant therapy, and these areas must be studied to improve quality of life and alleviate symptoms of these types of immune disorders.

REFERENCES

1. Chakinala MM, Kollef MH, Trulock EP. Critical care aspects in lung transplant patients. J Intens Care Med 2002; 17:8–33.
2. Wilson, Shannon, Shields, Stang. Drug Guide. Upper Saddle River, NJ: Prentice Hall Health, 2003.
3. Liversidge GG, Conzentino P. Drug particle size reduction for decreasing gastric irritancy and enhancing absorption of naproxen in rats. Int J Pharm 1995; 125:309–15.
4. Ostrander KD, Bosch HW, Bondanza DM. An in-vitro assessment of a nanocrystal beclomethasone dipropionate colloidal dispersion via ultrasonic nebulization. Eur J Pharm Biopharm 1999; 48:207–15.
5. Maia CS, Mehnert W, Schafer-Korting M. Solid lipid nanoparticles as drug carriers for topical glucocorticoids. Int J Pharm 2000; 196:165–7.
6. Maia CS, Gysler A, Mehnert W, Muller RH, Schafer-Korting M. Local tolerability of solid lipid nanoparticles for dermal use. Proc Int Symp Control Release Bioactive Mater 1999; 26:399–400.
7. Higaki M, Ishihara T, Izumo N, Takatsu M, Mizushima Y. Treatment of experimental arthritis with poly (D,L lactic/glycolic) acid nanoparticles encapsulating betamethasone sodium Phosphate. Ann Rheum Dis 2005; 64:1132–6.
8. Guzman M, Molpeceres J, Garcia F, Aberturas MR, Rodriguez M. Formation and characterization of cyclosporin loaded nanoparticles. J Pharm Sci 1993; 82:498–502.
9. Morand EF. Corticosteroids in the treatment of rheumatologic diseases. Current Opin Rheumatol 2000; 12:171–7.
10. Van Der Velden VHJ. Glucocorticoids: Mechanism of action and anti-inflammatory potential in asthma. Mediators Inflamm 1998; 7:229–37.
11. Faassen F, Kelder J, Lenders J, Onderwater R, Vromans H. Physicochemical properties and transport of steroids across CACO-2 Cells. Pharm Res 2003; 20:177–86.
12. Wang S, Zheng Y, Zhang W. Betamethasone dipropionate. Acta Crystal Struct Rep 2004; E60:O1063–4.
13. Duax WL, Cody V, Strong PD. Structure of the asthma drug beclomethasone dipropionate. Acta Crystal Struct Crystallog Cryst Chem 1981; B37:383–7.
14. Beclovent Aerosol for Inhalation: professional literature. Research Triangle Park, NC: Glaxo Smithkline Pharmaceuticals, 1998.
15. Rocca-Serra JP, Vicaut E, Lefrancois G, Umile A. Efficacy and tolerability of a new non-extrafine formulation of beclomethasone HFA-134a in patients with asthma. Clin Drug Invest 2002; 22:653–65.

16. Juniper E, Price DB, Stampone P, Creemers J, Mol S, Fireman P. Clinically important improvements in asthma-specific quality of life, but no difference in conventional clinical indexes in patients changed from conventional beclomethasone to approximately half the dose of extrafine beclomethasone dipropionate. Chest 2002; 121:1824–32.
17. Merisko-Liversidge E, Liversidge GG, Cooper E. Nanosizing: A formulation approach for poorly water soluble compounds. Eur J Pharm Sci 2003; 18:113–20.
18. Wiedmann TS, DeCastro L, Wood RW. Nebulization of nanocrystals: production of respirable solid-in-liquid-in-air colloidal dispersion. Pharm Res 1997; 14:112–16.
19. Sakagami M, Kinoshita W, Makino Y, Fujii T. Mucoadhesive BDP microspheres for powder inhalation - their unique pharmacokinetic/pharmacodynamic properties. Respir Drug Del 1998; 4:193–9.
20. Edsbacker S, Wollmer P, Nilsson A, Nilsson M. Pharmacokinetics and gastrointestinal transit of budesonide controlled ileal release (CIR) capsules. Gastroenter 1993; 104:A695.
21. Rodriguez M, Antunez J, Taboada C, Seijo B, Torres D. Colon specific delivery of budesonide from microencapsulated cellulosic cores: Evaluation of the efficacy against colonic inflammation in rats. J Pharm Pharmacol 2001; 53:1207–15.
22. Edsbacker S, Andersson T. Pharmacokinetics of budesonide (entocort EC) capsules for Crohn's disease. Clin Pharmaco 2004; 43:803–21.
23. Edsbacker S, Larsson P, Wollmer P. Gut delivery of budesonide, a locally active corticosteroid from plain and controlled-release capsules. Eur J Gastroenterol Hepatol 2002; 14:1357–62.
24. Waldrep JC, Arppe J, Jansa KA, Knight V. High dose cyclosporin A and budesonide liposome aerosols. Int J Pharm 1997; 152:27–36.
25. Waldrep JC, Gilbert BE, Knight CM, Black MB, Sherer PW. Pulmonary delivery of beclomethasone aerosol in volunteers. Chest 1996; 11:316–23.
26. Lobo JM, Schiavone H, Palakodaty S, York P, Clark A, Tzannis S. SCF-Engineered powders for delivery of budesonide from passive DPI devices. J Pharm Sci 2005; 94:2276–88.
27. Benameur H, Latour N, Schandene L, VanVooren JP, Flamion B, Legros F. Liposome-incorporated dexamethasone palmitate inhibits in-vitro lymphocyte response to a mitogen. J Pharm Pharmacol 1995; 47:812–17.
28. Suntres Z, Shek P. Liposomes promote pulmonary glucocorticoid delivery. J Drug Targ 1998; 6:175–82.
29. Suntres Z, Shek P. Prophylaxis against lipopolysaccharide-induced lung injuries by liposome-entrapped dexamethasone in rats. Biochem Pharmacol 2000; 59:1155–61.
30. Hickey T, Kreutzer D, Burgess DJ, Moussy F. Dexamethasone/PLGA Microspheres for continuous delivery of an anti-inflammatory drug for implantable medical devices. Biomat 2002; 23:1649–56.
31. Chorny M, Mishaly D, Leibowitz A, Domb A, Golomb G. Site-specific delivery of dexamethasone from biodegradable implants reduces formation of pericardial adhesions in rabbits. J Biomed Materials Res A 2006; 78A: 276–82.
32. Ho S, Clipstone N, Timmerman L, et al. Mechanism of action of cyclosporin A and FK506. Clin Immunol Immunop 1996; 80:S40–5.

33. Sandimmune Capsules and Oral Liquid: professional literature. East Hanover, NJ: Novartis Pharmaceuticals, 1999.

34. Neoral Soft Gelatin Capsules and Oral Liquid: professional literature. East Hanover, NJ: Novartis Pharmaceuticals, 1998.

35. Prograf Capsules: professional literature. Osaka, Japan: Fujisawa Pharmaceuticals, 2004.

36. Protopic ointment: professional literature. Osaka, Japan: Fujisawa Pharmaceuticals, 2002.

37. Morris RE. Prevention and treatment of allograft rejection in vivo by rapamycin: molecular and cellular mechanisms of action. Ann NY Acad Sci 1992; 1:68–72.

38. Molnar-Kimber KL. Mechanism of action of rapamycin (Sirolimus, Rapamune). Transplant Proc 1996; 28:964–9.

39. Kruijtzer C, Beijnnen J, Schellens J. Improvement of oral drug treatment by temporary inhibition of drug transporters or CYP 450 in the gastrointestinal tract and liver: an overview. Oncologist 2002; 7:516–30.

40. Ran Y, Zhao L, Xu Q, Yalkowsky S. Solubilization of Cyclosporin A. AAPS Pharm Sci Tech 2001; 2.

41. Dressman JB, Reppas, C. In vitro-in vivo correlations for lipophilic, poorly water soluble drugs. Eur J Pharm Sci 2000; 11:S73–80.

42. Balandrand-Pieri N, Queneau P, Caroli-Bosc. F, et al. Effects of Tauroursodeoxycholate Solutions on Cyclosporin A Bioavailability in Rats. Drug Metabol Distrib 1997; 25:912–16.

43. Sanchez A, Seoane R, Quireza O, Alonso MJ. In vivo study of the tissue distribution and immunosuppressive response of cyclosporin A-loaded polyester micro and nanospheres. Drug Deliv 1995; 2:21–28.

44. Mitruka SN, Pham SM, Zeevi A, et al. Aerosol cyclosporin prevents acute allograft rejection in experimental lung transplantation. J Thoracic Cardiovasc Surg 1997; 115:28–37.

45. Mitruka SN, Won A, McCurry KR, et al. In the lung aerosol cyclosporine provides a regional concentration advantage over intramuscular cyclosporine. Clin Lung Heart/Lung Transplant 2000; 19:969–75.

46. Soeda S, Akashi T, Maeda K, Kawagita T. Studies on the development of tacrolimus production. Seibutsu Kogaku Kaishi 1998; 76:389–97.

47. Yamashita K, Nakate T, Okimoto K, et al. Establishment of new preparation method for solid dispersion formulation of tacrolimus. Int J Pharm 2003; 267: 79–91.

48. Hancock BC, Zograffi G. Characteristics and significance of the amorphous state in pharmaceutical systems. J Pharm Sci 1997; 86:1–12.

49. Raghavan SL. Effect of cellulose polymers on supersaturation and in vitro membrane transport of hydrocortisone acetate. Int J Pharm 2000; 193: 231–37.

50. Honbo T, Kobayahi M, Hane K, Hata T, Ueda Y. The Oral Dosage form of FK-506. Transplant Proc 1987; 19:17–22.

51. Lamprecht A, Yamamoto H, Takeuchi H, Kawashima Y. Design of pH-sensitive microspheres for the colonic delivery of immunosuppressive drug tacrolimus. Eur J Pharm Biopharm 2004; 58:37–43.

52. Meissner Y, Pellequer Y, Lamprecht A. Nanoparticles in inflammatory bowel disease: particle targeting vs. pH sensitive delivery. Int J Pharm 2006; 316: 138–143.

53. Lamprecht A, Yamamoto H, Ubrich N, Takeuchi H, Maincent P, Kawashima Y. FK-506 microparticles mitigate experimental colitis with minor renal calcineurin suppression. Pharm Res 2005; 22:193–99.

54. Sehgal SN, Camardo JS, Scarola JA, Maida BT. Rapamycin (Sirolimus, Rapamune). Dialys Transplant 1995:482–487.

55. Grube E, Buellesfeld L. Rapamycin analogs for stent based local drug delivery. Herz 2004; 29:162–66.

56. Sorbera LA, Leeson PA, Castaner J. SDZ-RAD. Drugs Fut 1999; 24:22.

57. Simamora P, Alvarez JM, Yalkowsky SH. Solubilization of Rapamycin. Int J Pharm 2001; 213:25–29.

58. Sehgal SN, Molnar-Kimber K, Ocain TD, Weichman BM. Rapamycin: a novel immunosuppressive macrolide. Med Res Review 1994; 14:1–22.

59. Rapamune Tablets and Oral Liquid: professional literature. Philadelphia, PA: Wyeth Pharmaceuticals, 2004.

60. Wishart DS, Knox C, Guo AC, et al. Drug Bank: A comprehensive resource for in silico drug discovery and exploration. Nucleic Acids Res 2006; 34: D668–72.

61. Trexall Tablets: professional literature. Pomona, NY: Duramed Pharmaceuticals, 2005.

62. Liang L, Jackson J, Min W, Risovic V, Wasan K, Burt H. Methotrexate loaded poly (L-lactic) acid microspheres for intra-articular delivery of methotrexate to the joint. J Pharm Sci 2004; 93:943–55.

63. Wunder A, Muller-Ladner U, Stelzer E, et al. Albumin-based drug delivery as novel therapeutic approach for rheumatoid arthritis. J Immunol 2003; 170: 4793–801.

64. Kalariya M, Padhi B, Chougule M, Misra A. Methotrexate loaded solid lipid nanoparticles for topical treatment of psoriasis: formulation and clinical implications. Drug Del Tech 2004; 4:65–71.

65. Imuran Tablets: professional literature. Mississagua, Ontario: Glaxo-Smithkline Pharmaceuticals, 2005.

66. Report on Carcinogens, 11th edn. In: Services UDoHaH, ed. Vol. 11: Public Health Service, National Toxicology Program, 2004.

67. Zins BJ, Sandborn WJ, McKinney JA, et al. A dose ranging study of azathioprine pharmacokinetics after single-dose administration of a delayed-release oral formulation. J Clin Pharmacol 1997; 37:38–46.

68. Van Os EC, Zins BJ, Sandborn WJ, et al. Azathioprine pharmacokinetics after intravenous, oral, delayed release oral, and rectal foam administration. Gut 1996; 39:63–68.

13

Solid Dispersion Technologies

Dave A. Miller, James W. McGinity, and Robert O. Williams III

College of Pharmacy, University of Texas at Austin, Austin, Texas, U.S.A.

INTRODUCTION

Background

The oral route of drug administration is the most common and preferred method of delivery due to convenience and ease of ingestion. From a patient's perspective, swallowing a dosage form is a comfortable and familiar means of taking medication. As a result, patient compliance and hence drug treatment is typically more effective with orally administered medications as compared with other routes of administration, for example, parenteral.

Although the oral route of administration is preferred, for many drugs it can be a problematic and inefficient mode of delivery for a number of reasons. Limited drug absorption resulting in poor bioavailability is paramount amongst the potential problems that can be encountered when delivering an active agent via the oral route. Drug absorption from the gastrointestinal (GI) tract can be limited by a variety of factors with the most significant contributors being poor aqueous solubility and/or poor membrane permeability of the drug molecule. When delivering an active agent orally, it must first dissolve in gastric and/or intestinal fluids before it can then permeate the membranes of the GI tract to reach systemic circulation. Therefore, a drug with poor aqueous solubility will typically exhibit dissolution rate limited absorption, and a drug with poor membrane permeability will typically exhibit permeation rate limited absorption. Hence, two areas of pharmaceutical research that focus on improving the oral bioavailability of active agents include: (*i*) enhancing solubility and dissolution rate of poorly water-soluble drugs and (*ii*) enhancing permeability of poorly permeable drugs. This chapter focuses on the former, in particular, the use of solid dispersion technologies to improve the dissolution characteristics of poorly water-soluble drugs and in turn their oral bioavailability.

Numerous solid dispersion systems have been demonstrated in the pharmaceutical literature to improve the dissolution properties of poorly water-soluble drugs. Other methods, such as salt formation, complexation with cyclodextrins, solubilization of drugs in solvent(s), and particle size reduction have also been utilized to improve the dissolution properties of poorly water-soluble drugs; however, there are substantial limitations with each of these techniques. On the other hand, formulation of drugs as solid dispersions offers a variety of processing and excipient options that allow for flexibility when formulating oral delivery systems for poorly water-soluble drugs.

Much of the research that has been reported on solid dispersion technologies involves drugs that are poorly water-soluble and highly permeable to biological membranes as with these drugs dissolution is the rate limiting step to absorption. Hence, the hypothesis has been that the rate of absorption in vivo will be concurrently accelerated with an increase in the rate of drug dissolution. In the Biopharmaceutical Classification System (BCS) drugs with low aqueous solubility and high membrane permeability are categorized as Class II drugs (1). Therefore, solid dispersion technologies are particularly promising for improving the oral absorption and bioavailability of BCS Class II drugs.

With recent advances in molecular screening methods for identifying potential drug candidates, an increasing number of poorly water-soluble drugs are being identified as potential therapeutic agents. In fact, it has been estimated that 40% of new chemical entities currently being discovered are poorly water-soluble (2). Unfortunately, many of these potential drugs are abandoned in the early stages of development due to solubility concerns. It is therefore becoming increasingly more important that methods for overcoming solubility limitations be identified and applied commercially such that the potential therapeutic benefits of these active molecules can be realized.

Overview of Solid Dispersion Technologies

The term *dispersion* is typically used to describe a heterogeneous, multiphase system in which discrete domains of a discontinuous phase are distributed within a second, continuous phase (3). Although the term, dispersion, implies immiscibility of phases, the term, solid dispersion, as applied to drug delivery is often used generically to describe all systems in which an active agent is dispersed in an inert excipient carrier, including systems in which the drug may exist as a molecular dispersion (solid solution) in which there is not a discernable second phase. Thus, in this chapter the term, solid dispersion, will also be used generally to describe all systems in which an active agent is dispersed in an excipient carrier. With respect to the state of the drug in the systems, solid dispersions in this sense can include compositions in which the drug is dispersed as discrete domains of crystalline or amorphous drug, or as individual molecules within an excipient carrier. With respect to the complete drug–excipient composite, solid dispersions can be relatively large solid masses such as pellets, tablets, films, or strands; or they can exist as free flowing powders consisting of micro or nano-sized primary particles or aggregates thereof. The bulk state of the solid dispersion composition depends largely on the mode of processing.

Generally, solid dispersions are produced by solvent evaporation or melt processing methods; however, recently supercritical fluid and cryogenic freezing technologies have been demonstrated to produce fine particles of drug-excipient solid dispersions (4–6). By common solvent methods, the drug and the excipient(s) that comprise the carrier system are dissolved in a common solvent or cosolvent and subsequently the solvent is removed yielding a homogenous solid composite with the drug dispersed in the excipient matrix. The morphology of the drug in such systems depends largely on the miscibility of the drug with the excipient carrier, where greater miscibility tends toward greater amorphous drug content. Following solvent removal, the solid dispersion can then be milled to yield a fine powder which can then be filled into capsules or combined with tableting excipients and compressed.

Melt methods involve rendering at least one of the components of the drug–excipient mixture molten such that the components can be intimately mixed to form the solid dispersion. Typically the drug and the carrier are co-melted and homogenously mixed to produce the solid dispersion system. The extent of amorphous drug content in a solidified co-melt, as with solvent methods, depends largely on the miscibility of the drug and the carrier system in the molten state. In some cases only the carrier is melted and the drug particles are dispersed in the molten excipient system to form a solid dispersion of crystalline drug particles. Although melt methods were the first reported techniques used to produce solid dispersions, they waned in popularity with the advent of solvent methods due to various limitations of thermal processing. However, with the application of hot-melt extrusion to pharmaceutical processing, melt processing was revived as a result of improved processing efficiency and drug/excipient stability. Solid dispersions formed by melt processing can be milled into fine granules or powders and filled into capsules or compressed into tablets. The products of hot melt extrusion can be either strands or films. Strands can be directly cut into tablets, pelletized, or milled; while films only require cutting to the desired size.

Supercritical fluid technologies can produce solid dispersions in the form of engineered particles by utilizing supercritical fluids as antisolvents or rapidly evaporating solvents to force rapid precipitation of drug–excipient particles. Depending on the formulation and processing parameters, supercritical fluid-based processes can yield particles containing drug in an amorphous or crystalline form. Examples of such processes include the Gaseous antisolvent precipitation process (7), precipitation with compressed antisolvent (8), rapid expansion from supercritical solutions (9), and precipitation from gas-saturated solutions (10).

Generally, cryogenic technologies involve the rapid-freezing of single solvent or co-solvent based solutions containing drug and stabilizing excipients by either spraying on or into a cryogenic liquid, or applying the solution onto a cryogenic substrate. The frozen material is then lyophilized to remove the solvent by sublimation, thus yielding a freely flowing powder of high surface area that, in cases where excipient stabilizers are used, are solid dispersions of the active agent in an excipient carrier. Examples of these processes include spray-freeze-drying (11), spray-freezing into a halocarbon refrigerant vapor (12), spray-freezing into halocarbon refrigerant (13), spray freezing onto liquid nitrogen (14), spray-freezing into liquid (4), and ultra-rapid freezing (15).

The underlying principles for improving the dissolution properties of drugs by solid dispersion techniques are: (*i*) reducing particle size, (*ii*) altering the crystalline morphology, and (*iii*) intimately mixing the drug with hydrophilic excipients. By altering the bulk drug according to these principles, drug particle surface area is increased, the thermodynamic barrier to dissolution imposed by the crystal lattice is eliminated, and the wetting properties of the drug particles are enhanced. Each of the solid dispersion technologies mentioned above modify drug particles by one or more of these principles to improve the dissolution properties of the active molecule. For BCS Class II drugs where dissolution is the rate limiting step to absorption, enhancing the dissolution properties of a drug should correlate to improved oral drug absorption and overall therapeutic efficacy. In this chapter, the application of solid dispersion technologies to the oral delivery of BCS class II drugs is examined by reviewing published studies that detail the effects of solid dispersion systems on in vivo drug absorption. These studies demonstrate the current contributions and future potential of solid dispersion technologies toward improving the therapeutic outcomes by drug treatment for a variety of important medical conditions.

TREATMENT OF CARDIOVASCULAR DISORDERS

Solid dispersion technologies have been applied to several drug molecules that are indicated for the treatment of cardiovascular disorders. Several drugs belonging to classes of lipid lowering agents, anti-hypertensives, anti-angina, etc. demonstrate poor aqueous solubility resulting in dissolution rate limited absorption when administered orally. Hence, solid dispersion technologies have been utilized to enhance the dissolution properties of these drugs with the aim of improving in vivo drug absorption.

One such drug, fenofibrate, is a lipid lowering agent which is highly lipophilic and practically insoluble in water (0.1 µg/mL) (16). It is readily absorbed through the membranes of the GI tract, but is poorly bioavailable by oral administration due to slow and incomplete dissolution. Fenofibrate exhibits particularly low bioavailability when taken on an empty stomach (17). Sant et al. demonstrated the use of solid dispersion systems with polymeric micelles to improve the dissolution properties and enhance the oral absorption of fenofibrate (18). In this study, fenofibrate was encapsulated in a block copolymer consisting of poly(ethylene glycol) and poly(alkyl acrylate-co-methacrylic acid) by an oil-in-water emulsion method (19) and cast into films by evaporating the solvents under reduced pressure. The resulting film contained submicron particles consisting of fenofibrate dispersed within the hydrophobic acrylate blocks of the polymer micelles with the hydrophilic blocks forming an exterior shell. This delivery system exhibited pH dependant drug release as a result of the pendant carboxyl groups present on the hydrophobic chains. At pH below 4.7 these functional groups remain protonated, and hence neutrally charged promoting the formation of nanoaggregates in aqueous media by hydrophobic interaction. As the pH of the media was increased, the carboxyl groups became deprotonated rendering them anionically charged and more hydrophilic. Increased hydrophilicity of the polymer promoted deaggregation of the micellular nanoparticles followed by subsequent solubilization of the polymer leading to release of fenofibrate in a molecularly dispersed form.

This system was evaluated for in vivo drug absorption in Sprague–Dawley rats. The hypothesis tested with this study was that the pH-dependant aggregation of the nanoparticles would prevent rapid release of fenofibrate in the acidic environment of the stomach that could lead to drug precipitation and promote the release of drug in the small intestine to thereby improve bioavailability. The in vivo performance of the polymeric micelles was compared to bulk fenofibrate and a commercial formulation known as Lipidil Micro® which consisted of micronized fenofibrate with solubilization enhancers. From this study, it was determined that the polymeric micelle solid dispersion formulation exhibited the shortest T_{max}, highest C_{max}, and greatest AUC_{0-24h} when compared to the bulk powder and the Lipidil Micro formulation. The relative bioavailability of the polymeric micelle formulation was observed to be 156% and 15% greater than the bulk drug and the commercial formulation, respectively. The plasma concentration versus time curve from this study is shown in Figure 1.

The use of polymeric micelles in this study illustrated an advanced solid dispersion system in which the carrier polymers where designed to

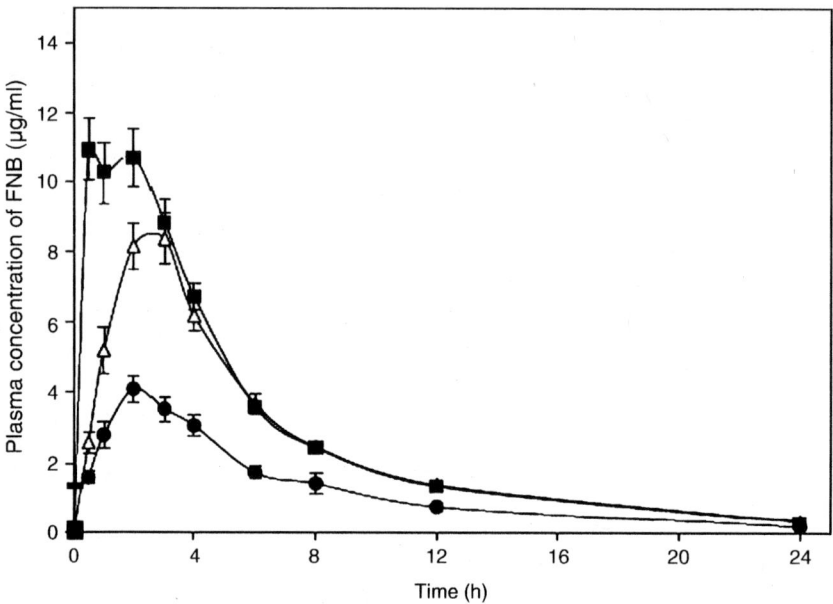

Figure 1   Plasma concentration vs. time curve of FNB after oral administration of
(■) pH-sensitive self-assemblies of PEG-b-P(nBA17-co-MAA17), (•) Lipidil MicroR
and (Δ) FNB powder to fasted Sprague–Dawley rats at a dose of 7.5 mg/kg. Mean
FSEM for $n = 6$.

form self-assembling nanoparticles by a simple O/W emulsion system, and
which provided pH-dependant drug release allowing for targeted release to
the small intestine. The oral delivery of fenofibrate from this unique solid
dispersion system was demonstrated to provide enhanced drug absorption in
vivo, and hence suggests the potential for improved treatment of hyperlipidemia with fenofibrate.

Recently, a study conducted by Ambike et al. demonstrated improved
absorption of simvastatin (SIM), a frequently prescribed cholesterol
lowering agent, from solid dispersions of amorphous SIM in stabilizing
excipients (20). The amorphous SIM solid dispersion particles were
produced by spray drying an organic solution of the drug and polyvinylpyrrolidone (PVP) along with Aerosil 200. Differential scanning
calorimetry (DSC) and powder X-ray diffraction (PXRD) confirmed that
SIM was present in a predominantly amorphous state for spray dried
dispersions in a 1:2:2 ratio of SIM:PVP:Aerosil 200. Tablets containing this
solid dispersion formulation exhibited a substantially accelerated in vitro
dissolution rate in pH 6.8 phosphate buffer. Hypolipidemic activity of the
1:2:2 SIM:PVP:Aerosil 200 solid dispersion was compared to bulk SIM in
healthy Wistar rats that were administered excess coconut oil to promote

hypercholesterolemia. The results of this study indicated enhanced absorption of SIM from the solid dispersion formulation over bulk SIM as the solid dispersion formulation substantially reduced total cholesterol and triglycerides while increasing HDL-cholesterol levels beyond that of the bulk powder. Hence, this study demonstrated the potential for enhanced treatment of hypercholesterolemia with SIM by formulation of the drug as an amorphous solid dispersion.

TAS-301 is a compound which is currently being developed as an antirestenosis drug for use following percutaneous transluminal coronary angioplasty. The water solubility of TAS-301 is approximately 20 ng/mL, and hence the drug has been shown to be very poorly bioavailable following oral dosing to fasted rats and dogs (21). In a study by Kinoshita et al. solid dispersions of TAS-301 were produced by melt-adsorption of the drug onto porous calcium silicate known commercially as Florite® RE (FLR) (21). The melt adsorption process used to produce the dispersions was conducted by both a small-scale batch heat-treating process, and in a continuous manner by the use of a twin-screw melt extruder. The crystallinity of TAS-301 in melt adsorbed formulations with a 1:2 drug to carrier ratio was reduced to concentrations below the limits of detection of both the DSC and PXRD instruments. The authors speculated that the significant reduction in drug crystallinity may be due to hydrogen bond formation between a lone carbonyl group on the TAS-301 molecule to the sylanol group on FLR during the melt adsorption process. The apparent solubility of TAS-301 from the melt-adsorbed formulations showed an approximately 20-fold improvement over the crystalline drug. The dissolution rate of TAS-301 was also shown to increase substantially when melt adsorbed onto FLR with 100% of the drug released in 15 min compared to less than 40% released with the crystalline drug and a physical mixture.

The in vivo absorption of the drug from the 1:2 TAS-301:FLR formulation produced by melt extrusion was evaluated using beagle dogs in both the fasted and fed states. The melt adsorbed formulation was dosed in capsules and compared to the crystalline drug in both a capsule and an aqueous suspension. This study revealed that plasma concentrations of TAS-301 with the melt adsorbed formulation were higher than both the crystalline TAS-301 suspension and capsule forms under all feeding conditions. The AUC_{0-12hr} was found to be 1.8 and 4.6 times greater with the melt-absorbed formulation than with the crystalline capsule formulation with the standard and rice-fed diets, respectively. Additionally the C_{max} of the melt-adsorbed product was 2.3 and 5.1 times greater with the amorphous formulation than with the crystalline formulation with the standard and rice fed diets, respectively. This study thus demonstrated that the increase in apparent solubility and dissolution rate by formulation of TAS-301 as an amorphous solid dispersion on the porous calcium silicate carrier correlated to enhanced drug absorption in vivo. Therefore, by overcoming the solubility limitations of the drug via formulation as an amorphous solid dispersion, greater systemic absorption and hence increased efficacy of restenosis treatment with TAS-301 via oral dosing can be achieved.

Verreck et al. conducted a formulation development study for a novel microsomal triglyceride transfer protein inhibitor known as R103757 in which a cursory evaluation of prospective delivery systems revealed the potential for enhanced drug efficacy by formulation as a solid dispersion (22). An initial evaluation of the physiochemical properties of the compound indicated possible poor oral bioavailability as indicated by poor aqueous solubility (0.5 μg/mL at neutral pH) and a relatively high estimated effective human dose (1 mg/kg). A proof-of-principle study was therefore conducted to verify that R103757 exhibited solubility limited absorption by evaluating bioavailability in dogs following oral administration of a fast disintegrating tablet containing crystalline drug, a solid dispersion formulation in which R103757 and hydroxypropyl methylcellulose (HPMC) (40:60 w/w) were coated onto inert sugar beads and filled into capsules, and an oral solution containing 10% hydroxypropyl-β-cyclodextrin (HP-β-CD) which served as a reference formulation. Therefore, this study evaluated absorption of the drug in a crystalline state and an amorphous state in relation to an oral solution reference. This in vivo study confirmed the assumption of dissolution rate limited absorption for R103757 as the tablets containing crystalline drug were unable to produce quantifiable plasma concentrations of the drug, while the un-optimized solid dispersion showed quantifiable plasma levels.

Based on the result of the proof-of-principle study, the researchers concluded that a solid dispersion would likely be the most efficacious formulation for oral delivery of the drug, and hence focused on optimizing the formulation by assessing three different solid dispersion platforms: (*i*) film-coated sugar beads, (*ii*) a glass thermoplastic system (GTS) and (*iii*) melt extrusion. The film coated beads were produced by spraying an organic solution of the drug and HPMC (40:60 w/w) onto sugar spheres in a fluidized bed apparatus. The GTS contained R103757 (11.1%), citric acid monohydrate (55.6%), HP-β-CD (27.8%), and HPMC (5.6%) and was formed by dissolution in heated ethanol followed by evaporation of the solvent in a vacuum oven. The gel residue that remained following evaporation of the solvent was then manually formed into cylinders and placed into hard gelatin capsules. Melt extrusion was conducted with a twin screw extruder to produce a solid dispersion formulation consisting of 25% drug and 75% HPMC. The extrudates were then milled, mixed with tableting excipients, and compressed into tablets. DSC and PXRD confirmed the amorphous nature of the drug in each of these formulations while in vitro dissolution testing in 0.1 N HCl demonstrated faster drug release from the melt extruded formulation over the two capsule formulations which performed similarly. The bioavailability of R103757 from these three formulations was then evaluated in a clinical trial with healthy male subjects in fed and fasted states. In the fasted state, the three solid dispersion formulations were compared to an oral solution containing 25% HP-β-CD. The results of these studies are shown in Figure 2. All of the solid dispersion platforms tested in this study exhibited relatively high bioavailability when considering that plasma concentrations for the crystalline material dosed to dogs were not quantifiable. In the fasted state, the relative bioavailability with respect to the oral solution was 27% for the film coated beads, 75% for the melt extruded tablet, and 97% for the GTS capsule. The GTS system provided greater AUC values than both the melt extruded tablets and the film coated beads in both the fed and fasted states.

In general, this study demonstrated the potential of solid dispersion systems for improving the oral bioavailability of poorly water-soluble drugs with solid state dosage forms, and how solid dispersion systems can be optimized with respect to production and formulation to provide maximum drug absorption enhancement. Specifically, this study showed that an advanced multi-component carrier system for amorphous drug provided equivalent bioavailability to an oral solution in a fasted state. This illustrates how solid dispersion systems can be tailored with respect to formulation and process technology to improve the in vivo absorption and overall therapeutic efficacy of a poorly water-soluble drug.

Dannenfelser et al. also reported the results of a formulation study involving an experimental microsomal triglyceride transfer protein inhibitor known as LAB687 (23). The compound was known to have low solubility (0.17 µg/mL) and high lipophilicity (Caco2 $P_{app} = 6.1 \times 10^{-4}$ cm/min) and was previously demonstrated to be poorly bioavailable in dogs from a dry blended formulation containing crystalline drug. Therefore, the objective of this study was to assess a solid dispersion formulation with a carrier system consisting of a water soluble polymer [polyethylene glycol 3350 (PEG)]

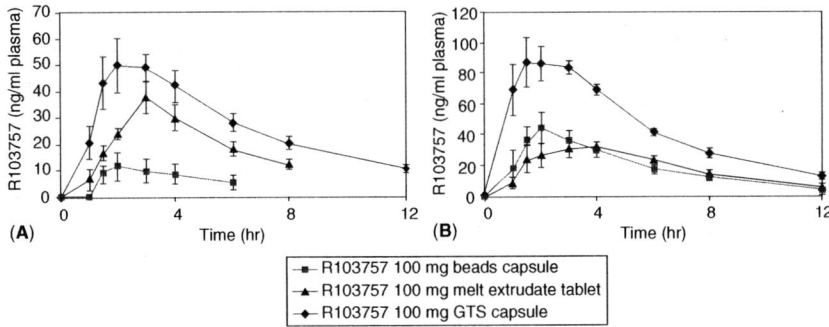

Figure 2 Results of the pharmacokinetic study with 18 healthy male subjects. Comparison of the R103757 100-mg melt extrudate tablet, the 100-mg bead capsule, and the 100-mg GTS capsule under (**A**) fasting and (**B**) fed conditions.

and a surface active agent (polysorbate 80) to improve the bioavailability of LAB687. The performance of the solid dispersion system was compared to a cosolvent-surfactant solution of the drug as well as a dry blend of the micronized drug with microcrystalline cellulose (MCC). For the production of the solid dispersion formulation, LAB687 (4% w/w) was first dissolved into the molten carrier consisting of a 3:1 mixture of PEG and polysorbate 80 at $65 \pm 5°C$. A 500 mL aliquot of this molten solution was then filled into hard gelatin capsules to produce the final dosage form. The dry blend formulation consisted of 50% micronized drug (mean particle size 4.9 μm) along with 49.8% MCC and 0.2% fumed silica. The cosolvent-surfactant solution contained 20 mg of drug per one milliliter of the cosolvent system which was made up of 10% propylene glycol, 45% Cremophor RH40, 35% corn oil glycerides, and 10% ethanol (w/w/w/v). An in vivo study was conducted in beagle dogs in which 50 mg of LAB687 was administered in one size 000 capsule for both the solid dispersion formulation and the dry blend formulations, and in two 00 capsules for the 20 mg/mL cosolvent surfactant solution. The results of the in vivo study are given in Table 1.

In this study, high inter-animal variability was observed with the dry blend formulation whereas the variability of the solid dispersion and cosolvent-solution was relatively low. Since the absolute bioavailability for LAB687 could not be determined, the relative bioavailabilities of each formulation were calculated based on drug absorption from the cosolvent-surfactant solution. The solid dispersion formulation showed 99.1% bioavailability with respect to the cosolvent-surfactant solution which was ten-fold greater than the dry blend formulation at 9.8%. Therefore, it was demonstrated by this study that delivering this experimental lipid lowering agent in the form of a solid dispersion substantially enhanced drug absorption in dogs, and hence suggests the potential for improved efficacy in human patients.

PAIN MANAGEMENT

Ibuprofen is one of the most commonly used non-steroidal anti-inflammatory drugs (NSAID) for the management of mild to moderate pain. It is very

Table 1 Mean Pharmacokinetic Parameters and Relative Bioavailabilities for the Three Formulation Approaches

Treatment: dose, formulation	T_{max} (h)	C_{max} (ng/mL)	AUC_{0-48h} (ng·h/mL)	F_{rel} (%)
50 mg, cosolvent-surfactant solution	1–2	1160	6960	100.0
50 mg, solid dispersion	1–2	803	6900	99.1
50 mg, dry blend (micronized drug)	1	128	681	9.8

slightly soluble in water (50 µg/mL) but highly permeable through physiological membranes (24). The bioavailability of ibuprofen from conventional dosage forms is typically high with approximately 80% of an oral dose absorbed from the GI tract (25). Although it is well absorbed, the onset of action of ibuprofen can be somewhat delayed due to its poor aqueous solubility. Since rapid onset of action is preferred with anti-inflammatory agents, overcoming the solubility limitations of the drug could improve the efficacy of ibuprofen and the management of acute pain. In a recent clinical study conducted by Klueglich et al. a solid dispersion technology known as Meltrex® was evaluated for improving pain therapy with ibuprofen by accelerating the onset of therapeutic action (26). Hot-melt extrusion is the basis of the Meltrex technology, and the process is used primarily to produce drug delivery systems that contain the active ingredient in a more readily soluble form (27). For this study, the Meltrex system was utilized to produce a molecular dispersion of ibuprofen in a hydrophilic polymer matrix such that the dissolution rate of the drug becomes dependant on the dissolution rate of the polymer matrix rather than the drug itself (28). Hence, by eliminating the solubility constraints that impede the dissolution rate of ibuprofen, it was hypothesized that the drug would be more readily absorbed following oral administration and the time to onset of therapeutic action would be reduced.

In a clinical study with healthy male subjects, the Meltrex ibuprofen formulation was compared to a traditional ibuprofen tablet as well as a tablet containing a chemical derivative of ibuprofen, ibuprofen lysinate. This derivative of ibuprofen has shown improved solubility, rate of absorption, and faster relief of pain over conventional ibuprofen (29). The ibuprofen plasma concentration versus time profile from this study can be seen in Figure 3. These results demonstrated the bioequivalence of the Meltrex formulation and the ibuprofen lysinate formulation based on $AUC_{0-\infty}$, C_{max}, and t_{max}. In the fasted state, the pharmacokinetic profile of the extrudate formulation was found to differ from the traditional ibuprofen tablet in that the C_{max} was 20% greater and the t_{max} occurred almost one hour earlier, yet the total AUC was approximately equal. Therefore, total ibuprofen exposure between the extrudate and traditional ibuprofen formulations was equivalent; however, absorption was much more rapid with the extrudate formulation. The authors therefore concluded that since the pharmacological effect of ibuprofen has been shown to be correlated with serum concentrations these in vivo results suggest that a more rapid onset of action was achieved with the extrudate formulation. This study demonstrated that formulation of ibuprofen as a molecular dispersion by the Meltrex technology successfully overcomes the solubility limitations of ibuprofen to provide more rapid absorption following oral administration than conventional ibuprofen tablets, and hence presumably allows for faster onset of therapeutic action.

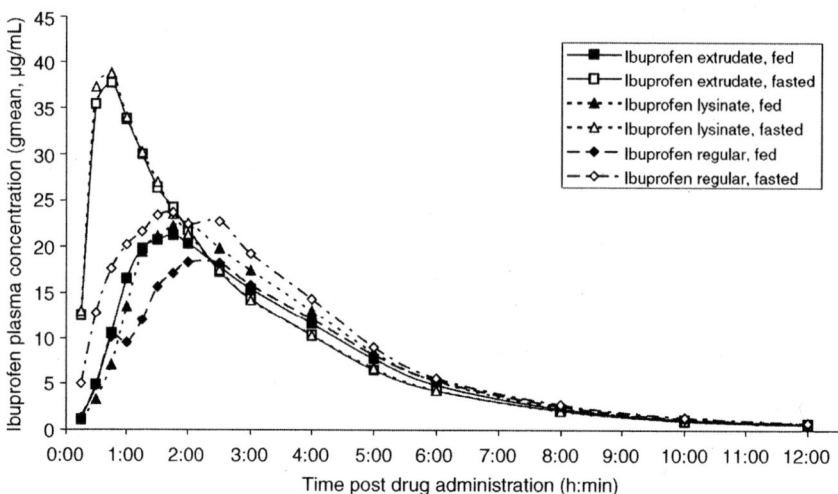

Figure 3 Geometric mean ibuprofen plasma concentrations following single oral administration of a 400-mg ibuprofen extrudate (*square*), lysinate (*triangle*), and regular (*diamond*) tablet under fed (*filled*) and fasted (*open*) conditions, respectively.

ABT-963 is a selective COX-2 inhibitor that is used for the treatment of pain and inflammation (30). The compound is very poorly water-soluble with a saturation solubility of 16 μg/mL in water at 25°C. A preliminary pharmacokinetic study in dogs conducted by Chen et al. revealed that the bioavailability ABT-963 was 24% from capsules containing the bulk drug, and absorption was determined to be dissolution rate limited (31). This preliminary pharmacokinetic evaluation prompted these researchers to evaluate a solid dispersion system as a potential solution to overcoming the solubility limitations of the compound that restricted its oral bioavailability. Hence, a melt method was adopted in which ATB-963 was dispersed in molten Pluronic F-68. Pluronic F-68 is a water soluble, non-ionic, surface active copolymer which has been used extensively in pharmaceutical preparations as a solubilizing agent and in solid dispersion systems to improve the solubility of active agents (32–34). In solubility studies performed by Chen et al. it was found that Pluronic F-68 had a substantial solubilizing effect on ABT-963, and therefore was selected as the carrier for the solid dispersion. The use of DSC and PXRD confirmed that the drug was in a crystalline state following production of the solid dispersions for all drug loadings. The presence of an active ingredient in a crystalline state is less common in solid dispersion formulations than the amorphous counterpart because for many drugs the thermodynamic stability of the crystal lattice is the primary obstacle to improving the drug's dissolution properties. However, with respect to stability, a crystalline dispersion would be preferred as amorphous solid dispersions are often thermodynamically metastable, and thus drug recrystallization and a corresponding decrease in dissolution rate can occur on storage. This example of ABT-963 as a crystalline solid dispersion in Pluronic F-68 is therefore unique as it was shown by dissolution testing to provide substantial improvement in the dissolution rate of the drug without a morphological change.

The Pluronic F-68 dispersion and a conventional immediate release (IR) wet granulation formulation were dosed to dogs in capsules with a solution of the drug in PEG 400 as a reference. The plasma concentration versus time curve from this study is shown below in Figure 4, and the PK parameters are shown in Table 2. The results of this in vivo study demonstrated that the ATB-963 solid dispersion formulation provided comparable drug plasma concentrations to the PEG 400 solution with a relative bioavailability of 94.1%. The traditional immediate release granulation formulation was found to have a much lower relative bioavailability of 46.5%. Additionally, the solid dispersion formulation was found to produce less inter-subject variability than the IR formulation. The increased plasma concentration implies improved therapeutic action with the solid dispersion formulation over the IR formulation. Hence, this study exemplifies the use of a crystalline solid dispersion system to improve

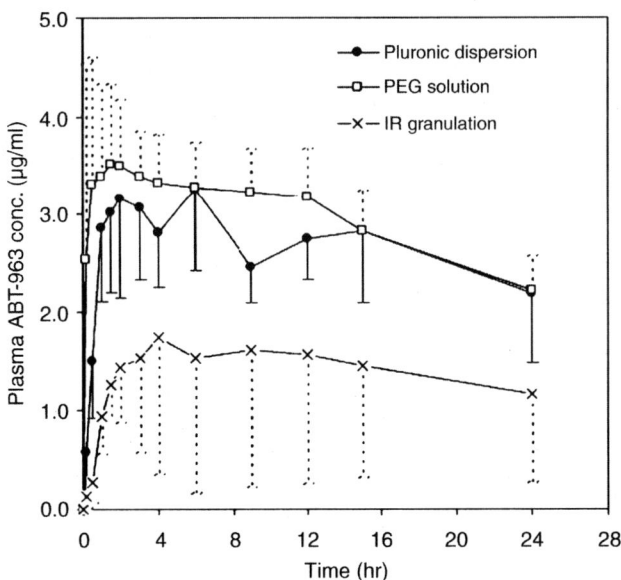

Figure 4   Plasma concentration of ABT-963 following oral administration of PEG solution and capsule formulations to fasted dogs.

Table 2 Pharmacokinetic Parameters of ABT-963 after Oral Dosing of 50-mg ABT-963 in Capsules and PEG Solution in Fasted Dogs

Formulation	Pharmacokinetic parameters (mean ± S.D., $n = 5$)				
	C_{max} (µg/ml)	T_{max} (h)	AUC (µg·h/ml)	AUC/D (µg·h/ml/mg/kg)	BA[a] (% relative)
IR capsule	1.86 ± 1.28	2.1 ± 1.1	34.2 ± 25.6	8.1 ± 5.3	46.5 ± 27.2
Solid dispersion[b]	3.62 ± 1.00	5.2 ± 5.8	64.6 ± 19.2	16.0 ± 4.1	94.1 ± 20.0[c]
PEG solution	4.00 ± 0.88	3.2 ± 5.0	70.1 ± 11.5	17.0 ± 2.4	

[a] Bioavailability relative to PEG solution.
[b] C_{max} and AUC of the solid dispersion formulation were normalized from 40 mg dose to 50 mg dose.
[c] Significantly different between the IT and solid dispersion formulations ($P < 0.05$).

the dissolution properties and oral absorption of a poorly water-soluble NSAID, ultimately resulting in improved management of acute pain.

Indomethacin (IND) is another example of an NSAID which has shown improved efficacy by formulation as a solid dispersion. Chowdary and Babu utilized a solvent evaporation technique to produce amorphous dispersions of indomethacin with hydroxypropyl cellulose-SL (HPC-SL), HPMC, and hydroxyethyl cellulose (HEC) (35). All formulations showed substantial improvement in dissolution rate, with the greatest improvement being with the HPC-SL formulation. In vivo studies were conducted in healthy fasted human volunteers with the IND-HPC-SL and IND-HPMC solid dispersion formulations using bulk IND as a reference. The results of this study demonstrated that faster absorption and higher serum levels were achieved with the solid dispersion formulations. The mean absorption rate constants (k_a) and AUC_{0-3hr} values for the IND, IND-HPMC, and IND-HPC-SL doses were 0.5159, 0.6879, 0.9016 h^{-1} and 1.72, 2.02, 2.44 $\mu g \cdot h/mL$; respectively. However, the $AUC_{0-\infty}$ values were similar for all formulations indicating that total IND exposure was similar for each formulation. Therefore, as with the ibuprofen solid dispersion system, formulation of IND as an amorphous dispersion in cellulosic polymers did not enhance overall bioavailability, but rather accelerated the onset of therapeutic action. A good correlation was found between the in vitro dissolution results and in vivo absorption as the IND-HPC-SL formulation exhibited the most rapid dissolution rate and had the greatest k_a value. This result further confirmed that the onset of therapeutic action of IND is dissolution rate limited. Therefore, this study indicated that the efficacy of orally administered IND in treating acute pain or inflammation is improved by formulation as a solid dispersion owing to an accelerated rate of absorption.

Piroxicam is another drug from the family of NSAIDs that is classified as a BCS class II drug. Oral piroxicam is typically indicated for treatment of osteoarthritis and rheumatoid arthritis and not for analgesia owing to slow and gradual absorption with a long elimination half-life (36). This pharmacokinetic profile provides prolonged therapeutic action; however, the onset of anti-inflammatory or analgesic action is delayed. Therefore, as with the previously discussed NSAIDs, treatment of acute pain and inflammation with piroxicam could be improved by formulating the drug in a manner that accelerates drug dissolution and promotes rapid absorption.

Yüksel et al. evaluated a solid dispersion of piroxicam in a carrier system composed of Gelucire 44/14 and Labrasol (37). The dispersion was produced by a melt method in which piroxicam was stirred into the molten carrier system and subsequently the molten dispersion was filled into hard gelatin capsules. The solid dispersion system of piroxicam was comparatively evaluated against bulk piroxicam and a commercially available tablet containing piroxicam-β-cyclodextrin complexes for in vitro drug release and in vivo absorption. Dissolution tests were conducted in various media and while the bulk piroxicam showed pH-dependant and incomplete dissolution, the solid dispersion formulation provided rapid dissolution of piroxicam (85% in 30 min) irrespective of pH. The commercial tablet also showed substantial improvement in the dissolution properties of piroxicam; however, some pH-dependence was found with a reduction in release rate at pH 4.5. The in vivo performance of these three piroxicam formulations was evaluated in healthy human volunteers. An equivalent of 20 mg of piroxicam was dosed to each subject with the bulk drug and solid dispersion dosed in hard gelatin capsules. This study revealed that the solid dispersion formulation provided the greatest maximum concentration (2.64 μg/mL) in the shortest time-to-peak (82.5 min), followed by the commercial tablet (2.44 μg/mL, 120 min) and the bulk drug (0.999 μg/mL, 144 min). These results indicated that the solid dispersion and commercial tablet formulations performed substantially better than the bulk drug with respect to rate and extent of absorption, with the solid dispersion formulation providing greater acceleration of absorption than the commercial tablet. With respect to overall absorption, the solid dispersion and commercial tablet formulations showed statistically equivalent total AUC values which were over two-fold greater than that of the bulk drug. Thus, in this case, the advanced formulation designs were able to improve the initial rate and extent of absorption, as well as total exposure over that of the bulk drug.

The studies discussed above illustrate how the use of solid dispersion formulation technologies can improve drug therapy for the management of acute pain and inflammation with BCS class II NSAIDs. In the studies reviewed, it was repeatedly demonstrated that poorly soluble drugs can be rendered more soluble via solid dispersion systems by the individual or synergistic effects of particle size reduction, morphology alteration, and improved particle wetting properties by intimate association with hydrophilic carrier systems. It is by these modifications that dissolution rates of the above mentioned NSAIDs can be dramatically improved. Since the absorption of these drugs form the GI tract is primarily limited by the rate at which they enter solution in the gastrointestinal lumen, solid dispersion systems directly promote absorption of these drugs by accelerating their dissolution rates. For acute pain management, accelerated absorption relates directly to the onset of therapeutic action, and hence solid dispersion formulations appear to provide more therapeutically effective oral doses of BCS class II NSAIDs than traditional formulations.

CANCER THERAPY

Solid dispersion technologies have also been explored for improving drug therapies for the treatment of various conditions associated with cancer.

Many anticancer drugs have poor aqueous solubility characteristics that preclude the efficacy of traditional oral dosage forms. Therefore, these drugs are commonly administered to the patient as parenteral injections. Since intrusive methods of drug administration are inconvenient and undesirable for the patient, there would be substantial benefit to the development of effective oral dosage forms with respect to ease of administration and patient compliance. Many researchers are therefore pursuing advanced formulation design strategies to develop solid oral dosage forms for cancer therapy drugs that provide equivalent bioavailability and therapeutic efficacy to the injectable dosage forms.

One such study was conducted by Etienne et al. in which the bioavailability of a new solid dispersion oral preparation of medroxyprogesterone acetate (MPA) was evaluated in breast and endometrial cancer patients (38). MPA has been used since the early 1960s for treatment in hormone-responsive tumors and is a primary hormonal treatment for advanced breast carcinoma. Oral administration of MPA has been demonstrated to be as effective as intramuscular injection (39); however, oral dosage forms are preferred due to convenience and the avoidance of harmful effects of frequent injections such as abscess formation around the site of injection. MPA is highly lipophilic drug and poorly water-soluble ($1\text{-}2\,\mu g/mL$) which results in a low extent of absorption ($< 10\%$). Hence, relatively high daily oral doses ($1000\,mg$) are required to elicit the therapeutic effect of the drug in patients. In addition to improving the bioavailability beyond that of the current oral dosage form, the authors of this study cited the cost of the active agent as a motivating factor for developing a more highly absorbed dose since greater than 90% of the administered drug passes out of the system as waste.

The solid dispersion formulation evaluated in this study was originally described by Carili (40). Specifically, the delivery system is an amorphous dispersion of MPA in crosslinked polyvinylpyrrolidone. In the in vivo study conducted by Etienne et al. sachets containing the solid dispersion formulation in an amount equivalent to $200\,mg$ of MPA were used for dosing (38). These sachets were evaluated with respect to a commercial MPA tablet ($500\,mg$) known as Farlutal. A single dose of both the sachet and the tablet formulation were administered twice daily in twelve hour intervals. Therefore, a 400-mg daily dose of the solid dispersion formulation was evaluated against a 1000-mg daily dose of Farlutal. It was found from this study that higher plasma levels were achieved with the $400\,mg$ sachets dose of the solid dispersion than with the 1000-mg dose of Farlutal. Specifically, the C_{max} values for the solid dispersion formulation were significantly higher than the Farlutal tablet. The solid dispersion formulation produced peak plasma concentrations of greater than $100\,ng/mL$ [minimum effective concentration necessary to elicit tumor response (41)] in 75% of patients. Additionally, the relative bioavailability of the solid dispersion formulation as evaluated by the AUC_{ss} measured between the two administration times was found to be approximately 3.5 times greater than that of the Farlutal formulation. The authors attributed the superior performance of the MPA solid dispersion formulation over the commercial tablet to the stable amorphous form of the drug and the hydrophilicity of the PVP carrier whose synergistic effects provided rapid drug dissolution. Therefore, formulation of the anti-cancer compound MPA as a solid dispersion was able to improve the efficacy of the twice daily dosing regime while concurrently reducing the administered dose. Hence, chemotherapy treatment with MPA may be improved with respect to efficacy as well as cost of treatment by formulation design as a solid dispersion.

An experimental anti-cancer agent known as HO-221 was the subject of a formulation development study and bioavailability evaluation conducted by Kondo et al. in which solid dispersion systems with different drug release characteristics were investigated (42). From initial studies with this compound these researchers identified that the oral absorption of HO-221 was dissolution rate limited owing to its aqueous insolubility (0.055 µg/mL) (43). Initial formulation work aimed at improving the bioavailability of HO-221 involved wet bead milling of the drug, but these preparations yielded only 5–20% oral bioavailability. This result prompted these researchers to investigate the use of advanced solid dispersion systems with various polymer carriers as a means of further improving the bioavailability of HO-221. Specifically, amorphous dispersions of the compound were produced by co-precipitation from a common solvent with PVP, copolyvidone, and an enteric polymer hydroxypropyl methylcellulose phthalate (HP-55). These coprecipitates were evaluated with regard to the physical state of the drug in the dispersion, in vitro drug release, and bioavailability in beagle dogs. XRD studies revealed that the drug was present in an amorphous state in each solid dispersion formulation. Dissolution studies in pH 6.5 buffer showed substantial supersaturation with each solid dispersion formulation with the order of maximum peak concentrations being HP-55 > copolyvidone > PVP. Enteric dissolution testing conducted for the HP-55 formulation revealed that release of HO-221 followed a typical enteric profile as no drug release was detected in pH 1.2 medium, and rapid release occurred following the pH shift to 6.5. The release profile of HO-221 from the HP-55 formulation following the pH shift was similar in rate and extent to the profile from the non-enteric dissolution testing in pH 6.5.

For the oral absorption studies, each of the coprecipitate formulations and the micronized drug were dosed to beagle dogs in capsules. The bioavailability of the micronized drug was found to be in the range of 9.2–16.6%. Despite significant differences in dissolution profiles, coprecipitates with PVP and copolyvidone produced very similar plasma profiles with each formulation resulting bioavailability values 3.5 times greater than the micronized drug. The absorption of HO-221 from the HP-55 formulation was found to be approximately two-fold that of the PVP and copolyvidone coprecipitates. The authors attributed this difference in drug absorption to the supersaturation of HO-221 in the small intestine versus the stomach. It was speculated that the immediate release and supersaturation of HO-221 in the stomach with the PVP and copolyvidone formulations resulted in reduced drug absorption as a result of the minimal surface area for absorption in the stomach and precipitation of supersaturated drug prior to passage into the small intestine. Whereas with the enteric HP-55 formulation, supersaturation of HO-221 was targeted to the small intestine where there is substantially greater surface area for drug absorption to occur. This hypothesis was confirmed by evaluating the bioavailability of the copolyvidone formulation following intraduodenal administration by which the drug plasma concentrations were found to slightly exceed that of the HP-55 coprecipitates. Hence, it was determined that localizing the release of supersaturated concentrations of HO-221 in the small intestine was the optimum delivery strategy for achieving extensive oral absorption of HO-221. This study thus demonstrated an advanced solid dispersion system which provided rapid drug release at supersaturated concentrations which could be targeted to the small intestine in order yield optimum oral bioavailability.

Another insoluble anti-cancer drug, lonidamine, has shown significant improvement in oral absorption by formulation as a solid dispersion. In a study by Palmieri et al. lonidamine solid dispersions in PEG 4000 and PVP K 29/32 were produced by spray drying of an organic solution of the drug and the polymer (44). XRD studies revealed the miscibility limits of lonidamine in PVP and PEG to be < 20% and < 10%, respectively. In vitro dissolution studies demonstrated that with both carriers the dissolution rate increased with decreasing drug content. Additionally, it was found that extensive levels of supersaturation were achieved with the PVP formulations yet not with the PEG formulations. The in vivo release kinetics of spray dried dispersions of lonidamine in PEG and PVP in a 1:9 drug polymer ratio were evaluated along with the bulk drug and lonidamine-cyclodextrin complexes. The pharmacokinetic data from this study can be seen in Table 3.

Table 3 Pharmacokinetics Data of the Solid Dispersions and Cyclodextrin Complexes with Lonidamine

	C_{max} (μg/mL)	T_{max} (min)	AUC (μg min/mL)
Lonidamine	57	90	364
Lon/β-cd 1:4	71.77	60	533
Lon/HP β-cd 1:4	63.7	60	674
Lon/PVP 1:9	60.9	60	632
Lon/PEG 4000 1:9	81.43	30	521

The results of this study indicated that the solid dispersion formulations were able to significantly improve the oral absorption of lonidamine over that of the bulk powder. The PVP formulation exhibited the most marked improvement in oral bioavailability of the solid dispersion formulations with a near two-fold increase in total AUC over the bulk drug. These results, as with the study involving HO-221, suggested that supersaturation of GI fluids is the primary cause for the absorption improvement observed with amorphous dispersions of highly insoluble drugs. The formulation of lonidamine as an amorphous dispersion in PVP was shown to produce supersaturated drug concentrations in vitro that translate in vivo to the supersaturation of GI fluids, thus providing a greater concentration gradient by which to drive drug absorption. In the case of traditional dosage forms in which the drug is present in its crystalline state, this extensive supersaturation of GI fluids is not possible. These advanced amorphous solid dispersion formulations are therefore essential to achieving adequate oral bioavailability for very poorly water-soluble BCS class II drugs.

HIV/AIDS TREATEMENT

Recently, Abbott Labratories introduced a new tablet formulation for Kaletra®, an anti-HIV protease inhibitor, as an alternative to the original soft gelatin capsule (SGC) formulation (45). This new tablet formulation is based on the previously discussed Meltrex technology which utilizes hot-melt extrusion to produce solid dispersion drug formulations. The Kaletra brand contains two insoluble active ingredients, lopinavir and ritonavir. Lopinavir is an inhibitor of the HIV protease that undergoes almost complete metabolism by CYP3A. Ritonavir inhibits CYP3A metabolism of lopinavir, and thus co-administration of these drugs has been demonstrated to increase plasma levels of lopinavir (46). Although the Kaletra SGC formulation proved to be an effective oral antiretroviral drug therapy, the regime required frequent dosing and administration with food. Additionally, the SGC formulation had to be stored in a refrigerated environment. The inconvenience of the SGC formulation, the potential for improved bioavailability/reduced food effect, and presumably reduced manufacturing costs motivated the development of the Meltrex based tablet formulation.

Utilizing the Meltrex system, 33% greater drug loading was achieved in tablets (200mg lopinavir/50mg ritonavir) than in the SGC formulation (133mg lopinavir/33mg ritonavir). The drug contained in the tablet formulation was found to be as readily soluble as drug contained in the SGC formulation, and hence on a dose-normalized basis the tablets were bioequivalent to the SGCs. Therefore, one-third fewer tablets were required to deliver the same therapeutic dose. Also, clinical studies determined that there are no significant differences in C_{max} and AUC with the Kaletra tablets between fed and fasted states, thus indicating negligible food effect on the absorption of the actives from the tablet formulations. Therefore, the Meltrex-based tablets provided the added convenience of dosing with or without food. Additionally, the Kaletra tablets were determined to be stable when stored at room conditions, thus offering another advantage over the SGC which must remain refrigerated.

This example of improving a commercial HIV drug therapy by the use of a melt extruded solid dispersion system represents the growing interest of the pharmaceutical industry in solid dispersion technologies as these systems become increasingly more viable. In this example, it was demonstrated that solid dispersion technologies can provide more effective and convenient oral drug therapies which are in many cases simpler to manufacture and more stable than the current alternatives.

Law et al. have also explored solid dispersion systems for improved oral bioavailability of ritonavir with the aim of improving the efficacy of the oral treatment of HIV (47). Amorphous dispersions of ritonavir in PEG at different drug loadings were prepared by a solvent evaporation-fusion method (48) and evaluated for in vitro and in vivo performance. In vitro dissolution studies were conducted in 0.1 N HCl with amorphous ritonavir solid dispersions in PEG at 10%, 20%, and 30% (w/w) drug loading as well as a physical mixture of 10% (w/w) crystalline ritonavir with PEG. The results of this dissolution study are shown below in Figure 5.

This figure illustrates the enhanced dissolution performance of the solid dispersion formulations over the physical mixture and the effect of drug loading on the rate and extent of ritonavir dissolution from the solid

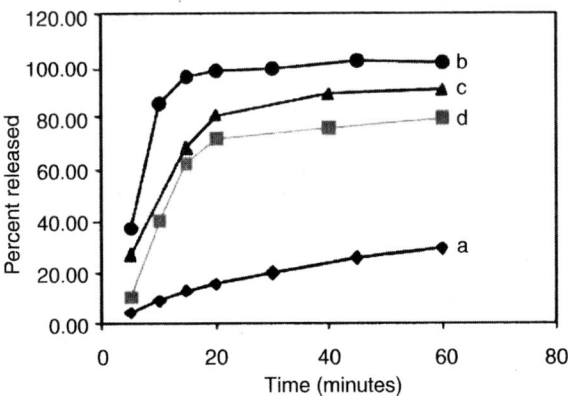

Figure 5 In vitro dissolution in 0.1N HCl of (*a*) physical mixture containing crystalline ritonavir–PEG at 10:90, and amorphous ritonavir in PEG solid dispersions at concentrations of (*b*) 10%, (*c*) 20%, and (*d*) 30% (w/w). Dissolution was determined by the USP I method (50 rpm, 378C).The data for 20% dispersion is an average of two runs; others are three runs.

dispersions. The in vivo performance of these compositions was then evaluated in beagle dogs by oral administration of the solid dispersion formulation in dry capsules. The ritonavir plasma concentration versus time curve from this study is shown in Figure 6.

From Figure 6, it is seen the extent of in vivo absorption of ritonavir closely follows the in vitro dissolution performance as ritonavir from the solid dispersion formulations is absorbed to a much greater extent than with crystalline ritonavir, and the extent of absorption from the solid dispersions is inversely proportional to drug loading. The 10%, 20%, and 30% (w/w) dispersions showed 22.04-, 17.98-, and 10.98-fold increases in average AUC over the crystalline drug, respectively. From the results of this study, it was concluded that absorption of ritonavir is concurrently improved with improved dissolution rate. Therefore, the enhanced in vivo performance of the solid dispersion formulations was directly attributed to

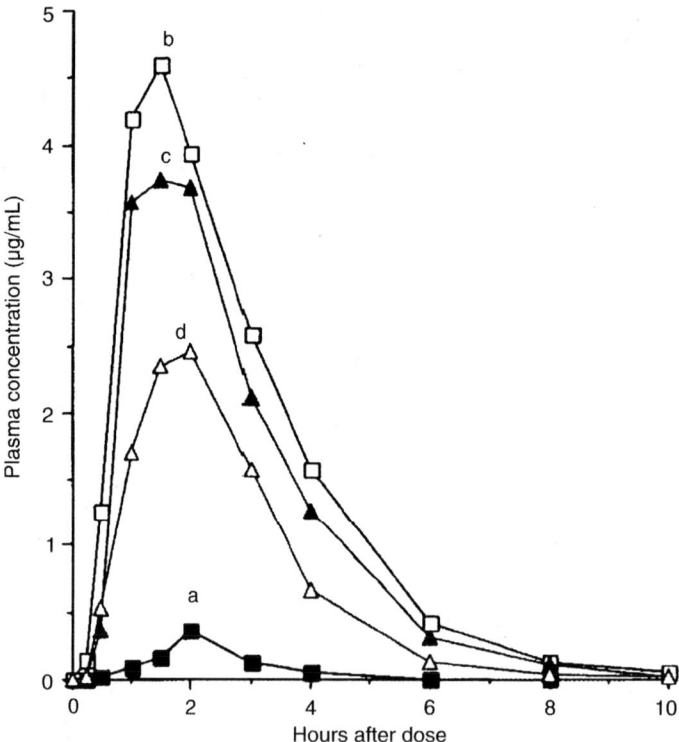

Figure 6 Mean plasma concentration–time profiles of ritonavir after a single oral dose to beagle dogs. Ritonavir was administered in (*a*) the crystalline form or the amorphous in PEG solid dispersions at concentrations of (*b*) 10%, (*c*) 20%, and (*d*) 30% (w/w).

higher drug concentrations in digestive fluids when dosed as an amorphous dispersion. Since ritonavir is a substrate and inhibitor of P-glycoprotein (Pgp) intestinal absorption could be limited by Pgp efflux (49). Therefore, improved absorption by supersaturation of digestive fluids may be due to the synergistic effects of an increased concentration gradient to drive absorption and saturation of Pgp resulting in decreased efflux. Irrespective of the mechanism of absorption improvement, this study clearly demonstrated improved absorption, and hence enhanced efficacy of oral ritonavir that is attributable to its delivery as a highly soluble amorphous solid dispersion.

In another similar study, De Jaeghere et al. evaluated pH sensitive micro and nanoparticle solid dispersions for improved absorption of a new HIV-1 protease inhibitor known as GCP 70726 (50). As with the previously described protease inhibitors, GCP 70726 is a poorly water soluble (0.12 µg/mL at pH 6.2) highly lipophilic compound (log P = 4.77). Therefore, poor oral bioavailability of the compound observed in preliminary studies was attributed to limited dissolution of the drug in the lumen of the GI tract. Toward the aim of improving the dissolution properties and oral absorption of GCP 70726, these authors evaluated a solid dispersion formulation consisting of amorphous drug dispersed in a poly(methacrylic acid-*co*-ethylacrylate) copolymer known as Eudragit® L100-55. This enteric polymer exhibits pH-dependant solubility in which the polymer becomes soluble above pH 5.5 owing to the ionization of carboxylic acid functional groups contained on the polymer. The pH solubility profile of the carrier system thus prevents drug release in the acidic environment of the stomach and promotes dissolution in the more neutral pH environment of the small intestine. The rationale for a solid dispersion system with a pH-dependant release profile is to concentrate the release of the molecularly dispersed drug to the key absorption site, that is, the upper regions of the small intestine. The authors expected that this mode of release would increase the probability of drug absorption by maximizing the exposure of transient supersaturated drug concentrations with the extensive surface area of the small intestine.

The solid dispersions of CGP 70726 in Eudragit L100-55 were produced by two processes: (*i*) emulsification-diffusion method in which a drug-polymer nanoparticle suspension was produced, rapidly frozen, and freeze-dried to yield a dry powder; (*ii*) spray drying a methanolic solution of the drug and the polymer to produce dry microparticles. X-ray diffraction confirmed the amorphous nature of the drug in both solid dispersion particle formulations. Particle size analysis revealed that the mean diameter of the particles formed by the emulsification-diffusion method was near 300 nm while the mean diameter of the spray dried particles was near 10 µm. The in vivo performances of the micro and nanoparticle formulations of GCP 70726 were evaluated in a single dose study with beagle dogs in both the fed and fasted states. An aqueous suspension of the crystalline drug was used as a reference; however, the resulting plasma levels were below the limit of quantitation of the analytical method. The pharmacokinetic parameters in the fed and fasted states for the solid dispersion systems are shown in Table 4.

In contrast to the crystalline drug, both groups of solid dispersion particles produced substantial plasma levels of CGP 70726 in both nutritional states; however, it was found that the presence of food reduced drug absorption. The authors suggested that the presence of food increased the gastric pH which caused premature release and partial precipitation of drug in the stomach. Contrary to expectations, the mean AUC of the microparticles was higher than that of the nanoparticles, but this discrepancy was found statistically insignificant. The improved absorption of the solid dispersion particles over the crystalline drug was attributed to rapid dissolution owing to the high specific surface area of the particles and the release of drug in a molecularly dispersed state following dissolution of the polymer carrier. In addition, targeting this rapid drug release to the extensive mucosal membrane surface area in the small intestine provided optimal opportunity for drug absorption. Similarly to previously discussed examples, this solid dispersion formulation was able to improve oral drug absorption by not only improving the dissolution properties of the drug, but also by targeting the delivery of the more soluble drug form to the optimum site for absorption in the GI tract.

ANTIFUNGAL TREATMENTS

Itraconazole is a synthetic triazole antifungal agent which exhibits activity against histoplasmosis, blastomycosis, and onychomycosis (51,52). The compound is a poorly water-soluble weak base, and thus is extremely

Table 4 Pharmacokinetic Parameters of CGP 70726 Incorporated in Eudragit L100-55 pH-Sensitive Particles after Oral Administration to Dogs; Mean 6 SEM ($n = 4$)

Formulation	$C_{max} \pm$ SEM (μmol/L)	T_{max} (h)	AUC 0-8 h $\pm$ SEM (μmol·h/L)
Fasted State			
Nanoparticles	1.62 ± 0.04	2	5.83 ± 0.77[a]
Microparticles	1.59 ± 0.32	2	7.83 ± 1.55
Fed State			
Nanoparticles	0.86 ± 0.21	1	2.00 ± 0.50
Microparticles	0.88 ± 0.33	2	4.40 ± 1.38

[a] Statistically different (Student test, $P < 0.01$).

insoluble at neutral pH (1 ng/mL) with increasing solubility in acid (4 µg/mL at pH 1). Itraconazole is very lipophilic (calculated log $P = 6.2$), and therefore shows excellent biological membrane permeability (53). These physiochemical parameters place itraconazole amongst the group of poorly soluble, readily permeable BCS class II compounds (1).

The leading commercial solid oral dosage form of itraconazole is Sporanox® capsules produced by Janssen Pharmaceutica. This formulation is a solid dispersion system produced by film coating sugar spheres with an organic solution of itraconazole and HPMC. By this process, a readily soluble, amorphous dispersion of itraconazole in HPMC is formed as a film on the surfaces of small spherical particulate substrates. These film coated pellets are then filled into hard gelatin capsules to produce a single unit, multiparticulate dosage form with substantially improved bioavailability over the crystalline drug (51).

In a recent study by Six et al. a clinical study was conducted to evaluate different solid dispersion formulations produced by hot-melt extrusion against Sporanox capsules (52). Amorphous solid dispersions containing itraconazole where produced with the following polymer carrier systems: (*i*) HPMC, (*ii*) Eudragit E 100 and (*iii*) Eudragit E 100-polyvinylpyrollidone-covinylacetate (PVPVA 64) in a 70:30 (w/w) ratio. All solid dispersion formulations evaluated in this study contained 40% (w/w) itraconazole in an amorphous state. In vitro dissolution testing was conducted with the three extrudate formulations and Sporanox in simulated gastric fluid. The results demonstrated that the Eudragit E 100 and Eudragit E 100-PVPVA formulations showed instantaneous dissolution of itraconazole with 80% release within the first 10 min while the HPMC extrudate and Sporanox formulations showed much slower dissolution rates with 80% release in 2 h.

The in vivo study with these formulations was conducted with healthy human volunteers. The pharmacokinetic parameters from this study are shown in Table 5. The relative bioavailabilities of the extrudate formulations with respect to Sporonox capsules were 102.9%, 77.0%, and 68.1% for the HPMC, Eudragit E 100, and Eudragit E 100-PVPVA 64 formulations, respectively. Interestingly, these in vivo results are the converse of the in vitro dissolution data. The authors proposed that this could be attributed the lack of polymer stabilization of supersaturated itraconazole in GI fluids by the rapidly dissolving formulations versus the more slowly dissolving HPMC-based formulations where the polymer remained in close proximity to dissolving itraconazole and thereby more effectively maintaining supersaturation. Drug absorption was similar to Sporonox in all cases, but was most similar in the case of the HPMC extrudate formulation. Despite the lack of improved in vivo performance the authors argued that the extrusion formulations were more attractive solid dispersion formulations than Sporonox capsules by citing the economic and environmental benefits of the continuous and solvent-free hot-melt extrusion process.

Table 5 Pharmacokinetic Parameters (±SD) of Itraconazole and Hydroxyitraconazole after Oral Administration of Four Different Formulations in Healthy Humans

Parameter	Sporanox ($n = 8$)	HPMC ($n = 8$)	Eudragit E100 ($n = 8$)	Eudragit E100-PVPVA64 ($n = 8$)
Itraconazole				
AUC_{0-72} (ng·h/ml)	1365.5 ± 619.9	1405.6 ± 778.2	1054.0 ± 583.9	928.9 ± 355.7
C_{max} (ng/ml)	115.8 ± 51.2	118.8 ± 53.7	77.3 ± 39.9	76.6 ± 23.4
T_{max} (h)	4.1 ± 1.2	3.7 ± 1.3	4.0 ± 1.8	3.9 ± 1.1
F (%)		102.9	77.0	68.1
Hydroxyitraconazole				
AUC_{0-72} (ng·h/ml)	3552.0 ± 2241.8	4033 ± 2631.7	3109.6 ± 1970.3	2625.6 ± 1401.2
C_{max} (ng/ml)	217.6 ± 106.7	231.0 ± 104.6	172.6 ± 89.6	164.7 ± 51.5
T_{max} (h)	4.6 ± 1.2	4.4 ± 1.1	4.5 ± 1.3	4.6 ± 1.2

Therefore, this study demonstrated a safe and efficient means of producing amorphous solid dispersion formulations of itraconazole with similar bioavailability to Sporanox. Moreover, these authors also found that a carrier system with slower dissolution properties could stabilize super-saturated concentrations of itraconazole to extend elevated drug levels in the intestinal lumen beyond that of rapidly dissolving carrier systems and thus increasing drug absorption. The Sporanox and HPMC extrudate formulations evaluated in this study demonstrate the benefit of solid dispersion systems that deliver a more soluble form of a poorly soluble drug in a modified release carrier to not only improve the dissolution properties of the drug, but also to optimally deliver a soluble form of the drug in the GI tract.

Albendazole is a drug with a wide spectrum of antihelminthic activity used to treat a variety of worm infestations. The poor water-solubility of albendazole limits oral bioavailability of the drug and leads to substantial inter-subject absorption variability (54,55). Similar to itraconazole, albendazole is more soluble in acidic media than in media of neutral pH. Consequently, the intestinal absorption of albendazole has been observed to be substantially lower in animal models with low gastric acidity than with those of high gastric acidity (56).

In a study conducted by Kohri et al. solid dispersion systems were investigated as a means of overcoming the poor solubility of albendazole in neutral pH environments to improve intestinal absorption (57). Amorphous solid dispersions were produced by a solvent evaporation method with carrier systems containing HPMC, HP-55, or a combination of HPMC and HP-55. A pH shift dissolution test method was used as an in vitro simulation of the transition from the stomach to the small intestine that the drug would undergo in vivo. The results of this dissolution study are shown in Figure 7.

This dissolution testing method revealed that for each formulation, with the exception of the HP-55 only formulation, albendazole dissolved rapidly in the pH 1.2 medium. Precipitation of dissolved albendazole occurred rapidly following the pH shift; however, the solid dispersion formulations, particularly those containing HP-55, maintained high drug concentrations at pH 6.5. The combination HPMC and HP-55 formulation showed the greatest inhibition of recrystallization following the pH shift, and hence was selected for in vivo evaluation. In vivo studies were conducted with two groups of rabbits, one group having normal gastric acidity and the other having low gastric acidity. A physical mixture of crystalline albendazole with lactose was used a as a reference. The plasma concentration versus time data for these two groups are shown in Figure 8.

For both groups, the C_{max} and AUC values were higher with the solid dispersion formulation than the physical mixture. The bioavailability of the solid dispersion formulation in the normal gastric acidity was almost 100%,

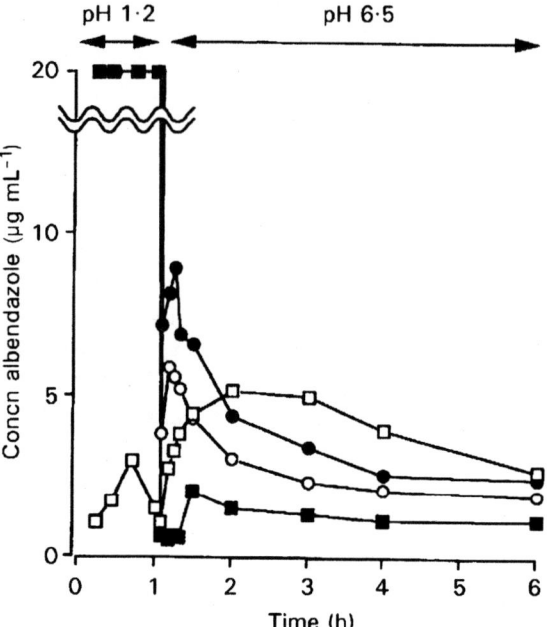

Figure 7 Dissolution behavior of albendazole from solid dispersions in media of pH 1.2–6.5: ■, physical mixture; ●, solid dispersion with hydroxypropylmethylcellulose and hydroxypropyl methylcellulose phthalate; O, solid dispersion with hydroxypropyl methylcellulose; □, solid dispersion with hydroxypropyl methylcellulose phthalate.

and in the reduced gastric acidity group the bioavailability of the solid dispersion formulation was 3.2 times that of the physical mixture. With the normal gastric acidity group, albendazole dissolved rapidly from both formulations; however, the presence of the polymers in the solid dispersion

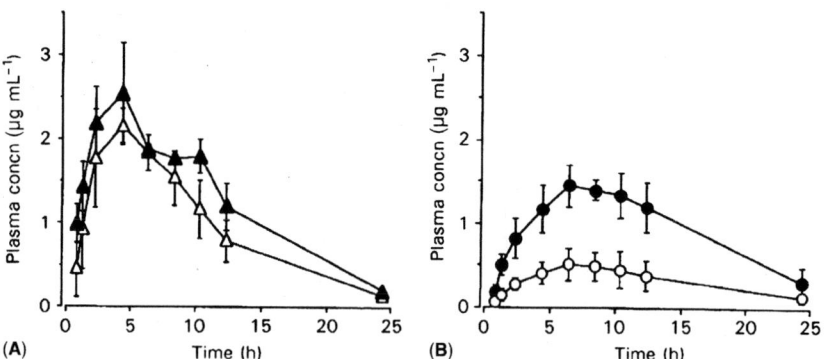

Figure 8 Mean plasma concentrations of albendazole sulphoxide after oral administration of physical mixture (△,O) and solid dispersion (▲,●) to normal acidity rabbits (**A**) and to low acidity rabbits (**B**) at a dose of 5 mg/kg. Each point represents the mean ± standard error of results from five rabbits.

formulation most likely prevented precipitation in the neutral pH environment of the small intestine, thus promoting greater drug absorption. With the low gastric acidity group, the authors approximated that drug dissolution with the solid dispersion formulation was four times that of the physical mixture. The enhanced solubilization of albendazole in a neutral pH environment coupled with the stabilizing effect of the polymers was proposed as the reason for improved drug absorption with the solid dispersion formulation for the low-gastric acidity group. Thus, the amorphous solid dispersion formulation proved to be a more effective delivery system for albendazole as the combination of a more soluble form of the drug with a polymeric carrier system that acted to stabilize supersaturated drug concentrations provided for extensive drug absorption that was less affected by pH than the crystalline drug formulation. Since the pH of the GI tract can vary substantially on both an intra and interpatient basis, the solid dispersion formulation would provide more effective treatment of albendazole overall considering the reduced pH dependency of drug absorption.

Griseofulvin is another antibiotic drug whose therapeutic efficacy has shown improvement when formulated as a solid dispersion. Griseofulvin is a systemic anti-fungal agent that is indicated for the treatment of fungal infections that commonly cause ringworm infestations of the hair, skin, and nails (58). Griseofulvin is poorly water-soluble, exhibits high biological membrane permeability, and is a relatively high dose drug (125–250 mg); and thus is classified as a BCS class II drug (59). Griseofulvin is a commonly mentioned drug in pharmaceutical papers that focus on solid dispersion technologies because it is one of the original drugs to be marketed as a solid dispersion formulation (Gris-PEG, Novartis).

Recently, Wong et al. demonstrated a crystalline solid dispersion of griseofulvin in poloxamer 407 produced by spray drying from an organic solution (60). The authors utilized both DSC and XRD to determine that spray drying did not alter the crystalline morphology of griseofulvin spray died alone or in combination with poloxamer 407. In vitro dissolution testing revealed that the griseofulvin-poloxamer 407 spray dried formulation exhibited a dramatically improved dissolution rate with 50% of drug released in 15 min, as compared to 18% and 7% for spray dried griseofulvin alone and the bulk drug, respectively. Since drug morphology was not altered by the spray drying process, this improvement was not due to improved apparent solubility, and thus was attributed to increased surface area due to the spray drying process and improved wettability by intimate mixing of the drug with poloxamer. In vivo absorption studies were conducted in rats with the bulk drug, spray dried griseofulvin, and spray dried griseofulvin-poloxamer 407 formulations dosed in capsules. Figure 9 illustrates the plasma concentration of griseofulvin versus time for each formulation.

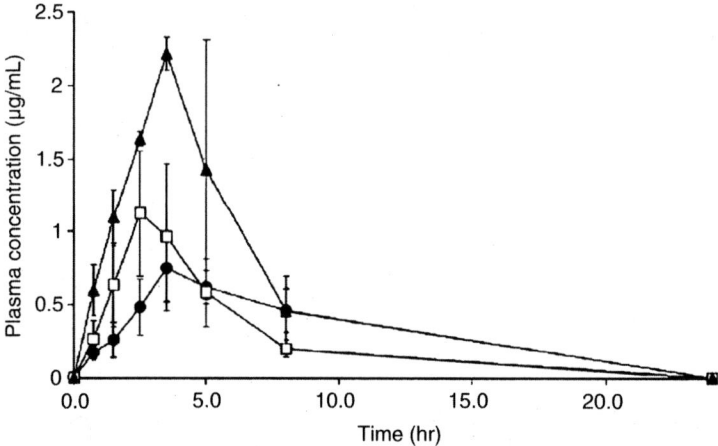

Figure 9 Mean griseofulvin plasma concentration-time profile following oral administration to rats (± S.D., $n = 4$). (●) Control; (□) Spray-dried; (▲) Spraydried + Poloxamer 407.

This study revealed that the spray dried griseofulvin-poloxamer 407 formulation provided the greatest drug absorption with significantly greater C_{max} and AUC values than the two non-dispersion formulations. Also, the absolute bioavailability of the griseofulvin-poloxamer solid dispersion formulation was determined to be approximately two-fold greater than the non-dispersion formulations. This study thus demonstrated that intimately mixing particles of griseofulvin with a polymeric surfactant greatly improved the wettability of the particles which resulted in an accelerated dissolution rate and ultimately doubled the in vivo absorption over the bulk drug.

IMMUNOSUPPRESSION

Tacrolimus is an immunosuppressive drug primarily indicated for the reduction of immune response following organ transplantation to reduce the risk of rejection. The solubility of tacrolimus in water is 1–2 µg/mL (61), and thus has shown low oral bioavailability (62). Solid dispersion systems for improved oral bioavailability of tacrolimus have recently been explored by Yamashita et al. (63). These researchers evaluated solid dispersion systems of tacrolimus with PEG 6000, PVP, and HPMC in order to identify the optimum carrier. Initially, the solid dispersion formulations were produced by dissolving tacrolimus and each polymer carrier in a cosolvent system consisting of dichloromethane and ethanol, and then removing the solvent in a vacuum dryer at 40°C. Solid dispersions produced by this process contained tacrolimus in an amorphous state as determined by DSC and XRD. Dissolution testing was conducted in pH 1.2 medium with tacrolimus in excess of the saturation solubility to determine which carrier polymer would yield the greatest extent of supersaturation. The results of this dissolution testing revealed that although each solid dispersion formulation showed equivalent peak supersaturation concentrations that were 25-times that of saturation (50 μg/mL), the HPMC formulation was the only solid dispersion that was able to maintain these elevated concentrations for 24 h. This result demonstrated that HPMC can effectively prevent the precipitation of supersaturated tacrolimus which presumably would provide enhanced in vivo absorption, and thus the tacrolimus-HPMC formulation was selected for oral bioavailability evaluation.

The initial bioavailability study was conducted in beagle dogs with the crystalline powder and the solid dispersion formulation dosed as a suspension. The blood concentration profiles of the crystalline drug powder and the solid dispersion formulation is shown in Figure 10. This study clearly indicated that the ability of the tacrolimus-HPMC formulation to

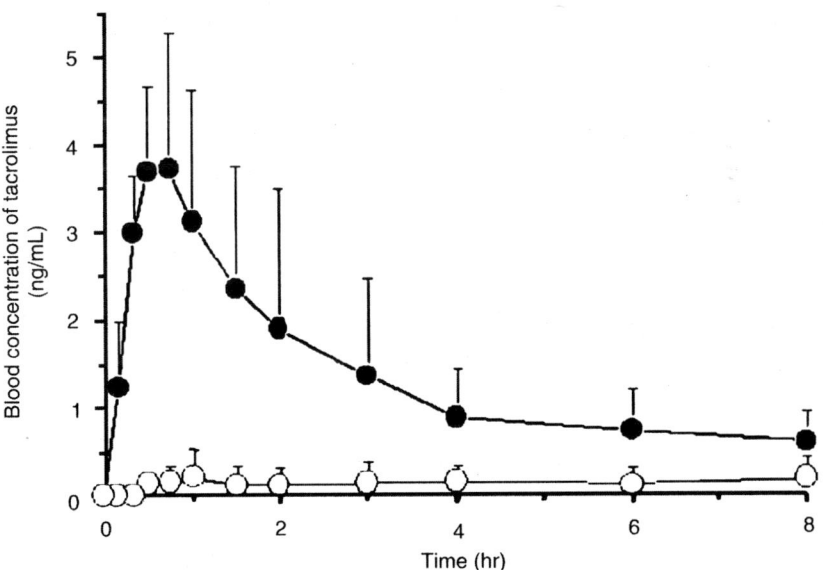

Figure 10 Blood concentration of tacrolimus after oral administration of SDF with HPMC to beagle dogs. (●) SDF of tacrolimus with HPMC; (○) tacrolimus crystalline powders. Values are expressed as the mean with a vertical bar showing SE of six animals. Each dosage form was administered at the dose of 1 mg as tacrolimus.

produce and maintain supersaturated drug levels during dissolution testing correlated well to in vivo drug absorption as the solid dispersion exhibited C_{max} and AUC values that were ten-fold higher than the crystalline powder.

Based on this result, the authors sought to optimize the solvent evaporation method by eliminating dichloromethane from the cosolvent system due to the health and environmental hazards associated with this solvent. HPMC is not soluble in ethanol, but is swellable. Therefore, by the new production method, tacrolimus was dissolved in ethanol, and this ethanolic solution was then used to swell the HPMC carrier. The solvent was then removed by the use of a vacuum oven at 40°C to form the dried solid dispersion. By the use of scanning electron microscopy, DSC, and XRD it was determined that tacrolimus was present in the HPMC carrier in an amorphous state. Dissolution testing and in vivo studies in monkeys revealed that the tacrolimus-HPMC solid dispersion prepared by the new method was equivalent to the solid dispersion prepared with the original cosolvent formulation. Therefore, in this study the authors demonstrated a solid dispersion formulation for tacrolimus that extensively supersaturated aqueous media and correspondingly improved oral absorption. By eliminating dichloromethane from the solvent system, the solvent evaporation method used to produce this tacrolimus solid dispersion was made considerably more feasible for commercial production. Therefore, this solid dispersion formulation of tacrolimus has substantial potential for improving immunosuppressive drug treatment and thereby decreasing the frequency of transplantation rejection.

Cyclosporin is another immunosuppressive agent that exhibits poor and variable absorption from the GI tract owing in part to its poor water solubility (7.3 µg/mL) (64). The leading solid oral dosage form for cyclosporin in the US is Neoral®, a microemulsion filled soft gelatin capsule formulation marketed by Novartis. As soft gelatin capsule dosage forms are relatively expensive, complicated to manufacture, and often have stability related issues (45,65); there would substantial benefit to replacing these dosage forms with conventional tablets or capsules that are bioequivalent. Toward this aim, Liu et al. evaluated a solid dispersion system of cyclosporin A in polyoxyethylene stearate (S40) as potential solid alternative to the liquid filled capsule (66). The solid dispersion formulation was prepared by a solvent-melt method in which cyclosporin A was dissolved in anhydrous ethanol, added to molten S40 at 65°C, and stirred until a majority of the solvent was evaporated. The molten mixture was then cooled at −18°C, dried under vacuum at 25°C for 24 h, and then pulverized. The results of XRD and DSC studies demonstrated that cyclosporine A was present in the dispersion formulation in an amorphous state. Dissolution testing in water showed substantial improvement in the rate and extent of cyclosporine release form the solid dispersion formulation over that of the crystalline drug powder and the physical mixture.

A bioavailability study conducted in rats showed very similar drug absorption with the solid dispersion formulation as with the Neoral soft gelatin capsule formulation. The pharmacokinetic parameters from this study are shown in Table 6. Statistical analysis of these results revealed no significant differences in the pharmacokinetic parameters between these formulations. Relative to the Neoral formulation, the bioavailability of the solid dispersion system was found to be 98.1%. This study therefore demonstrated the apparent bioequivalence of the solid dispersion formulation to the Neoral soft gelatin capsule. Considering the manufacturing cost of soft gelatin capsules and the associated stability issues (45,65), the solid dispersion system may be a more attractive alternative for the oral delivery of cyclosporine.

OTHER DRUG TREATMENTS

In a study by Vaughn et al. the bioavailability of danazol, a drug used in the treatment of endometriosis, was found to be enhanced in correspondence with in vitro supersaturation from solid dispersion nanoparticles (67). Specifically, high potency, high surface area particles of amorphous danazol dispersions in PVP K15 were prepared by two novel nanoparticle production methods known as evaporative precipitation into aqueous solution (EPAS) (68) and spray freezing into liquid (SFL) (4). The danazol nanoparticles produced by SFL were determined to be completely amorphous with a primary particle size of 30 nm, whereas the EPAS nanoparticles were found to be partially crystalline with a primary particle size of 500 nm. Dissolution testing revealed that the SFL particles produced 33% supersaturation relative to the physical mixture at peak concentration and remained supersaturated for 90 min. The EPAS particles exhibited 27%

Table 6 Pharmacokinetic Parameters of CyA after Oral Administration of CyA Solid Dispersion and Sandimmun Neoral® to Wistar Rats

Pharmacokinetic parameter	CyA solid dispersion	Sandimmun Neoral®
C_{max} (ng/ml)	2348.65 ± 495.69	2557.38 ± 555.09
T_{max} (h)	2.00 ± 0.67	1.71 ± 0.67
AUC_{0-60} (ng/mlh)	40283.99 ± 5203.16	41021.10 ± 6239.87
K_a (h^{-1})	1.56 ± 0.57	1.54 ± 0.44
K_{10} (h^{-1})	0.079 ± 0.017	0.085 ± 0.020
$T_{1/2 \beta}$ (h)	21.24 ± 5.02	21.84 ± 6.78
MRT_{0-60} (h)	17.96 ± 1.88	17.97 ± 1.29
F (%)	98.20	-

Mean $\pm$ SD ($n = 10$)

supersaturation relative to the physical mixture at peak concentration and remained above saturation for 60 min. The reason for the discrepancy in the level and duration of supersaturation between the SFL and EPAS particles was explained by the differences in the drug morphology. As the EPAS formulation was partially crystalline, there is not only less amorphous drug content to drive supersaturation, but also the crystalline particles would act as nucleation sites to promote precipitation of danazol from supersaturated solution. These particle compositions along with the physical mixture and a commercial danazol formulation were evaluated in vivo using mice. The two nanoparticle formulations exhibited greater C_{max} values than the physical mixture and the commercial dosage form with the EPAS formulation having a slightly greater C_{max} than the SFL formulation. The SFL particles showed substantially greater AUC than the other formulations with the EPAS formulation showing similar AUC to the commercial product followed by the physical mixture. A plot of the danazol serum concentrations versus time for each formulation is shown in Figure 11. These results demonstrated the direct correlation between the extent of supersaturation of danazol in vitro and in vivo drug absorption. Therefore, by formulating danazol in a highly soluble form such as amorphous nanoparticles, the oral bioavailability and thus therapeutic efficacy can be substantially improved.

Solid dispersion systems have been demonstrated to improve the oral absorption of several other drugs used to treat a variety of conditions. Due to the similarity of the solid dispersion technologies used to produce the delivery systems for these drugs to those already described, the studies will be only briefly discussed. Yakou et al. produced solid dispersions of phenytoin, an anticonvulsant drug, in PEG 4000 by a fusion method (69). Phenytoin was determined to be amorphous in the dispersions and the in vitro dissolution rate

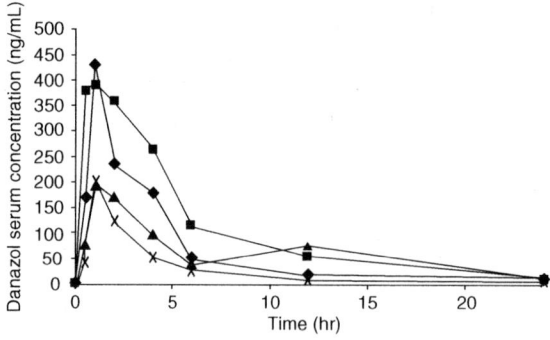

Figure 11 Oral bioavailability of Danazol in a mouse model for the SFL composition (Danazol:PVP-K15 1:1) (■), EPAS composition (Danazol:PVP-K15 1:1) (♦), physical mixture (Danazol:PVP-K15 1:1) (*), and commercially available Danazol (▲).

was substantially improved over the crystalline drug powder and the physical mixture. Bioavailability assessment of the solid dispersion formulation was conducted in human volunteers along with phenytoin crystals and a physical mixture. The mean AUC and peak plasma concentration of the solid dispersion formulation were both found to be greater than the physical mixture and crystalline drug. Thus, from this study it was determined that a solid dispersion system offers clinical advantages of rapid drug release and excellent bioavailability that may provide improved efficacy of oral phenytoin.

Doherty et al. produced amorphous solid dispersions of a poorly water soluble diuretic drug, frusemide by solvent evaporation with PVP (70). The amorphous solid dispersion formulation showed marked dissolution improvement over crystalline frusemide in a variety of media. This in vitro dissolution performance correlated well to the in vivo performance of the solid dispersion formulation as a significant reduction in the time to maximum effect was found relative to the crystalline drug in human subjects. Thus, the solid dispersion formulation was found to enhance the therapeutic effect of frusemide by decreasing the time to onset of action.

Glibenclamide (Glyburide) is a hypoglycemic agent used in the treatment of diabetes mellitus. As glibenclamide is only slightly soluble in water, its dissolution characteristics limit oral absorption and lead to high variability. Tashtoush et al. examined solid dispersion formulations with Gelucire 44/14 and PEG 6000 produced by the fusion method for enhancing dissolution rate and bioavailability of glibenclamide (71). The results of these studies revealed more rapid and extensive in vitro drug release from the solid dispersion formulations than the commercial product Daonil® with the PEG formulation performing slightly better than the formulation with Gelucire. In vivo testing in human volunteers revealed that the greatest bioavailability was achieved with the solid dispersion formulation with PEG ($AUC_{0-\infty}=$ 1035.7 ng·hr/mL) followed by the Gelucire ($AUC_{0-\infty}=$ 680.8 ng·hr/mL) and Daonil ($AUC_{0-\infty}=$ 432.1 ng·hr/mL). Therefore, it was demonstrated that the oral bioavailability and hence efficacy of glibenclamide can be substantially improved by formulation as a solid dispersion in a hydrophilic carrier.

CONCLUDING REMARKS

The studies discussed in this chapter illustrate a variety of solid dispersion technologies that improved the dissolution properties, and consequently the oral absorption of BCS class II drugs. It was found that accelerating the dissolution rate of BCS class II drugs concurrently accelerated the absorption rate since these drugs exhibit dissolution rate limited absorption. Moreover, the work of several researchers discussed herein made evident that supersaturation of GI fluids was the direct cause for improved absorption and overall bioavailability. This is a benefit unique to solid dispersion systems as extensive supersaturation cannot be achieved with traditional dosage forms containing crystalline drug.

The studies discussed in this chapter have also demonstrated that solid dispersion systems can provide numerous additional benefits to oral drug therapy beyond improving bioavailability. Solid dispersion formulations were demonstrated to accelerate the onset of action for drugs such as NSAIDs where immediacy of action is crucial to relieving acute pain and inflammation. For anti-cancer drugs in particular, solid dispersion systems were shown to provide bioavailable oral dosage forms which could be substituted for standard injections to improve patient comfort and compliance. Solid dispersion systems were also found to reduce food effects on drug absorption, thus increasing the convenience of drug therapy as the need for some drugs to be taken with food was eliminated. It was also demonstrated that a solid dispersion-based dosage form allowed for greater drug loading per dose and improved stability over a soft gelatin capsule formulation which thereby improved the convenience of the drug therapy by reducing the dosing regime and eliminating the need for refrigerated storage. Additionally, the improved absorption efficiency demonstrated for solid dispersion systems allows for a reduction in the content of active agent per dose, thus decreasing the cost associated with these drug therapies. Finally, it was demonstrated that solid dispersion systems can be produced utilizing functional carriers that offer the added benefit of targeting the release of highly soluble forms of poorly water-soluble drugs to an optimum site for absorption. These benefits highlighted in this chapter demonstrate the current contributions and future potential of solid dispersion systems toward improving drug therapies for a variety of important medical conditions whose treatment involves poorly water-soluble drugs.

These studies represent decades of investigation and continued improvement of solid dispersion technologies with regard to drug release capabilities, stability, processing efficiency, and overall feasibility. However, despite the promise of solid dispersion technologies for improving drug delivery, there are only a few commercial products currently on the market that are produced using solid dispersion technologies (72). Previous authors have attributed this to stability or processing issues; however, in recent years the understanding of the solid state stability of drugs in these systems has increased markedly such that formulation parameters are now known that eliminate storage instability (27). Additionally, with the advent of continuous manufacturing processes such as hot-melt extrusion, the complications of manufacturing solid dispersions are diminishing. Therefore, the obstacles that once prevented the successful commercialization of solid dispersion formulations are becoming less prohibitive. Hence, it is believed that the current lack of marketed commercial products merely represents the time lag between conceptualization and maturation of solid dispersion technologies in which increased knowledge of solid dispersion technology has now made it a viable commercial manufacturing option. Thus, it is expected that in the near future the number of commercialized products that are based on solid dispersion technologies will increase markedly. This is further evidenced by the steady increase in solid dispersion research in recent years, particularly the number of clinical studies evaluating the potential for improved drug therapy via the use of solid dispersion technologies. This chapter has highlighted many pre-clinical and clinical in vivo studies to illustrate the current contributions of solid dispersion technologies and the future prospects for improving oral drug therapies with poorly water-soluble drugs.

REFERENCES

1. Amidon GL, Lennernas H, Shah VP, Crison JR. Theoretical basis for a biopharmaceutical drug classification: the correlation of in vitro drug product dissolution and in vivo bioavailability. Pharm Res 1995; 12(3):413–20.
2. Lipinski CA. Avoiding investment in doomed drugs, is poor solubility an industry wide problem? Curr Drug Dis 2001; 4:17–9.
3. Tabibi SE, Rhodes CT. Disperse Systems. In: Banker GS, Rhodes CT, eds. Modern Pharmaceutics, 3rd edn. New York, NY: Marcel Dekker, Inc.; 1996: 299–331.
4. Rogers TL, Hu JH, Yu ZS, Johnston KP, Williams III RO. A novel particle engineering technology: spray-freezing into liquid. Int J Pharm 2002; 242(1–2): 93–100.
5. Hu J, Johnston KP, Williams III RO. Nanoparticle engineering processes for enhancing the dissolution rates of poorly water soluble drugs. Drug Dev Ind Pharm 2004; 30(3):233–45.
6. Leuner C, Dressman J. Improving drug solubility for oral delivery using solid dispersions. Eur J Pharma Biopharm 2000; 50(1):47–60.
7. Reverchon E. Supercritical antisolvent precipitation of micro- and nano-particles. J Supercrit Fluids 1999; 15(1):1–21.
8. Palakodaty S, York P. Phase behavioral effects on particle formation processes using supercritical fluids. Pharm Res 1999; 16(7):976–85.
9. Matson DW, Peterson RC, Smith RD. Production of fine powders by the rapid expansion of supercritical fluid solutions. Adv Ceramics 1987; 21:109.
10. Dixon DJ, Johnston KP, Bodmeier RA. Polymeric materials formed by precipitation with a compressed fluid antisolvent. AIChE J 1993; 39:127–39.
11. Mumenthaler M, Leuenberger H. Atmospheric spray-freeze drying: a suitable alternative in freeze-drying technology. Int J Pharm 1991; 72(2):97–110.
12. Briggs AR, Maxwell TJ, inventors; Process for Preparing Powder Blends. US Patent 3,721,725. 1973.
13. Dunn DB, Masavage GJ, Sauer HA, inventors; Method of Freezing Solution Droplets and the Like Using Immiscible Refrigerants of Differing Densities. US Patent 3,653,222. 1972.
14. Hebert PF, Healy HS, inventors; Production Scale Method of Forming Microparticles. US Patent 5,922,253. 1999.

15. Evans JC, Scherzer BD, Tocco CD, et al. Preparation of nanostructured particles of poorly water soluble drugs via a novel ultra-rapid freezing technology. Poly Mater Sci Eng (2003) 2003; 89:742.

16. Law D, Wang W, Schmitt EA, Qiu Y, Krill SL, Fort JJ. Properties of rapidly dissolving eutectic mixtures of poly(-ethylene glycol) and fenofibrate: the eutectic microstructure. J Pharm Sci 2003; 92:505–15.

17. Guivarc'h P-H, Vachon MG, Fordyce D. A new fenofibrate formulation: results of six single-dose, clinical studies of bioavailability under fed and fasting conditions. Clin Ther 2004; 26(9):1456–69.

18. Sant VP, Smith D, Leroux J-C. Enhancement of oral bioavailability of poorly water-soluble drugs by poly(ethylene glycol)-block-poly(alkyl acrylate-co-methacrylic acid) self-assemblies. J Control Release 2005; 104 (2):289–300.

19. Sant VP, Smith D, Leroux J-C. Novel pH-sensitive supramolecular assemblies for oral delivery of poorly water soluble drugs: preparation and characterization. J Control Release 2004; 97(2):301-12.

20. Ambike AA, Mahadik KR, Paradkar A. Spray-dried amorphous solid dispersions of simvastatin, a low Tg drug: in vitro and in vivo evaluations. Pharm Res 2005; 22(6):990–8.

21. Kinoshita M, Baba K, Nagayasu A, et al. Improvement of solubility and oral bioavailability of a poorly water-soluble drug, TAS-301, by its melt-adsorption on a porous calcium silicate. J Pharm Sci 2002; 91(2):362–70.

22. Verreck G, Vandecruys R, De Conde V, Baert L, Peeters J, Brewster ME. The use of three different solid dispersion formulations – melt extrusion, film-coated beads, and a glass thermoplastic system – to improve the bioavailability of a novel microsomal triglyceride transfer protein inhibitor. J Pharm Sci 2004; 93(5):1217–28.

23. Dannenfelser R-M, He H, Joshi Y, Bateman S, Serajuddin ATM. Development of clinical dosage forms for a poorly water soluble drug I: Application of polyethylene glycol-polysorbate 80 solid dispersion carrier system. J Pharm Sci 2004; 93(5):1165–75.

24. Levis KA, Lane ME, Corrigan OI. Effect of buffer media composition on the solubility and effective permeability coefficient of ibuprofen. Int J Pharm 2003; 6:49–59.

25. McNeil-Motrin®. (ibuprofen suspension, oral drops, chewable tablets, caplets) prescribing information. 56 ed. Montvale, NJ: Medical Economics Company Inc; 1994.

26. Klueglich M, Ring A, Scheuerer S, et al. Ibuprofen extrudate, a novel, rapidly dissolving ibuprofen formulation: relative bioavailability compared to ibuprofen lysinate and regular ibuprofen, and food effect on all formulations. J Clin Pharmacol 2005; 45(9):1055–61.

27. Lewis J. Solid Solutions for Insoluble Substances. Drug Delivery Report 2006; Spring/Summer:62–3.

28. Breitenbach J, Mägerlein M. Melt-Extruded Molecular Dispersion. New York, NY: Marcel Dekker; 2003.

29. Geisslinger G, Dietzel K, Bezler H, Nuernberg B, Brune K. Therapeutically relevant differences in the pharmacokinetical and pharmceutical behavior of

ibuprofen lysinate as compared to ibuprofen acid. Int J Clin Pharmacol Ther Toxicol 1989; 27:324–8.

30. Bell R, Black L, Harries R, et al. The discovery and characterization of selective cyclooxygenase-2 inhibitors. Inflamm Res 2000; 49:S101.

31. Chen Y, Zhang GGZ, Neilly J, Marsh K, Mawhinney D, Sanzgiri YD. Enhancing the bioavailability of ABT-963 using solid dispersion containing Pluronic F-68. Int J Pharm 2004; 286(1–2):69–80.

32. Collett JH, Popli H. Poloxamer. Washington, DC: American Pharmaceutical Association; 2001.

33. Shin SC, Cho CW. Physicochemical characterizations of piroxicam-poloxamer solid dispersion. Pharm Dev Tech 1997; 2:403–7.

34. Passerini N, Albertini B, Gonz′alez-Rodr′1guez ML, Cavallari C, Rodriguez L. Preparation and characterization of ibuprofen-poloxamer 188 granules obtained by melt granulation. Eur J Pharm Sci; 2002:71–8.

35. Chowdary KPR, Babu KVVS. Dissolution, bioavailability and ulcerogenic studies on solid dispersions of indomethacin in water-soluble cellulose polymers. Drug Dev Ind Pharm 1994; 20(5):799–813.

36. Tagliati CA, Kimura E, Nothenberg MS, Santos SRJC, Oga S. Pharmacokinetic profile and adverse gastric effect of zinc-piroxicam in rats. Gen Pharmacol 1999; 13:67–71.

37. Yüksel N, Karatas A, Ozkan Y, Savaser A, Ozkan SA, Baykara T. Enhanced bioavailability of piroxicam using Gelucire 44/14 and Labrasol: in vitro and in vivo evaluation. Eur J Pharma Biopharm 2003; 56(3):453–9.

38. Etienne MC, Milano G, Rene N, et al. Improved bioavailability of a new oral preparation of medroxyprogesterone acetate. J Pharm Sci 1991; 80(12):1130–2.

39. Pannuti F, Martoni A, Di Marco AR, Piana E. Eur J Cancer 1979; 15:593–601.

40. Carli F, inventor Farmitalia Carlo Erba S.p.A. Pat. # 4,632,828, assignee. Pharmaceutical composition US. 1986.

41. Johnson JR, Priestman IJ, Fotherby K, Kelly KA, Priestman SG. An evaluation of high-close medroxyprogesterone acetate therapy in women with advanced breast cancer. Br J Cancer 1984; 1984(50):363–6.

42. Kondo N, Iwao T, Hirai K, et al. Improved oral absorption of enteric coprecipitates of a poorly soluble drug. J Pharm Sci 1994; 83(4):566–70.

43. Kondo N, Iwao T, Kikuchi M, et al. Pharmacokinetics of a micronized, poorly water-soluble drug, HO-221, in experimental animals. Biol Pharm Bull 1993; 16 (8):796–800.

44. Palmieri GF, Cantalamessa F, Di Martino P, Nasuti C, Martelli S. Lonidamine Solid dispersions: in vitro and in vivo evaluation. Drug Dev Ind Pharm 2002; 28 (10):1241–50.

45. Breitenbach J. Melt extrusion can bring new benefits to HIV therapy: the example of Kaletra tablets. Am J Drug Deliv 2006; 4(2):61–4.

46. Abbott_Laboratories. KALETRA® (lopinavir/ritonavir) tablets (lopinavir/ritonavir) oral solution [prescring information; online]. Available from URL: http://wwwrxabbottcom/pdf/kaletratabpipdf2005;[Accessed December 14, 2006].

47. Law D, Schmitt EA, Marsh KC, et al. Ritonavir-PEG 8000 amorphous solid dispersions: in vitro and in vivo evaluations. J Pharm Sci 2004; 93(3):563–70.

48. Law D, Krill SL, Schmitt EA, et al. Physicochemical considerations in the preparation of amorphous ritonavir/PEG 8000 solid dispersions. J Pharm Sci 2001; 90(8):1015–25.

49. Aungst BJ, Nguyen NH, Bulgarelli JP, Oates-Lenz K. The influence of donor and reservoir additives on Caco-2 permeability and secretory transport of HIV protease inhibitors and other lipophilic compounds. Pharm Res 2000; 17: 1175–80.

50. De Jaeghere F, Allemann E, Kubel F, et al. Oral bioavailability of a poorly water soluble HIV-1 protease inhibitor incorporated into pH-sensitive particles: effect of the particle size and nutritional state. J Control Release 2000; 68(2):291–8.

51. Janssen LP. Sporanox (Itraconazole Capsules) Prescribing Information. Available from URL: http://wwwsporanoxcom2006; Janssen Pharmaceutica N.V.(Beerse, Belgium).

52. Six K, Daems T, de Hoon J, et al. Clinical study of solid dispersions of itraconazole prepared by hot-stage extrusion. Eur J Pharm Sci 2005; 24(2-3): 179–86.

53. Peeters J, Neeskens P, Tollenaere JP, Van Remoortere P, Brewster M. Characterization of the interaction of 2-hydroxypropyl-B-cyclodextrin with itraconazole at pH 2, 4 and 7. J Pharm Sci 2002; 91:1414–22.

54. Marriner SE, Morris DL, Dickson B, Bogan JA. Pharmacokinetics of albendazole in man. Eur J Clin Pharmacol 1986; 30:705–8.

55. Jung H, Hurtado M, Sanchez M, Medina MY, Sotelo J. Clinical pharmacokinetics of albendazole in patients with brain cysticercosis. J Clin Pharmacol 1992; 32:28–31.

56. Kohri N, Yamayoshi Y, Iseki K, Sato N, Todo S, Miyazaki K. Effect of gastric pH on the bioavailability of albendazole. Pharm Pharmacol Commun 1998; 4: 267–70.

57. Kohri N, Yamayoshi Y, Xin H, et al. Improving the oral bioavailability of albendazole in rabbits by the solid dispersion technique. J Pharm Pharmacol 1999; 51(2):159–64.

58. Maheswaran AM, ed. Mosby's Drug Consult. 16 edn. St. Louis, Missouri: Mosby, Inc.; 2006.

59. Lindenberg M, Kopp S, Dressman JB. Classification of orally administered drugs on the World Health Organization Model list of Essential Medicines according to the biopharmaceutics classification system. Eur J Pharma Biopharm 2004; 58(2):265–78.

60. Wong SM, Kellaway IW, Murdan S. Enhancement of the dissolution rate and oral absorption of a poorly water soluble drug by formation of surfactant-containing microparticles. Int J Pharm 2006; 317(1):61–8.

61. Hane K, Fujioka M, Namiki Y, et al. Physico-chemical properties of (−)-1R,9S,12S,13R,14S,17R,18E,21S,23S,24R,25S,27R)-17-allyl-1,14-dihydroxy-12-[(E)-2-[(1R,3R,4R)-4-hydroxy-3-methoxycyclohexyl]-1-methylvinyl]-23,25-dimethoxy-13,19,21,27-tetrametyl-11,28-dioxa-4-azatricyclo[22.3.1.04,9] octacos-18-ene-2,3,10,16-tetrone hydrate (FK-506). Iyakuhin Kenkyu 1992; 23: 33–43.

62. Honbo T, Kobayashi M, Hane K, Hata T, Ueda Y. The oral dosage form of FK-506. Transpl Proc 1987; 19:17–22.

63. Yamashita K, Nakate T, Okimoto K, et al. Establishment of new preparation method for solid dispersion formulation of tacrolimus. Int J Pharm 2003; 267 (1–2):79–91.

64. Ismailos G, Reppas C, Dressman JB, Macheras P. Unusual solubility behaviour of cyclosporin A in aqueous media. J Pharm Pharmacol 1991; 43: 287–9.

65. Gibson M. Product Optimisation. In: Gibson M, ed. Pharmaceutical Preformulation and Formulation: A Practical Guide from Candidate Drug Selection to Commercial Dosage Form. Boca Raton, FL: CRC Press; 2001.

66. Liu C, Wu J, Shi B, Zhang Y, Gao T, Pei Y. Enhancing the Bioavailability of Cyclosporine A Using Solid Dispersion Containing Polyoxyethylene (40) Stearate. Drug Dev Ind Pharm 2006; 32(1):115–23.

67. Vaughn JM, T. MJ, Crisp MT, Johnston KP, Williams RO, III. Supersaturation produces high bioavailability of amorphous danazol particles formed by evaporative precipitation into aqueous solution and spray freezing into liquid technologies. Drug Dev Ind Pharm 2006; 32(5):559–67.

68. Sarkari M, Brown J, Chen XX, Swinnea S, Williams RO, Johnston KP. Enhanced drug dissolution using evaporative precipitation into aqueous solution. Int J Pharm 2002; 243(1–2):17–31.

69. Yakou S, Umehara K, Sonobe T, Nagai T, Sugihara M, Fukuyama Y. Particle size dependency of dissolution rate and human bioavailability of phenytoin in powders and phenytoin-polyethylene glycol solid dispersions. Chem Pharm Bull 1984; 32(10):4130–6.

70. Doherty C, York P. The in-vitro pH-dissolution dependence and in-vivo bioavailability of frusemide-PVP solid dispersions. J Pharm Pharmacol 1989; 41(2):73–8.

71. Tashtoush BM, Al-Qashi ZS, Najib NM. In vitro and in vivo evaluation of glibenclamide in solid dispersion systems. Drug Dev Ind Pharm 2004; 30(6): 601–7.

72. Serajuddin ATM. Solid disperison of poorly water-soluble drugs: early promises, subsequent problems, and recent breakthroughs. J Pharm Sci 1999; 88(10):1058–66.

Index